The Physics of
SPORTS

Michael Lisa
The Ohio State University

THE PHYSICS OF SPORTS

Published by McGraw-Hill Education, 2 Penn Plaza, New York, NY 10121. Copyright © 2016 by McGraw-Hill Education. All rights reserved. Printed in the United States of America. No part of this publication may be reproduced or distributed in any form or by any means, or stored in a database or retrieval system, without the prior written consent of McGraw-Hill Education, including, but not limited to, in any network or other electronic storage or transmission, or broadcast for distance learning.

Some ancillaries, including electronic and print components, may not be available to customers outside the United States.

ISBN 978-0-07-351397-3
MHID 0-07-351397-0

Senior Vice President, Products & Markets: *Kurt L. Strand*
Vice President, General Manager, Products & Markets: *Marty Lange*
Vice President, Content Design & Delivery: *Kimberly Meriwether David*
Managing Director: *Thomas Timp*
Brand Manager: *Thomas Scaife, Ph.D.*
Director, Product Development: *Rose Koos*
Product Developer: *Eve Lipton*
Marketing Manager: *Nick McFadden*
Director, Content Design & Delivery: *Linda Avenarius*
Program Manager: *Faye M. Herrig*
Content Project Managers: *Jane Mohr, Tammy Juran, and Sandra M. Schnee*
Design: *Studio Montage, St. Louis, MO*
Content Licensing Specialists: *DeAnna Dausener*
Cover Image: *Bubble chamber image: © Fermilab; baseball player: Donald Miralle/Getty Images*. On the cover: The physics behind a curveball creates a suprising mathematical relationship with the motion of subatomic particles produced by cosmic rays and large accelerators. This relationship is discussed in Chapter 5.
Compositor: *Laserwords Private Limited*

All credits appearing on page or at the end of the book are considered to be an extension of the copyright page.

Isaac Newton images: © Pixtal/agefotostock RF

www.mhhe.com

ABOUT THE AUTHOR

Originally from New Jersey, Mike Lisa began his career at the National Superconducting Cyclotron Laboratory at Michigan State University, where he earned his Ph.D. in experimental nuclear physics in 1993. He continued his research on high-energy nuclear reactions at Lawrence Berkeley National Lab in Berkeley, California, as a postdoctoral fellow. In 1996, he joined the faculty at Ohio State University, where he has continued research in large-scale experiments at the Alternating Gradient Synchrotron and the Relativistic Heavy Ion Collider at Brookhaven National Lab (BNL) on Long Island, NY, and the Large Hadron Collider at CERN in Europe. In 2010, Dr. Lisa began teaching the Physics of Sports at Ohio State, and it quickly became an instructional focus. He and his wife Laura have three children. He enjoys long-distance cycling and cheering for the Cleveland Cavaliers.

(© Suzanne Gipson)

To Laura, Oscar, Sammy, and Anna.

BRIEF CONTENTS

McGraw-Hill Smartbook xv
Preface to the Student xvi
Preface to the Instructor xviii
Acknowledgments xxiii

Part I Primary Chapters 1

1   Warm-Up: Basic Concepts 2
2   Racing, Mathematically 14
3   Net Force: Dwight Howard Illustrates 39
4   Punts, the Fosbury Flop, and Other Projectile Motions 80
5   Curveballs, Foul Shots, and Bent Kicks 105
6   Game Changers: Collisions in Sports 145
7   Energy in Sports: Bursts of Power 186
8   Energy and Timing in Elastic Equipment 224
9   The Physics of Cycling 244
10 Twisting Athletes in Flight 302

Part II Supplementary Chapters 321

11 Lines of Action on the Line of Scrimmage: The Torque Wars   322
12 A Barry Bonds Home Run 326
13 The Pole Vault 335
14 Is It Better to Run through First Base or to Dive? 351
A   Unit Conversions 360
B   Tables of Relevant Physical Properties 362
Further Reading 367
Answers to Odd-Numbered Problems 372
Index 376

CONTENTS

McGraw-Hill Smartbook xv

Preface to the Student xvi

Preface to the Instructor xviii

Acknowledgments xxiii

Part I Primary Chapters 1

1 Warm-Up: Basic Concepts 2

1.1 Quantifying the World of Sports
Units, conversions, scientific notation 3
  1.1.1 Scientific Notation 3
  1.1.2 Units and Conversions 3

1.2 When We Don't Have Exact Numbers
Estimation, typical scales 5
  1.2.1 Estimation with Photos and Videos 6
  1.2.2 Typical Scales 7

1.3 The Center of Mass
Center of mass 8
  1.3.1 Some Facts 9

Problems 11

2 Racing, Mathematically
1D kinematics 14

2.1 Phelps in Beijing
Speed, velocity, position, and graphs 15
  2.1.1 Speed 15
  2.1.2 Position and Graphs 15
  2.1.3 Displacement and Average Velocity 16
  2.1.4 Vectors and Scalars 17
  2.1.5 Instantaneous Velocity 18

2.2 Bolt in Berlin
Acceleration, constant-acceleration kinematics 20
  2.2.1 Exploding Off the Block 20
  2.2.2 Constant Acceleration Kinematics 23
  2.2.3 The Signs a and v 24

2.3 Rope-Climbing and Diving
Vertical motion and gravity 27
  2.3.1 The 800-lb Gorilla: Gravity 28
  2.3.2 Freefall Time 29
  2.3.3 Landing Velocity 31
  2.3.4 The Symmetry of a Vertical Leap, Hang Time 31

Collected Equations 33

Problems 34

3 Net Force: Dwight Howard Illustrates
Forces, dynamics **39**

3.1 How Things Interact: Forces 40
3.2 The Physics of a Dwight Howard Dunk 40

- 3.2.1 Waiting for the Pass
  Weight, ground reaction force, equilibrium, free-body diagrams, Newton's first law   40
- 3.2.2 Spec Sheet
  Weight and mass   42
- 3.2.3 The Launch
  Newton's second and third laws; dynamics of a jump; dynamics with nonconstant forces; ground reaction force   45
- 3.2.4 The Flight
  Free-fall dynamics, revisited   52
- 3.2.5 The Landing
  Cushioning limitations, GRF revisited   53

3.3 Sideways Traction
Friction 55

- 3.3.1 Static and Kinetic Friction   56
- 3.3.2 Anti-Lock Brakes and Traction Control   57

3.4 More-Complex Situations
2D dynamics and applications 59

- 3.4.1 Dwight Howard Takes a Quick Step
  Vectors   59
- 3.4.2 Football Tryouts
  Two moving bodies; how to increase friction   62
- 3.4.3 Increasing Friction in Race Cars   64
- 3.4.4 Ball Throw Speed
  Important application   65

3.5 "Imaginary Forces" in Sports 67

- 3.5.1 Imaginary Pushes on Dale Earnhardt Jr.
  Sensation of force in a noninertial frame   67
- 3.5.2 Two Nontraditional Olympic Events
  Potential to mistakenly interpret a reaction force   68
- 3.5.3 Discus Throw
  Centripetal and "centrifugal" forces   69

Collected Equations 75

Problems 76

4 Punts, the Fosbury Flop, and Other Projectile Motions
Projectile motion **80**

4.1 The Math: Simpler Than You Think
2D kinematics 81

4.2 Football Punt: Range, Hang Time, and Compromise
Range, hang time **83**

- 4.2.1 A Quick Example   84
- 4.2.2 The Optimum Angle   84
- 4.2.3 Punting Strategy and Compromises   86

4.3 Shot-Put
The range when the starting and ending heights are not the same   **88**

- 4.3.1 Optimum Shot-put Launch Angle   89

4.4 Human Projectiles
Parabolic motion of center of mass with changing body shape **91**

- 4.4.1 The Blake Griffin Ballet   91
- 4.4.2 Dick Fosbury's Flop   92
- 4.4.3 Bob Beamon's Long Jump   94

Collected Equations **99**

Problems **99**

5 Curveballs, Foul Shots, and Bent Kicks
Aerodynamic forces **105**

5.1 Overview
Four forces; use of approximations **106**

- 5.1.1 Approximations   106

5.2 Immersion in Fluid: Buoyancy
Buoyant force **107**

- 5.2.1 Buoyancy in Air   109

5.3 Moving Through Fluid Drag
Drag force, drag coefficient, terminal velocity **110**

- 5.3.1 A Cartoon Image of Drag   110
- 5.3.2 Important Complication 1: Speed Dependence of C_D 111
- 5.3.3 Important Complication 2: C_D and Surface Roughness   112
- 5.3.4 And More Complications...   113
- 5.3.5 Terminal Velocity   114

5.4 Sideward Forces from Asymmetries **115**

- 5.4.1 The Swing of a Cricket Ball
  Sideward force from asymmetric surface roughness   116
- 5.4.2 Bending a Ball's Flight
  Magnus force   117

5.5 Aerodynamic Forces, One at a Time
Examples **119**

- 5.5.1 Curveballs and Subatomic Physics
  Magnus as a centripetal force; unexpected analogy   120
- 5.5.2 Do We Need a Computer?
  Constant-force approximation   122

| | | 5.5.3 | A Simple Formula for a Curveball |
| | | | *Sideward deflection over a short part of the spiral*   123 |

        5.5.4 Roberto Carlos's "Impossible" Free Kick
Real-life spiral example; aerodynamic-dominated versus gravity-dominated ball sports   124

        5.5.5 John Paxson, Master of Forces
Systematic breakdown of a basketball shot   126

5.6 More Complicated Aerodynamics in Sports
Semiquantitative analysis of more complex situations **129**

        5.6.1 Knuckling
Effects of fluctuating orientation and drag   129

        5.6.2 Tilting into the Wind: Discus
Non-Magnus lift   130

        5.6.3 Human Wings: Ski Jumps
Lift with adjustable tilt   133

        5.6.4 Making the World Safer for Javelin Spectators
Changes in javelin design and rules   133

        5.6.5 Not So Fast! Polyurethane Swimsuits
Drag effects in water and rule changes   134

5.7 Not All Air Is Created Equal
Variations in air density and the effect on sports **135**

        5.7.1 Rocky Mountain (Natural) High
Altitude and air density   135

        5.7.2 Hot Days are (not) a Drag
Temperature and air density   138

        5.7.3 It's Not Just the Heat—It's the Humidity
Humidity and air density effects   139

        5.7.4 Storm Fronts   139

Collected Equations **139**

Problems **140**

6 Game Changers: Collisions in Sports
Collisions and momentum **145**

6.1 What a Collision Is and How to Think About It
Momentum, impulse **145**

6.2 The Physics of a Football Tackle
Completely inelastic collisions, conservation of momentum **147**

        6.2.1 The Energy of a Crunch
Kinetic energy, energy lost   149

        6.2.2 Helmet Design
Impulse, relation to force   151

        6.2.3 Forcing a Runner out of Bounds
2D inelastic collisions   154

6.3 Gentler Pursuits: Bowling
Elastic collisions; also isolated and nonisolated systems **156**

- 6.3.1 Beginner's First Roll: Head-on Collision
  1D elastic collisions   156
- 6.3.2 Birthday Party Bowling
  Importance of an isolated system when using momentum conservation   161
- 6.3.3 Off-Center Hits: Converting a Lily
  2D elastic collision   162
- 6.3.4 Off-Center Billiards Shots
  Special case: 2D elastic collisions for $m_1 = m_2$ 165
- 6.3.5 Beyond Two Dimensions: The Upward Hop of the Pin
  Impulse and again an isolated system   165

6.4 A Happy Medium: Dribbling and Driving
Partially inelastic collisions **167**
- 6.4.1 The Sad, Short Life of the NBA's Synthetic Ball
  Coefficient of restitution   169
- 6.4.2 Pádraig Harrington's Drive and Swinging Harder
  Inelastic collision with finite-mass objects; COR variation with speed   172

6.5 Off-Center Hits: Spinning the Ball
Qualitative introduction to torque and spin **177**
- 6.5.1 Bounce Pass
  Torque from friction; translation and rotation motion   178
- 6.5.2 Diving Shot
  Aerodynamic and collisional aspects of topspin   179
- 6.5.3 Backspin on a Golf Shot
  Aerodynamic and collisional aspects of backspin   179

Collected Equations **182**

Problems **183**

7 Energy in Sports: Bursts of Power
Energy, power, work, efficiency, elasticity **186**

7.1 Bouncing Basketball: The Whole Process
Conversion of energy; elastic and gravitational energy **186**
- 7.1.1 Heat in Basketball: Not Just for Miami
  Kinetic to thermal energy   186
- 7.1.2 Energy During the Bounce
  Elastic potential energy   187
- 7.1.3 Details of Energy During the Rise
  Gravitational potential energy   188

7.2 Efficiency
Various definitions **191**
- 7.2.1 The Efficiency of a Basketball Bounce   191
- 7.2.2 The Efficiency of a Golf Drive   192
- 7.2.3 Heat Death   193

7.3 The Athlete: The Energetic Starting Point
Chemical energy and its conversion **194**

7.3.1 The Source of Energy and Its Flow
Food energy, Calories 194

7.3.2 The Human Engine I: Energy Conversion
Biochemistry of food processing; energy storage 196

7.4 Keeping Score: Energy Accounting in Sports
The energy conservation concept and its application to athletes 199

7.4.1 The Water Analogy 200

7.4.2 How Useful is the Energy Conservation Concept Really?
In-principle versus practical utility of the concept 201

7.5 Uncle Rico's Hopes Dashed
Work, power, and the human engine 203

7.5.1 Work
Work–energy theorem; connection to forces 203

7.5.2 Power 205

7.5.3 The Human Engine II: Power
Caloric conversion rates 206

7.6 Behdad Salimikordsiabi's Clean and Jerk
Quantitative analysis of power lift; details of motion 211

7.6.1 Work During a Lift 211

7.6.2 The Snatch and Clean and Jerk Techniques
Details of a complicated set of moves 213

Collected Equations 219

Problems 220

8 Energy and Timing in Elastic Equipment
Stiffness, timing, elastic energy storage 224

8.1 The Physics of Archery I: Energy Storage and Transfer
Hooke's law, efficiency of energy transfer 224

8.1.1 The Arrow's Energy 225

8.1.2 The Bow's Energy
Hooke's law 225

8.1.3 Bow and Arrow Efficiency
Energy "loss" depends on system details 226

8.2 The Physics of Archery II: Fire Power
Oscillation frequency and period 227

8.3 The Physics of Archery III: Archer's Paradox
Stiffness of extended rod, timing details 229

8.3.1 Oscillations of an Arrow
Characterizing the flexibility of a rod 230

8.3.2 How the Archer's Paradox Works
Buckling, matching timing, details 231

8.4 Zdeno Chára's Slap Shot: Fast Storage, Faster Release
Energy storage and collisions 234

8.5 Bungee-Jumping Brides and Quadratic Equations
Extended example with tension; quadratic equation 237

          8.5.1   Dangling Above the Water
                  Tension as an equilibrating force   237
          8.5.2   How Low Will He Bounce?
                  Nontrivial energy example and the quadratic equation   238

   Collected Equations **240**

   Problems **241**

9 **The Physics of Cycling**
   Sustained power generation, rolling friction, more aerodynamics, power balance, rotational dynamics, torque **244**

   9.1   **Input to the Bike: Sustained Human Power** **246**
         9.1.1   Caloric Power Requirements for Long-Term Effort
                 Metabolic equivalent task (MET) ratings   246
         9.1.2   Oxygen Uptake, VO_2max, and Power
                 Definition of VO_2max; rider efficiency   248
         9.1.3   Power-to-Weight Ratio
                 PWR   252

   9.2   **Power Output**
         Power, force, and velocity **253**
         9.2.1   Hills
                 Gravitational force along a slope   253
         9.2.2   Rolling Resistance
                 Dissipation during rolling   255
         9.2.3   Wind Drag
                 Drag in still and moving air   256
         9.2.4   The Bicycle Power Equation
                 Iterative solution to complicated equations   260
         9.2.5   Cycling Versus Other Modes of Transport
                 Comparison of power requirements   263
         9.2.6   Drafting
                 Aerodynamics of more than one object   265

   9.3   **Talansky Drives the Bike**
         Torque and rotational motion **272**
         9.3.1   Rolling
                 Connection between linear and rotational motion   272
         9.3.2   Drivetrain I: Gears
                 Geometry of chain rings, importance of cadence   274
         9.3.3   That Annoying 2π: Angular Velocity
                 Angular velocity, the radian   279
         9.3.4   Drivetrain II: Chain Action
                 Torque with fixed right-angle lever arm   280
         9.3.5   Drivetrain III: Chain Reaction
                 Clarifying Newton's third law   283
         9.3.6   Drivetrain IV: Pushing the Pedals
                 Line of action; torque for forces at an angle   284

9.3.7 Rolling Friction, Revisited
Torque nature of "rolling friction," lever arm when line of action is not tangent to edge 287

9.3.8 Back to Basics: The Wheel Again
Moment of inertia, angular acceleration, rotational kinetic energy 290

Collected Equations 293

Problems 296

10 Twisting Athletes in Flight
Angular motion with changing moment of inertia, rotation about fixed axes and in free space, conservation of angular momentum 302

10.1 Human Rotation
Anatomical axes and moments of inertia 303

10.2 Backward Giant Circle
Torque, energy, rotation around fixed axis 305

10.2.1 Torques and Spin Rate
Torque changes with lever arm 305

10.2.2 Maximal Force at the Bottom of the Swing
Conservation of energy with rotational motion; centripetal force 306

10.2.3 Swinging to Speed Up 309

10.2.4 Dismount
Angular momentum and conservation 309

10.3 Figure Skating: Spinning on Ice
Angular momentum and work done by "internal" force 312

10.4 Rotational Action and Reaction
Angular momentum of different body parts 315

10.4.1 Acrobatics of a Long-Jumper, Revisited
Rotational action/reaction about the transverse axis 315

10.4.2 Throwing, Kicking, Twisting
Rotation and counterrotation along the longitudinal axis 316

10.4.3 Balance Beam
Rotation and counterrotation about the anteroposterior axis 317

Collected Equations 318

Problems 319

Part II Supplementary Chapters 321

11 Lines of Action on the Line of Scrimmage: The Torque Wars 322

12 A Barry Bonds Home Run
Ball-bat collisions 326

12.1 Ball–Bat Collision: Speeds, Impulse, Force 326

12.2 Batted Ball Speed (BBS) 329

	12.3	**Focus on the Bat** 330
		Collision with an extended object
		12.3.1 Bonds's Swing 330
		12.3.2 The Bat as an Extended Object
		Sweet spot, vibrations, effective mass 331
	Collected Equations 334
13	**The Pole Vault** 335
	13.1	**Origins** 335
	13.2	**The Modern Event** 335
		13.2.1 Performance Progression 335
		13.2.2 Contributions to Height 337
	13.3	**Pole Vault 101: Energy Flow** 338
		13.3.1 Energy-Based Estimate of Vaulting Height 338
		13.3.2 What Matters in the Simple Calculation 340
	13.4	**Pole Vault 102: Beyond Energy Flow** 341
		13.4.1 Maximizing Initial Energy: Carry Weight 342
		13.4.2 Minimizing Inelastic Energy "Loss" 344
		13.4.3 Fully Exploiting the Energy: Flexibility and Timing 347
		13.4.4 Work Done by the Athlete 349
14	**Is It Better to Run through First Base or to Dive?** 351
	14.1	**The Story according to *Sport Science*** 351
	14.2	**Too Close to Call** 353
	14.3	**Diving Speed** 354
		14.3.1 "50% Deceleration" 355
		14.3.2 Newton's First Law and Air Drag 355
	14.4	**What's Really Happening: Torque and Impulse** 356
	14.5	**Other Issues** 358
	14.6	**Concluding Remarks** 359
A	**Unit Conversions** 360
B	**Tables of Relevant Physical Properties** 362
Further Reading 367
Answers to Odd-Numbered Problems 372
Index 376

MCGRAW-HILL SMARTBOOK

Powered by the intelligent and adaptive LearnSmart engine, SmartBook is the first and only continuously adaptive reading experience available today. Distinguishing what students know from what they don't, and honing in on concepts they are most likely to forget, SmartBook personalizes content for each student. Reading is no longer a passive and linear experience but an engaging and dynamic one, in which students are more likely to master and retain important concepts, coming to class better prepared.

SmartBook includes powerful reports that identify specific topics and learning objectives students need to study. These valuable reports also provide instructors with insight into how students are progressing through textbook content and are useful for identifying class trends, focusing precious class time, providing personalized feedback to students, and tailoring assessment.

How does SmartBook work? Each SmartBook contains four components: Preview, Read, Practice, and Recharge. Starting with an initial preview of each chapter and key learning objectives, students read the material and are guided to topics for which they need the most practice based on their responses to a continuously adapting diagnostic. Read and practice continue until SmartBook directs students to recharge important material they are most likely to forget so as to ensure concept mastery and retention.

McGraw-Hill Connect® Physics

McGraw-Hill Connect® Physics to accompany *The Physics of Sports* offers online electronic homework, an eBook, and resources for both instructors and students. Instructors can create homework with easy-to-assign problems from the text. This feature also offers the simplicity of automatic grading and reporting.

PREFACE TO THE STUDENT

Welcome to a study of the physics of sports! Regardless of whether you have traditionally loved technical and scientific subjects in the past, we hope you will enjoy learning some of the important concepts at play in athletics. They are equally important to a basic understanding of our increasingly complex world, which relies partly on young leaders like yourself for input.

This is a course in physics. We will discuss physics concepts (force, momentum, acceleration) and equations relating them. We will not focus on mathematical derivations, but will use math to apply underlying scientific principles to real-life situations found in sports.

If you are an athlete, it is unlikely that completing this course will greatly improve your game; for that, you should listen to your coach, not your physics professor. Your coach will tell you what works, how to position your body, what exercises best prepare you for competition. Through years of practice, play, and coaching, your body already "knows" what works in sports. In this course, we will learn the science behind what your body knows intuitively. We will find some surprises—cases in which your intuition fails or comes to the wrong conclusion. And we will be able to make meaningful and interesting comparisons between the physical quantities in sports and those found elsewhere. For example, how do the forces involved in a hard tackle compare to the crushing force of a train car on the rail? Once you have a deeper knowledge of the physics underlying the game, your appreciation and enjoyment as a player or spectator will be heightened.

In addition to learning the science of sports itself, we hope you gain an appreciation of how science is applied to complicated problems. (In his book *The Physics of Baseball*, Bob Adair wittily but accurately notes, "The physics of baseball isn't rocket science. It's much harder.") The first aspect of any scientific approach is quantification: distances, forces, and coefficients of restitution all must be well defined and quantified; the job of science is to relate these quantities, explain their behavior, and predict what happens next in any situation.

The second aspect, especially important for complicated topics, is approximation. Your peers majoring in science and engineering are taking introductory physics courses, too. In those courses, the professor almost always starts off saying that she or he will ignore the presence of air when formulating a problem. Such an approach would hardly be appropriate in any serious treatment of sports! You, as the student, would justifiably feel that the professor at the front of the room was blathering on about something that has nothing to do with the "real world" of sports.

In this book, we aim to blather on about things relevant to sports, including "messy" concepts such as the effects of air and inelasticity. However, detailed, absolutely correct treatment of effects such as these is often extremely complicated, involving advanced mathematics and computer modeling, things we will not cover in this course. So, are we stuck with the "baseball without air" discussion found in the engineer's introductory physics class? No. Occasionally we will ignore air in order to focus on another point, but we will not pretend it is not there. Instead, we will use the tools we develop and treat the problems approximately. The capacity to model and approximate is probably the most powerful tool in the scientist's kit. It allows her to make meaningful conclusions and predictions even when the conditions are not 100% known and her treatment of the intricate processes is somewhat simplified. (Indeed, in the world outside of an introductory textbook, this is the situation all the time!)

You, the students of this course, might not become scientists or athletes. However, you are becoming educated citizens who vote and participate in an increasingly technological society with increasingly complex and pressing problems. It is also a society increasingly filled with disparate voices proclaiming or attacking scientific "truths" on charged, complicated issues. As a citizen and leader in that society, you should have a sense of how scientists can make real, meaningful statements even when their understanding or treatment of a situation is incomplete. In the years to come, you may or may not remember who came out on top in last year's March Madness. Either way, let's see if we can garner something even more out of a scientific treatment of its plays.

Game on!

PREFACE TO THE INSTRUCTOR

This book is intended as a textbook for a one-semester or one- to two-quarter undergraduate course for students not necessarily intending to major in physical science, engineering, or a related field. At many major institutions, all students are required to take at least one physical science. Some may not look forward fulfilling this requirement; indeed, some may really fear physics, though they are more likely to say they "hate" it instead. It is hoped that a student's natural interest in athletics and the direct relevance to concrete material will bridge the gap for students turned off by the seemingly abstract stuff covered in many undergraduate physics courses.

There are a large and growing number of excellent books on physics and sports; a list of some of my favorites may be found at the end of this book. Why, then, write another? Well, these books, well written, educational, and often entertaining, are simply not *text*books. Important physics concepts—force, velocity, and torque—come into the discussion, interesting facts are given, and occasionally a formula is applied. However, the focus is typically on conveying interesting physics-related facts about a particular sport, rather than developing a general appreciation and facility for scientific reasoning, using sports examples as the vehicle of transmission.

I have taught the Physics of Sports at the university level for several years and have spoken with colleagues at various institutions who have attempted similar courses. Rather than a lack of student interest or the inherent complexity of the subject matter, a chief obstacle to the success of such courses is the lack of a textbook. Such a book should provide the general framework for an integrated course, though timely supplemental material of local interest works especially well in this type of course, particularly in the American system in which universities often have large associated sports teams. The textbook must provide logical chapter structure, cross-referencing, examples, and end-of-chapter homework problems. The excellent sports-science popularizations mentioned above are simply not designed to develop in the students the skills required to solve problems and do their own calculations; homework assignments simply don't fit well.

Even if they've never read the aforementioned books, students are likely to have watched one of the many television or online shows that present a scientific treatment of some aspect of sports. The shows feature popular sports figures and excellent graphics and video. Occasionally the scientific treatment is wrong. More frequently it is partly wrong (though often in unimportant ways). Almost always the shows are interesting. And almost never is the scientific treatment complete. By this last statement, I mean that viewers come away from the program with the sense that something has been taught or explained to them, when in reality something has been *told* to them. There is a difference. Without some basic instruction in physical concepts—instruction that is neither the purpose nor the focus of these programs—the viewer actually benefits little from the show. Your job is to provide the basic instruction, and this book is intended to help. In your course, you may want to present and discuss these television or online shows. They provide excellent and interesting opportunities for discussion and education. What might be considered as flashier "junk food" competition for your more rigorous instruction can be instead an aid.

Academically, the material covered is mostly classical kinematics and mechanics, the Newtonian science of motion. (Some elementary biochemistry is needed and covered when energy and power are discussed in chapters 7 and 9.) Basic concepts (for example, acceleration, force, torque, momentum, and energy) and interrelationships between them are taught largely through examples relevant to sports. Considerable care went into selecting appropriate examples. In typical introductory mechanics courses, students are continually told to ignore air and friction; indeed, the effects of air might never be treated at any point. This may work for beginning engineering students, but to the typical student taking a course on the physics of sports as a science requirement, a discussion of "golf on the moon" only confirms his suspicion that physics is a contrived, irrelevant study.

On the other hand, most "realistic" examples are quite realistic indeed, characterized by a complicated interplay of many effects of approximately equal importance, impossible to calculate directly, and difficult to measure. The majority of examples that upon first thought seem perfect, fall into this category. I have tried to identify cases that are sufficiently compelling and realistic, but yet may be used to make a crisp point in the science under discussion. The overlap region in the Venn diagram of these two categories is surprisingly thin.

Scientific reasoning involves math, including simple algebra and graphs, and an appreciation of numbers, including estimation of orders of magnitude. The mathematics in this book are basic; no calculus is required. In fact, the quadratic equation is covered only in one section, which may be skipped. However, it differs from many algebra-based undergraduate courses in that when algebra is not sufficiently powerful to solve a given problem, we resort to approximate solutions, as in chapter 5 on aerodynamics. Equations that cannot be solved analytically can often be solved iteratively, as in section 9.2.4. Simple tools like these are powerful and familiar to the scientist. Beginning students have less trouble with these tools than one might initially guess, and there is real value in the realization that quantitative understanding can be gained even if the problem is not sterile and textbook-tidy.

Any teacher soon realizes how sorely mathematic and analytic skills are lacking in many highly educated and intelligent young students. These intelligent young college students may or may not become engineers or scientists, but they will become the leaders and voters of tomorrow. The issues they will be facing are more complex and science-related than ever before, and in many cases the stakes may have profound implications for the near-term future of our species. Few among them will be prepared to fully understand these highly complex issues from first principles. Like most of us, they will need to rely on various sources, some with vested interests and more interested in influencing opinion than in conveying the truth. Both the individual and society as a whole benefit immensely when the former has even a basic sense of the scientific approach.

Content, Structure

There are several possible ways to structure a book on the physics of sports. For the physics teacher, one natural path to follow would be along the lines of a standard physics sequence, the development of each topic (for example, forces) building on the previous topics (for example, kinematics). The points could be illustrated by use of sports examples, for example, in the chapter on momentum that contains discussions on football tackles, baseball bunts, etc. We have followed this route in the past, with some success. However, for the students the discussion can sometimes become rather stilted or contrived. Because the chapter only focuses on a given concept, only a particular aspect of each sports situation is highlighted. It becomes repetitive.

There is also some allure to discussing one sport at a time—a series of chapters on football, another series on tennis. This allows for a smooth narrative perfectly suited to the sports science popularizations on the market. However, in a science course, the physics concepts do need to be developed pedagogically at some point. If the class starts with a focus on tennis, then force, momentum, etc., will need to be developed during the tennis discussion. If the next sport to be discussed is football, then it makes little sense to develop all the physics topics from scratch, as if the tennis discussion had never occurred. Continuing along these lines, it becomes clear that a certain ordering of sports would be required in such an approach. This seems insufficiently flexible in a one-semester course; by the time you've gone through the whole sequence for one sport, you may have no time to cover another sport your students find more compelling.

I have adopted something of a hybrid approach, designed to balance the desire to be compelling with the need for pedagogy. The first part of the book is composed of 10 "primary" chapters. It is expected that these chapters will be covered more or less in order. They focus on a few specific situations in athletics, chosen to naturally develop a physics topic. Boxed examples or short subsections from other sports are brought in to illustrate the physics topics further.

The general narrative of the book focuses on sports, even if science pedagogy lies beneath. By its nature, this approach leads to some atypical topical structures. For example, the instructor may be shocked to find the concept of "power" defined in a subsection of a section about a fictional movie character, rather than power having its own chapter or sharing one with energy. However, the narrative is more natural this way. To understand *why* Uncle Rico can't throw a football over a mountain, we *need* to know something about power.

The primary chapters are followed by four "supplementary" chapters. These focus on specific sports situations involving several physics concepts simultaneously. These sports situations would not be good choices as the basis of developing any of the concepts initially. It is assumed that these ideas have been covered previously (in the primary chapters), so the emphasis is on magnitude calculations and on identifying the most appropriate tool to understand a given situation. There are boxed questions and examples sprinkled throughout these chapters, but there are no end-of-chapter homework questions.

The supplementary chapters represent an array of directions the instructor might want to take. Chapter 11 looks in detail at the torques involved in a single photograph. Chapter 12 covers details of an historical event (Barry Bonds's 756th home run). Chapter 13 discusses several physical concepts in a single sport (the pole vault). Chapter 14 dissects the message of one episode of a popular television show on the science of sports, highlighting physics misconceptions that go beyond sports.

The instructor should use (or not use) these chapters as he or she sees fit. It may be best is to use one of these directions as "inspiration" and instead cover something interesting from last week's game of the school's team. Indeed, a detailed frame-by-frame analysis of a recent game downloaded from YouTube can draw students' interest as little else can. Another option is to generate a supplement based on recent sports-science articles that frequently appear in *ESPN Magazine* or *Sports Illustrated*. The hope is that, after covering the primary chapters, the instructor is free to find the most compelling situations and *apply* the physics that has previously been developed.

The Human Element

In laying out any book, decisions about boundaries must be made. I have limited discussion of biomechanics and kinesiology, deciding that their study may not illustrate general scientific principles, but rather would be extremely specific to a given joint or motion (such as a swim stroke). However, once torque and force are covered in the primary chapters, an instructor should have little problem supplementing with a biomechanic example.

In contrast, I found that at least a very basic discussion of biochemistry is necessary for any reasonable discussion of energy and power in sports. These topics are not typical in a traditional course on mechanics, and the instructor at first may be tempted to skip them. (Indeed, even after consideration, he may decide to skip them.) But I find that the students appreciate the connection between the food they eat and the power they generate. They already of course know that there is a connection; our treatment spells out details they may not have considered. Also it helps make clear the well-known phenomenon of the ability to produce powerful bursts of effort, but only for a short time.

If the instructor accepts that the biochemistry is an important part of the course, she or he may nevertheless be annoyed that its discussion is broken into three parts. Wouldn't it be neater to keep them all together? Well, yes, but the students may feel bored and wonder why this is being taught in a physics class. The material is distributed as it is in order to keep a close connection to the larger discussion that motivates covering it.

Finally, discussion of many of these nontypical topics, such as VO_2max, metabolic equivalents (MET), power-to-weight ratio (PWR), etc., can be annoying in terms of all the units and might seem rather "too detailed." This may be so, but students who read about sports or watch televised sports shows with any science in them at all hear these terms all the time. It is important for the material in this book to make connection with the articles they'll read in *ESPN Magazine*. Hence, the fact that VO_2max is measured sometimes in liters per minute and sometimes in milliliters per minute per kilogram needs to be mentioned and laid out.

Units

To the horror of some of my colleagues, I do not rigidly adhere to SI units and in fact tend to use Imperial units (lb, ft, etc.) in many discussions. Anybody who has taught an outreach course to a clientele of American nontechnical majors will understand why. It is particularly important to connect with the student's existing knowledge base; quantifying forces in newtons would seem—would *be*—contrived. Context would be lost.

Sports in particular are often tied strongly to a system of units; discussing play on a 91.74-m American football field will immediately draw the focus away from the science at hand to a pointless discussion of units. On the other hand, discussing the 10-m platform dive or 100-m dash requires familiarity with SI. To further complicate the picture, speed is often quantified in miles per hour.

As physicists, we could simplify the discussion with a tidy and strict adherence to SI units, but we would be doing our students no favors. First and foremost, we'd be wasting a precious opportunity to make a connection to the outreach student; the course would instantly become less compelling and more "academic" (read: boring). We would be ignoring the context of their everyday lives, hence adding credence to the perception (strong among the target audience) that science is irrelevant in the "real world." It would also deprive them of the experience of working simultaneously with two systems of units, a reality for twenty-first-century Americans, like it or not.

For quantities such as force, length, and power, the situation is annoying but straightforward. The question of what to do with mass is less clear. Here, Americans would seem at a distinct disadvantage, as the Imperial unit of slug is not commonly used. The SI unit of kilogram is familiar even to most Americans; the newton as a unit of force is not. (I was a bit surprised to find that even for the international students, the unit of force with which they are most familiar is the pound (lb); many had never heard of the newton.)

To address the mass-unit issue in the Imperial system, some authors turn to the "pound-mass" and "pound-force" solution, as Tim Gay does in his excellent book

The Physics of Football. (In this system, 1 pound-mass is the mass of an object that weighs 1 lb on Earth's surface; that is, 1 pound-mass = $\frac{1}{32}$ slug.) While not strictly necessary, it is natural in this approach to redefine Newton's second law to read $F = \frac{1}{32}ma$, as Gay does. Another approach is to characterize an object's mass by its weight (in whatever units) divided by g (in whatever units).

Both of these approaches are workable, but are a bit awkward and can worsen the perennial confusion between mass and force (particularly weight). Braving the ridicule of colleagues who have never taught an outreach course on sports, I follow Brancazio (*SportScience*) and other authors and boldly use slug. Simply put, it is the natural unit of mass in an admittedly unnatural system.

In a way, its use can turn the situation into an actual advantage for the American student, as the very novelty of the slug unit helps focus and clarify the distinction between mass and force in discussion.

ACKNOWLEDGMENTS

Developing and writing a new type of book like this one has been an adventure of surprising scope. I could not have written it without the help and support of many people, some of whom I'd like to thank here.

First, I'd like to thank the hundreds of students who have taken the Physics of Sports course over the past several years here at Ohio State. These bright students inspired me with their questions and interest, and they drove the development of this book. Also at OSU, I'd like to thank Prof. Terry Walker, who somehow conned me into taking on the Physics of Sports course, and Prof. Evan Sugarbaker for his administrative support of it. Thanks also to Dr. Thomas Barrett for interesting discussions and providing several end-of-chapter problems for the later chapters.

I am grateful to Prof. Daniel Cebra, who wanted to provide a unique course offering to the students at the University of California, Davis. It is a huge effort to develop an entirely new undergraduate course and an act of trust to be the first to adopt a new text. Thanks to him and them for insightful feedback and advice.

I have benefited greatly with colleagues from around the country who generously shared their enthusiasm and expertise, including Prof. Wolfgang Bauer of Michigan State University; Prof. Tim Gay of the University of Nebraska; Prof. Eric Goff of Lynchburg College; Prof. Alan Nathan of the University of Illinois; and Prof. David Wilson of the Massachusetts Institute of Technology. I was especially delighted when Prof. Nathan, baseball physics expert, agreed to author a supplemental chapter (chapter 12) on the topic.

The people at McGraw-Hill have been fantastic to work with, and only through their professional skill could a project like this come to fruition. Many thanks are due to Michael Lange, Nick McFadden, Tomm Scaife, and Thomas Timp. Special thanks to Eve Lipton and Jane Mohr, who have been absolutely instrumental. Over a year's worth of phone meetings, Eve provided patient (but not too patient!) editorial and logistical guidance, and Jane smoothly coordinated the production process. Without their hard work, this book would certainly not exist. Many thanks to Ramya Thirumavalavan, who very professionally managed the production needs of the text, and to Patti Scott for careful proofreading and suggestions for improvement. Photo researchers Danny Meldung and Julie De Adder identified and coordinated many photos. Dharma Lingam and the art team did a wonderful job with the figures.

My greatest appreciation goes to my wife and children, scholars and athletes all. Many, many weekends and nights were consumed by this book during the past three years. My name goes on the cover, but my wife, Laura, sacrificed equally much. Thanks to Oscar and Sammy for teaching me practically everything I know about sports and for infecting four-fifths of our household with a passion for the Cleveland Cavaliers. And many thanks to Anna, for her scholarly inspiration and for putting up with it all.

Reviewers

Thanks to reviewers who provided feedback on early drafts of the book:

Prof. John Di Bartolo *Polytechnic Institute of New York University*
Prof. Alan Nathan *University of Illinois*
Prof. Milun Rakovic *Grand Valley State University*
Dr. Michael Randall *the University of Wisconsin*
Prof. Lloyd Smith *Washington State University*

Primary Chapters

PART I

1 Warm-Up: Basic Concepts

The workings of the physical world can be precisely formulated and predicted in mathematical terms. These powerful concepts lead to a deeper understanding of the connection between the various aspects of athletics. The first step is to quantify these aspects.

Scientists and coaches quantify an athlete's ability through vertical jumps and sprint times, as well as with accelerometers, power meters, and other devices. (© *Science Photo Library/Corbis RF*).

The most important numbers are not on the players' uniforms.

Drew Brees of the New Orleans Saints is an awesome quarterback. Any course—physics or otherwise—that claims to prove the opposite is probably nonsense. But can we convey in words exactly how good he is? We can use general adjectives—he's smart; he's quick; and he's strong. But that doesn't get us very far. We need to use numbers to quantify his talent. We could point to his 2000 season at Purdue, where he rushed for a total of 512 yd in 95 carries, or his Superbowl winning season with the 13-3 New Orleans Saints. Or perhaps the 468 completed passes (of 657 attempts) for a NFL record 5476 passing yards in the 2011 season. But really, that can say as much about his (great) team and (awestruck) opponents, as about Brees himself.

A coach will want to know Brees's weight (207 lb), how quickly he can run 40 yd (a very respectable 4.67 s), and perhaps the force his arms can exert in a bench press (275 lb). The physics we will learn will allow us to use numbers like these to find the forces involved in a tackle, the angle at which Brees must throw a pass, the range of a ball hit with an aluminum or wood bat, the effect of backspin on a three-point shot. This information is all numerical, so we need to be able to identify relevant physical aspects and quantify them.

This short chapter introduces some of the important concepts we'll need in order to translate an athletic situation into numbers. We will also take a first look at the center of mass (also known as the center of gravity), something crucial to a quantitative understanding of an athlete. This concept is usually introduced much later in an introductory physics course, but it plays such an important role in the physics of sports that we'll need it almost immediately.

1.1 Quantifying the World of Sports

In our scientific study of sports, we will use mathematical relationships to understand the movement and interactions between players and objects. Everything needs to be quantified—speed, weight, distance, and energy. Here, we quickly cover a few of the concepts we'll need to get us started.

1.1.1 Scientific notation

In the scientific study of the world around us, numbers can range from the minuscule to the enormous. Some examples:

radius of the nucleus of the gold atom = 0.0000000000000059 m

Drew Brees's weight = 207 lb

distance of a marathon = 1,660,032 in.

number of stars in the universe = 9,000,000,000,000,000,000,000

Scientific notation is used to handle these awkward numbers more easily. It essentially "counts the zeros," rewriting the number as something times a power of 10. For example:

radius of the nucleus of the gold atom = 5.9×10^{-15} m

Drew Brees's weight = 2.07×10^2 lb

distance of a marathon = 1.66×10^6 in.

number of stars in the universe = 9×10^{21}

Notice that because $10^{-2} = \frac{1}{10^2} = \frac{1}{100} = 0.01$, a negative exponent moves the decimal point to the left.

If you were to enter the number of stars into your calculator, you might type something like **9E21**, but calculators differ. Play around and make sure you know how to enter numbers like this on yours.

As it turns out, the need for scientific notation in our study of sports will not be great. This is because we generally will not quantify the length of a marathon in terms of inches, but instead will use miles (26.2 mi) or kilometers (42.2 km). Likewise, the distance required for a first down in football is 0.0057 mi = 5.7×10^{-3} mi, but we would more naturally describe it as 10 yd.

1.1.2 Units and conversions

Quantification requires more than numbers, of course. If we hear of a person who ran 100, we don't know whether he's an Olympic 100-m sprinter, a football player returning a ball 100 yd down the field, or an insane distance runner covering 100 mi.

Quantifying sports will require frequent unit conversions.

It's in fact an absolute rule that any number representing a physical quantity (length, time, force, and so on) must include a unit of measure.[1]

We will need to be conversant in both the **Imperial (English) unit system** and the **SI (Système International) unit system**, also known as the metric system. Imperial units are much more common when describing American sports—we don't usually talk about a 91.7-m football field, even though this is the same as a 100-yd field—whereas SI units dominate most international or foreign sports. In both systems, there are units that characterize large distances (such as miles in Imperial and kilometers in SI) and smaller distances (such as inches and centimeters). It is often necessary to know how many centimeters there are in a kilometer, or inches in a mile.

At the time of this writing, Burma, Liberia, and the United States are the only three nations that have not adopted SI as the official system of measurement. By sticking with the Imperial system, we Americans have made it somewhat hard on ourselves:

Imperial (British)	SI (Metric)
1 mi = 5280 ft	1 km = 1000 m
1 ft = 12 in.	1 m = 100 cm
1 lb = 16 oz	1 kg = 1000 g
1 gallon = 8 pints	1 liter = 1000 ml

To convert units in the SI system, we use the **SI prefixes**. There are 20 of these, but we'll only be using three:

kilo	k	$1000 = 10^3$
centi	c	$0.01 = 10^{-2}$
milli	m	$0.001 = 10^{-3}$

So, 1 mg = 0.001 g (1 milligram = $\frac{1}{1000}$ gram). Obviously, the relationships between units in Imperial system are more complicated.

Appendix A lists various **conversion factors** between different units. For example, 1 mi = 1.61 km. It can be useful to write this another way:

$$1 \text{ mi} = 1.61 \text{ km} \quad \rightarrow \quad 1 = \frac{1.61 \text{ km}}{1 \text{ mi}}.$$

When converting a quantity from one unit to another, it is worthwhile to remember that "**you can always multiply anything by one.**" As an example, let's say a pitcher throws a blazing fastball at $v_{ball} = 100$ mph, and we need to know that speed in meters per second $\left(\frac{m}{s}\right)$ to answer a question about the physics of sports. We can just multiply 100 mph by one, three times:

$$v_{ball} = 100 \, \frac{\text{mi}}{\text{hr}} \times \frac{1.61 \text{ km}}{1 \text{ mi}} \times \frac{1000 \text{ m}}{1 \text{ km}} \times \frac{1 \text{ hr}}{3600 \text{ s}}$$
$$= \frac{100 \times 1.61 \times 1000}{3600} \, \frac{\text{m}}{\text{s}}$$
$$= 44.7 \, \frac{\text{m}}{\text{s}}.$$

Mathematically, units divide and multiply just like numbers. Unit conversion is most easily expressed by "canceling" units in a fraction.

The point is that units multiply, divide, and cancel in fractions in just the same way that numbers do. This is important in any mathematical science (like physics), and we will see this repeatedly in the rest of the book. Do yourself a favor and always include units in any analysis of a sports situation.

---

[1] Purely mathematical numbers like π don't represent physical quantities, so they have no units.

EXAMPLE 1.1 Which "one" to use?

Both are called "football" by their fans, but American football and Association Football (soccer) are quite different games. Soccer is one of the few sports for which the dimensions of the pitch (or "field," as it's usually called in the United States) are not precisely specified. Rather, a range of allowed dimensions are specified by FIFA's "Law 1." The length of the pitch can be anywhere from 90 to 120 m, while its width must be between 45 and 90 m.

Could a FIFA-endorsed football pitch be laid out on the turf of an American football field whose dimensions are 300 ft by 160 ft?

To make the comparison, we need to convert the soccer field dimensions into Imperial units. Using the conversion found in appendix A,

$$1 \text{ m} = 3.28 \text{ ft} \quad \rightarrow \quad \begin{cases} 1 = \frac{1 \text{ m}}{3.28 \text{ ft}} = 0.305 \, \frac{\text{m}}{\text{ft}} \\ 1 = \frac{3.28 \text{ ft}}{1 \text{ m}} = 3.28 \, \frac{\text{ft}}{\text{m}} \end{cases}$$

So, we see that there are *two* possible values for "one" that we can use. Which is the "right" one?

As it turns out, they both are right, so relax. Let's use the first one to convert the shortest possible length of a soccer field:

$$L_{\text{soccer field}} = 90 \text{ m} \times 0.305 \, \frac{\text{m}}{\text{ft}} = 27.45 \, \frac{\text{m}^2}{\text{ft}}.$$

As a matter of fact, this *is* the true length of the field—it is not "wrong." However, it is in units that aren't very helpful to us.

Let's try the other value for "one":

$$L_{\text{soccer field}} = 90 \, \cancel{\text{m}} \times 3.28 \, \frac{\text{ft}}{\cancel{\text{m}}} = 295.2 \text{ ft}.$$

Just as the numerical parts of the equations multiply and divide, so do the units. By taking care of this, we find the useful information we are looking for: a soccer field should be at least 295.2 ft long, just fitting on an American football field in the long dimension.

The minimum width of the field is

$$W_{\text{soccer field}} = 45 \, \cancel{\text{m}} \times 3.28 \, \frac{\text{ft}}{\cancel{\text{m}}} = 147.6 \text{ ft},$$

a bit less than the width of an American football field.

So, an American football stadium can host a regulation soccer match, simply by re-lining the field.

1.2 When We Don't Have Exact Numbers

In this book, we will look at concrete examples of sports situations and analyze them scientifically. As mentioned, this means that we will need numerical information about the situation: How fast was the ball going? How high did he jump? What was the angle between his leg and the floor? What was the force on the ball? However, except for special cases set up for educational television shows, there is no scientist who measures these things as they happen. It will be up to us to estimate the information, often from photographs or by frame-by-frame analysis of videos.

Because our analyses will often rely on these approximate numbers for input, our conclusions will be approximately correct. For example, based on slightly inaccurate information about a pole vaulter's approach speed, we may predict that he will clear 16 ft 10 in., when in reality he clears only 16 ft 8 in. The important part of our study of the physics of sports will be to develop the scientific *principles* behind athletics.

There are two tools we will use repeatedly to put sports situations into numbers that we can analyze scientifically. The first is estimation, usually based on photographic or video records of the event. The second is the proper use of scales characteristic of athletics—the typical speed of a football player, the typical stride of a sprinter, and so on—when a key piece of information about a sports situation is unknown.

These important tools are used constantly by professional physicist studying the natural world. Interestingly enough, they are seldom emphasized or taught in an introductory physics course for science or engineering majors, where the situations are significantly more idealized and less "messy" than the real world of sports.

1.2.1 Estimation with photos and videos

The modern age provides a huge boon to any study of the physics of sports, with the wide variety of free, relatively high-quality video clips available online. Usually the day after any given event, key plays are posted at sites like YouTube, and one needn't be an expert to perform an analysis of a sports situation. There are many free or low-cost software tools that will enable you to download the videos and step through them frame by frame. This allows us to quantify the motion of a person or object, and much of the study of sports is the study of motion.

An example is shown in figure 1.1, which shows five frames from a downloaded video of tennis player Roger Federer delivering one of his flawless serves. How fast is he serving, in this clip? To answer this, we need to know how far the ball travels in a given amount of time.

Standard videocameras and most posted videos have a frame rate of $r = 30$ frames per second (fps); this information is usually encoded in the movie file and is easily displayed in standard players like QuickTime. The movie used in figure 1.1 was taken with a high-speed camera and corresponds to $r = 120$ fps.

We need this information to measure the time of the motion. In this case there are 120 frames in 1 s, so the time between subsequent frames is $\frac{1}{120}$ s = 8.33 ms. The ball and racket make contact in the top frame near the elbow of the lady in pink (as seen from our vantage), after which the ball moves to the right. By the fifth frame, the ball has been in the air for 4×8.33 ms = 33.3 ms.

We can estimate how far it has traveled in this time by using other objects for scale. The 6 ft 1 in. Federer provides a convenient choice. The yellow arrows in the figure compare the ball's flight distance to Federer's height. Based on this, we estimate that the ball has traveled roughly 6 ft. (We could do a more careful analysis—with a ruler, for example.) The ball's speed is therefore

$$\text{speed} = \frac{6 \text{ ft}}{0.0333 \text{ s}} \times \frac{1 \text{ mi}}{5280 \text{ ft}} \times \frac{3600 \text{ s}}{1 \text{ hr}} = 123 \text{ mi/hr}.$$

Indeed, Federer regularly delivers his victims serves at speeds of 120 and even 130 mph.

This was an example of a simple video analysis. You may perform more-sophisticated ones as your study progresses. All cases, however, will involve estimations of position and time.

Section 1.2   *When We Don't Have Exact Numbers*   **7**

Figure 1.1. Five frames from a high-speed video of a serve by Swiss tennis star Roger Federer. He makes contact with the ball in the first frame. The fifth frame is blown up to the right. (© *Game by Frame Tennis*).

1.2.2 Typical scales

Sometimes we will want to analyze a situation in which some crucial information not only is not precisely known, it is not even recorded in a photo or video. What do we do then?

In such situations, we should be comfortable filling in these blanks with "typical values." For example, in studying some situation, we may need to know the height at which a basketball is released during a jump shot in the NBA. If we don't know

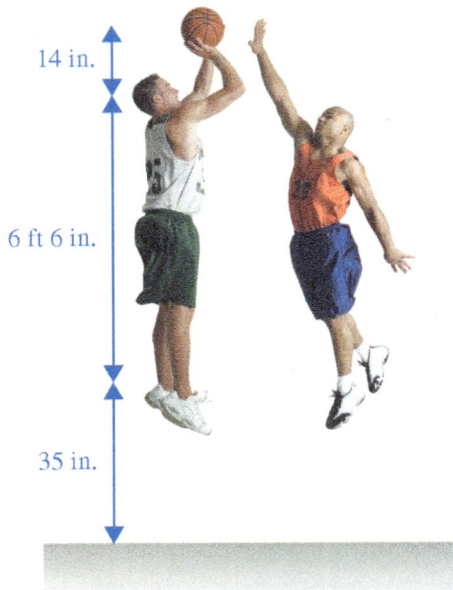

Figure 1.2
"Typical" sizes of a basketball player taking a jump shot. (© *Rubberball/Getty Images RF*).

the precise value and don't have a photo from which we can get an estimate, we just have to fill in the blank ourselves.

We might say that the typical NBA player is about 6 ft 6 in. A good vertical jump might be 35 in. (about the same as the inseam for a 6 ft 6 in. man) and his arm might extend about 14 in. above his head, during a jump shot. See figure 1.2 for a visualization. In this case, the release height of the ball is easy to find:

$$\text{release height} = 35 \text{ in.} + 6 \text{ ft} \times \frac{12 \text{ in.}}{1 \text{ ft}} + 6 \text{ in.} + 14 \text{ in.}$$
$$= 35 \text{ in.} + 72 \text{ in.} + 6 \text{ in.} + 14 \text{ in.} = 127 \text{ in.} = 10 \text{ ft } 7 \text{ in.}$$

(Notice how the "feet" part of the player's height needed to be converted to inches here.)

It may be that you are not familiar with the typical height of a basketball player or a good vertical leap. In this case, peruse the Internet or magazines to get some idea. As for the 14 in. extension of the arm above the head? Grab a tape measure and do the measurement on yourself—that's not unscientific!

Knowledgeable fans may point out that a player usually doesn't jump as high as he possibly can, when shooting a jump shot. Therefore, a player whose vertical is listed at 35 in. may rise only 31 in. during a shot. There are a number of such potential inaccuracies, but they are usually relatively small.

The point is that you should feel comfortable filling in missing information with "reasonable" numbers, based on real-life situations.

1.3 The Center of Mass

Roger Federer flexes his body like a spring, leaps up, and crushes a serve. German goalkeeper Oliver Kahn sails horizontally through the air to block a brutally curving free kick. American pole vaulter Jenn Suhr executes precise and complicated choreography as she strains for precious inches, more than 16 ft in the air. San Francisco's Ted Ginn Jr. jukes his way through a rushing onslaught of defenders as he returns a punt. A diver launches from a platform (figure 1.3).

Movement defines sports, and the complexity of the movement achievable by the human body leads to the immense richness of athleticism appreciated by both participants and fans. In addition, the equipment itself—from the flexible pole used by Suhr to the crazily bouncing football grabbed by Ginn—further challenge the scientist who wishes to quantify and analyze the motion in a sports situation.

For the first several months in an introductory physics class for scientists or engineers, books and professors teach about the physics of "point particles" (tiny objects of zero size, something found only in subatomic physics) or at best totally rigid objects that cannot change shape. This obviously will not do for us. Right at the outset, we need to develop a concept usually covered much later in a course for scientists or engineers—the center of mass.

In our discussion of motion, we often discuss how "the position" of an object changes with time. For our hypothetical point particle, "the position" is a straightforward concept. But what about a baseball, which has 108 double-stitches all at different positions and following different trajectories as they revolve about the center of the spinning ball even as the ball curves through space? Your intuition probably tells you that the position of the *center* of the ball is the most appropriate thing to associate with "the position" of a baseball. And your intuition is correct.

Figure 1.3
What happens next? One thing is for sure—she's not going to hold that pose! She is going to "fall" due to the relative position of her center of mass and her base of support. (© *Getty Images RF*).

Figure 1.4. The center of mass sits at the average position of the matter comprising an object or person.

The center of a baseball or of any spherical ball is the "average" position of the matter comprising the ball. This average position is called the **center of mass**, or c.m., of the ball. Some other bodies are shown in figure 1.4. Due to its obvious symmetry, a balanced barbell has a center of mass that is located on the bar itself, midway between the weights. If one side has more weight than the other, then the matter is, on average, located toward the end with more weight.

One of the most important assets of the best athletes is their ability to precisely control the position of their center of mass. The human body is wonderfully malleable, and it turns out that the "average position" of the matter in one's body can even be located where there is no matter at all! This is similar for a donut: clearly by symmetry, the "average position" of a donut's mass lies at its empty center. Further examples can be seen in figure 1.5, taken from a gymnastic coach's handbook.

 The center of mass is the average position of the atoms making up a body.

1.3.1 Some facts

We will discuss many factors involving the center of mass, sometimes called the center of gravity,[2] as it applies to different athletic events. We will not delve into the details of finding the center of mass mathematically, which usually involves integral calculus. Rather, we'll start here with a few essential facts.

- The center of mass of a typical man standing at rest is just above the height of his belt buckle.

- Due to the different mass distribution, the center of mass of a woman is about 2 in. lower than that of a man of the same height.

- The position of a person's center of mass changes as she changes her body shape, as illustrated in figures 1.4 and 1.5.

- When discussing the forces acting on a person or an object, as we'll do in chapter 3, it is the reaction of the center of mass that matters most.

- A person is stable—that is, able to maintain her position without falling over—if her center of mass is above her base. The *base* is simply the area surrounded

---
[2] Technically, the center of gravity and the center of mass are not precisely the same, but there is no meaningful difference for our discussion of sports.

Figure 1.5. The center of mass for a gymnast in several common positions. From Tony Smith's *Gymnastics: A Mechanical Understanding*, a 1982 guide for gymnastic coaches.

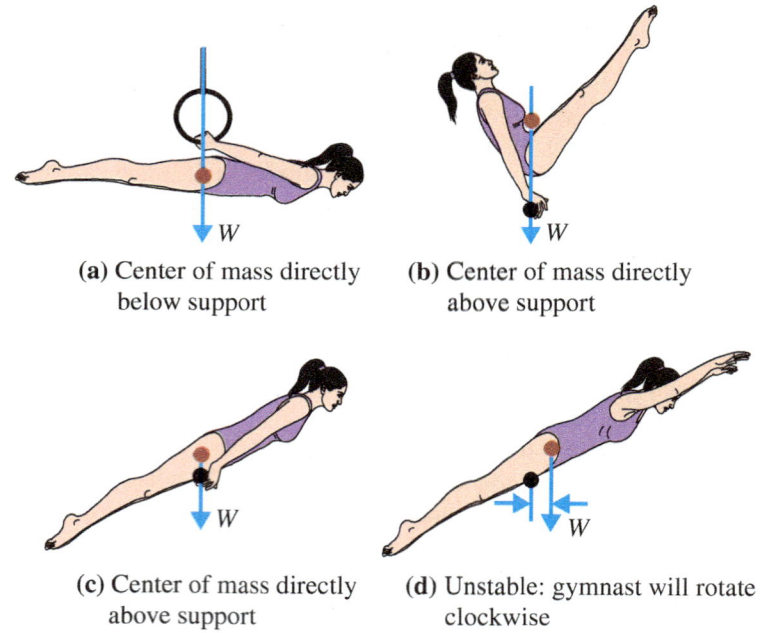

Figure 1.6. A gymnast is stable—i.e., he can remain stationary for as long as his strength holds out—if his center of mass lies directly above or below the point of support. The point of support is the same in (c) and (d), but in (d) the gymnast has raised his arms, moving his center of mass to the right. From *Gymnastics: A Mechanical Understanding*, by Tony Smith.

by her points of support on the ground. For example, in figure 1.4 the woman's base is the area between her feet; she is not in danger of toppling, because her center of mass is above this area. In contrast, in figure 1.3 there is no chance that the diver could maintain—or regain—her balance in this position.

- If a person has a *single* point of support, he is stable only if his center of mass lies directly above or directly below the point of support. See figures 1.6 and 1.7 for examples.

If person's center of mass is not above his base of support on the ground, he topples over.

(a)   (b)

Figure 1.7. Performers of various stripes intuitively know and exploit the laws of physics to strike dramatic poses. Compare these with the diver's pose in figure 1.3. (a) A member of the Ohio State University marching band performs during a matchup between the Buckeyes and the Nebraska Cornhuskers, October 6, 2012. Through precise placement of hands, baton, and knees, the center of mass is placed directly above the points of support (the feet). (© *David Dermer/Diamond Images/Getty Images*). (b) Elvis's legendary balance came from his instinctive understanding of Center of mass. (© *Metro-Goldwyn-Mayer/Sunset Boulevard/Corbis*).

- While an athlete's various body parts may execute beautifully complex motions while he soars through the air (think of the basketball player making a dunk, a pole vaulter flexing his back), his center of mass will always follow the extremely simple trajectories we'll study in chapters 2 and 4.

- For a freely spinning body—a diver performing somersaults and twists, or a football toppling end-over-end as it soars through the air—the rotation is always around the center of mass.

We shall have much more to discover about this topic throughout the book. For chapter 2, keep in mind that when I talk about "the position" of a person or an object, I am referring to the center of mass.

Problems

1. Express the following distances in meters, using scientific notation:
   (a) the length of a football field
   (b) the distance of a marathon (26 mi + 384 yd)
   (c) a good basketball player's vertical leap (40 in.)

2. Choose the slowest among the following speeds:
   (a) 220 ft/s
   (b) 100 mph
   (c) 70 m/s
   (d) 300 km/hr
   (e) 70 yd/s

3. An excellent time for the 1500-m race is 3 min 31.4 s. Express this time in (a) minutes and (b) seconds.

4. Europeans often brag that they drive 100—that is, 100 km/hr. How fast is this in
    (a) mph?
    (b) ft/s?

5. Express Terrelle Pryor's weight (233 lb) in Newtons. (See appendix A for conversion factors.)

6. Figure P1.6 shows a woman touching her toes.
    (a) Why does she instinctively angle her legs backward away from vertical?
    (b) Try, yourself, to touch your toes (or reach for them, anyway) with your legs straight and vertical. To make sure they remain vertical (and they don't tilt backward like the woman's), stand with your heels against a wall and then bend over. Your body and instinct will teach you this important point more convincingly than will any book.
    (c) In January, 2008, the *New York Times* ran a story about UNC-Asheville's basketball player Kenny George, who is 7 ft 7 in. tall! Asheville has a contract with Nike, who made 12 pairs of size-26 shoes for George.
    The formula for American men's shoe size is
        Size = 3 × length in inches − 24
    If the woman in the figure had Kenny George's feet, would she be able to maintain her balance if she did not angle her legs backward, but kept them vertical?

7. (a) In horsepower, what is the power expended by a 65-W lightbulb? (See appendix A for conversion factors.)
    (b) How many 65-W lightbulbs does it take to expend as much power as a 0.5-hp lawnmower?

8. In physics, Hooke's Law describes the force F exerted by a spring when it is stretched a distance x. Force is measured in Newtons (N), and distance is measured in meters (m). For a spring with "spring constant" $k = 10$ N/m, which of the following formulas are *definitely wrong*, based solely on the units involved?
    (a) $F = kx$
    (b) $F = kx^2$
    (c) $F = x/k$
    (d) $F = x^2/k$
    (e) $F = kx/2$
    (f) $F = kx^2/2$
    (g) $F = k\sqrt{x}$

9. As I will discuss in chapter 2, the acceleration of a runner is measured in ft/s^2. Below are some formulas for the distance x a runner travels in some time t. Based solely on units, which of them are *definitely wrong*?
    (a) $x = at$
    (b) $x = a\sqrt{t}$
    (c) $x = a^2 t$
    (d) $x = at^2$
    (e) $x = at^2/2$
    (f) $x = a/t$
    (g) $x = a/t^2$

Figure P1.6
To maintain a stable pose, this woman must keep her center of mass above her feet. (© *Stockbyte/PunchStock RF*).

Section 1.3

10. What might be a reasonable estimate for the number of strides an Olympic athlete takes in the 100-m dash?

11. In the horizontal direction, about how far is the diver's center of mass from her feet, in figure 1.3?

12. (a) In miles per hour, how fast does the typical marathon runner run?
    (b) In miles per hour, what is the typical speed for a fastball pitch in the major leagues?
    (c) About how far can a college track athlete run in one minute?

13. How many American football fields (ignore the endzones), laid end-to-end, make a mile?

14. Europeans always tell their height in centimeters. What is yours, in centimeters?

15. Estimate the number of times a basketball is dribbled (hits the floor while being dribbled) during one 12-min quarter of an NBA game. Describe how you arrived at your estimate.

16. In one variant of rugby (rugby union), when the ball goes out of bounds ("out of touch" in rugby terms), the ball is thrown in in what is known as a line-out. This is more or less equivalent to a throw-in in soccer. In a line-out, teams can lift players into the air to help them intercept the ball. (In rugby league, the other variant, line-outs are not used. Instead, when the ball goes out of touch, a scrum—one of those giant huddle-looking things with both teams pushing against each other—is held near the spot that the ball left the field.)

    In figure P1.16a, Leinster (blue uniforms) have captured the ball on a line-out.

    (a) Where is the center-of-mass of the three-person configuration?
    (b) The player on the left appears to be bearing most of the weight of his teammate. If the player on the ground to the right would suddenly vanish, how would the center of mass of the remaining two (taken together) differ from that of the three-player combination?
    (c) Again, if the player on the ground to the right would suddenly vanish, could the remaining two players maintain their pose without toppling over? (Strength is not the issue—assume the players are infinitely strong.) Explain your answer.
    (d) Now look at figure P1.16b, Assume that the athletes are infinitely strong. If the cheerleader on the right would suddenly vanish, could the remaining two maintain their pose without toppling over? Explain your answer.

17. (a) In figure 1.7a, the drum major extends his hand with the baton as far forward as possible. Is that important for his balance?
    (b) Would this part of the routine be easier with a heavy baton or a light one?
    (c) Would this maneuver look more dramatic with his hat on (as shown) or off? (This is an opinion.)
    (d) Would this maneuver *be* more difficult with his hat on or off? (This is not an opinion.)

18. Acceleration a is measured in m/s^2, and speed v in m/s. If $Z = v/a$, then is Z a time, a distance, a speed, an acceleration, or something else?

Figure P1.16a

29 March 2014; Devin Toner, of Leinster, and Paul O'Connell of Munster, contest a lineout. Celtic League 2013/2014, Round 18, Leinster v. Munster, Aviva Stadium, Lansdowne Road, Dublin. (© *Brendan Moran/Sportsfile/Corbis*).

Figure P1.16b

A trio of cheerleaders strike a pose. (© *Rubberball/Getty Images RF*).

Racing, Mathematically

Complex interactions between athletes and the environment determine intricate motions in sports. In this chapter we develop the mathematical language of motion, the first step toward a scientific understanding of the physics behind sports.

The motion of a snowboarder carving virgin snow on the edge of the board can be mathematically formulated in terms of his position, velocity, and acceleration, each of which change with time. (© *Purestock/Superstock RF*).

Remember, it's not how fast you get to the finish line, it's . . . well, actually it **is** *how fast you get to the finish line.*

—Dan Lipinski

In ranking draft picks, football coaches pay great attention to a player's time in the 40-yd dash. Focus on this particular distance derives from the average range of a punt; the average punt in the NFL is a little longer than 40 yd, and has a hangtime of 4.5 s. So if a player on the punting team could dash 40 yd in 4.5 s, he could arrive at the receiver at the moment when the receiver catches the ball, forcing a fair catch in the typical punt. (Of course, defending blockers might have different ideas!) In any event, the 40-yd dash is one standard benchmark for football players in all positions.

But what, exactly, does a 40-yd time quantify? Does it give an indication of a player's "juking" ability? (And how do we quantify a "juke"? More on that later.) Is the player with a faster 40-yd time able to exert more force? Is he more powerful? If a player has a 6-s 40-yd time, does that mean he can do 20 yd in 3 s? or 80 yd in 12 s?

In order to understand this standard benchmark and what it really tells us about a player's ability, we will need to understand quantities such as speed, velocity, and acceleration, as well as the relationships between them. As we discuss these quantities, we'll need to take care: terms such as *speed* and *velocity* are used interchangeably in everyday conversation, but in a scientific discussion of sports they have distinct and precise meanings. In fact, throughout this book we will find terms (*momentum*, *energy*, etc.) that appear in both science and colloquial vernacular, and we'll see the importance of careful consideration of meanings.

Much of sports is defined by motion. In this chapter, we will be putting motion into the language of mathematics. Scientifically, this *description* of motion is called **kinematics**. Later on, we will get to physical formulas and relationships that *explain* motion and how to control it; that is the more interesting area of **dynamics**—of energy and slapshots, collisions and power. All of it builds crucially on the simple yet powerful concepts and relationships we develop in this chapter.

To get us started, we will focus in this chapter on straight-line (or one-dimensional) kinematics.

2.1 Phelps in Beijing

2.1.1 Speed

In the 2008 Summer Olympics in Beijing, American Michael Phelps thrilled the world by winning a record eight gold medals. Among the seven world's records he set was that for the 400-m individual medley, in which two 50-m lengths each are swum in butterfly, backstroke, breaststroke, and freestyle (Phelps used the front crawl for the freestyle). His split times are shown in table 2.1.

An important quantity for any race is the athlete's *speed*. The *average speed* is written as

$$\text{average speed} = \frac{\text{distance traveled}}{\text{time elapsed}}. \qquad (2.1)$$

So, the average speed of Phelps's second length is

$$\frac{50 \text{ m}}{54.92 \text{ s} - 25.60 \text{ s}} \approx 1.705 \text{ m/s}.$$

Notice that to get the time elapsed, we often take the difference between two numbers, in this case the difference between the starting and ending times of the leg. It is common to use the capital Greek letter delta (Δ) to denote quantities that are the difference between two numbers, so we will usually use the symbol Δt to denote "time elapsed."

2.1.2 Position and graphs

It is often useful to make a *graph* to denote the motion of an object or athlete. The graph of Phelps's swim is shown in figure 2.1, and it contains more information than simply the split times. On the horizontal axis is the time in seconds, and on the vertical is Phelps's position. The position in one-dimensional motion is usually given as the symbol x (which is why we are careful not to call the horizontal axis in the graph the "x-axis"). In sports, we often talk about motion—how an object's position changes with time—so let's pause for a moment here to make three important points about position. They are illustrated by figure 2.2, which shows Phelps's position at some point in the race.

Table 2.1. Split times for Michael Phelps's gold-medal-winning 400-m individual medley in the 2008 Olympics in Beijing.

Leg of race	Cumulative time
Butterfly out	0:25.60
Butterfly return	0:54.92
Backstroke out	1:25.50
Backstroke return	1:56.49
Breaststroke out	2:31.26
Breaststroke return	3:07.05
Freestyle out	3:35.99
Freestyle return	4:03.84

16 Racing, Mathematically

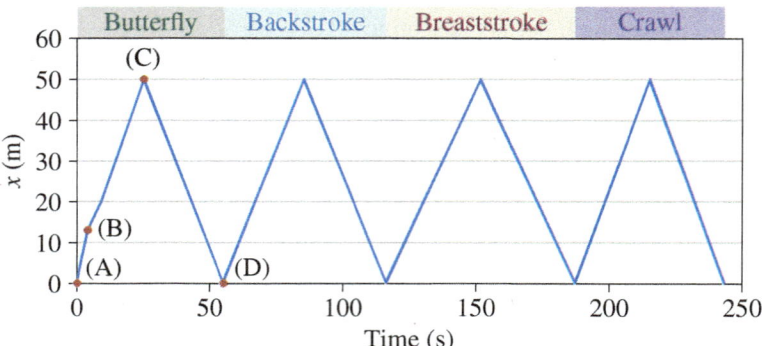

Figure 2.1. Michael Phelps's world-record-setting 400-m medley race in the 2008 Summer Olympics. Phelps's displacement relative to the edge of the pool is plotted on the vertical axis, versus time on the horizontal.

1. Any position must always be specified *relative* to an origin; the position at the origin is defined to be $x = 0$. In the graph shown in figure 2.1, we have set the origin at the edge of the pool where the race started, so that $x = 50$ m is the position of the opposite side of the pool. This seems natural enough, but sometimes we might want to use the camera sitting at the bottom of the pool as the origin. In that case, when Phelps starts the race, $x = -25$ m, and when he kicks off at the other end, $x = +25$ m. Either one is fine, but as we'll see in our examples, once we decide on an origin, we need to stick with it.

2. This immediately raises the second point: the position can be negative! When we choose an origin, we also have to choose which direction is positive. In our figures and examples, we'll usually make left the negative and right the positive direction.

3. By x, we are indicating only the horizontal *component* of Phelps's position. His height above the camera on the floor or below the diving block is insignificant. We'll deal with other components later on.

2.1.3 Displacement and average velocity

What is usually important is the change in position, or the *displacement*. Again we use the symbol Δ and define an athlete's displacement when he travels starting at point 1 and ending at point 2 by

$$\Delta x = x_2 - x_1 . \tag{2.2}$$

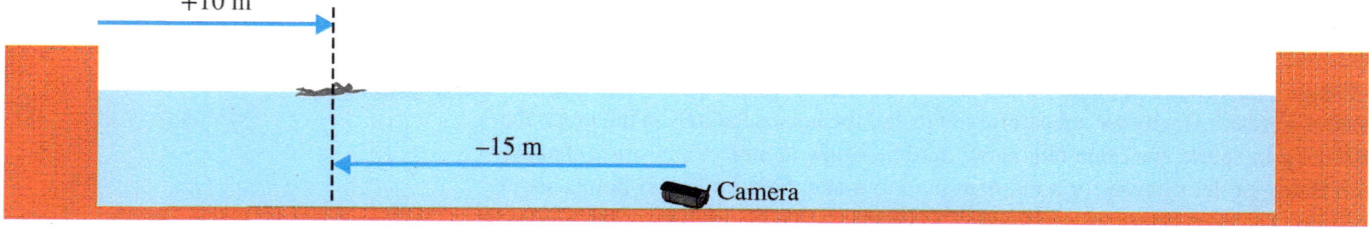

Figure 2.2. A not-to-scale illustration of Michael Phelps at some point in the race. If the diving block is used as the origin, his horizontal position is $x = +10$ m. If the camera at the center of the pool is used, it is $x = -15$ m.

Let's look again at Phelps's second length. In that case, if we use the diving block as the origin,

$$\Delta x_{(\text{2nd length})} = 0 \text{ m} - (+50 \text{ m}) = -50 \text{ m}.$$

If we use the camera at the bottom of the pool as the origin, then

$$\Delta x_{(\text{2nd length})} = -25 \text{ m} - (+25 \text{ m}) = -50 \text{ m}.$$

That's a nice thing about displacement—it's independent of the choice you make for the origin.

The *average velocity* is just the displacement for a particular part of the journey, divided by the time elapsed during the journey.

$$\bar{v}_x = \frac{\Delta x}{\Delta t}. \tag{2.3}$$

The bar above the v indicates "average," and the subscript x indicates that this velocity relates to horizonal motion.

The units for velocity are the same as those for speed—usually mph or m/s—and the two quantities are closely related. However, they are not the same. For example, the average velocity for Phelps's second length is

$$\bar{v}_{x,(\text{2nd length})} = \frac{-50 \text{ m}}{54.92 \text{ s} - 25.60 \text{ s}} \approx -1.705 \text{ m/s}.$$

This is the same as his average speed for that length, but with a minus sign.

2.1.4 Vectors and scalars

We need to pause here to discuss a couple of mathematical terms. We'll get back to the race in a moment.

Displacement and velocity are examples of **vector** quantities. They have a *magnitude*, which is always a positive number: the magnitude of Phelps's average velocity in the second length is 1.705 m/s.

Vectors also have a *direction*—they can point left, right, northwest, up-and-right, and so on. Because of this, we will usually write vector quantities with arrows over them; for example, we can write average velocity as $\vec{v}$. To indicate the magnitude of a vector in an equation, we put bars around it: $|\vec{v}|$.

When we're talking about just the horizontal *component* of a vector, we get rid of the arrow and use the "x" subscript. (For the vertical component, we use the "y" subscript, as we'll see soon.) In this chapter, we are discussing motion in one dimension only, so we'll only need individual components.

The *sign* indicates the direction of motion. The minus sign on the horizontal component of the average velocity above tells us that Phelps traveled to the left in that part of the race.

The distance traveled can never be negative, so the average speed is always a positive number—it does not depend on the direction of travel. Your car's speedometer tells you the car's speed—it does not register a negative number when traveling south and a positive one when traveling north! Quantities like this that do not depend on direction are called **scalars**. In our discussions, everything that is not a vector is a scalar. Examples that we will discuss include energy, power, and mass.

We shall have more to say about vectors in chapter 2.

What do we have so far? Is the average speed just the magnitude of the average velocity? For our example of Phelps's second length, yes; but this is not always the case, as we discuss next.

2.1.5 Instantaneous velocity

Consider the entire race, which lasted 243.84 s. Imagine the look Phelps would give you if, immediately after his record-setting race, you explained that his average velocity for the race was zero! You would be right, however, because his initial and final positions were the same ($x_2 = x_1$), so his displacement is $\Delta x = 0$. Of course, although his *displacement* was indeed zero, the *distance traveled* was not, so his average speed was (400 m)/(243.84 s) ≈ 1.64 m/s ≈ 3.7 mph.

As Michael would vehemently point out, the *average* velocity obviously doesn't tell the whole story. His coach, standing at the diving block the whole time, had the same average velocity as Phelps—zero! Clearly, Phelps's velocity is changing during the race. On lengths 1, 3, 5, and 7, his average velocity is positive (he swims to the right in our diagram), and on the even-numbered lengths it is negative.

This is clear from figure 2.1, and a direct connection can be made between the graph and the velocity. In fact, the average velocity for any part of the journey is simply the slope of the line connecting the endpoints of the journey. The slope of the line connecting points C and D is simply -1.705 m/s. A line with negative slope slants down.

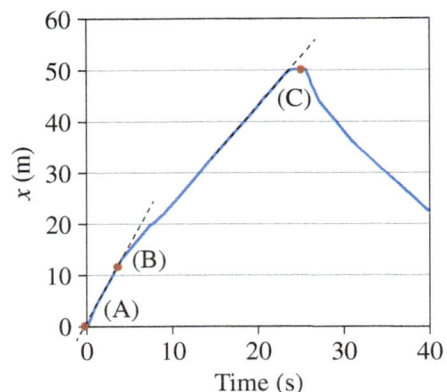

Figure 2.3
A blown-up portion of the graph shown in figure 2.1. Two tangent lines are drawn and discussed in the text.

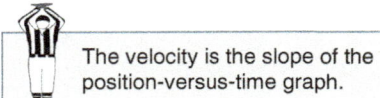

But the graph isn't just a collection of straight lines—for example, what is happening between points A and B? Figure 2.3 focuses on the first portion of the graph. Clearly, in the first 5 s or so, the curve is steeper than it is later on. This reflects the fact that Phelps launched himself from the block at high speed before the water slowed him after a few seconds. Before Phelps hits the far edge of the pool, the slope—and hence the velocity—is still positive, but it decreases as he approaches point C. After point C, the slope and velocity are negative.

In fact, by now it should be clear that the slope of the curve can be different at every single point on the curve—there may be no straight section at all! In this case, the slope of a tangent line—a line that just "kisses" the curve at a given point—gives us the **instantaneous velocity**, the velocity *at* a given point in time.

What is the velocity right *at* point C? There, the tangent to the curve is flat: the slope is zero, so the velocity is zero. When an athlete reverses direction (here, by pushing off the wall), he is momentarily at rest.

If you look closely at figure 2.3, you will notice that the curve just to the right of point C is steeper than that to the left. This is due to Phelps's legendary proficiency at the "dolphin kick" just after kicking off from the wall. Here, the swimmer undulates his body like a dolphin while still under water, avoiding the increased drag of swimming at the surface.

Clearly, graphs of motion contain a huge amount of information, much more than average speeds and velocities. The instantaneous velocity changes moment to moment. We will soon learn that these changes are due to *forces* exerted on and by athletes; that's when it really gets interesting! To get to that discussion, in the next session we develop the mathematics of changing velocities—that is, of acceleration.

Section 2.1   Phelps in Beijing   19

EXAMPLE 2.1 *The slowest stroke*

Without analyzing the split times on table 2.1, can we identify the stroke that slows Michael Phelps the most?

Understanding the relationship between velocity and graphs, this is straightforward. A quick glance at figure 2.1 tells us that the slopes of the curves during his breaststroke lengths are smaller than the others. Therefore, the magnitude of his velocity (his speed) is least when he is doing the breaststroke.

EXAMPLE 2.2 *Sound and the football line*

You probably know that sound travels more slowly than light.

During a storm, the light flash and booming sound generated by a lightning strike are created simultaneously, but you hear the thunder after observing the flash. On the other hand, sound moves pretty fast—760 mph. Can the delay matter for athletic events?

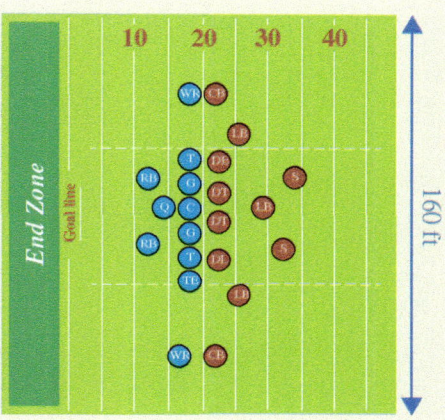

The diagram shows a typical football formation. The quarterback (shown as "Q") calls out a signal. At some point in the signal, the ball is hiked from the center ("C") to the quarterback. Only at this point are the players allowed to cross the line of scrimmage. Players on the offense know the signal beforehand, so they can anticipate it and jump across the line precisely at the moment the ball is hiked. Because the defense must react to the ball's movement, their motion is delayed by human reaction time, about $0.1 \text{ s} = 100 \text{ ms}$.

What about the wide receivers ("WRs") on the offense? Even though they know the signal to hike the ball, there is still a delay in the sound reaching their ears. How long is that delay? Can it matter at all?

First, we look at the sketch and estimate the distance that the sound must travel between the quarterback and the wide receiver. A reasonable estimate is about 60 ft.

Because we have our distance in feet, we'll convert the units for the speed of sound into feet per second:

$$\text{average speed} = 760 \, \frac{\cancel{\text{mi}}}{\cancel{\text{hr}}} \times \frac{5280 \text{ ft}}{1 \cancel{\text{mi}}} \times \frac{1 \cancel{\text{hr}}}{3600 \text{ s}} = 1115 \, \frac{\text{ft}}{\text{s}}.$$

Then, we rewrite equation 2.1 as

$$\Delta t = \frac{\text{distance the sound traveled}}{\text{average speed}} = \frac{60 \cancel{\text{ft}}}{1115 \cancel{\text{ft}}/\text{s}} = 0.053 \text{ s} = 53 \text{ ms}.$$

Interestingly, this is more than half of the human reaction time, so it may be about as important as the lag time experienced by the defense.

The next time you watch a game, pay attention to the wide receivers ready at the line of scrimmage. Sometimes they are looking downfield, relying only on sound to tell them when to jump; in this case, they suffer from the 53-ms delay described above. On the other hand, if they rely *only* on sight, they'd suffer from the 100-ms human reaction time, which is worse. When they are watching the ball, their brain probably combines both sound and sight information to give minimal delay.

> **Question**
>
> The graphs in figures 2.1 and 2.3 have been simplified to help direct our discussion, but what if we put in more detail?
>
> The butterfly and breaststroke are quite dynamic strokes, with the swimmer surging ahead with a powerful thrust, then slowing before the next thrust. What would the graph look like, if we included the details of each thrust?

2.2 Bolt in Berlin

2.2.1 Exploding off the block

In sports, the most interesting action occurs when an athlete's or a ball's velocity *changes*. A basketball, initially shot up, slows until it reaches the peak of the arc, then accelerates down toward the hoop. Football running backs receive the handoff, speed up through a hole, and are suddenly stopped by a devastating hit from a linebacker (whose velocity also changes). To understand and describe this type of motion, we need to mathematically define **acceleration**, which is the rate at which velocity changes.

Now, because velocity is a vector, a change in velocity can mean a change in its magnitude (its speed), or a change in direction, or both. Because of this, a race car going around a circular track at constant speed is actually accelerating! We'll soon get to motion in more than one dimension (such as a car racing around a track), but for now we continue to focus on motion along a straight line. We start with the definition of the **average acceleration**,

$$\bar{a}_x = \frac{\Delta v_x}{\Delta t} = \frac{v_{x,2} - v_{x,1}}{t_2 - t_1}. \tag{2.4}$$

Here, $v_{x,2}$ and $v_{x,1}$ are the instantaneous velocity at two points (t_2 and t_1) in time.

The similarity between equations 2.3 and 2.4 is clear. In fact, the similarity goes farther: just as we discussed the instantaneous velocity as the slope of the curve of position versus time (x versus t), the **instantaneous acceleration** is the slope of the curve of velocity versus time (v_x versus t).

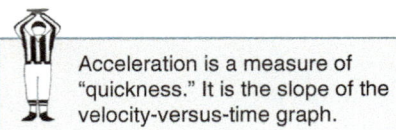
Acceleration is a measure of "quickness." It is the slope of the velocity-versus-time graph.

Let's see a concrete example.

Figure 2.4 shows the average speed for world-record holders in races of different lengths. It has a logarithmic scale on the horizontal axis. We won't go into the details of a log scale, except to say that it is a convenient way to put races of *very* different lengths on the same graph.

Several things are notable about this graph. The first is the remarkably universal nature of the trend: men are a bit faster than women, and average speeds in 2013 were a bit more than in 1957, but the curves are nearly parallel for all three cases. When we discuss power, we will describe how the human body's storage and consumption of food energy dictates the shape of these curves. Secondly, for races of more than a mile, the average speed drops little—from about 16 mph for the mile run, to 12 mph for a marathon. Finally, even though one expects the average velocity to drop as the race gets longer, it at first seems strange that the average speed for the 100-m dash is

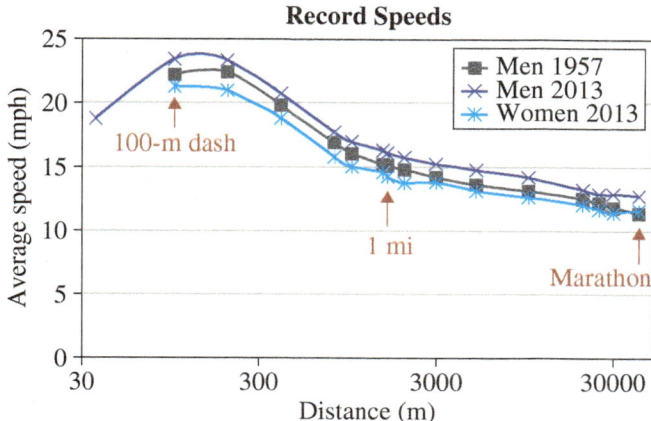

Figure 2.4. Average speeds for world-record races ranging from 100 to 42,195 m (marathon) are based on official times. The 1957 times were compiled by Peter Brancazio in *Sport Science* (1984), who first produced a similar graph. The single point at 36.7 m is based on the unofficial 4.32-s time of University of Michigan quarterback Denard Robinson in the 40-yd dash.

a bit *less* than for the 200-m dash. Records on the 50-m or 60-m dash vary, but they are reasonably represented by one of the fastest (if unofficial) reported 40-yd-dash times, by Denard Robinson. Apparently, very short sprints have lower average speeds than longer ones do.

The reason, of course, is that it takes some time for a human to "get up to speed" after the gun goes off (figure 2.5). Those first few seconds are insignificant for the marathoner or even the miler. But for sprinters they are the most important moments. Table 2.2 lists the 10-m split times for Usain Bolt's record-smashing 100-m sprint, set at the IAAF World Championships in Berlin, 2009. These splits are plotted in figure 2.6. After a brief reaction time,[1] Bolt increases his speed, as indicated by steepening slope of the x-vs.-t curve in figure 2.6. After about 4 s, Bolt approaches his maximum speed, at which point the slope stops changing and the curve resembles a straight line with a slope of 12.2 m/s ≈ 27.3 mph.

Table 2.2. Official IAAF 10-m split times for Usain Bolt's world record 100-m sprint, set in Berlin, 2009.

Cum. time (s)	Distance (m)	Cum. time (s)	Distance (m)
0	0	5.47	50
0.146	0 (reaction time)	6.29	60
1.89	10	7.10	70
2.88	20	7.92	80
3.78	30	8.75	90
4.64	40	9.58	100

Figure 2.5
IAAF regulations require a 0.1-s delay between the starting sound and runners pushing off the block. **Top:** (© *Ingram Publishing RF*). **Bottom:** (© *2009 Jupiterimages Corporation RF*).

---

[1] It is usually assumed that no human can react to the gun in less than 100 ms. Sensors in the starting block detect when a sprinter pushes off. Even if he pushes off *after the gun has fired*, if he does so in less than 100 ms, it is ruled a false start.

Forty-yard times reported for football players are often "hand times," which do not include these reaction times. Therefore, to perform a fair comparison one should probably add about 0.15 s to reported 40-yd times.

 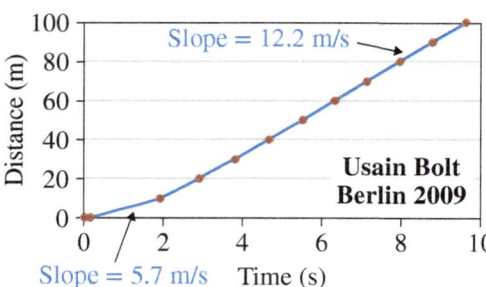

Figure 2.6. The world's fastest man. Usain Bolt watches the clock as he crosses the finish line in the 2009 IAAF World Championships in Berlin. Like most sprinters, he decelerates somewhat over the final several meters, but some wonder whether Bolt could have pushed harder through the finish. The right panel charts his official split times, listed in table 2.2. (© *Wolfgang Rattay/Reuters/Corbis*).

Figure 2.4 tells us that no human being even comes close to this value for his *average* speed, for a race of *any* length. Bolt's average velocity for his world-record 100-m race is

$$\bar{v}_x = \frac{\Delta x}{\Delta t} = \frac{100 \text{ m}}{9.58 \text{ s}} = 10.44 \text{ m/s} \approx 23.35 \text{ mph}.$$

A more intuitive feeling for the development of Bolt's sprint can be gotten by plotting the slope of the *x*-vs.-*t* curve—that is, the velocity. Figure 2.7 shows two types of velocity graph. On the right is a graph we might see in a sports magazine, in which velocity is plotted versus distance. (Actually, it is the *average* velocity, over the 10-m split.) Here it is easy to see that even the most explosive sprinter doesn't reach top speed until he's run about 40 m. Obviously, football's 40-yd sprint is all about acceleration.

On the left is the "physicist's way": velocity is plotted as a function of *time*. We see clearly that a world-class sprint is composed of three phases. The first phase is one of high acceleration: in the first 2.4 s, Bolt has gone from rest to 10.1 $\frac{\text{m}}{\text{s}}$. His acceleration is

$$\bar{a}_{x,\text{early}} = \frac{\Delta v_x}{\Delta t} = \frac{10.1 \text{ m/s} - 0}{2.4 \text{ s}} = 4.21 \frac{\text{m/s}}{\text{s}} = 4.21 \text{ m/s}^2.$$

Notice the new unit for acceleration.

In the second phase, from about 2.5 to 5 s, the slope of the v_x-vs.-*t* curve drops, even as the slope of the *x*-vs.-*t* curve grows. (Ask yourself: Does this make sense?) Bolt's acceleration is decreasing, though his speed is increasing.

Bolt's third and final phase is the cruising phase, beginning about 5 s after the starter's gun. When he finally reaches his maximum speed of 27.6 mph, Bolt is certainly moving (faster than ever, in fact!), but his acceleration is zero. The velocity graphs of figure 2.7 reveal what is imperceptible on the position graph of figure 2.6 or in a listing of split times: the slope of the *v*-versus-*t* graph is slightly negative after about *t* = 6.3 s—Bolt is slowing somewhat. Such end-of-sprint slowing is not uncommon among runners, as it is impossible for the human body to maintain speeds above 22 mph for more than a few seconds. Nevertheless, those who regularly watch Bolt's races often find themselves wondering whether his times could be even lower if he would finish the race before beginning his celebration or looking at the race clock as he runs by.

The velocity graph reveals that Bolt, like many sprinters, actually slows somewhat toward the end of the race.

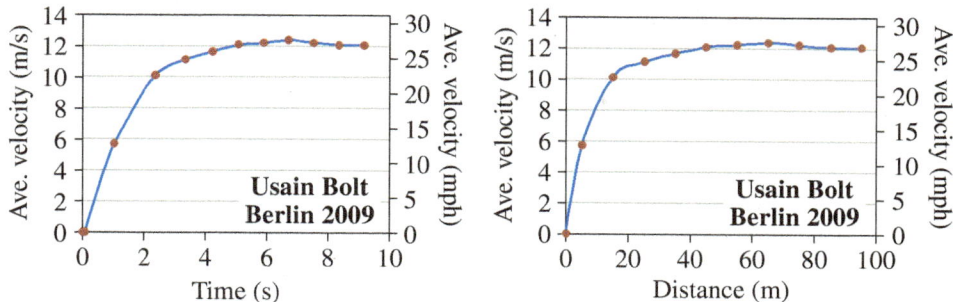

Figure 2.7. Velocity graphs for Usain Bolt's record-setting 100-m dash. On the left is the "physicist's style" of plotting velocity versus time; the acceleration is the slope of this curve at any point. On the right is a common graph found in sports newspapers and magazines, of velocity versus *distance*.

2.2.2 Constant acceleration kinematics

Clearly, the motion of an athlete can be extremely complicated: not only her position (x) but also her velocity (v_x) and acceleration change moment to moment! A full treatment of all possibilities requires advanced math and often a computer. However, we will find that we can get amazingly far by focusing on the constant acceleration case. The mathematics we will develop for constant acceleration motion is simple yet powerful; it will be used repeatedly in this book. It will even be of some use when the acceleration is *not* constant, as in Michael Phelps's 400-m medley or the juking of a football player before getting tackled.

However, we will need some practice with the simple case first, so for the rest of this chapter we are going to assume that the acceleration is constant. In this special and important case, the acceleration is the same as the average acceleration, so we'll write a_x without the overbar seen in equation 2.4.

The following constant acceleration kinematic equations relate position, velocity, and acceleration:

$$\Delta v_x = a_x \Delta t \qquad v_x = v_{x,0} + a_x t \qquad (2.5)$$

$$v_x^2 = v_{x,0}^2 + 2a_x \Delta x \qquad v_x^2 = v_{x,0}^2 + 2a_x x \qquad (2.6)$$

$$\Delta x = v_{x,0}\Delta t + \tfrac{1}{2}a_x(\Delta t)^2 \qquad x = x_0 + v_{x,0}t + \tfrac{1}{2}a_x t^2 \qquad (2.7)$$

Don't worry—there are only three equations above, not six! We have written each equation in two ways, to help avoid the confusion that sometimes arises when students first start to use these equations.

As in all of our discussions so far, for the equations on the left, Δv_x, Δx, and Δt are the difference in velocity, position, and time from the beginning of the interval of motion being considered, to the end. For the equations on the right, the time at the beginning of the interval is set to 0, and at the end, to t; therefore, Δt may simply be written t. Also, the positions at the beginning and end of the interval are defined as x_0 and x, and the initial and final velocities are $v_{x,0}$ and v_x.

Students often get overwhelmed by the choice of the three equations above, and wonder: "How do I know which one to use?" For any given problem, consider the information you have, such as x, $v_{x,0}$, or Δt, and the information you want, such as v_x. Then, identify the equation that relates all of these variables, but does *not* contain a

Trial and error is a perfectly fine way to choose among equations 2.5–2.7; you won't get the wrong answer.

If the acceleration is constant, *x*-vs.-*t* makes a parabolic graph, and *v*-vs.-*t* linear graph.

variable that you neither have nor care to find. For example, if you do not know Δt and you are not trying to find Δt, then probably you won't want to use equation 2.5 or 2.7.

If it still seems confusing, and it might at first, keep this in mind: you cannot get the wrong answer with *any* of them, so relax. You can even pick one at random, if you like. The worst that will happen is that you'll find that the one you've chosen is useless—for example, you want to find position (x), but you've chosen equation 2.5, which doesn't have x in it. Or you've chosen equation 2.6, but realize that you have been given time t in your homework problem, but not x. Before long, you will grow accustomed to these equations and will know which one to pick.

EXAMPLE 2.3 *SuperBolt*

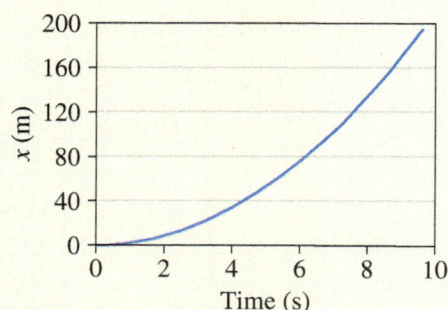

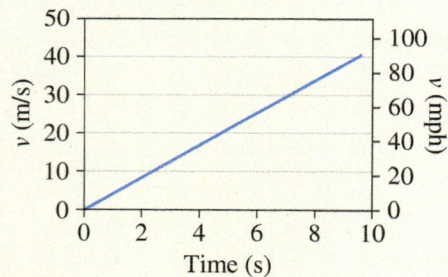

In the first phase of his sprint, Usain Bolt accelerated at 4.21 m/s². What if he could maintain this rate of increasing velocity? What would be his velocity and displacement after 9.58 s?

Let's look first at his position, x, as a function of time. As we'll usually do, we'll say the starting line is the origin ($x_0 = 0$), so his position x is the same as his displacement Δx. According to equation 2.7,

$$x = \Delta x = (0 \text{ m/s})(9.58 \text{ s}) + \tfrac{1}{2}(4.21 \text{ m/s}^2)(9.58 \text{ s})^2 = 193.2 \text{ m}.$$

This is considerably more than the 40 m his coach told him to run. One imagines the scene in *Forrest Gump*, where the crowd holds up signs saying "Stop, Forrest!"

The x-vs.-t graph of this motion, shown here, is a *parabola*; that is, the position grows quadratically with time. The slope—the velocity—keeps increasing as a function of time. In fact, as the other graph here shows, it grows linearly. The slope of v_x-vs.-t is constant (it is a straight line), which makes sense because the slope of v_x-vs.-t is acceleration, and we're assuming constant acceleration here! It's easy to figure out Bolt's velocity at the end of 9.58 s, using equation 2.5:

$$v_x = 0 \text{ m/s} + (4.21 \text{ m/s}^2)(9.58 \text{ s}) = 40.3 \text{ m/s} \approx 90 \text{ mph}!$$

2.2.3 The signs *a* and *v*

In our discussion of the sprinter, his velocity was a positive number, and so was his acceleration. However, there are important situations in sports where acceleration and velocity point in opposite directions. There can be some confusion about how to treat these cases mathematically, so let's briefly discuss an example.

Consider young running back Chris "Beanie" Wells of the Arizona Cardinals, racing down the field with the football in the positive direction at 16 mph (7.15 m/s). Running directly at him at 12 mph is the fearsome Ray Lewis, linebacker for the Baltimore Ravens. Clearly, Lewis's goal is to change Beanie's velocity—to accelerate him; in fact, he wants to change Beanie's velocity to zero, as quickly as possible!

In common language, one might say Lewis wants to "decelerate" Beanie, but we will avoid introducing a new term to our scientific description of motion. Acceleration quantifies *any* change in velocity, regardless of direction.

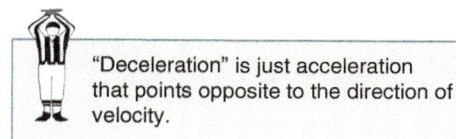

"Deceleration" is just acceleration that points opposite to the direction of velocity.

Wells's and Lewis's official weights are listed at 238 lb and 250 lb, respectively. After we've discussed forces and collisions, we'll be able to understand, scientifically, that the resulting collision between Lewis and Wells can generate more than half a ton of force on Wells (!) and an average acceleration of -24.7 m/s^2 during the tackle. Note that the acceleration is negative because we said Wells's velocity is positive and Lewis accelerates Wells in the opposite direction.

How much time does it take Lewis to bring Beanie to a dead stop ($v_x = 0$) once he's made contact? We know v_x, $v_{x,0}$, and a_x, and we'd like to know Δt. Looking at our three kinematic equations, we choose equation 2.5, because it has these quantities and does not have x, which we neither know nor care to find at this point. Only a bit of algebra is needed:

$$v_x = v_{x,0} + a_x \Delta t \quad \rightarrow \quad \Delta t = \frac{v_x - v_{x,0}}{a} = \frac{0 - 7.15 \text{ m/s}}{-24.7 \text{ m/s}^2} = 0.29 \text{ s}.$$

Lewis brings Wells from 16 mph to a dead stop in a little more than a quarter of a second—ouch!

We may also ask how much farther forward Beanie got (that is, his displacement, Δx) before he was brought to a dead stop ($v_x = 0$) by Lewis. Looking at our choice of equations, we note that we know $v_{x,0}$, v_x, and a_x, and we'd like to know Δx. An equation with those variables in it is equation 2.6. We only need to use a little bit of algebra:

$$v_x^2 = v_{x,0}^2 + 2a \Delta x \quad \rightarrow \quad \Delta x = \frac{v_x^2 - v_{x,0}^2}{2a_x} = \frac{0 - (7.15 \text{ m/s})^2}{2(-24.7 \text{ m/s}^2)} \approx 1 \text{ m}.$$

So Beanie gained an extra 1.1 yd due to his forward momentum—not bad against someone like Ray Lewis.

Wells's velocity and position are plotted in figure 2.8. Once again, his velocity is linear in time, and his position depends quadratically on time. Because $a_x < 0$, the v_x-vs.-t line slants down, and the concavity of the x-vs.-t parabola is concave down.

Notice what happens if Lewis continues to accelerate Wells after 0.29 s. Wells's velocity becomes negative and his displacement starts going down—Lewis has brought Wells to a stop and is now pushing him backward! Of course, if Wells had progressed beyond the line of scrimmage, this wouldn't affect the play, since the ball is placed at the point of farthest forward motion. This fact doesn't always deter players like Lewis, however.

To make the point more generally, take a look at figure 2.9. There, we see a ballplayer sliding into third base. His velocity is initially negative (i.e., to the left) but is eventually zero. The lower photo shows a hockey player accelerating a puck that was initially moving to the right, to move even faster to the right. In both cases, a plot of velocity versus time is shown, and you should note the similarity between them. We have not yet discussed forces, but we shall see that accelerations are caused by forces, and in the plot, the force acts on the player and puck in the time window marked Δt. The acceleration due to the force is simply the slope of the line in this time window. In both cases this slope is positive, reflecting the fact that the acceleration points to the right.

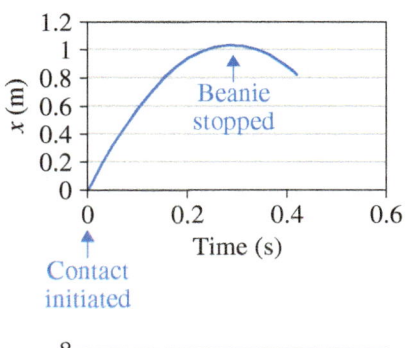
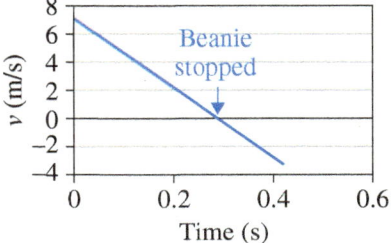

Figure 2.8
Beanie Wells's position and velocity, after being tackled by Ray Lewis at $t = 0$.

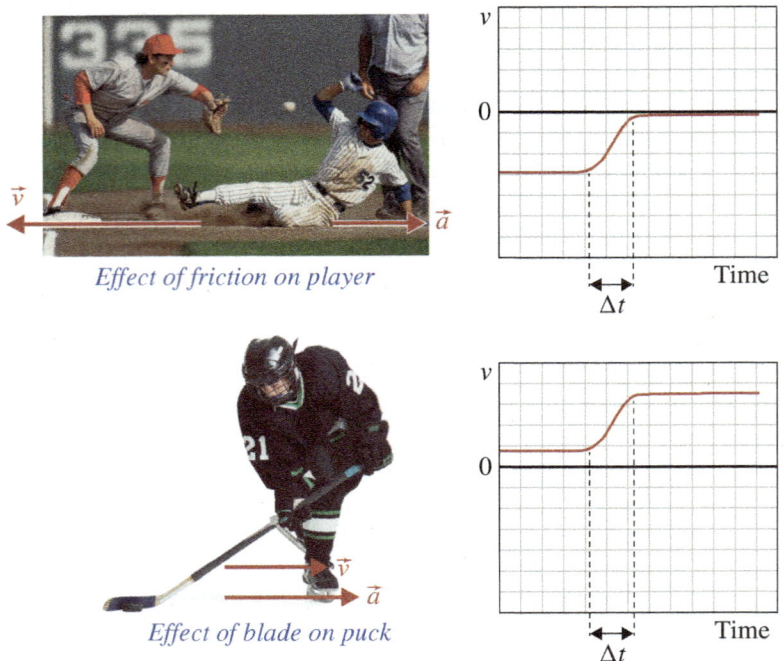

Figure 2.9. The velocity-versus-time curves of a ballplayer sliding into third base and a hockey player accelerating a puck have an important similarity. **Top:** (© David Madison/Getty Images RF). **Bottom:** (© Rubberball/Getty Images RF).

EXAMPLE 2.4 *Sliding into third*

Consider the ballplayer shown sliding into third base in figure 2.9. From years of experience, players instinctively know that they can expect an acceleration of about 20 ft/s² when sliding on the clay of the basepath, though few would quantify it as such. Before the slide, the player shown achieved a speed of 19 ft/s (about 13 mph). How far from the bag should he begin his slide, if he wants to reach third base but not slide past it (and risking getting tagged out)?

Paying attention to signs, what do we know? We want him to come to rest, so $v_x = 0$. His initial velocity $v_{x,0}$ is negative because he is *moving* to the left, but his acceleration a_x is positive because he is *accelerating* to the right. To find the player's displacement Δx during the slide, we use equation 2.6:

$$v_x^2 = v_{x,0}^2 + 2a_x \Delta x \quad \rightarrow \quad \Delta x = \frac{v_x^2 - v_{x,0}^2}{2a_x} = \frac{0 - (-19 \text{ ft/s})^2}{2 \times 20 \text{ ft/s}^2} = -9 \text{ ft.}$$

That we obtained a negative displacement might seem surprising at first, but it only means that during the slide the player's position changed to the left—in the direction of third base. Indeed, we should be *really* surprised if it were positive!

Question

What if, in the previous example, the ballplayer was running half as fast (9.5 ft/s) when he began his slide? Would he slide half as far? Why or why not? It is very important for ballplayers to understand instinctively the relationship between their speed and the distance they want to slide, and it's not as simple as one might initially think.

2.3 Rope-Climbing and Diving

So far we've discussed the mathematical formulation of horizontal motion. Naturally, vertical motion is crucial in sports as well: in the same complex in which Phelps was setting world records through blazing horizontal motion, Australian Matthew Micham took gold on the 10-m platform, launching a stunning backward dive with $2\frac{1}{2}$ somersaults and $2\frac{1}{2}$ twists. These motions are quite complicated, and we shall discuss some of the physics behind them later on. Just as we did with Phelps's swim, we ignore the complicated motion of individual body parts and continue to focus on the motion of a body's average position—its center of mass.

When we talk about position in the vertical direction, we'll use the symbol y. If there is constant acceleration, the equations of motion are

$$\Delta v_y = a_y \Delta t \qquad v_y = v_{y,0} + a_y t \qquad (2.8)$$

$$v_y^2 = v_{y,0}^2 + 2a_y \Delta y \qquad v_y^2 = v_{y,0}^2 + 2a_y y \qquad (2.9)$$

$$\Delta y = v_{y,0} \Delta t + \tfrac{1}{2} a_y (\Delta t)^2 \qquad y = y_0 + v_{y,0} t + \tfrac{1}{2} a_y t^2. \qquad (2.10)$$

Now, those should look familiar! They are identical to equations 2.5–2.7 but involve velocity v_y and acceleration a_y in the vertical direction.

Straight-line races in the vertical direction are not as common as sprints or swim meets in the horizontal. However rope-climbing, a familiar component of the grueling regimen faced by recruits entering the military, periodically emerges as a competitive sport. Until 1936 it was an Olympic event.

In 1954, UCLA's Don Perry set the NCAA record for the 20-ft climb, with an astonishing time of 2.8 s. This is about as fast as this author was able to pull a loose 20-ft rope between his legs on the ground! Figure 2.10 shows Garvin Smith, a contemporary of Perry, in a strobe photo of a competitive climb. Climbers began from a sitting position, with their outstretched hands about 4 ft above the floor: $y_0 = +4$ ft if we use the floor at the origin. Their outstretched hands reach a plate at a final position of $y = +20$ ft. The displacement of their center of mass is essentially the same as the displacement of their outstretched hand: $\Delta y = y - y_0 = 16$ ft. The average velocity then, is

$$\bar{v}_y = \frac{\Delta y}{\Delta t} = \frac{16 \text{ ft}}{2.8 \text{ s}} = 5.71 \text{ ft/s} \times \frac{1 \text{ m}}{3.28 \text{ ft}} = 1.74 \text{ m/s}.$$

Vertical motion is described by the same formulas as used for horizontal motion.

Interestingly, Perry's average speed was almost identical to that of Michael Phelps in the water in Beijing. The sign of $\bar{v}_y$ is positive, reflecting our choice of "up" as the direction of increasing vertical position. All of our equations would work equally well if we chose "down" as the positive-y direction, but this can get confusing; unless said explicitly otherwise, $v_y > 0$ will mean an upward-directed velocity.

Look again at the photo of Smith in figure 1.20. The strobe light went off at equal time intervals Δt. Smith's displacement in the first time interval (Δy_1) is clearly larger than his displacement in subsequent ones, reflecting the fact that climbers begin with an explosive pull and later climb at a smoother, slower rate. While his displacement in the second interval is smaller than it is in the first interval, it is larger than it is in the third: $\Delta y_1 > \Delta y_2 > \Delta y_3$. His velocity is changing with time. Thus we see that this is another case in which the velocity and acceleration point in opposite directions: while moving upward ($v_y > 0$), Perry is accelerating *downward* in the photo, $a_y < 0$.

28 *Racing, Mathematically* Chapter 2

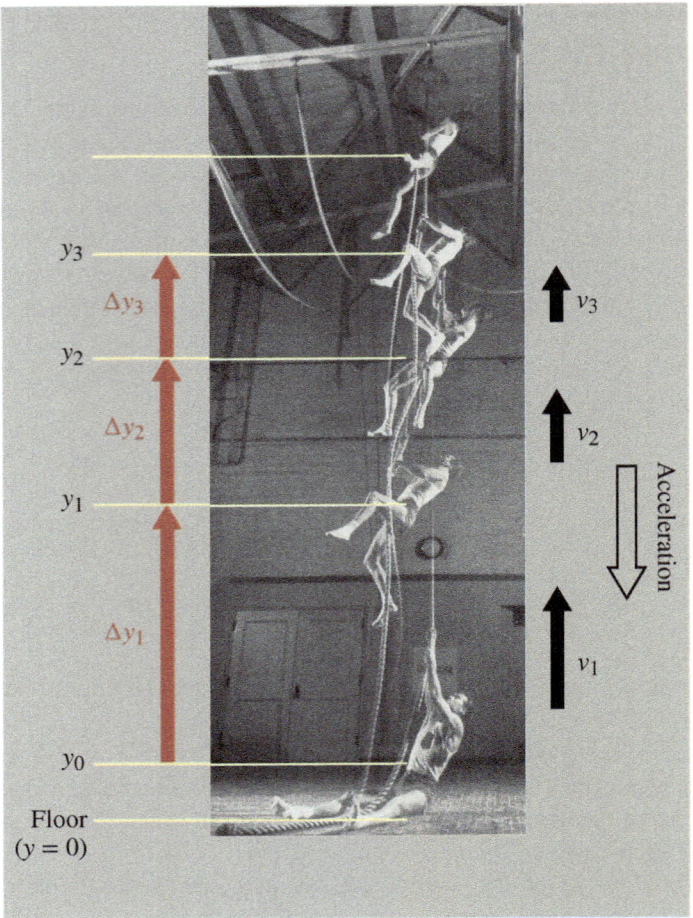

Figure 2.10. A stroboscopic shot of Cal State LA's Garvin Smith circa 1950. The climber begins from a seated position, with his hands outstretched about 4 ft above his head. He can kick his legs but cannot use them on the rope or to push off of the floor. A metal disk coated with ink sits 20 ft above the floor. (25-ft and 8-m climbs were also common.) Ink on the climber's fingers served as proof that he had completed the climb. The approximate height of Perry's center of mass at each strobe shot is indicated. (© *Peter Stackpole/The LIFE Picture Collection/Getty Images*).

> **Question**
>
> In our discussion above, we said that the displacement of Smith's hands during the climb is the same as the displacement of his center of mass. This is approximately true, but not exactly. Look again at the photo. Is the displacement of his center of mass greater than, or less than, that of his hands? By about how much?

2.3.1 The 800-lb gorilla: gravity

As I've mentioned and will discuss in detail later, accelerations are caused by forces. Different forces play roles in different sports, but the gravitational force is central to all. Even in the purely horizontal Canadian sport of curling, gravity is important, as it determines the force of friction (we'll come to this later, too) that slows the stone sliding across the ice. If you ever curl on the moon, be sure not to push as hard as you would on Earth!

Soon we will be discussing other forces too, including air drag and lift. However, for the rest of this chapter we'll discuss the situation where gravity is the only important force around. So long as an object isn't moving too fast and is sufficiently dense, it is okay to ignore the forces from air. An excellent example is a human in freefall over a short distance.

Consider Perry after he's completed his climb—his race is over, and he catches his breath. If he releases the rope, gravity takes over and he begins to accelerate downward: $a_y < 0$. How long will he be in the air before his feet hit the floor, and how fast will he be going at impact?

If nothing else interferes, all people and objects experience the same downward acceleration due to gravity.

To answer these questions, we need to know something about gravity. It turns out that so long as we are talking about athletic events within a few miles of the Earth's surface, gravity is especially simple. In fact, if gravity is the only relevant force, then the acceleration of any object or person is constant: $a_y = -g$, where

$$\text{approximate:} \quad g = 9.8 \text{ m/s}^2 = 32 \text{ ft/s}^2 = 22 \, \frac{\text{mph}}{\text{s}}$$
$$\text{standard:} \quad g = 9.80665 \text{ m/s}^2 = 32.17405 \text{ ft/s}^2. \tag{2.11}$$

In almost all of our discussions, we will use the approximate values listed above.

Equation 2.11 gives the "typical" value of g. Later on, we will discuss the fact that this acceleration is a little bit different at different locations on the planet, and that this can have small but real consequences for sports.

2.3.2 Freefall time

At the top of the rope, Perry's outstretched hand is 20 ft above the floor. Assuming that he is about 6 ft tall, his feet are 7 ft lower than his hand, or 13 ft above the floor. So, his displacement for his journey down is $\Delta y = -13$ ft. To figure out how much time he spends in the air, we'll use equation 2.10, with $a_y = -g$ and setting his initial velocity $v_{y,0} = 0$ because he's at rest just at the instant of letting go:

$$\Delta y = 0 - \tfrac{1}{2} g (\Delta t)^2 \quad \rightarrow \quad \Delta t = \sqrt{\frac{2 \Delta y}{-g}} = \sqrt{\frac{2 \times (-13 \text{ ft})}{-32 \text{ ft/s}^2}} = 0.9 \text{ s}.$$

It takes Perry a bit less than a second to hit the floor. Take a moment to understand how all the minus signs worked out here.

In fact, we'll often want to know how long it takes something to fall, so let's take what we did and write down a useful formula. The time it takes to fall (beginning at rest) from a height h is

$$T_{\text{fall}} = \sqrt{\frac{2h}{g}}. \tag{2.12}$$

While Perry fell 13 ft ≈ 4 m in 0.9 s, Australian diver Micham was in freefall from about three times as high—11.5 m.[2] The time Micham had available to execute his $2\frac{1}{2}$ somersaults and $2\frac{1}{2}$ twists was then

$$T_{\text{fall, Micham}} = \sqrt{\frac{2 \times 11.5 \text{ m}}{9.8 \text{ m/s}^2}} \approx 1.5 \text{ s}.$$

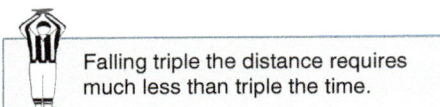

Falling triple the distance requires much less than triple the time.

Notice that we needed to use consistent units: because Micham's original height h was in meters, we expressed g in m/s².

Notice something else important: Although Micham fell almost three times as far as Perry did, he was not in the air three times as long as Perry. Instead he was in the air about $\sqrt{3}$ as long—or 75% longer. The reason for this is clear if we look at figure 2.11: it takes Micham 0.9 s to fall about 4 m, just as it does Perry. However, because his speed grows with time, it takes only 0.6 s to fall the remaining 7 m. This

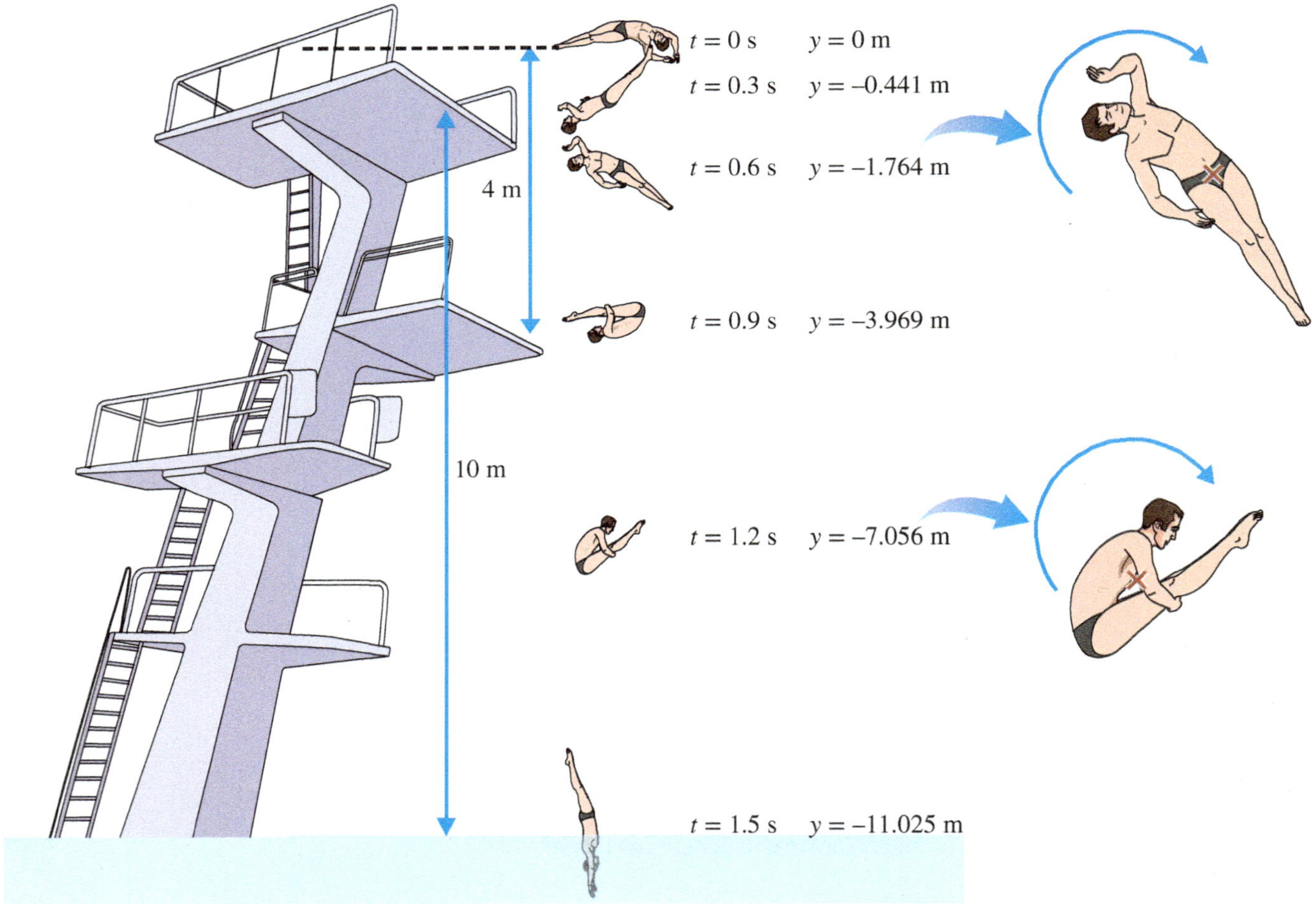

Figure 2.11. A sketch based on video footage of Matthew Micham's gold-winning 10-m platform dive at the 2008 Olympics. The height of Micham's center of mass is shown at equal time intervals. On the right, blowups show his body's orientation and spin at two times; notice how the position of his center of mass changes in relation to his body. It is the center of mass that follows the kinematics of equations 2.8–2.10.

fact is reflected in the fact that Δy is proportional to the square of the time (t^2), rather than time (t) itself.

Because of this unalterable fact, a diver's final maneuvers take up much more "vertical space" than his initial ones. From long experience, Micham knows that "twisting" along his long axis is most attractive when done over a short distance. In his Olympic dive sketched in figure 2.11, he performed $2\frac{1}{2}$ twists in the upper 2.5 m ($\Delta y = -2.5$ m). His $2\frac{2}{2}$ somersaults took roughly the same amount of time, but are much more graceful spread over the next 8 m or so.[3] Artistic choices made in judged events must reconcile aesthetics with the laws of physics.

Finally, we remember that our kinematic equations apply to Micham's center of mass—the "average position" of all the atoms that compose his body. The other parts of his body undergo all sorts of strange motion and are under his control. However, once he steps off the platform, there is nothing at all that he can do to change the trajectory of his center of mass.

2.3.3 Landing velocity

To find Perry's velocity at the moment his feet hit the floor, we multiply his acceleration ($-g$) with the amount of time that's passed since he let go of the rope, given by equation 2.12:

$$v_{\text{land}} = -g \times T_{\text{fall}} = -g \times \sqrt{\frac{2h}{g}} = -\sqrt{2gh}. \tag{2.13}$$

$$v_{\text{land,Perry}} = -\sqrt{2\left(32\,\frac{\text{ft}}{\text{s}^2}\right)(13\text{ ft})} = -\sqrt{832\,\frac{\text{ft}^2}{\text{s}^2}} = -28.8 \text{ ft/s},$$

or about 20 mph. If he flexes his legs upon landing (as we discuss in chapter 3), a fit athlete can usually sustain such a landing without injury.

It is well that Micham "lands" in the water, as his speed is $\sqrt{2\left(9.8\text{ m/s}^2\right)(11.5\text{ m})}$ = 15 m/s ≈ 33.6 mph.

2.3.4 The symmetry of a vertical leap, hang time

Perhaps no other situation in sports is as common, or as important to understand, as the flight of an object or a person between launch from and return to the ground. When gravity is the only relevant force, analysis is made much easier by the symmetry of this motion. Let's see how this works, by reversing the freefall of Don Perry that we've just discussed.

What if Perry were to jump straight up from the floor at $v_{y,0} = 28.8$ ft/s? How high would he go, and how long would it take to reach his peak? At his peak, his vertical velocity is $v_y = 0$ (otherwise that's not his peak, is it?), so we can use equation 2.9 and find the vertical distance traveled from the floor to the peak, h:

$$v_y^2 = v_{y,0}^2 + 2a_y \Delta y \quad \rightarrow \quad h = \Delta y = \frac{v_y^2 - v_{y,0}^2}{2a_y} = \frac{0 - (28.8\text{ ft/s})^2}{2\left(-32\text{ ft/s}^2\right)} = 13 \text{ ft}.$$

---

[3] It is also much easier to transition to a straight entry, from a somersault than from a twist. This will become clearer in chapter 10.

32 *Racing, Mathematically*

The general formula relating peak height to launch velocity is

$$h = \frac{v_{y,0}^2}{2g}. \tag{2.14}$$

So, we find that if Perry leaps from the ground at 28.8 ft/s, he will reach a height of 13 ft and (as we discussed in section 2.3.3) come crashing to the ground at the same speed—28.8 ft/s. (Now, we know that no athlete can jump straight up 13 ft! That's because none can leap from the ground at 28.8 ft/s.) It's generally true that $v_{\text{launch}} = v_{\text{land}}$.

In the absence of air or other forces, the time it takes an object to rise from the ground to its peak is the same as the time required to fall back to the ground.

It should not surprise you, either, to learn that the time a body spends on the way up to the peak is the same as the time spent on the way back down (T_{fall}). From football punts to jump shots, it is common to discuss the **hangtime** of a projectile—the time between its launch and landing on the ground. The hangtime is just double the fall time, and we can write down a formula we'll find quite handy:

$$T_{\text{hang}} = 2\frac{v_{y,0}}{g} = 2\sqrt{\frac{2h}{g}}. \tag{2.15}$$

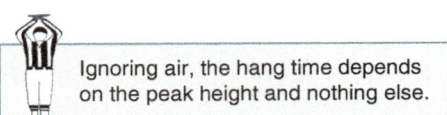
Ignoring air, the hang time depends on the peak height and nothing else.

Here, $v_{y,0}$ is the vertical component of the velocity as the ball/player leaves the ground.

height (h)	hangtime (T_{hang})	launch/land speed (v_0)
40 in.	0.91 s	14.5 ft/s (9.9 mph)
13 ft	1.8 s	28.8 ft/s (19.6 mph)
32 ft 10 in. (10 m)	2.9 s	45.7 ft/s (31.2 mph)

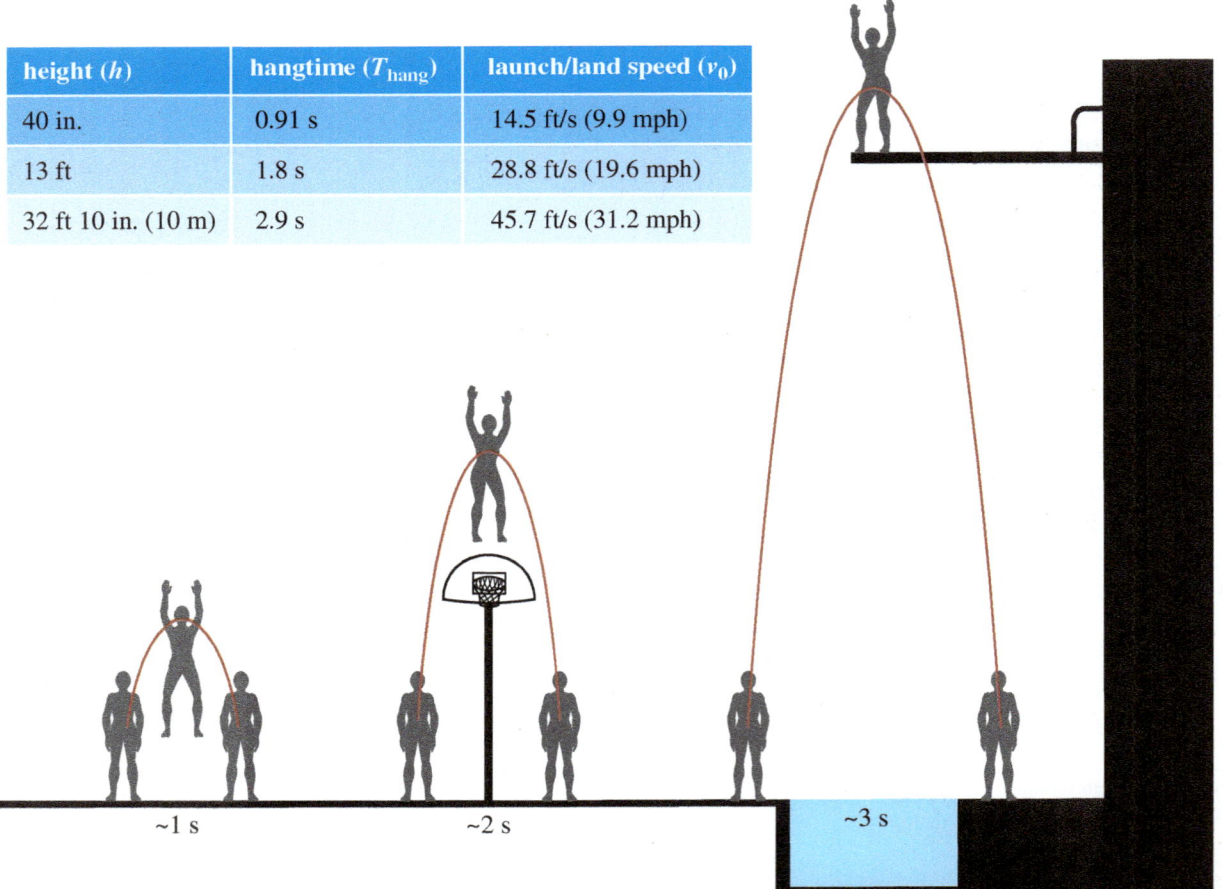

Figure 2.12. A realistic 40-in. vertical leap is compared with two hypothetical leaps. Their hangtimes are approximately 1, 2, and 3 s, respectively.

Equation 2.15 tells us something important: the hang time of a person or object is determined entirely by the height of his/its trajectory. We've mentioned the freefall of Don Perry from 13 ft and of Matthew Micham from the 10-m tower dive. Figure 2.12 sketches leaps from the ground to these unrealistic heights, together with a more realistic (if still impressive) 40-in. vertical leap possible for the best basketball players. The latter can achieve a hang time of almost one full second. On the other hand, occasional claims that players have 2-s hang times are clearly rubbish—this would require their *feet* to sail above the backboard. (As we shall discuss later, combining horizontal motion with the leap does not change this fact, though it can help generate the illusion of "flying.")

EXAMPLE 2.5 *A quick wrist shot*

A hard wrist shot can send an initially stationary puck flying at 55 mph, or 24.6 m/s. How does its acceleration while in contact with the stick compare to the acceleration of an object in freefall?

Evaluating the information we have, $v_x = 24.6$ m/s and $v_{x,0} = 0$, and looking at our equations 2.5–2.7, we realize that we don't know enough to answer this question. If we want to find a_x using equation 2.5, we need to know Δt. If we use equation 2.6, we need Δx. Using equation 2.7 would require both Δx and Δt.

To make progress, we need to use the tool of estimation. In the photo, the arrow indicates the approximate distance over which the stick is in contact with the puck. This is the distance Δx, over which the acceleration is occurring. Comparing the length of the arrow to the player's height (assumed to be typical, 6 ft), we can estimate $\Delta x \approx 2$ ft = 0.61 m. This is only an estimate; reasonable estimates might vary from 1 ft to 3 ft, but one would never use $\Delta x = 7$ ft.

Now our job is easy: we can use equation 2.6 to find

$$v_x^2 = v_{x,0}^2 + 2a_x \Delta x \rightarrow a_x = \frac{v_x^2 - v_{x,0}^2}{2\Delta x} = \frac{(24.6 \text{ m/s})^2 - (0 \text{ m/s})^2}{2 \times 0.61 \text{ m}} = 495 \text{ m/s}^2.$$

One way to compare this with the acceleration due to freefall is by taking a ratio:

$$\frac{a_x}{g} = \frac{495 \text{ m/s}^2}{9.8 \text{ m/s}^2} \approx 50 \rightarrow a_x = 50g.$$

(© Royalty-Free/Corbis)

When an objects accelerates at 9.8 m/s² (or 32 ft/s²), we say that it accelerates with 1 *g* (pronounced "gee"). This is true whether the object is falling under gravity or due to something else. With a good wrist shot, the player is able to accelerate the puck at 50 *g*'s, impressive indeed.

Collected Equations

$$\text{average speed} = \frac{\text{distance traveled}}{\text{time elapsed}}. \quad (2.1)$$

$$\Delta x = x_2 - x_1. \quad (2.2)$$

$$\bar{v}_x = \frac{\Delta x}{\Delta t}. \quad (2.3)$$

$$\bar{a}_x = \frac{\Delta v_x}{\Delta t} = \frac{v_{x,2} - v_{x,1}}{t_2 - t_1}. \tag{2.4}$$

$$\Delta v_x = a_x \Delta t \qquad v_x = v_{x,0} + a_x t \tag{2.5}$$

$$v_x^2 = v_{x,0}^2 + 2a_x \Delta x \qquad v_x^2 = v_{x,0}^2 + 2a_x x \tag{2.6}$$

$$\Delta x = v_{x,0} \Delta t + \tfrac{1}{2} a_x (\Delta t)^2 \qquad x = x_0 + v_{x,0} t + \tfrac{1}{2} a_x t^2 \tag{2.7}$$

$$\Delta v_y = a_y \Delta t \qquad v_y = v_{y,0} + a_y t \tag{2.8}$$

$$v_y^2 = v_{y,0}^2 + 2a_y \Delta y \qquad v_y^2 = v_{y,0}^2 + 2a_y y \tag{2.9}$$

$$\Delta y = v_{y,0} \Delta t + \tfrac{1}{2} a_y (\Delta t)^2 \qquad y = y_0 + v_{y,0} t + \tfrac{1}{2} a_y t^2. \tag{2.10}$$

$$\begin{aligned}\text{approximate:} &\quad g = 9.8 \text{ m/s}^2 = 32 \text{ ft/s}^2 = 22 \tfrac{\text{mph}}{\text{s}} \\ \text{standard:} &\quad g = 9.80665 \text{ m/s}^2 = 32.17405 \text{ ft/s}^2.\end{aligned} \tag{2.11}$$

$$T_{\text{fall}} = \sqrt{\frac{2h}{g}}. \tag{2.12}$$

$$v_{\text{land}} = -g \times T_{\text{fall}} = -g \times \sqrt{\frac{2h}{g}} = -\sqrt{2gh}. \tag{2.13}$$

$$h = \frac{v_{y,0}^2}{2g}. \tag{2.14}$$

$$T_{\text{hang}} = 2\frac{v_{y,0}}{g} = 2\sqrt{\frac{2h}{g}}. \tag{2.15}$$

Problems

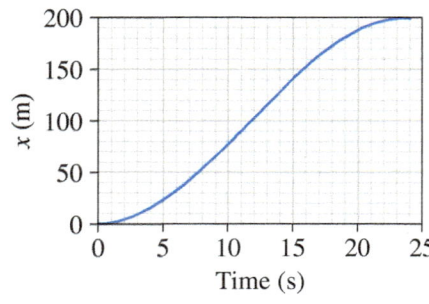

Figure P2.2

1. What is the contact time between the puck and the stick in example 2.5?

2. In figure P2.2, a runner's position is plotted as a function of time.
   (a) Graphically estimate her velocity at $t = 3$ s.
   (b) For the whole run, what is her average speed, in m/s?
   (c) On two grids, roughly sketch her speed as a function of time, and her acceleration as a function of time. Include numbers on the axes.
   (d) Is this run near Olympic standards?

3. A baseball is thrown straight up with a velocity of 30 m/s. Indicating both the magnitude and direction of the velocity,
   (a) what is the ball's velocity 2 s after being thrown?
   (b) what is the ball's velocity 4 s after being thrown?

4. A runner is moving at 5 m/s at $t = 2$ s, and by $t = 5$ s, she moves at 11 m/s in the same direction. What is her average acceleration?

5. A lot of confusion about "floating" comes from the fact that a big jumper spends so much time near the very peak of his jump.
   (a) If a basketball player jumps 36 in. straight up, for how long is he in the air?
   (b) For how long is he within 9 in. of the top of his jump?
   (c) What fraction of the time is he in the top 9 in.?

6. L.A. Dodgers centerfielder Matt Kemp is one of the fastest players in baseball today, near the top of stolen base lists. He's been clocked at 22 mph, not too far from Bolt's peak speed, shown in figure 2.7. Kemp rounds second and is racing toward third at 24 mph.
   (a) If he maintains a constant speed, how long does it take for him to cover the final 12 ft to third?
   (b) Of course, coming into third base at full speed means that the runner will likely overrun the base. Therefore, the base coach signals Kemp to slide the final 12 ft. He does so, decelerating at a constant rate, coming to rest precisely on the base. What is the magnitude of his acceleration? How does this compare to the acceleration due to gravity?
   (c) By sliding, how much more time has Kemp taken to reach the bag? That is, how long does it take him to cover the final 12 ft when sliding, and how does that compare with your answer to part (a)?

7. A distance runner completes 1 mi in 5 min. What is his average velocity, in m/s?

8. A Ferrari Scaglietti has an acceleration of 5.9 m/s^2, which we assume constant. How far down the track has it gone, 5 s after it's jumped off the starting line?

9. When an object accelerates at 9.8 m/s^2 (or 32 ft/s^2), we say it accelerates with 1 g. This is true whether the object is falling under gravity, or due to something else. Using the constant-acceleration approximation we've discussed in this chapter, how many g's are involved in the following situations?
   (a) Cedar Point's Top Thrill Dragster (an awesome ride, by the way): According to the park, it goes 0–120 mph in 3.8 s on the horizontal track, before turning up vertically.
   (b) A football kickoff: The ball is initially at rest on the tee. After being in contact with the kicker's foot for only 8 ms, it leaves the kicker's foot at 85 ft/s.

10. In figure P2.10 are four sketched graphs of position versus time.
    (a) Which velocity-versus-time graph (on the right) corresponds with each position-versus-time graph?
    (b) Which of the velocity-versus-time plot(s) could correspond to motion under a constant, nonzero acceleration?

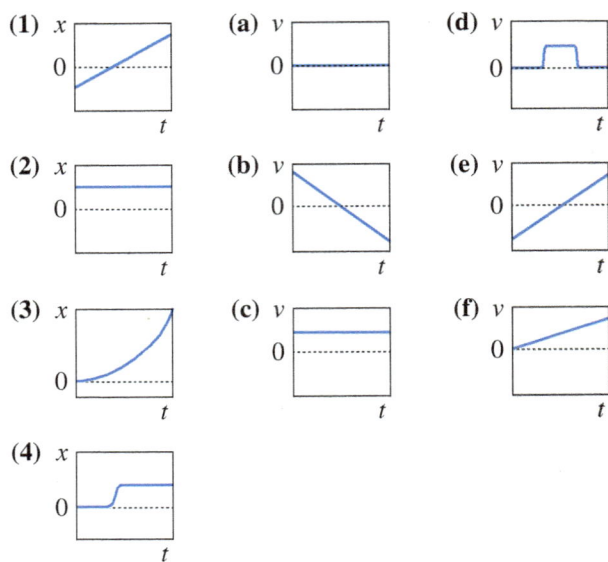

Figure P2.10

11. Unaware that you have taken a physics class, a major league pitcher claims that he can throw a baseball from the ground, straight up, as high as the Empire State Building, which is 1250 ft tall.
    (a) At what speed would he have to throw the ball, to accomplish this?
    (b) Approximately how fast *can* a major league pitcher throw a ball?
    (c) Using your answer from part (b), how high would the ball go, if thrown straight up?

12. There are many legends of basketball players with 2-s hang times. This would mean 1 s on the way up, and 1 s coming back down.
    (a) How high would his feet be above the floor, at the peak of the jump?
    (b) At what speed would the player come crashing back down onto the hard floor?
    (c) Is such a jump plausible?

13. Hockey players have to react fast in a face-off. If the ref throws down the puck from waist level at 18 ft/s, how long does it take to hit the ice? How does this compare to human reaction time?

14. Denard Robinson has one of the fastest 40-yd-dash times, with an average speed of 18.94 mph. As we said in section 2.2, though, 40-yd times are often "hand-measured," and do not include human reaction time as track-and-field events do. If an additional 150 ms were added to Robinson's time, to account for reaction time, what would his average speed be, in miles per hour?

15. In Olympic diving, the time one spends in the air determines the number of somersaults and twists the athlete can attempt. On the 3-m springboard, the diver's center of mass first rises by 1 m, then falls to the water. On the 10-m platform dive, the diver more or less falls vertically, with very little initial rise.
    (a) What is your impression: which diver spends more time in the air?
    (b) How long is the 10-m platform diver in the air?
    (c) How long is the 3-m springboard diver in the air?

16. Referring to table 2.1 and figure 2.1, what was Michael Phelps's average speed on the backstroke portion of his 2008 400-m medley? What was his average velocity on that portion?

Section 2.3   *Rope-Climbing and Diving*

17. Baseball players report that pitches are faster in Denver due to the lower air density there. But can drag really be a noticeable effect for such a short flight? In chapter 5, we will be discussing the effects of air, but let's get a feel for these effects right now. Consider a 95-mph fastball thrown at 60 ft 6 in. from mound to plate. Don't worry about the vertical motion of the ball—we just care about horizontal.

    (a) If the ball travels at constant velocity, how long does it take to reach the plate?

    (b) Air drag tends to "decelerate" the ball at about 32 ft/s^2 (or 9.8 m/s^2), which is (rather coincidentally) 1 g. If we account for drag, how long does the ball's journey take?

    (c) To get a feeling of whether air drag really matters, compare the difference between your answers to (a) and (b) with the time a bat takes to cross the plate. Is the difference between your answers to (a) and (b) much larger than this time, much smaller than this time, about the same? To answer this question, you need to know that the bat is typically moving at about 70 mph as it crosses the plate and that the plate is about 1 ft long.

18. We will study aerodynamic forces in chapter 5, but you already know that air drag will slow a table tennis ball speeding through the air. Professionals like Jörgen Persson and fellow Swede Jan-Ove Waldner have had some epic matches. In one, Persson smashed a ball at 47 mph at his edge of the table. By the time it had crossed to the other side of the 9-ft table, it had slowed to 40 mph.

    (a) In g's, what was the acceleration of the ball due to drag?

    (b) How long did the ball take to cross the table?

    (c) How does your answer in (b) compare to typical human reaction time?

    (d) How much time was added due to the air drag? That is, how much more quickly would the ball have crossed the table if it had continued at 47 mph on its entire trip?

    (e) In order to have enough time to react, Walder was not standing at the edge of the table, but 15 ft behind it. Assuming the same acceleration as you found in (a), how long does it take the ball to reach him, once Persson hits it?

    (f) Make motion graphics: sketch the velocity and position of the ball as a function of time.

19. In November 2011, Miami Dolphins running back Reggie Bush was fined $7500 for "excessive celebration" after a touchdown when he purposely slid 7 yd in the wet endzone, coming to rest right in front of a TV camera. He was moving at 25 ft/s (8.3 yd/s) when he began his slide.

    (a) What was his average acceleration?

    (b) How long (in seconds) did his slide last?

20. As clearly as you can, draw x-t and v-t graphs for a basketball player running the full length of the court—from one basket to the other—at 19 mph for a breakaway layup, then, over the course of 1 s, turning around and finally moving back to play defense under the hoop at 16 mph. You need to put numbers on all axes.

21. In his record-setting sprint in Berlin, Usain Bolt's split time at the 20-m mark was 2.88 s.

    (a) Accounting for the fact that he didn't start moving until $t = 0.146$ s, due to reaction time, what was his average acceleration in this interval? Give your answer in m/s and g.

    (b) How does this compare with a Ford Mustang, which goes from 0 to 60 mph in 5 s?

22. If a race is begun with a starting pistol like the one in figure 2.5, the runner nearest the gun might in principle have an advantage over the other runners, because the

sound reaches his ears before it reaches theirs. To prevent even this small unfairness, modern meets use a system of speakers behind the starting blocks of all competitors that sound simultaneously to start the race. But does it *really* matter? Let's see.

(a) A track lane is 4 ft wide. If the starter with the pistol stands on the inside of the track (next to lane 1), then how much sooner does the runner in lane 1 hear the report of a gun, as compared to the runner in lane 9?

(b) How does this time difference compare with, say, the amount by which world records are broken?

Net Force: Dwight Howard Illustrates

Now that we've learned how to describe motion mathematically, it is time to understand what *controls* the motion of balls and athletes and, well, everything else. The various physical elements of the game—players, balls, sticks, the floor—interact with each other by exerting **forces** on one other. There are many types of forces in sports. They speed up, slow down, or change the direction of the object being acted on. Control the forces and you control the motion—and the game.

The Orlando Magic's Dwight Howard, right, drives to the basket against the Chicago Bulls' Joakim Noah during the third quarter of an NBA basketball game in Chicago on Wednesday, December 1, 2010. The Magic won 107–78 (© *Nam Y. Huh/AP Photo*). In this chapter, we'll discuss the forces between athletes and objects that control the flow of the game.

> *When I dunk, I put something on it. I want the ball to hit the floor before I do.*
> —Darryl Dawkins, NBA player

As his opponents can attest, one can learn a lot about basketball by watching 6-ft-11-in. center Dwight Howard. We can learn a lot of physics, too, by studying one of Howard's deadly vertical slam dunks. We will spend much of this chapter analyzing less than 2 s of action, in which Howard crouches, builds up explosive momentum as he straightens his legs and swings his arms up, slams the ball down through the hoop, and finally lands, knees bent to cushion the shock.

To analyze the details of this action, we will need the mathematics of motion we discussed in chapter 2. We also need to develop the concepts of a changing center of mass, force, weight versus mass, all three of Newton's laws, and the free-body diagram. To get Howard into the paint, we'll need a basic understanding of vectors and the force of friction. Clearly, there's a lot of physics involved when Howard scores!

Once we have these important concepts under control, we'll take a look at circular motion and "imaginary forces" as they apply to curling, discus, and other sports. Let's get started.

3.1 How Things Interact: Forces

In chapter 2, we learned how to put motion into the language of math once we were given the acceleration. However, all of sports hinges on the *cause* of acceleration—objects (or people) accelerate due to their interactions with other objects (or people). A golf ball accelerates due to its interaction with a club; a baseball player accelerates as he slides into third base due to his interaction with the ground; a tennis ball accelerates (slows down) as it interacts with the air it passes through; a diver accelerates downward as she interacts gravitationally with the Earth.

In a scientific approach to sports, we quantify all of these interactions in terms of **forces**. A force is basically a push or a pull that one object exerts on another. Forces are vectors, which means they have (1) a magnitude, quantifying the strength of the push or pull; (2) a direction, because one can be pushed left, right, up, or down; and (3) a unit of measure.

The unit of force is pound (lb) in the Imperial system and Newton (N) in the SI system. Conversions between these units are as follows:

$$1 \text{ lb} = 4.45 \text{ N} \qquad 1 \text{ N} = 0.225 \text{ lb}. \tag{3.1}$$

As we'll discuss in a moment, an object's weight can be measured in pounds or Newtons. You're probably less familiar with the latter unit, so it might help to remember that the typical apple weighs 1 Newton.

It is fair to say that sports is all about mastering forces to control motion.

3.2 The Physics of a Dwight Howard Dunk

3.2.1 Waiting for the pass

We begin with Howard waiting for the pass under the hoop. He is momentarily at rest—a rare state for him. What is especially important for physics is that he is not *accelerating*. When someone is not accelerating, we say he is in **equilibrium**.

An object in equilibrium is not *accelerating* ($|a| = 0$). However, it may be *moving* with constant velocity.

As we will see repeatedly, scientific understanding of a sports situation often starts with an analysis of the forces at play. The first step in this process is identifying all of those forces.

You should always start with gravity—gravity *always* acts on all objects and people, regardless of whether they are flying through the air or sitting on the floor. The force of gravity on an object is called its **weight**, $\vec{W}$. For Dwight Howard, $W_y = -265$ lb; the subscript y indicates that we are here interested in the vertical component of the force (in fact, of course, weight has no horizontal component) and the negative sign means that the force points vertically down.

As we mentioned in chapter 2, if gravity is the only force acting on something, then that thing must accelerate downward at 32 ft/s². Because Dwight is in equilibrium, some other force is also in play, counteracting gravity. Of course, we know that this has something to do with the floor he is standing on. In fact, the floor is exerting an upward-pointing **normal force** on each of Dwight's feet.

We will discuss only a few types of forces in this book; these are listed in table 3.1. Normal forces are among the most important. A normal force is a "pushing" force that occurs when two objects are in contact with each other. In geometry, "normal" is another word for "perpendicular"; normal forces are perpendicular to the plane between the two objects where they touch. For example, the normal force on Howard's feet is perpendicular to the floor (and to the bottom of his shoes). Any time we see two objects in contact, it's a good bet that they are exerting normal forces on each other.

Table 3.1. The types of forces we will discuss in this book.

Weight	The downward pull of gravity
Normal	The push exerted by an object in contact with another
Tension	The pulling force of a rope, muscle, tendon, and so on
Friction	The resistance to sliding when two objects are in contact
Fluid forces	Discussed in detail in chapter 5:
	• Buoyancy: the upward force when *in* a fluid (air or water)
	• Drag: the retarding force when *moving through* a fluid
	• Magnus: the sideways force when *moving and spinning* in a fluid

In sports physics, the upward normal force exerted by the ground is often called the **ground reaction force** (GRF). We'll use this name and and the symbol $\vec{G}$ to designate the force, to help distinguish it from other normal forces that will come into play later on.

Figure 3.1 shows a **free-body diagram** (FBD) of Howard as he waits for the pass. An FBD is simply a sketch of an object (or person) showing all forces acting *on* that object. As I will repeatedly stress, only forces acting *on* Dwight Howard belong in an FBD of Dwight Howard. If, as he tends to do, *Howard* exerts a force on *another* player, then that force belongs in an FBD of the other player. Forces are drawn as arrows: the direction of the arrow shows the direction of the force, and the length of the arrow indicates the strength of the force.

When drawing a free-body diagram to analyze a situation, be sure to draw only the forces of other objects acting *on* the object, not those that the object itself exerts.

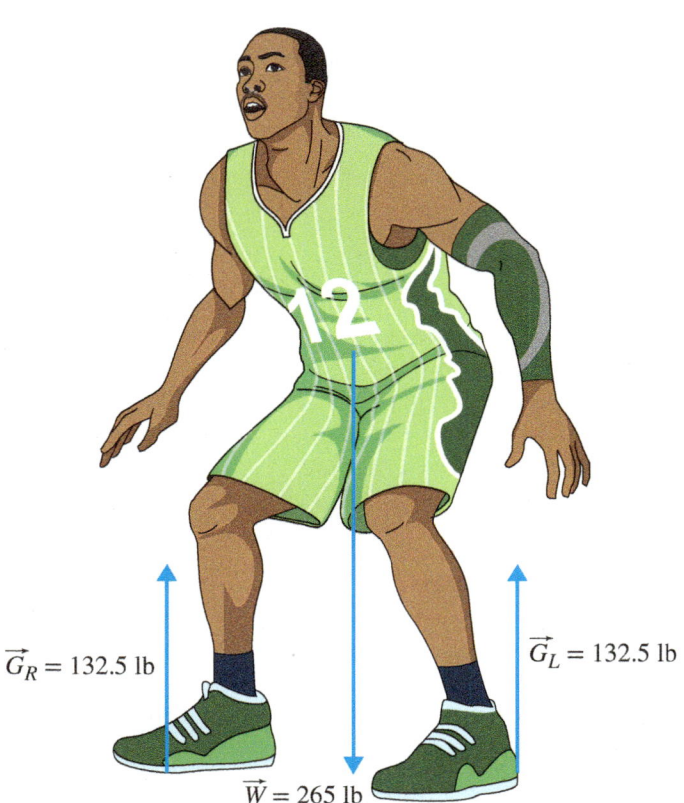

Figure 3.1. Howard waiting for the pass. This is a free-body diagram (FBD) showing all forces acting on Howard: the downward force of gravity and the upward forces from the floor on his feet. At this moment, the forces cancel each other out, so Howard is not accelerating—he is in equilibrium.

In addition, one should usually draw the arrow starting (or sometimes ending, if the diagram gets too crowded) at the point of application of the force. For example, the arrows for the normal force should begin at Howard's feet). This aspect of FBDs won't be very important for us until we discuss torque later on.

The normal forces on Howard's feet clearly push upward, but how do we know the magnitudes of these forces? An important clue is in his posture: to be equally ready to spring left or right, Howard takes the perfect position, with his center of mass directly between his two feet. It doesn't take a whole lot of mathematical analysis to figure out that this means that the normal forces on Howard's two feet are of equal magnitude: $G_{R,y} = G_{L,y}$.

The forces are equal, but what *are* they? You can probably guess the answer, but it turns out that Howard's stance is a perfect example of one of the most important scientific laws, formulated in 1687, **Newton's first law**.

Newton's First Law

The *net* force on a player in equilibrium is zero.

$$\sum_i \vec{F}_i = \vec{F}_1 + \vec{F}_2 + \vec{F}_3 + \cdots = 0 \leftrightarrow \vec{a} = 0 \qquad (3.2)$$

The summation symbol $\sum_i$ just means we are adding up all the forces. The sum of all the forces is called the **net force**.

The calculation for Howard in equilibrium is then

$$W_y + G_{R,y} + G_{L,y} = 0$$
$$-265 \text{ lb} + G_{R,y} + G_{L,y} = 0$$
$$G_{R,y} + G_{L,y} = 265 \text{ lb}.$$

So, the upward force on each of Howard's feet is (265 lb)/2 = 132.5 lb.

Notice how pounds (lb) are used to quantify all forces, not only weight.

For the rest of this chapter, we won't consider the left and right foot separately. The symbol G_y will refer to their sum.

Equilibrium does not always mean "at rest"

Just like Howard in figure 3.1, a hockey puck sitting at rest on the ice is obviously in equilibrium (figure 3.2). What about one that glides across "frictionless ice" (a magical substance that exists only in the physicist's mind) at a constant velocity? Well, because $\vec{v}$ is constant, there is no acceleration, so this puck is also in equilibrium.

Don't make the mistake of equating "equilibrium" with "at rest." A puck with no net force acting on it will move at constant velocity. This is not obvious in everyday observation, since a puck on *real* ice will slow slightly due to the unbalanced forces of friction and air resistance.

3.2.2 Spec sheet

As he waits for the pass, Howard is in equilibrium because the various forces on his body "cancel out"—the net force is zero. He can accelerate only when there is an *im*balance of forces. It doesn't take a whole lot of thought to realize that the *direction*

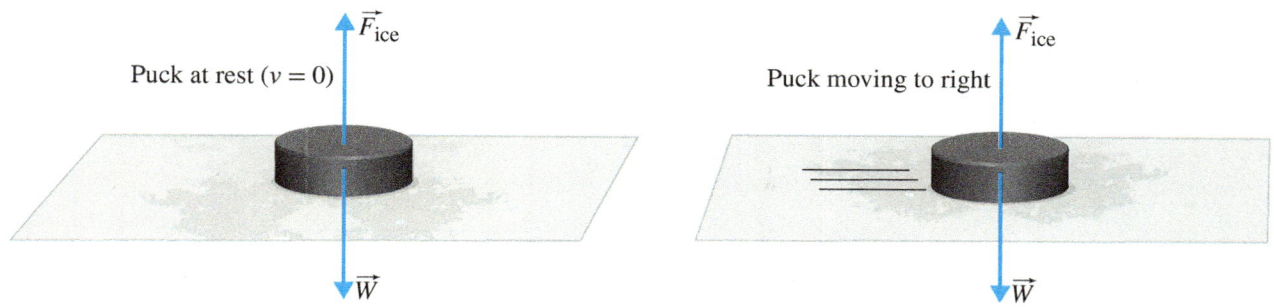

Figure 3.2. A puck at rest is in equilibrium, as is one moving to the right at constant speed.

of the resulting acceleration is the same as the direction of the force imbalance—that is to say, the direction of the net force. For Howard to get off the ground, the net force on him will need to point upward.

However, to get any kind of reasonable acceleration, the net force will need to be large, due to Howard's large **mass**. Mass is a property of any object (including human bodies) and quantifies the object's tendency to resist acceleration. Due to their smaller mass, the bodies of average-size physics professors are less able to resist acceleration than is Dwight Howard—a push from LeBron James that might nudge Howard out of the lane would knock a poor professor across the court.

Because mass and weight are strongly related, students often confuse them; do yourself a favor and don't be such a student. Weight is a force: it is a vector with a direction and like all forces is due to the interaction with some other object; in this case that other object is the planet Earth. Mass is not a vector, and mass does not come from the interaction with anything. It is a measure of how much matter is in an object. Basically, an object's mass is determined by the number and type of atoms making up the object.

The elements carbon (C), oxygen (O), and hydrogen (H) account for almost all (95%) of the human body, by mass. So, Dwight Howard reacts less to the push by LeBron James because his body has more C, O, and H atoms than the professor's body.

The relationship between mass and weight is simple:

$$W = mg. \tag{3.3}$$

The equation just gives the magnitude of the weight force; we already know that $\vec{W}$ points downward.

The g in equation 3.3 is the same as given in equation 2.11 in chapter 2. There, we mentioned that g is a little different at various places on Earth: it is smaller in Mexico City (host of the 1968 Olympic games) than in Helsinki, Finland (1952 Olympics). Dwight's weight will be larger if he steps off a plane in Helsinki than if he lands in Mexico City. In fact, his weight will be zero if he ends up in deep space (even Howard can't jump this high, however). In all of those places, though, his body will contain the same number of atoms: his mass—and his resistance to force—will be the same.

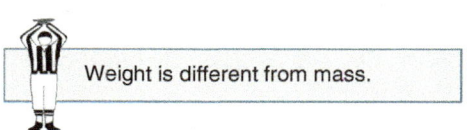
Weight is different from mass.

Hopefully now it is clear that, although they are related, mass and weight are quite different things. A clear case of this is a failed get-rich-quick scheme in example 3.1.

Like most other quantities, mass is measured with different units in the SI ("metric") and Imperial systems. Most Americans are comfortable with the Imperial force unit of pound and much less familiar with the Newton. On the other hand, you are likely familiar with the SI unit of mass, the **kilogram** (kg), and haven't even heard of the customary unit for mass, which is the **slug**. (Reportedly, the name refers to the fact that a massive object reacts "sluggishly" to an applied force.)

To give some feeling for the units, a 150-lb person has a mass of about 70 kg, or 5 slug.

We will find the following relationships between the units quite handy:

$$1 \text{ N} = 1 \text{ kg} \times \text{m/s}^2.$$
$$1 \text{ lb} = 1 \text{ slug} \times \text{ft/s}^2.$$
(3.4)

Because Dwight Howard's mass is a property of Dwight Howard himself, and not one that depends on where he travels, his spec sheet should really list his mass, not his weight. His weight is reported to be 265 lb, or 1179.25 N. We can easily find his mass with equation 3.3:

$$m_{\text{DH}} = \frac{W_{\text{DH}}}{g} = \frac{265 \text{ lb}}{32 \text{ ft/s}^2} = \frac{265 \text{ slug} \times \text{ft/s}^2}{32 \text{ ft/s}^2} = 8.28 \text{ slug}.$$

$$m_{\text{DH}} = \frac{W_{\text{DH}}}{g} = \frac{1179.25 \text{ N}}{9.8 \text{ m/s}^2} = \frac{1179.25 \text{ kg} \times \text{m/s}^2}{9.8 \text{ m/s}^2} = 120.3 \text{ kg}.$$

EXAMPLE 3.1 *Gold: Buy high and sell low!*

(© Glow Images RF)

Speculators buy gold when they feel it's hit a low price, to sell it later at a higher price—the "buy low and sell high" approach. They are buying and selling at different points in *time*, and taking a risk that the price of a fixed quantity of gold will change.

However, gold is often sold by the ounce—a measure of its weight. As we've discussed, the weight of a gold coin is different in different cities; these differences are well known and will not change. Therefore, a trading strategy of buying and selling by weight at different points in *space* would be no risk at all. Shocking traditional traders, we should follow a "buy high and sell low" policy—buy gold coins in Denver where they weigh less, and sell them in New York where they weigh more.

Say gold costs $1100/oz on a given day. If we buy a pound (16 oz) of gold coins in Denver (g_{Denver} = 9.796 m/s^2 = 32.1309 ft/s^2) and sell them in New York ($g_{\text{New York}}$ = 9.803 m/s^2 = 32.1538 ft/s^2), what will be our profit?

It's easy to figure out our cost:

$$\text{Buying Price} = \frac{\$1100}{\text{oz}} \cdot (16 \text{ oz}) = \$17{,}600.$$

When we bring our purchase to New York, how much will it weigh and what will we get for it? Well, even though the weight of the gold will be different in the two cities, the mass will be the same:

$$m = \frac{W_{\text{Denver}}}{g_{\text{Denver}}} = \frac{1 \text{ lb}}{32.1309 \text{ ft/s}^2} = 0.031123 \text{ slug}.$$

The weight of our purchase in New York is

$$W_{\text{New York}} = mg_{\text{New York}} = (0.031123 \text{ slug})(32.1538 \text{ ft/s}^2) = 1.0007 \text{ lb}$$

$$= 1.0007 \text{ lb} \times \frac{16 \text{ oz}}{1 \text{ lb}} = 16.0114 \text{ oz}.$$

We'll be able to sell it for

$$\text{Selling Price} = \frac{\$1100}{\text{oz}} \cdot (16.0114 \text{ oz}) = \$17612.54.$$

Section 3.2   The Physics of a Dwight Howard Dunk   **45**

We've made $12.54, which doesn't even cover the cost of transportation. Of course, if we'd bought 1000 lb of gold in Denver, we'd make a profit of more than $12,000.

Although the profit margin is small, this appears a surefire way to make money! Sadly, however, the scheme won't work. Although gold was once sold by weight and is still sold by the ounce, the "troy ounce" is now used for precious metals and is defined in terms of *mass*: 1 troy ounce = 31.1034768 gram. This is about 95 billion trillion atoms of gold, regardless of whether it is in New York or Denver.

3.2.3 The launch

Howard is reported to have a 40-in. vertical leap, a claim we can verify by studying the photo of him kissing the rim in figure 3.3. This is an impressive vertical for anyone,

Figure 3.3. Dwight Howard kisses the rim during training camp in 2005. Knowing his height and the height of a rim, we can accurately estimate his vertical jump. In the photo, he is 37 in. above the ground and his lips are about 2 in. from the hoop. In this or other attempts, Howard's lips reportedly do touch the rim, so his vertical is certainly at least 39 in. Claims of a 40-in. vertical seem quite believable. (© *NBAE/Getty Images*).

but especially unusual for a 265-lb player; as we'll see, the forces required for such a leap are formidable. Before we get to the forces, though, let's consider the motion involved in the launch.

Kinematics of the jump

As we saw in chapter 2, Howard must leave the ground at 9.9 mph in order to achieve a 40-in. vertical. His acceleration from 0 mph to 9.9 mph cannot happen instantly. Instead, it is the result of the imbalance of forces generated as the center explodes upward from a crouching position, as shown in figure 3.4. That figure is busy, so let's take some time to discuss it.

Athletic motions of the human body are usually very complex, with arms, legs, and torso moving in different directions. Recall, however, that the most important motion is the motion of the center of mass—the "average" position of the person's mass. We will pay special attention to the center of mass, indicated by the red "X" in the figure. Remember that once Howard's feet leave the floor, the only force on him is his weight, and the motion of his center of mass is then determined—there is nothing he can do about it.

Any upward acceleration of the center of mass must be achieved in the interval between panels (a), where $v_{y,0} = 0$, and (c), where $v_y = 9.9$ mph. The acceleration during this interval will be different at different instants in time, but for now we will figure out the average acceleration. This lets us use our constant-acceleration

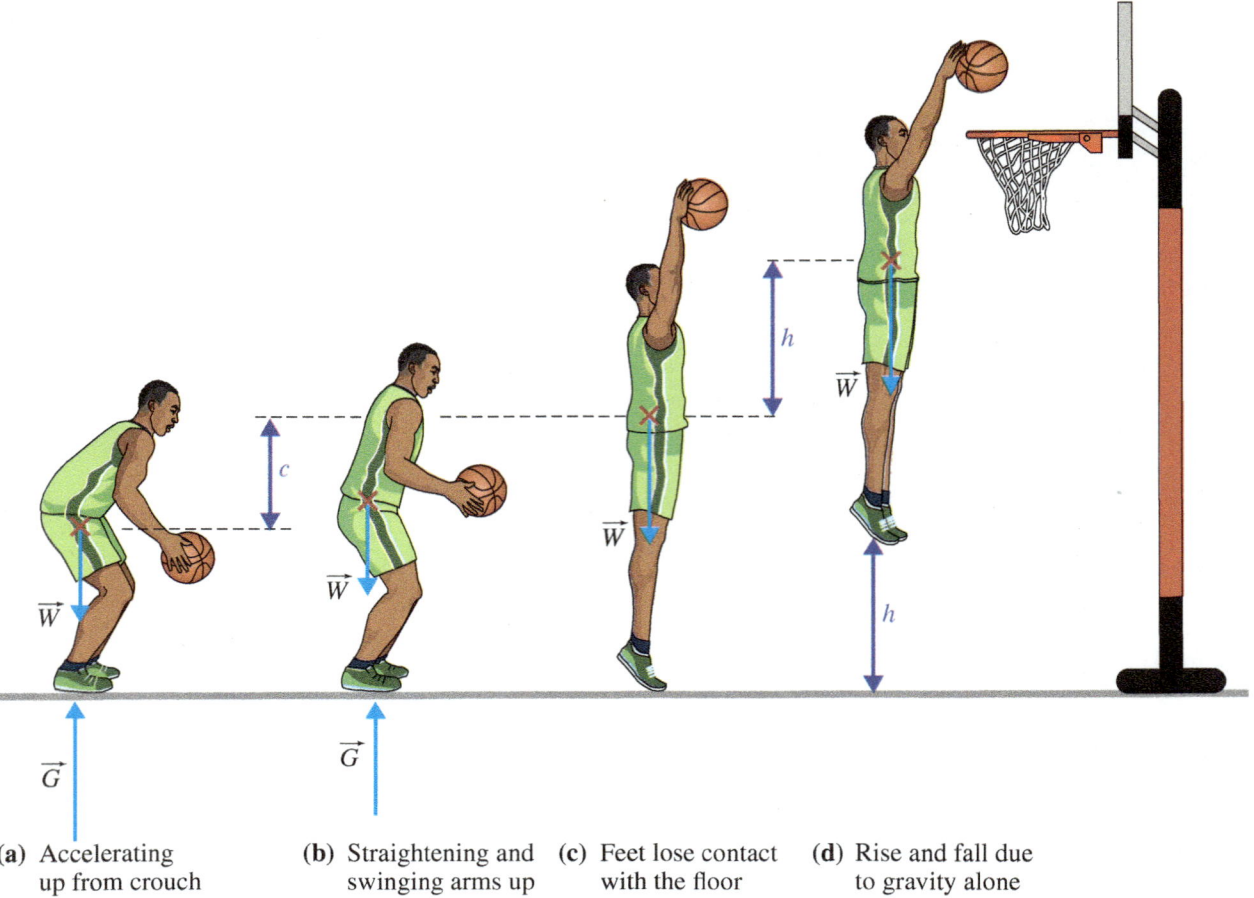

Figure 3.4. The mechanics of Dwight Howard's vertical leap. In panels (a) and (b), the magnitude of the normal force from the floor is greater than Howard's weight. Hence, the net force points upward and so does Howard's acceleration. Once contact with the floor is broken, the only force acting on Howard is his weight, and he instantly begins accelerating downward.

equations from chapter 2. In particular, let's use equation 2.9 to relate the acceleration to the rise in Howard's center of mass; this is labeled c in the figure.

$$v_y^2 = v_{y,0}^2 + 2a_y \Delta y \rightarrow a_y = \frac{v_y^2 - v_{0,y}^2}{2\Delta y}$$

$$\rightarrow \boxed{a_{y,\text{launch}} = \frac{v_{y,\text{launch}}^2}{2c}}. \tag{3.5}$$

From video analysis of a Dwight Howard vertical leap, one can estimate that Howard's center of mass rises about 20 in. before his feet leave the floor: $c = 20$ in. $= 1.67$ ft. (We'll say more about how this is achieved, shortly.) If he wants to leap 40 in., $v_{y,\text{launch}} = 9.9$ mph $= 14.6$ ft/s, so we can estimate Howard's average acceleration as

$$a_{y,\text{launch}} = \frac{(14.6 \text{ ft/s})^2}{2 \cdot 1.67 \text{ ft}} = 63.8 \text{ ft/s}^2.$$

Howard is thrust up at about two g's!

Imbalance of forces

Howard achieves an upward acceleration during the jump due to an *im*balance of the forces on him. In particular, as indicated in the figure, the upward-pointing GRF is larger than the downward-pointing force of gravity (his weight). The precise relationship between this force imbalance and Howard's acceleration is given by **Newton's second law**.

Newton's Second Law

The acceleration of the center of mass of an object or player is the ratio of the *net* force on the object and its mass.

$$\vec{a}_{\text{c.m.}} = \frac{\sum_i \vec{F}_i}{m} = \frac{\vec{F}_1 + \vec{F}_2 + \vec{F}_3 + \cdots}{m} \tag{3.6}$$

It is crucial to understand that the forces in equation 3.6 are the forces *on* Howard. They do *not* include the force that Howard's legs exert on the floor—that force can only accelerate the floor, not Howard. The only relevant forces are those shown in the FBD—the GRF and the weight. Howard's weight is always the same, regardless of whether he is in the air, on the ground, in equilibrium, or accelerating. So, the only force we need to figure out is $\vec{G}$, which only has a vertical component, G_y. The calculation for the jump is

$$a_{y,\text{launch}} = \frac{G_y + W_y}{m} \rightarrow G_y = ma_y - W_y$$

$$= (8.28 \text{ slug})(+63.8 \text{ ft/s}^2) - (-265 \text{ lb})$$

$$= 528 \text{ lb} + 265 \text{ lb}$$

$$= 793 \text{ lb}.$$

That's a lot of force!

It is so easy to get confused by signs, so take just a minute to notice that in equation 3.6 all of the forces are *added*, not subtracted. Pay attention to what we've done above, adding the y-components of the forces and remembering that because the force of weight points downward, W_y has to be negative. We also explicitly included the sign on the upward acceleration, just to be clear.

The origin of the GRF

You may have heard the admonition to "pull yourself up by your bootstraps." This may be good advice to the young person who needs to work hard to get ahead, but as a physical concept it is nonsense. Try to pull up your bootstraps and all you'll end up with are tight boots. A law of nature is that no person or object can accelerate himself by exerting a force on himself.

Dwight Howard's acceleration is caused by forces exerted *on* Dwight Howard *by* other things (the floor and the Earth). In particular, he was accelerated because the GRF changed. Howard only accelerated upward when *the floor pushed on him* with an upward force stronger than 265 lb.

Well, how nice of the floor to do that for Dwight! But how did it "know" when Dwight wanted it? The answer is given by **Newton's third law**.

Newton's Third Law

When body 1 exerts a force on body 2, then body 2 exerts an equal and opposite force on body 1.

$$\boxed{\vec{F}_{1,2} = -\vec{F}_{2,1}} \qquad (3.7)$$

These forces are often called an "action-reaction pair."

This law—the final of Newton's three—tells us that the upward GRF exerted *by* the floor *on* Howard is the "reaction" force to Howard's downward normal force *on* the floor. So, Howard's upward acceleration clearly is strongly related to the downward push his legs exert on the floor (you already knew that), but that force can only accelerate the floor. The reaction force is a different force—it is exerted on a different object (Howard) and creates a different acceleration (71 ft/s^2).

GRF: Peak force

We have noted that the *average* force exerted by Howard's legs during the jump is 793 lb—impressive indeed! However, this force is not constant during the jump—it grows with time and, depending on the jumper, can show some complicated wiggles. Figure 3.5 shows the GRF on a nonprofessional jumper, as he executes a jump. For the first second or so, the jumper is stationary and not accelerating.[1] He is in equilibrium; thus, the GRF is equal to his weight, 175 lb. He begins his jump at about $t = 3.5$ s (corresponding to panel (a) of figure 3.4), and his feet leave the ground at about $t = 3.9$ s (panel (c) of figure 3.4).

The average force during this interval is about 250 lb, but the *peak* force occurs at about $t = 3.8$ s and is 350 lb, 40% higher. It is reasonable to assume that Dwight Howard's peak force is likewise 40% higher than his average force of 793 lb—about 1100 lb, more than half a ton!

GRF: Swinging the arms

Why do players usually raise their arms during take-off, rather than lifting their arms before starting to jump, or lifting them while in the air? There is a physics-based

---

[1] Of course you remember that "stationary" ($|\vec{v}| = 0$) doesn't *necessarily* mean $|\vec{a}| = 0$, right?

Section 3.2   The Physics of a Dwight Howard Dunk   49

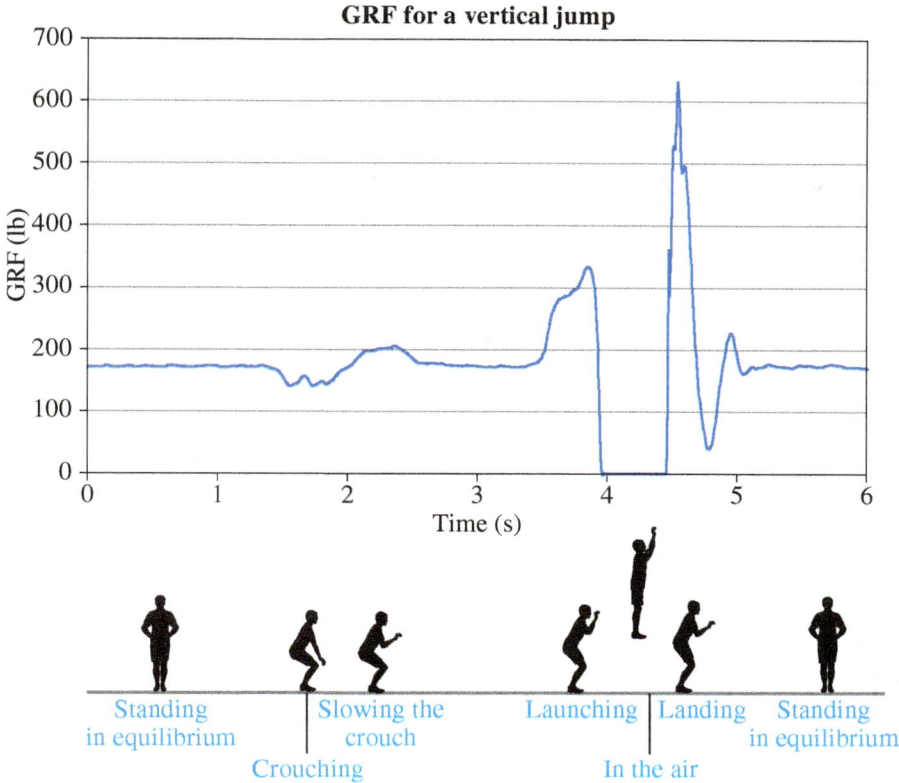

Figure 3.5. The GRF versus time, for a jumper whose lack of jumping prowess steered him toward an academic career (the author). The GRF is measured by a force plate—essentially a bathroom scale that can be read out via computer.

benefit to swinging them up during the jump, beyond simply the "natural" feeling of jumping in this way. We can understand this by reviewing the physics of the jump.

What are the ingredients that enter into maximizing the vertical height h? By combining equations 2.14 and 3.5, we can write down a simple formula that gets right to the point:

$$h = \frac{F_{\text{leg},y} - W_y}{W_y} \cdot c = \frac{a_{y,\text{launch}}}{g} \cdot c = (g\text{'s of acceleration})\, c. \qquad (3.8)$$

So, if the average acceleration during the launch is $2\,g$, then the rise of one's center of mass after the jump is twice as much as its rise before the jump.

It is no surprise that more acceleration—more g's—means a higher jump. To achieve that, we need legs as strong as possible (to generate a large GRF reaction) and mass as little as possible (to minimize resistance to the force). Naturally, there is tension between these requirements, as increased muscle mass leads to increased strength. Many of the best jumpers in field events (high jump, long jump) develop strong thighs through lifting and other strengthening exercises but are relatively slender on top; the upper body is basically dead weight for jumping. That said, big men in basketball—including 7-ft-1-in. Shaquille O'Neal, who weights 300+ lb—regularly achieve verticals in the mid 30-in. range.

What may be more surprising is that c, the rise in height *before* leaving the ground, is equally important as the acceleration—increase either one by 20%, and you jump 20% higher. How does one increase c? The most obvious way is to start in a lower crouch, as in panel (a) of figure 3.4. There is a natural limit to this,

 The depth of the crouch—the distance that the center of mass rises during the jump—is as important as the force generated.

however: physiologically the muscles in the thigh can exert the greatest explosive force when stretched by about 20% of their resting length.

Another way to increase c is by swinging the arms from a low to a high position during the jump, *while the feet are still on the floor*. Remember from section 1.3 that raising the arms raises the center of mass, even if nothing else in the body moves. On figure 3.4, Howard's center of mass rises from below his belt line in panel (a), to above his belt line in panel (c), even while his belt line itself was accelerated upward. The 4- to 5-in. contribution coming from swinging the arms is about 25% of the total center of mass rise of $c = 18$ in.; equation 3.8 tells us that this adds 10 in. to Howard's jump.

According to our discussion so far, Howard can be out of equilibrium even if he never straightens from his crouch. Simply by accelerating his arms, he causes his center of mass to accelerate—the forces on him do not "cancel out." Since Howard's weight $\vec{W}$ never changes, this means that the normal force from the floor, the GRF, depends on what his arms are doing.

Figure 3.6 shows the effect when our academic jumper swings his arms. At first, he is in equilibrium, and the GRF is equal to his weight, 175 lb. At about $t = 1.5$ s, the GRF is *larger* than 175 lb; this implies an upward acceleration of the jumper's center of mass—he is accelerating his arms upward. At $t \approx 2$ s, the net force on the jumper points downward, implying that his center of mass is accelerating downward as he accelerates his arms.

Accelerating your arms up while jumping can increase your vertical by 20% or more.

It is interesting to note that moderately vigorous swings of the arms change the GRF by about 25 lb, a not-insignificant contribution to the overall force of the jump. Swinging one's arms while jumping from a crouch is important—it increases both c and $a_{y,\text{launch}}$ in the crouch-jump equation 3.8.

> **Question**
>
> In figure 3.6, the GRF is alternately greater than and less than the person's weight. When the GRF is less than the person's weight of 175 lb,
>
> - can you say in which direction the hands are *moving*? If so, are they moving up or down?
> - can you say in which direction the hands are *accelerating*? If so, are they accelerating up or down?

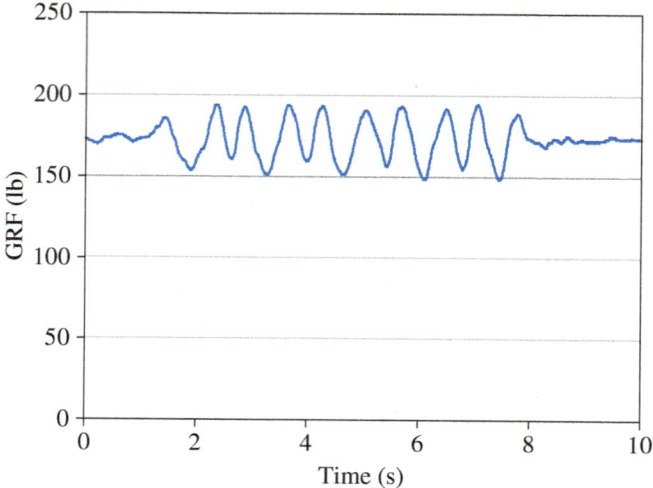

Figure 3.6. The GRF versus time for a person who never springs with his legs, but only swings his arms.

Section 3.2   The Physics of a Dwight Howard Dunk

EXAMPLE 3.2 *How much did the jumper crouch?*

From the graph of GRF, we can tell almost everything there is to know about the jump. Using only the chart shown in figure 3.5, estimate how high the jumper rose (h) and how far he crouched (c).

Once the jumper leaves the ground, only gravity is at play, and we can directly relate the hang time (about 0.5 s, estimating from the figure) to the height jumped, using equation 2.15:

$$T_{\text{hang}} = 2\sqrt{\frac{2h}{g}} \rightarrow h = \frac{gT_{\text{hang}}^2}{8} = \frac{(32 \text{ ft/s}^2)(0.5 \text{ s})^2}{8} = 1 \text{ ft} = 12 \text{ in.}$$

This is a nice rule of thumb to keep in mind: a half-second hang time means a 1-ft jump.

Now, how about the crouch? As with many problems in this book, there is more than one way to solve this problem—if you find a solution, don't worry about whether you solved it in the way you "should" have. In both of the methods used here, however, we will need to find the launch acceleration.

First method

In the first method, we start by finding the jumper's mass, which is easy enough by looking at the first 0.5 s where he is standing still. His weight is 175 lb, so $m = (175 \text{ lb})/(32 \text{ ft/s}^2) = 5.47$ slug.

Next we estimate the average GRF during the time interval between $t \approx 3.4$ s and 3.8 s at about 275 lb. At this point, we need to be careful: Newton's second law tells us the acceleration is determined by the force *imbalance*—the *net* force:

$$a_{y,\text{launch}} = \frac{G_y + W_y}{m} = \frac{275 \text{ lb} + (-175 \text{ lb})}{5.47 \text{ slug}} = 18.3 \text{ ft/s}^2,$$

about 0.6 g.

Finally, we use the jump formula of equation 3.8 to find c:

$$h = \frac{a_{y,\text{launch}}}{g} \cdot c \rightarrow c = \frac{hg}{a_{y,\text{launch}}} = \frac{(1 \text{ ft})(32 \text{ ft/s}^2)}{18.4 \text{ ft/s}^2} = 1.75 \text{ ft} = 21 \text{ in.}$$

Our jumper is no Dwight Howard—his center of mass was raised more with his feet on the ground than when they were in the air!

Second method

Here, we estimate $a_{y,\text{launch}}$ by estimating the change in velocity during the jump and the time over which that change happened.

Because the jumper starts from rest, the change in velocity is just his launch velocity, which we can figure out using equation 2.14:

$$h = v_{y,\text{launch}}^2/2g \rightarrow v_{y,\text{launch}} = \sqrt{2gh} = \sqrt{2(32 \text{ ft/s}^2)(1 \text{ ft})} = 8 \text{ ft/s}^2.$$

From the graph of figure 3.5, we see that the GRF surpasses the jumper's weight at about 3.4 s; this is the beginning of upward acceleration—the beginning of the jump. His feet leave the ground at about 3.8 s, so $\Delta t = 0.5$ s, and his acceleration is

$$a_{y,\text{launch}} = \frac{\Delta v_y}{\Delta t} = \frac{8 \text{ ft/s}^2}{0.4 \text{ s}} = 20 \text{ ft/s}^2.$$

The first thing we notice is that this value for acceleration is not identical to the value of 18.4 ft/s² we came up with above, but this is not a problem nor is it surprising. These are estimates based a rough (not to say sloppy) evaluation of the graph.

At this point we could simply use the jump formula (equation 3.8) as we did in method 1, but for variety's sake we'll use instead equation 2.9 from chapter 2:

$$v_y^2 = v_{y,0}^2 + 2a_y\Delta y \rightarrow c = \Delta y = \frac{v_y^2 - v_{y,0}^2}{2a_y} = \frac{(8\text{ ft/s})^2 - 0}{2\left(20\text{ ft/s}^2\right)} = 1.6\text{ ft} = 19.2\text{ in.}$$

This is not precisely the same number we got with the previous method, again because our estimates were rough. However, the conclusion is precisely the same: the jumper is no Dwight Howard.

GRF: Tucking

Look one last time at figure 3.5. The jumper begins accelerating upward at about $t \approx 1.5$ s—this is when the upward-pointing GRF is larger than his weight, and the net force points upward. Before this, between $t = 1$ and 2 s, the GRF is *smaller* than the weight, implying that the net force on the jumper points down. Newton's second law tells us that here, the jumper's center of mass must be accelerating downward. This is the interval over which the jumper is crouching and accelerating his arms downward, in anticipation of the leap.

3.2.4 The flight

Once Howard is in the air, the situation becomes much simpler. The only force on him is his weight, $\vec{W}$, and this does not change; we've dealt with this situation in section 2.3. Howard has taken off with $v_{y,\text{launch}} = 9.9$ mph, and there is *nothing* he can do to alter the trajectory of his center of mass. Unless another player exerts a force on him during a foul (something Howard is quite accustomed to), his center of mass will rise 40 in. and return to its original height in 0.91 s.

While the flight of Howard's center of mass is simple and unchangeable, however, the motions of his individual body parts—hands, feet, and so on—are not. What are the effects of his body's contortions while he is in the air? As we've discussed already in some detail, when an athlete changes his body's configuration (positions of arms, legs, etc.), he changes the location of his arms and feet *relative to the position of his center of mass*.

Some basketball players have the misconception that raising their feet by bending their knees after lift-off will increase the height of their jump (see figure 3.7). Whether on the floor or in the air, by bending, Howard raises his center of mass relative to his belt buckle and his head by perhaps 3 in. But this is the same as saying that bending his knees *lowers* the height of his head relative to his center of mass by 3 in. Since Howard's center of mass will reach a peak height of 40 in. no matter what he does with his body, by bending his knees Howard reduces the peak height of his head—and the ball in his outstretched hands—by 3 in.

Bending the knees, once airborne, *decreases* the height achieved by the jumper's head and the ball in his upstretched arms.

> **Question**
>
> In figure 3.3, Dwight Howard's lips will just barely touch the rim at the peak of his jump. If he left the ground at the same speed, but lifted his arms while in the air, would his lips fail to reach the rim, or would they crash into the rim? Why?

Figure 3.7. If he leaves the floor at 9.9 mph, Howard's center of mass reaches a peak height of 40 in. In the jump to the right, he bends his knees. This causes his *feet* to go higher than they would have otherwise, but the peak height of his head and the ball are reduced.

3.2.5 The landing

At about $t = 4.5$ s, the jumper in figure 3.5 lands. His initial velocity is down $v_{y,0} < 0$ (in fact, as we discussed in chapter 2, he lands at the same speed with which he took off) and eventually he comes to rest ($v_y = 0$), so his average acceleration must be *upward*. Newton's second law says that the force imbalance must point upward, meaning the upward-pointing GRF must be larger than the downward-pointing weight. This is what happens at first—the force of the ground on the jumper's legs exceeds 600 lb, after which there are some complicated oscillations that eventually die away.

To see why the forces on a jumper's legs are so much higher during landing, we turn again to equation 2.9:

$$v_y^2 = v_{y,0}^2 + 2a_y \Delta y \quad \rightarrow \quad a_y = \frac{v_y^2 - v_{y,0}^2}{2\Delta y}.$$

Δy is the change in height of the center of mass during the landing or launching process. During the launch, the jumper uncrouches and swings his arms. Upon landing, however, he does not crouch, so Δy is less and a_y (and GRF) is more.

You can easily verify that gradually slowing yourself by crouching as you land will "soften" your landing. In so doing, you've increased Δy and decreased GRF.

This underscores the importance of bending one's knees upon landing; neglecting to do so results in truly staggering forces and injury.

But what about basketball shoes and their high-tech soles—can't they absorb the shock? Regardless of the technology, basic physics puts a fundamental limit on the degree to which shoes can reduce the force of landing. That limit is determined by the thickness of the sole, which can be as much as 2 in.

Upon landing, the sole of the shoe compresses as the shoe exerts an upward force on the person, accelerating his center of mass upward. As we've just seen, a smaller acceleration (hence a smaller net force) is required if the acceleration occurs over a

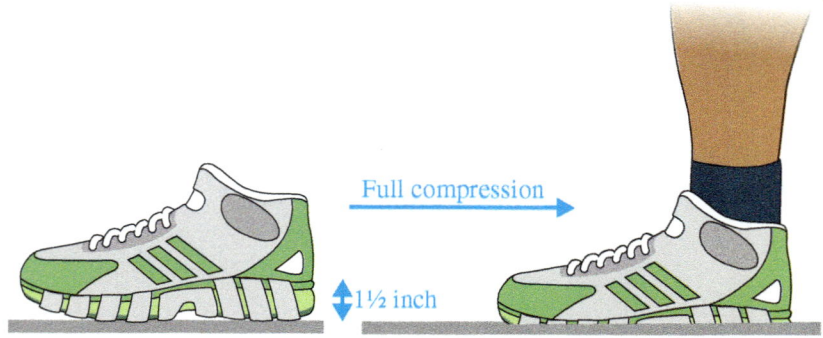

Figure 3.8. Howard's Adidas shoes have a sole about 1.5 in. thick. If a player locks his knees and remains perfectly rigid, then his center of mass moves only as much as the soles compress. This distance cannot be more than the original thickness of the soles.

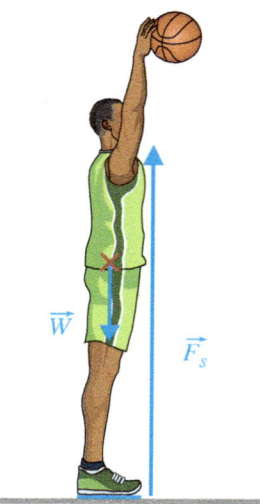

Figure 3.9
Free-body diagram for Dwight Howard as he lands. He is moving downward, but accelerating upward. The forces *on* Howard are his weight $\vec{W}$ and the upward force of his shoes $\vec{F}_s$. The shoes are sketched in blue. We do *not* include forces *on* the shoes (such as the force of the floor) in a FBD of Howard.

longer distance $|\Delta y|$. Independent of the shoe's technology, compression of the sole can only change the height of the center of mass at most by the overall thickness of the sole.

Say Dwight Howard locks his knees upon landing from his 40-in. vertical, relying on the 1.5-in.-thick soles of his Adidas TS Beasts to cushion the fall (see figure 3.8). What would be the average force exerted by the shoe on his legs?

To figure out his average acceleration ("deceleration" in this case), we use kinematics and put numbers into the equation we wrote above. As his shoes hit the ground, Howard is falling at 14.6 ft/s, and he stops falling ($v_y = 0$) when his soles are maximally compressed, $\Delta y = -1.5$ in. $= -0.125$ ft.

$$a_y = \frac{v_y^2 - v_{y,0}^2}{2\Delta y} = \frac{0 - (14.6 \text{ ft/s})^2}{2(-0.125 \text{ ft})} = 853 \text{ ft/s}^2,$$

more than 25 *g*'s!

The free-body diagram of Howard is shown in figure 3.9, where the upward-pointing force from the soles of his shoes, $\vec{F}_s$ is stronger than his downward-pointing weight. Newton's second law tells us

$$a_y = \frac{F_{s,y} + W_y}{m} \quad \rightarrow \quad F_{s,y} = ma_y - W_y$$

$$= (8.28 \text{ slug})(853 \text{ ft/s}^2) - (-265 \text{ lb})$$

$$= 7328 \text{ lb}!$$

In fact, 1.5 in. is unrealistic as a maximum that a sole could compress; using a more probable value like 0.5 in. would result in proportionally higher forces on Howard's legs. Clearly, it is crucial to bend one's knees upon landing!

 Physics places an unbreakable limit on the cushioning effect of soles.

EXAMPLE 3.3 *The force of a wrist shot*

In example 2.5, we estimated that a hard 55-mph shot from the ice required a puck acceleration of 495 m/s². How much force was exerted on the puck during the shot?

Here, the only horizontal force is the push $\vec{P}$ of the stick, so it is easy to find the force using Newton's second law.

$$a_x = \frac{P_x}{m} \quad \rightarrow \quad P_x = ma_x = (0.17 \text{ kg})(495 \text{ m/s}^2) = 84.2 \text{ N}.$$

In Imperial units, this is (84.2 N)(0.225 lb/N) = 19 lb, roughly the weight of a 9-month-old boy.

Curlers are known for after-match "festivities" and might come up with strange ways of having fun. If the hockey shot described above were given to a curling stone instead of a puck, how fast would the stone come off the stick? Looking up the stone's mass in appendix B, we find

$$a_x = \frac{P_x}{m} = \frac{84.2 \text{ N}}{18.2 \text{ kg}} = 4.6 \text{ m/s}^2,$$

which we can use in kinematic equation 2.6:

$$v_x = \sqrt{v_{x,0}^2 + 2a_x \Delta x} = \sqrt{0 + 2(4.6 \text{ m/s}^2)(0.61 \text{ m})} = 2.37 \text{ m/s} = 5.3 \text{ mph}.$$

Question

In using the "same shot" to move the stone, we kept the same force and shot length as for the puck. Would anything be *different* about the shot on the stone versus on the puck?

3.3 Sideways Traction

A thorough analysis of Dwight Howard's dunk has illustrated the three Newtonian laws that drive all of the physics we'll discuss in this book. We saw that the forces involved in Howard's leap vary with time in a complicated way, but that by analyzing the *average* forces and accelerations, we can understand the most important aspects of the process.

Now we will apply these laws to another force crucial to sports: friction. The two forces we've discussed so far, GRF and gravity, are an athlete's primary tools for accelerating vertically. Horizontal acceleration can come from a collision (just ask a victim of NFL linebacker Ray Lewis!), but more commonly one relies on friction.

To test a lineman's strength and technique, a coach may have him push a football sled, such as the one in figure 3.10. Clearly, the stronger he is, the better the lineman will perform. But what other factors come into play? Does the player's weight matter? (Yes.) If the player is overwhelmingly strong, is there a limit to how much weight he can push? (Yes and no.) Can the player adjust his technique to perform better? (Yes.) A simple study of friction allows us to answer these questions quantitatively.

We said that normal forces are generated whenever two objects come into contact. Frictional forces are generated when two objects are in contact and slide or

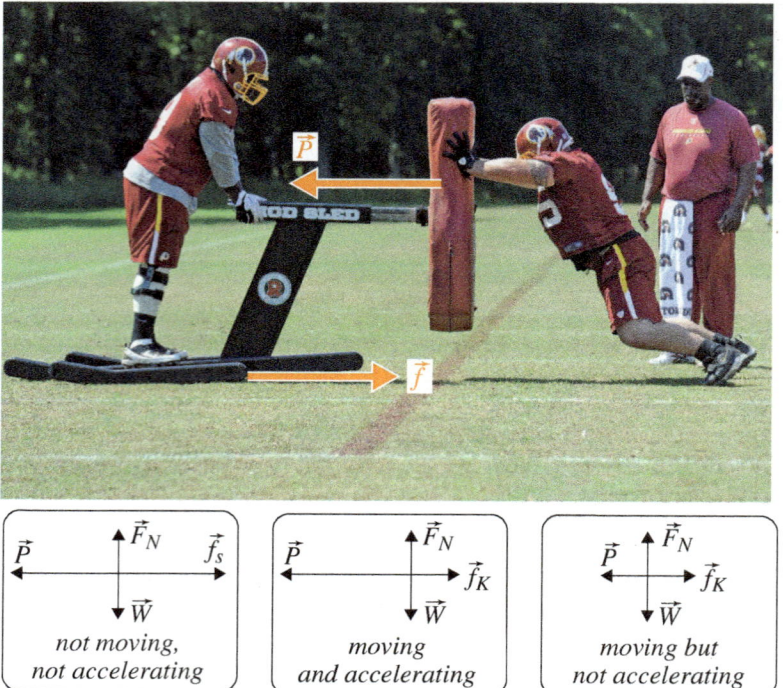

Figure 3.10. Washington Redskins nose tackle Chris Neild drives the sled with force $\vec{P}$ while guard Delvin Johnson weighs it down in training, May 2012. The tendency to slide is opposed by static friction (f_S) if the sled does not move, or kinetic friction (f_K) if it does. Because kinetic friction is smaller than static friction, the same push $\vec{P}$ will cause an acceleration, due to the force imbalance. (© *John Middlebrook/NewSport/Corbis*).

"try to slide" against each other. Either way, the force is directed parallel to the surfaces and opposes the slide or attempted slide.

3.3.1 Static and kinetic friction

In figure 3.10, Chris Neild pushes horizontally on a football sled with force $\vec{P}$. The sled and teammate David Johnson together probably weigh about 600 lb. This is a lot of weight, but keep in mind that Neild is not trying to *lift* 600 lb—he is trying to move it horizontally. Even if the Johnson+sled combination weighed 10,000 lb, Neild's push would accelerate it if no *horizontal* force canceled $\vec{P}$. Undergraduate physics majors often discuss what happens on "frictionless ice." On such ice, the merest push would start the Johnson+sled combination sliding, no matter what its weight.

In the real world, of course, friction opposes Neild's efforts. It may oppose him so much that the sled doesn't slide at all; if so, then we are talking about the **static friction force**, $\vec{f}_S$. The direction of $\vec{f}_S$ is clear—it always opposes the "attempted" slide.

The magnitude of static friction varies, depending on the push attempting to slide an object; in fact, $|\vec{f}_S|$ will be exactly the magnitude required to prevent a slide, up to a maximum limit, $|f_{S,\max}|$. If Neild is not pushing at all ($|\vec{P}| = 0$), then there is no frictional force. If he pushes with 20 lb of force, a 20-lb static friction force arises, and the sled does not move. When he pushes even harder with 50 lb, the static friction increases to 50 lb in response. At some point, however, static friction can grow no further; when Neild pushes with a force greater than some limit, the sled will start to slide.

 Static friction opposes any force that "attempts" to slide an object over a surface. If the force overcomes the static friction, sliding occurs and kinetic friction appears.

Two things determine the maximum value of the static friction force. The first is the force with which the two surfaces are being squeezed together; in the football sled case, this is the normal force from the ground $|\vec{F}_N|$. (In this case, $\vec{F}_N$ is the vertical component of the GRF, but later on we'll also be discussing friction with surfaces other than the ground.) Not surprisingly, the more tightly the ground presses against the sled, the stronger the static friction force can get.

The second factor is the **coefficient of static friction**, written μ_S. This is just a number quantifying the "stickiness" of the surfaces, and it has to be measured. There are tables of static friction coefficients for just about every pair of surfaces you can imagine; a quick online search reveals hundreds. Table B.4 in appendix B lists some coefficients relevant to sports. Tacky-surface pairs—like basketball shoes on an acrylic-coated basketball court—have large values, $\mu_S \approx 1.1$ or more. Slippery pairs, like skates on ice, have static friction coefficients $\mu_S \approx 0.1$ or less.

Together, these two factors determine the maximum static friction force,

$$f_{S,\max} = \mu_S |\vec{F}_N| \quad ; \quad |\vec{f}_S| \leq f_{S,\max}. \tag{3.9}$$

For the type of sled in the photo on a dry grass field, a typical value for the coefficient of static friction is $\mu_S = 0.5$. Therefore, $f_{S,\max} = 0.5\,(600\text{ lb}) = 300\text{ lb}$; if Neild can push harder than this, he'll get the sled moving.

Let's say Neild manages to generate $|\vec{P}| = 305$ lb. Does that mean that the force imbalance in the horizontal direction is 5 lb, that Neild's 305-lb push being opposed by a friction force of 300 lb? No. In fact, once static friction is overcome, the friction force that opposes the slide suddenly becomes significantly *less* than 300 lb. You've probably experienced this yourself: you push harder and harder to get something to slide until it suddenly "gives," perhaps causing you to stumble forward. You find that *it takes less force to keep something sliding than it takes to start it sliding in the first place*. This is because the **kinetic friction force** replaces static friction once sliding begins, and kinetic friction is a weaker force.

Whereas static friction opposes an "attempted slide," the force of kinetic friction is always directed so as to oppose a slide in progress. Like static friction, kinetic friction is stronger when the two surfaces are pressed more tightly together (i.e., when $|\vec{F}_N|$ is larger), and a **coefficient of kinetic friction** quantifies the stickiness of a pair of surfaces. Even the formula is very similar:

$$|\vec{f}_K| = \mu_K |\vec{F}_N|. \tag{3.10}$$

The force of kinetic friction is always less than the maximum limit of the static friction force.

Using the value listed in table B.4 in appendix B, the force of kinetic friction is $\mu_K |\vec{F}_N| = 0.4\,(600\text{ lb}) = 240$ lb, considerably less than the 300-lb force Neild had to overcome to start the slide.

3.3.2 Anti-lock brakes and traction control

Modern anti-lock braking systems (ABS) on passenger cars make use of the fact that $\mu_S > \mu_K$, i.e., that static friction is stronger than kinetic friction. In a sudden panic-stop situation, it is natural for the driver to step hard on the brakes. In older cars (or on your bicycle), this can lock the wheels; kinetic friction then slows the car as the rubber slides against the road. Example 3.4 discusses such a case.

However, the bottom of a *rolling* tire is instantaneously at rest with respect to the road, as indicated in figure 3.11. The object is moving and accelerating, but the

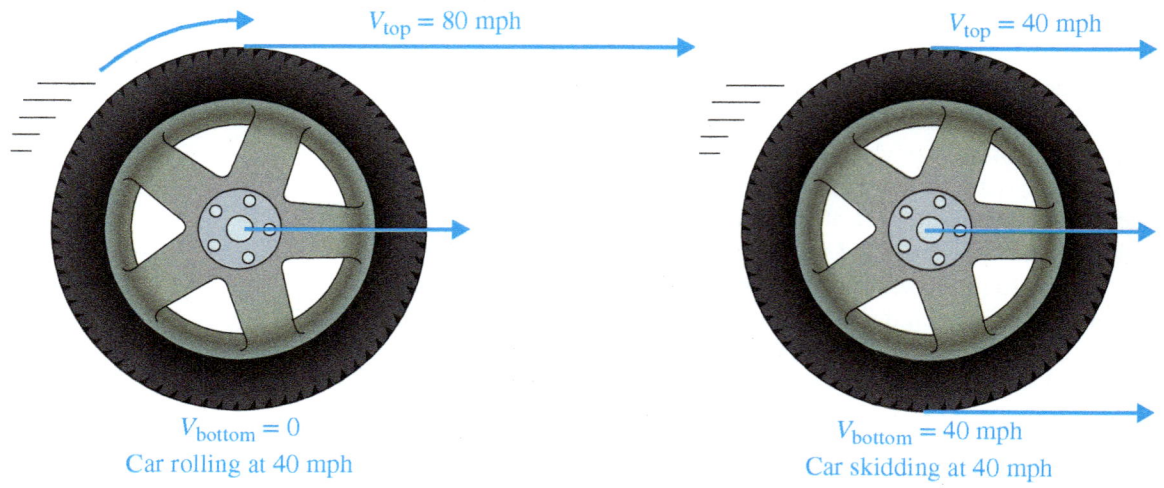

Figure 3.11. A car is traveling at 40 mph. If its wheel is rolling, the part of the tire touching the ground is instantaneously at rest.

surfaces are not sliding against each other. Therefore, the force of static friction can be used for braking. The trick is to push against the road up to the limiting value for static friction, but no harder. In a controlled situation, this can be done by a driver who knows his car and brakes well and has experience with various road conditions. For reliable quick stops in panic situations however, ABS systems employ computer-controlled actuators based on speed, road conditions, and sensors that detect when the wheel is about to lock.

EXAMPLE 3.4 *Busted?*

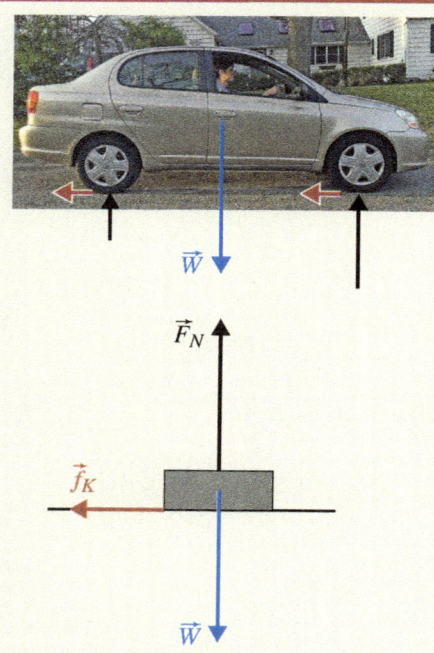

(Photo courtesy of Mike Lisa)

A driver without ABS panics, slams on the brakes, and leaves a 50-m skid mark before coming to rest. To the police officer who witnessed the whole thing, the driver says that he was traveling under the 55 mph limit. Sensing that the officer knows some physics, he points out that his economical Toyota Echo weighs only 2015 lb, so the stopping force of friction is smaller than that of a heavy car. Does he get a ticket?

In this problem, we have the constant force of kinetic friction. Therefore, the acceleration is constant, and we can use the ever-useful kinematic equations 2.5–2.7. All of those equations require us to know the acceleration, so we'll use Newton's second Law.

A Toyota Echo brakes to slow down. The simplified diagram on the right ignores the fact that the normal force on the front tire must be larger than that on the rear tire, for reasons we will discuss in chapter 9.

A free-body diagram is shown above. The fact that the normal force ($\vec{F}_N$) and friction force ($\vec{f}_K$) are distributed among the four tires is not important, and we can treat the car as a single sliding block. Because the Echo is not accelerating vertically,

$$a_y = \frac{\sum F_{y,i}}{m} = 0 \quad \rightarrow \quad |\vec{F}_N| = |\vec{W}| = mg.$$

The horizontal acceleration is due purely to kinetic friction:

$$a_x = \frac{\sum F_{x,i}}{m} = \frac{-\mu_K |\vec{F}_N|}{m} = \frac{-\mu_K mg}{m} = -\mu_K g = -0.8\,(9.8 \text{ m/s}^2) = -7.84 \text{ m/s}^2.$$

With the acceleration in hand, its time to solve the kinematics. We know the car's final velocity, $v_x = 0$, and the distance over which the acceleration occurred, $\Delta x = 50$ m. Equation 2.6 will give us the car's initial velocity $v_{x,0}$:

$$v_x^2 = v_{x,0}^2 + 2a_x \Delta x \rightarrow$$

$$v_{x,0} = \sqrt{v^2 - 2a_x \Delta x} = \sqrt{0 - 2\left(-7.84 \text{ m/s}^2\right)(50 \text{ m})} = 28 \text{ m/s} = 62.6 \text{ mph}.$$

Busted.

Question

What about the driver's point that the slowing force of friction is small on his car, due to its low weight? Why didn't that matter? Would a heavy car have skidded a longer or shorter distance?

3.4 More-Complex Situations

In our detailed analysis of Dwight Howard's leap and Chris Neild pushing the sled, we've actually covered a remarkable amount of physics. Now let's see how to apply these concepts to more-complicated situations.

In our analyses, we'll need to deal with multiple free-body diagrams within a single situation and with a little trigonometry.

3.4.1 Dwight Howard takes a quick step

In figure 3.12, Dwight Howard takes a quick step along the baseline. Clearly, he is accelerating to the left (as seen by the viewer). In fact, we're going to assume he's accelerating *purely* to the left—not up or down. Based solely on the photo, we can estimate how *much* he is accelerating.

First step: Free-body diagram

To arrive at interesting numerical answers, we need to translate the physical situation into mathematics. The free-body diagram is our crucial translation tool. The FBD (1) contains only those forces *on* Howard and (2) contains *all* forces on Howard.

When marking forces on a FBD, it's a good idea to begin with the object's weight, $\vec{W}$; you can't go wrong, and you know in which direction it points. For step two, we locate all objects with which Howard is in contact. In this case his left foot touches the floor, so we might expect a normal force and possibly friction. (The third and final step is to consider any forces due to air or water, but these are not important here.)

As usual, the floor exerts an upward normal force on Howard, $\vec{F}_N$. Let's carefully consider the frictional force between Howard (i.e., his shoe) and the floor. In this play, Howard did not slip, so although he is *moving*, his shoe is not sliding across the floor. Therefore, the horizontal force the floor exerts on him is *static* friction.

As for its direction, we know that $\vec{f}_S$ must point to the left. Some students will consider (correctly) that Dwight is pushing to the right, and so (incorrectly) draw $\vec{f}_S$ pointing right. This is a very common and understandable mistake. But remember

Even if the center of mass of an object is *moving*, the frictional force is static if there is no *slipping*.

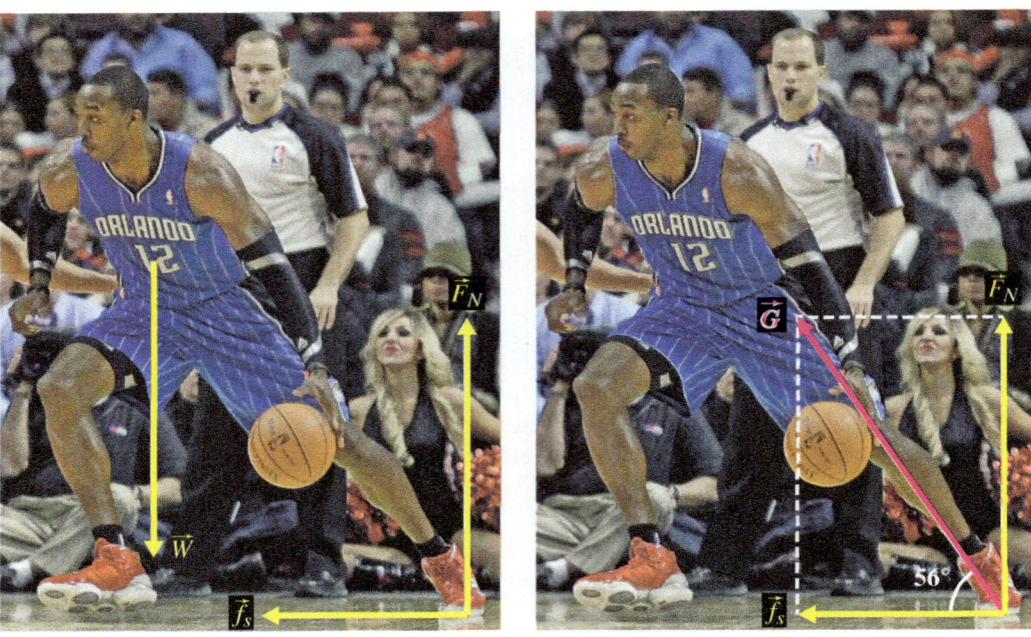

Figure 3.12. On the left is the FBD, showing all forces on Howard. On the right, the GRF is analyzed. **Left** and **Right:** (© *Nam Y. Huh/AP Photo*).

Take care to distinguish forces acting *on* an athlete from those that the athlete exerts.

our rule: "No person can accelerate himself." Dwight Howard accelerates to the left because *the floor* exerts a force on him to the left. The forces that Howard exerts on other things (like the floor) do not belong in our FBD of Howard. If $\vec{f}_S$ were directed to the right, the force imbalance would be to the right and he'd accelerate to the right.

Second step: Translation into equations

Now that we've got a handle on all forces, we can mathematically analyze them. We'll use Newton's second Law to relate the forces to Howard's acceleration. Remember that he's accelerating purely to the left, so $a_y = 0$.

$$\text{y-direction: } W_y + F_{N,y} = -|\vec{W}| + |\vec{F}_N| = ma_y = 0$$
$$\rightarrow |\vec{F}_N| = |\vec{W}| = 265 \text{ lb}$$
$$\text{x-direction: } f_{S,x} = ma_x \quad \rightarrow \quad a_x = \frac{f_{S,x}}{m}$$

The static friction force is usually *less* than its maximum possible value.

To find Howard's acceleration, we need to know the magnitude of the static friction force, $|\vec{f}_S|$. Careful—don't make the common mistake of assuming that $|\vec{f}_S| = \mu_S|\vec{F}_N|$! Remember: $\mu_S|\vec{F}_N|$ is the *maximum* value $|\vec{f}_S|$ can take. In general, $|\vec{f}_S|$ will be smaller than this maximum.

To make progress with this analysis, we need to know something about the optimum way to accelerate and something about vectors.

If you stand with your leg vertically straight down, you can exert a sideways force on the floor, generating a static friction force of the floor on you. But the most efficient way to generate a sideward force is to tilt your leg and then use your most powerful muscles to spring directly along the line your leg makes, as Howard is doing in figure 3.12.

Using a protractor, we find that his leg makes a 56° angle with the floor. He pushes out, and the ground reaction force pushes back directly along his leg. The GRF, $\vec{G}$, has two components—a vertical one ($\vec{F}_N$) and a horizontal one ($\vec{f}_S$). To understand how these components combine to accelerate Howard, we need to learn a little more about vectors.

Vector components and trigonometry

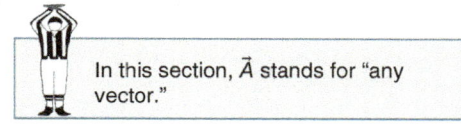

The trigonometry we'll need in this book is very basic; it relates the components of a vector. Figure 3.13 shows a vector, $\vec{A}$, that makes an angle θ with respect to the horizontal x-axis. A_x and A_y are the horizontal and vertical components of $\vec{A}$, and they are related to its magnitude as follows:

$$A_x = |\vec{A}| \cos\theta \quad (3.11)$$

$$A_y = |\vec{A}| \sin\theta \quad (3.12)$$

$$|\vec{A}| = \sqrt{A_x^2 + A_y^2} \quad (3.13)$$

$$\theta = \tan^{-1}\left(\frac{A_y}{A_x}\right) \quad (3.14)$$

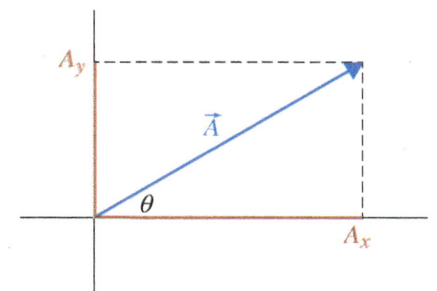

Figure 3.13
A vector $\vec{A}$ is decomposed into its horizontal and vertical components, A_x and A_y.

Completing the analysis

Applying these vector relationships to the situation of figure 3.12, we find the magnitude of the GFR:

$$|\vec{F}_N| = |\vec{G}| \sin 56° \quad \rightarrow \quad |\vec{G}| = \frac{F_N}{\sin 56°} = \frac{265 \text{ lb}}{0.83} = 319.3 \text{ lb}.$$

Now it's easy to find the magnitude of the static friction force:

$$|\vec{f}_S| = |\vec{G}| \cos 56° = (319.3 \text{ lb})(0.56) = 178.8 \text{ lb}.$$

Finally, we return to our Newton's second law equation from above, to find Howard's acceleration:

$$a_x = \frac{f_{S,x}}{m} = \frac{178.8 \text{ lb}}{8.28 \text{ slug}} = 21.6 \text{ ft/s}^2.$$

This is more than half a g! Howard is as deadly accelerating along the baseline (or into Joakim Noah) as he is under the basket.

Before we leave our analysis of this situation, look again at figure 3.12. Imagine Howard attempting a quick sideward acceleration wearing only socks. Clearly, he'd slip if he pushed too hard. In problem 10 at the end of this chapter, you'll figure out exactly how "sticky" Howard's shoe needs to be, to allow him to perform the move shown in the figure.

> **Question**
>
> There were a lot of forces and components in our analysis of Dwight Howard's quick step! It can be easy to get lost. To help you focus, answer this question: How hard is Howard pushing with his leg, and in what direction?

3.4.2 Football tryouts

Let's build on our knowledge and see how to analyze a situation involving all of Newton's laws, vector concepts, and two moving bodies.

A coach is trying out linemen. His test is to repeatedly increase the weight on a football sled like the one in figure 3.10, to find the maximum weight that the player can get going and *keep* going at a constant speed, pushing straight ahead. 280-lb Jerome has been in the weight room all year and is as strong as an ox. He's also been studying basic physics, so he knows a few tricks.

Like you, Jerome knows that the force required to *get* the sled moving will be greater than that needed to keep it going—he needs to exert a very strong force just for a moment at the beginning, to overcome static friction. As we've discussed, very strong forces can be achieved for very short times, in sudden collisions. Remembering figure 3.5, Jerome knows that if he can accelerate a little bit—build up a head of steam—and then violently crash into the sled, he can exert perhaps 600 lb. Because $\mu_S = 0.6$, that's strong enough to get a 1000-lb sled moving.

This is the reason linemen need to explode and then crash violently into their opponent—or the sled.

How heavy a sled can Jerome *keep* moving? To figure this out, we start by drawing *two* free-body diagrams, one for Jerome and one for the sled. Remember—restrict an FBD to a single object, and include only the forces acting *on* that object.

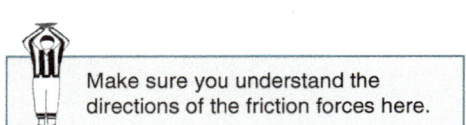

The FBDs in figure 3.14 translate the physical situation of Jerome and the sled into mathematics. Jerome and the sled are moving to the right, so the kinetic friction force on the sled $\vec{f}_{k,S}$ points left. (Notice the two subscripts there: k means "kinetic" and S means "sled.") Meanwhile, the force of *static* friction *on* Jerome $\vec{f}_{s,J}$ points right.

$\vec{P}_S$ is the force on the sled exerted by Jerome, and $\vec{P}_J$ is the associated "reaction" force exerted *on* Jerome. They are different forces, acting on different objects, related by Newton's third law:

$$\vec{P}_S = -\vec{P}_J \quad \rightarrow \quad |\vec{P}_S| = |\vec{P}_J|.$$

To impress the coach, Jerome needs to maximize $|\vec{P}_S|$, which is the same as maximizing $|\vec{P}_J|$.

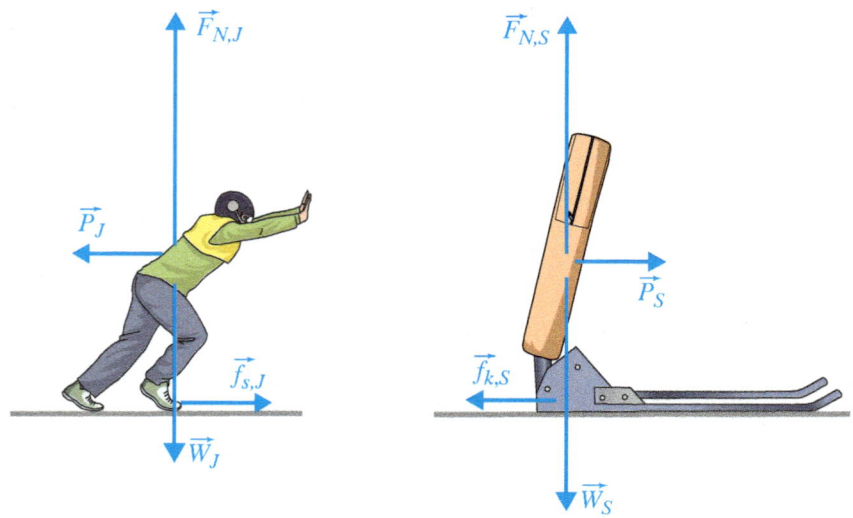

Figure 3.14. Separate FBDs for a lineman and the sled he is pushing.

The coach wants Jerome to keep the sled moving at a constant velocity. Therefore, even though there are many forces in the problem, Newton's second law tells us that the *net* force on Jerome is zero—he is in equilibrium. The equations for Jerome's forces are:

y-direction: $F_{N,J,y} + W_{J,y} = 0 \quad \rightarrow \quad F_{N,J,y} = -W_{J,y} = -(-280 \text{ lb}) = 280 \text{ lb}$

x-direction: $P_{J,x} + f_{s,J} = 0 \quad \rightarrow \quad |\vec{P}_J| = |\vec{f}_{s,J}|.$

Because we want as large a $|\vec{P}_J|$ as possible, we want $|\vec{f}_{s,J}|$ to be as large as possible. Equation 3.9 tells us the how large the static friction can be, so

$$|\vec{P}_J| = f_{s,J,\max} = \mu_S |\vec{F}_{N,J}| = 0.5\,(280 \text{ lb}) = 140 \text{ lb}.$$

The value $\mu_S = 0.5$ corresponds to Jerome's flat-soled shoes and the grass.

Now we know the maximum possible push Jerome can give to the sled. Let's work on the sled's equations, to find out how much weight Jerome can push. The sled is also in equilibrium, so its equations are:

y-direction: $|\vec{F}_{N,S}| = |\vec{W}_S|$ (which is what we're trying to find)

x-direction: $|\vec{P}_S| = |\vec{f}_{k,S}| = \mu_K |\vec{F}_{N,S}| = \mu_K |\vec{W}_S|$

$$\rightarrow \quad |\vec{W}_S| = \frac{|\vec{P}_S|}{\mu_K} = \frac{140 \text{ lb}}{0.4} = 350 \text{ lb}.$$

So, Jerome will be able to push a 350-lb sled. To push a heavier one, he would need to increase $|\vec{P}_J|$; but this would overwhelm the maximum static friction pushing him forward, and his feet would slip.

Improving performance

Is there *anything* Jerome could do to improve his performance on this test? Getting stronger will not help. His performance is limited by the friction between his shoes and the ground, not his strength; he could get 10 times stronger and he'd still max out at 350 lb.

Increasing his weight would help, because it would increase the normal force of the ground on him and hence the static friction on his shoes. The extra nonmuscle weight that many linemen carry is often just as valuable as muscle! (This extra weight helps in another way, too—it means they are more massive and thus harder for their opponent to accelerate, according to Newton's second law.)

Having studied physics, however, Jerome can use his head and not rely exclusively on his stomach to eat his way onto the team. Instead of pushing only *forward*, he should push partially *upward*. This does two things. Firstly, it decreases the ground's normal force on the sled, $|\vec{F}_{N,S}|$, thereby reducing the kinetic friction he has to overcome. Secondly, it *increases* the normal force $|\vec{F}_{N,J}|$ exerted on Jerome, increasing his own static friction.

This is one of the main reasons coaches tell their linemen to keep low. In figure 3.15, Rams defensive end Robert Quinn demonstrates this important principle as he comes in low against a block by Byron Stingily of Tennessee. The force Stingily exerts on Quinn, $\vec{P}_Q$, is partially directed downward, increasing the ground's normal force on Quinn ($|\vec{F}_{N,Q}|$) and hence the friction pushing him forward. Conversely, the ground's normal force $|\vec{F}_{N,S}|$ on Stingily is reduced by the partially upward force $\vec{P}_S$ Quinn exerts on him.

Figure 3.15. St. Louis Rams defensive end Robert Quinn comes in low against a block by Tennessee Titans tackle Byron Stingily in Edwards Jones Dome, August 20, 2011. By coming in low, Quinn exerts a force on Stingily that is directed partially upward, reducing $|\vec{F}_N|$. Conversely, the *down*ward force on Quinn *in*creases the ground's normal force on him. **Left, Middle, Right:** (© *Mark J. Peters/ZUMA Press/Corbis*).

To put it another way, the normal force on Quinn is *larger* than his weight, while the normal force on Stingily is *less* than his weight.

Finally, of course Jerome could trade his flat shoes for cleats, to increase μ_S and the force of static friction pushing him forward.

So, in addition to brute strength, we see that friction considerations provide several important guidelines for a lineman:

- Wear cleats.

- Explode into your opponent, to overcome static friction and deal with the weaker kinetic friction.

- Be massive, both to increase the static friction on you and to reduce the acceleration your opponent can generate.

- Stay low, to exert an upward force on your opponent.

We shall have more to say about the matchup between Quinn and Stingily later when we discuss torque.

Question

In figure 3.15, why are the two forces at the players' feet not "equal and opposite"?
Can you identify *any* pair of forces in the figure that are of equal magnitude and point in opposite directions?

3.4.3 Increasing friction in race cars

Race cars also require a tremendous amount of friction to accelerate and turn without skidding. As with all friction, the strength of the force is completely determined by

Figure 3.16. **Left:** At high speeds, inverted wings on the front and rear of an IndyCar generate a "lift" force similar to that created by an airplane's wings, but directed downward. **Right:** At 200 mph, the car could cling to an inverted road! The car's weight ($\vec{W}$) still pulls it down, but now the lift force ($\vec{F}_L$) on the wings push the car *up* into the road. The road pushes down on the car with the normal force $\vec{F}_N$, so the car can still turn and brake. **Left** and **Right:** (© *Glow Images RF*).

the surface stickiness and the normal force. Racing tires are carefully formulated to optimize frictional coefficients and properties like heat resistance and mass. The only remaining ingredient to optimize is the normal force.

Jerome could increase the normal force on his feet by gaining weight. Indeed, lashing a few cinderblocks to the side of an IndyCar would increase the normal force and hence the friction, but this is obviously not a good idea, as it just gives us more mass to accelerate and covers up the sponsor logos.

In the 1960s, racer and engineer Jim Hall developed a beautiful solution to increase the normal force without significantly increasing the car's mass when he introduced the movable wing for race cars. Endless innovations on the design have been occurred since, and refinements continue today. Modern Indy and Formula One cars feature wings and advanced spoilers that behave like inverted aircraft wings: at high speeds they generate a "lift" force $\vec{F}_L$, but directed *down*ward, as indicated in the left panel of figure 3.16.

 Clever use of aerodynamics can press the car more tightly to the ground, increasing friction without increasing the car's weight.

The car now pushes down on the road with the force of its weight plus the lift force. Newton's third law demands that the road exerts a normal force of equal magnitude.

Amazingly, at high speeds (around 200 mph) the lift force can be three times greater than the car's weight of about 1600 lb. As suggested in the right panel of figure 3.16, this means that even if the car and road were turned upside-down, the car would still be pressed "up" against the road!

3.4.4 Ball throw speed

Soon we will discuss what happens to a thrown ball after it leaves your hand, but for now consider this: Which could you launch at a higher speed, a baseball or a golf ball? Likely, you'll answer that the golf ball—which is three times lighter than the baseball—will leave your hand at higher speed. And you'd be right. However, the difference in speed may not be as large as you think, and it's worthwhile to understand why.

To throw the ball at maximum speed, you exert a force F_{max} to accelerate the ball over some distance Δx, at which point your arm is fully stretched out and you

 Especially for light balls, much of the force you apply is spent accelerating the mass of your hand.

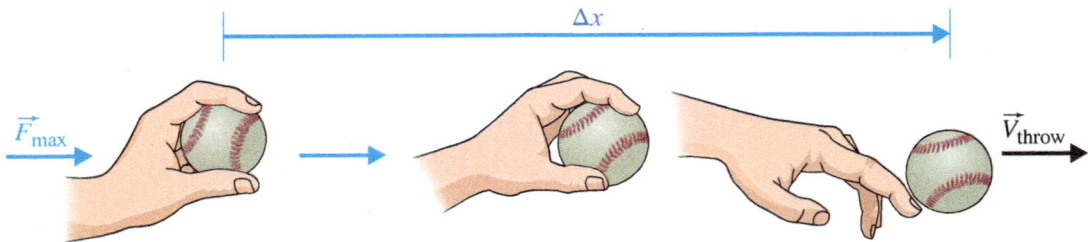

Figure 3.17. A simplified picture of a baseball throw. A force F_{max} is applied to the ball and hand together, over a distance of Δx.

release the ball at speed v_{throw}. If all of that force were exerted on the ball, then $a_{ball} = F_{max}/m_{ball}$ for the golf ball would be three times higher than for the baseball.

But here's the important part: *The force you exert needs to accelerate the ball and your hand, together.* See figure 3.17. So, if you exert your maximum force F_{max}, the acceleration of the hand-ball group[2] is

$$a_{\text{ball and hand}} = \frac{F_{max}}{m_{ball} + m_{hand}},$$

which is obviously smaller than F_{max}/m_{ball}.

We can use equation 2.6 to find the throw speed for a ball:

$$v_x^2 = v_0^2 + 2a_x \Delta x \rightarrow \boxed{v_{throw} = \sqrt{2\frac{F_{max}}{m_{ball} + m_{hand}} \Delta x}}. \quad (3.15)$$

Table 3.2 lists some maximum throw speeds for a variety of balls, assuming typical values $F_{max} = 75$ lb, $\Delta x = 0.75$ m, and a 0.5-kg hand. Although a table tennis ball is 55 times less massive than a baseball, it can only be thrown 13% faster! The throw speed begins to vary only when the object becomes more massive than the hand. For lighter objects like baseballs and golf balls, most of the force is spent accelerating the hand itself.

Table 3.2. Throw speeds for objects with widely varying masses, for an athlete who can exert a 75-lb force over 75 cm, according to the simple treatment described in this section.

Ball	Mass	Acceleration (m/s)²	Throw speed
Table tennis	2.7 g	670	31.5 m/s (70 mph)
Golf	45 g	618	30.3 m/s (68 mph)
Baseball	150 g	518	27.7 m/s (62 mph)
Clay brick	2.7 kg	105	12.5 m/s (28 mph)
16-lb shot	7.46 kg	42	7.9 m/s (18 mph)

---

[2] In this section, we are not discussing the complete physics of overarm throwing, because the additional complications are not important here. Physicists call our discussion an "effective treatment" of the problem, and equation 3.15 an effective equation. A more detailed discussion is found in Rod Cross's book *The Physics of Baseball and Softball*.

3.5 "Imaginary Forces" in Sports

It's happened to all of us: you're at a party and talking about the centrifugal force that presses you into your seat as the roller coaster makes a loop, when suddenly some know-it-all claims there is no such thing as centrifugal force. He starts talking about "centripetal force" instead, and how that's quite different. And soon everybody is talking about physics.

No? That hasn't happened to you? Hmm, maybe that's why physicists don't get invited to many parties...

Nevertheless, this is often a confusing topic, and it is worthwhile to get it right once and for all. As it turns out, because so many sports involve circular motion, clearing up the confusion is key to understanding many athletic situations.

Nobody knows the difference between centripetal and centrifugal forces so well as the discus thrower, even if she usually doesn't talk about it in physics terms. As is often the case, the knowledge is instinctive. Let's make it intellectual as well.

To get us started, let's quickly discuss a few examples of "imaginary forces"—forces that people often imagine acting in certain situations.

3.5.1 Imaginary pushes on Dale Earnhardt Jr.

When NASCAR racer Dale Earnhardt Jr. takes off from the starting line, he feels himself pressed back into his seat with a force of more than 150 lb. (End-of-chapter problem 3 asks you to find the exact number.) But wait—is there *really* a force pushing him back? Table 3.1 lists all forces relevant for us, and none of them are good candidates for this. Gravity always points down, and there is no strong wind *inside* the car to push Earnhardt backward. The tension in his seat belt is unimportant (he'd feel the "force" even without it), as is any small friction between his pants and the seat.

In fact, figure 3.18 shows all the forces on Earnhardt, and none of them points backward! Earnhardt sinks into the seat because the seatback is exerting a 150-lb normal force on him. Thanks to Newton's third law, Earnhardt pushes backward on the seatback with a normal force of equal magnitude; but note, *this is not a backward force on Earnhardt*—nothing is "pushing him back."

 When you accelerate forward in your car, there is no force pushing you back in your seat.

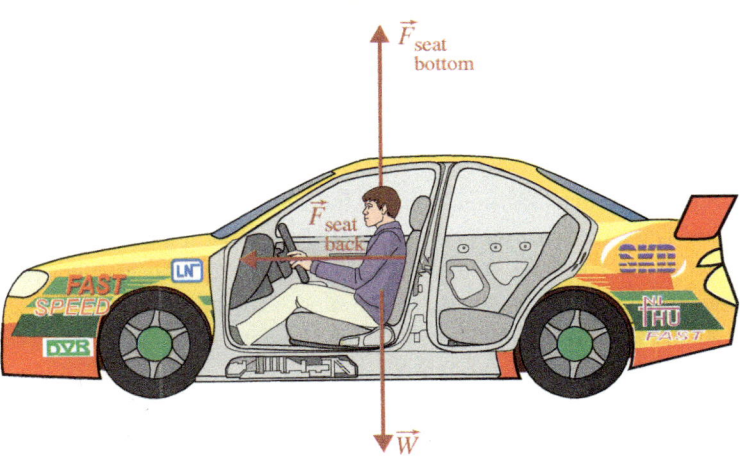

Figure 3.18. Dale Earnhardt Jr. accelerates along a straightaway. While he experiences a *sensation* of "heaviness" pulling him backward, there is in fact no backward force acting on him.

An Important Note

This business of "imaginary forces" can be confusing, and free-body diagrams like figure 3.18 are perfect for clarifying the situation. Let me warn you against a very common mistake: Do not include on such a diagram the "force of acceleration." There is no such force. The only forces you should worry about are listed in table 3.1. Draw all of those, and the imbalance will give the acceleration.

The resulting sensation of a "heaviness" pushing a driver backward is sometimes called an "imaginary force." If Earnhardt's car were *not* accelerating, then only a real, nonimaginary force could push him back into his seat.

Question

If Earnhardt slams the brake to slow down suddenly, what are the forces on him and which way do they point? (Draw a free-body diagram.) Is he "thrown forward" against his safety harness?

An important precursor: A turn in Earnhardt's car

Now consider Dale Earnhardt Jr. rounding a bend on the course. Heck, you paid good money for this book, so let's consider *yourself* driving car #88. After all, you know what it feels like to take a corner.

NASCAR races are always run counterclockwise around the track, so the cars are always turning left around the bends. Think about it—when you turn to the left at high speed in Earnhardt's car, do you feel "thrown" to the left, or to the right? Go through the forces listed in table 3.1; which force is throwing you?

If you identify carefully all of the forces acting on you, you will find an imbalance. This means, of course, that you are accelerating. But hey!—why does the same thing happen if you are taking the turn at constant speed?

It turns out that even if your speed is constant, your *velocity* is not constant (remember—velocity includes direction, and if you are turning, then velocity is not constant). *If you are not moving in a straight line, you are accelerating.* We'll get to the math of this in a moment.

3.5.2 Two nontraditional Olympic events

In the previous section, we looked at forces *on* an accelerating athlete and how he might conjure up "imaginary forces" to account for his sensations.

In this short section, we'll look at two cases where an athlete accelerates something *else*. In these cases, the athlete does not experience sensations he mistakenly attributes to imaginary forces acting *on himself*. However, he may still invoke some nonexistent forces *on the object*, to describe the situation.

Curling

Curling became an official Olympic event at the 1998 games in Nagano, Japan. The sport resembles the recreational game of shuffleboard, but it's played on a "curling sheet" of finely smoothed ice. In a smooth initial lunge, a curler accelerates a 40-lb stone from rest to a speed of about 1.5–2.5 mph. (The curler then slides with the stone for about 30 more feet, making slight adjustments to the stone's direction and giving it a slow spin before releasing it.)

The free-body diagrams for the stone and the curler are shown in figure 3.19. The action-reaction pair of pushes between the two are highlighted, and we see that the stone is pushing rightward on the curler. However, that does *not* mean that something is pushing rightward on the stone. (Inevitably there is some small amount of kinetic friction between the stone and the ice, but hopefully it is intuitively clear that the stone would push back on the curler even on "perfect" frictionless ice.)

This was a relatively easy example. Few would feel compelled to conjure up an imaginary rightward force on the stone, in order to explain the force of the stone on

 The backward-pointing force that the stone exerts on the curler does not arise from something pushing the stone backward.

Figure 3.19. Canada vice-skip Kaitlyn Lawes delivers a stone during the semifinal of the Women's Curling competition between Great Britain and Canada from the Ice Cube Curling Centre, Coastal Clustre—XXII Olympic Winter Games in Sochi Russia, 2014. Separate free-body diagrams are shown for the stone and curler. **Left** and **Right:** (© *Craig Mercer/ActionPlus/Corbis*).

the curler. Clearly, *an object can exert a pushing force on an athlete, even if nothing is pushing the object into the athlete*. The force arises from the object's acceleration and Newton's third law.

Tug-of-war

You may be familiar with it as a barbecue pastime, but tug-of-war was an Olympic sport in the early 20th century and is part of the World Games today. Several yearly tournaments are regulated and organized by the Tug of War International Federation (TWIF).

In official tournaments, teams consist of eight pullers, but to keep it simple, imagine participating in a one-on-one tug-of-war. During the match, you feel the rope pull away from you as your opponent pulls on his end. He, of course, is digging his heels into the ground to generate the maximum amount of friction possible, just as you are. If he stands there, straining but not accelerating in either direction, then the pull that you feel is simply equal to the resisting force of the ground on him. This is the situation in the top panel of figure 3.20.

In the bottom panel, we put your opponent on ice, and you can easily accelerate him toward you. In fact, $a_2 = P_{2,1}/m_2$. In this case, *even though there is no resistive force on your opponent*, you feel a tug on the rope. So, we can again conclude that *an athlete can feel a pull on a rope, even if nothing is pulling the object at the other end away*. The force arises from the object's acceleration and Newton's third law.

> **An Important Note**
>
> The free-body diagram of the curler on the right panel of figure 3.19 is a little more complicated than other ones we've seen, so it could use a word of explanation. Driven by the technique of "throwing" the stone, curlers wear two quite different shoes on their feet.
>
> The front foot should slide easily under the athlete, supporting his weight but doing little else. Nevertheless, there will always be some backward-pointing kinetic friction on this foot. The soles of slider shoes are often teflon or plastic to make μ_k as small as possible.
>
> The "hack foot shoe" is worn on the back foot and has a rubber sole to grip the ice as the curler pushes off. (Often the shoe is simply a slider shoe with a removable rubber "anti-slider" cover.) As in the examples we've discussed earlier, it is this *forward*-pointing *static* friction force that generates the acceleration of the stone-curler system. Both the curler and the stone have a leftward pointing force imbalance.

> **Question**
>
> In the top panel of figure 3.20, are the forces from the ropes felt by the two contestants equally strong (i.e., $|\vec{P}_{1,2}| = |\vec{P}_{2,1}|$)? How about in the bottom panel?

3.5.3 Discus throw

The curler accelerating a stone and the fortunate tug-of-war contestant whose opponent stands on ice, both feel real (not imaginary) forces from the stone and the

 Curlers, tug-of-warriors, and discus throwers all feel real pushing and pulling forces from the stone/opponent even if no additional force is acting on the object.

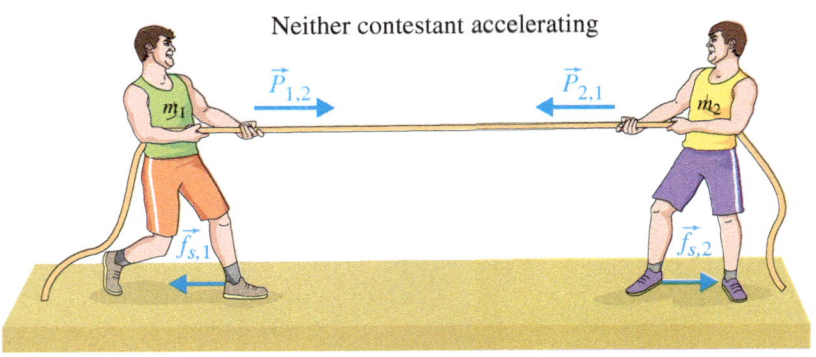

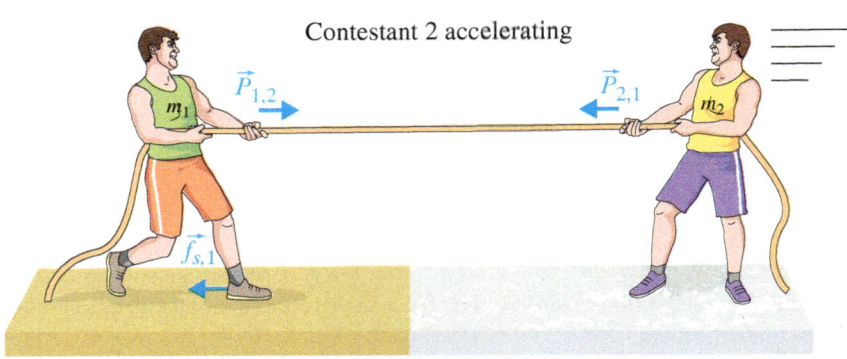

Figure 3.20. Contestants 1 and 2 in a tug-of-war. In the top panel, neither contestant is accelerating. On the bottom, contestant 2 gets no grip on the "perfect ice," so is accelerated toward contestant 1.

opponent. However, they would be mistaken if they concluded that some *other* force was pushing the stone into the curler or pulling the opponent away in the tug of war—*that* would be the "imaginary" force.

This is the crux of the matter in the famous centripetal-centrifugal confusion, and nobody has a clearer insight into the truth than the discus thrower.

Several factors go into a good discus throw, including foot placement, torque generated by the torso, angle of release, and the special aerodynamics of the discus itself. However, it is obvious that the athlete wants to release the discus (1) with the maximum possible speed and (2) in the proper direction. Let's focus on that second requirement.

Figure 3.21 shows a view of a thrower on the discus field from above. The region in front of him where the discus must land is only 35° wide; otherwise the discus goes into the safety net. At which point—**A** or **B**—should he release the discus in order to make a valid throw? The answer may be obvious, but let's do a quick analysis.

When the thrower is spinning, he feels a very real and strong outward tug from the discus at the end of his outstretched arm. *What is causing this tug?* Sometimes people attribute it to an outward-directed **"centrifugal force"** on the discus that the athlete needs to counteract by pulling the discus inward. Intuitively, that seems natural enough.

However, this is an imaginary force. There is no outward-directed "centrifugal" force on the discus at all! The free-body diagram for the discus is shown in figure 3.22. Two forces act on the discus: the tension $\vec{T}$ from the athlete's arm, and its weight $\vec{W}$. The tension generally follows along the length of the arm and is of course

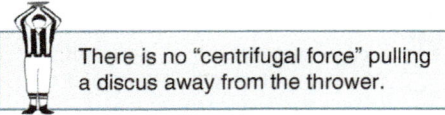

There is no "centrifugal force" pulling a discus away from the thrower.

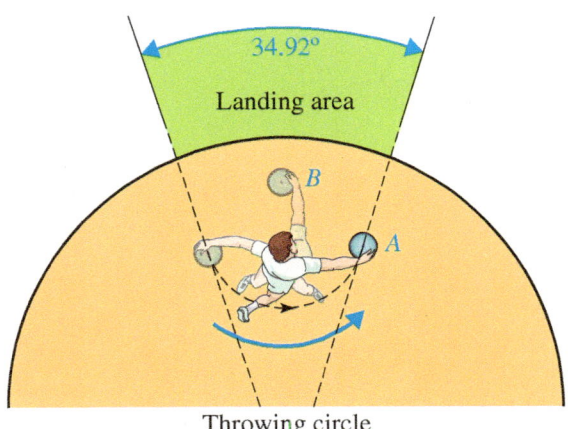

Figure 3.21. A bird's-eye view. The rapidly spinning discus thrower has a rather limited angular range for a valid throw. At which position, **A** or **B**, should he release the discus?

a pulling force. The vertical component of $\vec{T}$ cancels the discus's weight, and the net force on the discus is horizontal, as the athlete *pulls* the discus toward himself. Clearly there is a force imbalance on the discus, so it must be accelerating.

In fact, *when any person or object moves in a circle, a component of the acceleration points in toward the center of the circle.* (If the motion is at a constant speed, then the inward component is the only component.) This inward component is the **centripetal acceleration**, and it is stronger when the speed is higher (obviously) and when the circle's radius is smaller (perhaps less obviously):

$$a_c = \frac{v^2}{r}. \qquad (3.16)$$

Not surprisingly, the force that produces the centripetal acceleration is the **centripetal force**. It always points *inward*, and its strength is given by

$$|\vec{F}_c| = m\frac{v^2}{r}. \qquad (3.17)$$

Figure 3.22. The force that the athlete exerts on the discus is given by the tension in his arm, as he pulls it. The net force on the discus is in the horizontal direction, in toward the athlete. **Left** and **Middle:** (© *Rubberball/Getty Images RF*).

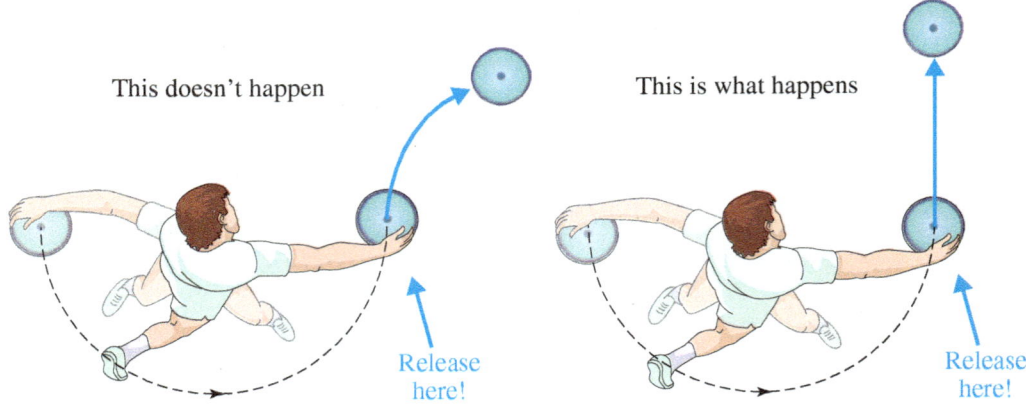

Figure 3.23. If the outward-directed "centrifugal force" were real, then the discus would accelerate outward upon release, as shown on the left. Because there is no such force, the discus does not accelerate at all (at least in the horizontal direction) but flies in a straight line, as shown on the right. (Inspired by a similar figure in *Physics*, 6th edition, by Giancoli.)

Now we can answer the question above: The athlete is accelerating the discus by exerting a force on it, so the tug that *he* feels is simply the corresponding "reaction force." This is a very real force, whose magnitude we'll estimate in example 3.5. The "imaginary" force is the one that we might mistakenly envision *on* the discus, pulling it away from the athlete. Again, nothing is pulling it away.

With this in mind, we can answer the question we asked previously. Clearly, a thrower instinctively knows to release the discus around position *A* in figure 3.21. As soon as he releases the discus, there will be no horizontal force on it, and it will travel in a straight line into the landing area.

This fact is emphasized again in figure 3.23. Yes, I know I'm hammering this point home repeatedly, but it's so easy to remain confused on the forces involved in circular motion, and I'm hoping that some of these pictures will stick in your head.

Centripetal forces have two important hallmarks: (1) They are always perpendicular (at right angles) to the object's velocity, and (2) they change only the *direction* of the velocity, not the speed $|\vec{v}|$. This means that if the thrower wants to increase the discus's speed, he'll have to exert other forces in addition to the tension acting as the centripetal force.

Another force?

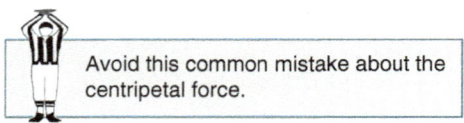
Avoid this common mistake about the centripetal force.

The centripetal/centrifugal issue and "imaginary forces" are hardly on a par with Einstein's theory of Relativity, but they can be confusing. If you've got it under control, then you are *way* ahead of most people—congratulations! Now I want to make sure you steer clear of another common confusion. Ready? Here it is:

> The centripetal force is not another force, to be added to the list in table 3.1. Rather, when one of the forces on that list is used to make something move in a circle (or part of a circle, like a curve), then it is a centripetal force.

The discussion just above already showed how tension can be a centripetal force. Figure 3.24 illustrates three more on our list as centripetal forces. In chapter 5 we'll discuss how fluid forces can be centripetal and give rise to curveballs.

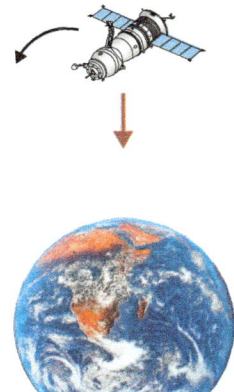  

Figure 3.24. Any of the forces we've discussed can be centripetal forces. From left to right: The circular orbit of a satellite is due to gravity; the normal force from the track forces the bobsled to turn; frictional forces are necessary to turn a bike. Make sure you understand the direction of the centripetal forces (arrows). The objects accelerate in the direction of the arrows. **Left:** (© *Stockbyte/Getty Images RF*); **Middle:** (© *Image Source/Getty Images RF*); **Right:** (© *Glow Images RF*).

EXAMPLE 3.5 *How hard does Stephanie Brown Trafton pull?*

American Stephanie Brown Trafton is one of the world's elite field athletes, winning the gold medal in discus in the 2008 Olympic games with a distance of 62.77 m (206 ft).

For such a toss, how large is the tension in Stephanie Brown Trafton's arm, just before she releases the discus? We'll just worry about the horizontal component of the tension, ignoring the small vertical component that compensates for the downward weight of gravity on the discus.

The tension in this case is a centripetal force *on the discus*, so equation 3.17 tells us we'll need to know the mass of the discus, the speed of the discus, and the radius of the discus's circular path. According to table B.3 in appendix B, $m = 1$ kg.

The radius of the discus's circular path is approximately the distance between the center of the thrower's shoulderblades and her wrist. A rough estimate of this distance, for the 6-ft-4-in. Brown Trafton, is 80 cm.

Finally, we need to know the discus's speed. Naturally, this speed grows as Brown Trafton accelerates it along its path,[3] so we'll focus on the speed toward the end of her throw. Video analysis reveals that she executes her final half-circle in 0.2 s, so a full circle would take $\Delta t = 0.4$ s. To get the speed, we need to know the distance the discus travels in that time, but this is easy—it is just the radius of its circular path, $2\pi r$! So, the speed is

$$v = \frac{2\pi r}{\Delta t} = \frac{2\pi (0.8 \text{ m})}{0.4 \text{ s}} = 12.6 \text{ m/s}.$$

Putting it all together, we can find the tension in her arm:

$$|\vec{T}| = |\vec{F}_c| = m\frac{v^2}{r} = (1 \text{ kg})\frac{(12.6 \text{ m/s})^2}{0.8 \text{ m}} = 198 \text{ N} = 44.6 \text{ lb}.$$

That little 2.2-lb discus is exerting 20 times its own weight on her arm!

> **Question**
>
> In Brown Trafton's throw discussed in example 3.5, how many *g*'s is the discus's centripetal acceleration? If you were experiencing the same acceleration, what would be the force on *you*? Would you survive one of those Hell Hole rides (where you stick to the wall of a rotating cylinder) with this centripetal acceleration?

EXAMPLE 3.6 *Earnhardt around the bends*

In 2000, Dale Earnhardt Jr. earned his first NASCAR Sprint Cup win at the Texas Motor Speedway, with an average speed over 130 mph. Although it's not recorded, it is safe to assume that his speed on the turns was lower than this; let's say it was 110 mph = 161.4 ft/s. At that speed, what is the centripetal force on him?

The curves at the approximately oval-shaped Texas Motor Speedway have a circular radius of 700 ft, so Earnhardt's centripetal acceleration is

$$a_c = \frac{v^2}{r} = \frac{(161.4 \text{ ft/s})^2}{700 \text{ ft}} = 37.2 \text{ ft/s}^2,$$

more than 1 *g*!

Earnhardt's weight is listed at 165 lb, so his mass is $m = (165 \text{ lb})/(32 \text{ ft/s}^2) = 5.16$ slug, so the centripetal acceleration on him is

$$F_c = ma_c = (5.16 \text{ slug})(37.2 \text{ ft/s}^2) = 192 \text{ lb}.$$

This 192-lb force on Earnhardt is being exerted by the car door and his safety harness, and *it is pointing in, toward the center of the curve*. Despite what his intuition might tell him, there is no force pushing Earnhardt "out." Such outward-pointing "centrifugal" forces are examples of imaginary forces.

Can we *ever* use the "centrifugal force" concept?

You know, a discus thrower could be forgiven for thinking that the discus is experiencing a centrifugal force. After all, she's pulling hard just to keep the discus from separating from her body. We know that the discus is simply trying to go in a straight line; nothing is "pulling it away." But from her point of view, because she is spinning, it seems the discus is trying to move "out," away from her.

We have discussed in detail what is really going on. There is no "centrifugal force" pulling the discus away from Brown Trafton. For this reason, many physics professors (and that bore at the party I mentioned on page 67) sneer at anybody who mentions a "centrifugal force." However, so long as you carefully keep in mind what is really going on, the concept of a centrifugal force can be useful, if you're looking at things from the standpoint of the rotating observer.

The concept arises in many sports studies. For example, detailed treatments of a golf swing almost always treat the arm-club system as a double pendulum; the lower pendulum pivots at the wrist mostly due to "centrifugal force."

"Centrifugal forces" are also the natural way to talk about, well, centrifuges. When a mixture (say, uranium) is spun in a centrifuge, the denser components (say, ^{238}U) become concentrated at the edge due to the larger "centrifugal force" on them.

In these contexts, the term and concept "centrifugal force" is used regularly, even by some of the very physicists who may jump down your throat if *you* use the term. They always use the term with implied quotation marks, however, because really the concept is a useful shorthand for other effects, as we've now discussed in detail. These physicists are afraid that you couldn't possibly understand the subtleties of this useful concept, but I have more confidence in you. So long as you use quotation marks as I have done to remind yourself that it is not a "real" force, you have the permission of this physicist to use the concept and term "centrifugal force," keeping in mind that you are looking at things from the point of view of an observer rotating with the system. Avoid using the concept to solve homework problems quantitatively.

Collected Equations

$$1 \text{ lb} = 4.45 \text{ N} \qquad 1 \text{ N} = 0.225 \text{ lb} \tag{3.1}$$

$$\sum_i \vec{F}_i = \vec{F}_1 + \vec{F}_2 + \vec{F}_3 + \cdots = 0 \leftrightarrow \vec{a} = 0 \tag{3.2}$$

$$W = mg \tag{3.3}$$

$$\begin{aligned} 1 \text{ N} &= 1 \text{ kg} \times \text{m/s}^2 \\ 1 \text{ lb} &= 1 \text{ slug} \times \text{ft/s}^2 \end{aligned} \tag{3.4}$$

$$a_{y,\text{launch}} = \frac{v_{y,\text{launch}}^2}{2c} \tag{3.5}$$

$$\vec{a}_{\text{c.m.}} = \frac{\sum_i \vec{F}_i}{m} = \frac{\vec{F}_1 + \vec{F}_2 + \vec{F}_3 + \cdots}{m} \tag{3.6}$$

$$\vec{F}_{1,2} = -\vec{F}_{2,1} \tag{3.7}$$

$$h = \frac{F_{\text{leg},y} - W_y}{W_y} \cdot c = \frac{a_{y,\text{launch}}}{g} \cdot c = (g\text{'s of acceleration}) \, c \tag{3.8}$$

$$f_{S,\text{max}} = \mu_S |\vec{F}_N| \; ; \quad |\vec{f}_S| \leq f_{S,\text{max}} \tag{3.9}$$

$$|\vec{f}_K| = \mu_K |\vec{F}_N| \tag{3.10}$$

$$A_x = |\vec{A}| \cos\theta \tag{3.11}$$

$$A_y = |\vec{A}| \sin\theta \tag{3.12}$$

$$|\vec{A}| = \sqrt{A_x^2 + A_y^2} \tag{3.13}$$

$$\theta = \tan^{-1}\left(\frac{A_y}{A_x}\right) \tag{3.14}$$

$$v_{\text{throw}} = \sqrt{2\frac{F_{\max}}{m_{\text{ball}} + m_{\text{hand}}}\Delta x} \tag{3.15}$$

$$a_c = \frac{v^2}{r} \tag{3.16}$$

$$|\vec{F}_c| = m\frac{v^2}{r} \tag{3.17}$$

Problems

1. When a hockey player takes a shot from deep in his own zone toward the other end of the rink 200 ft away, one hardly notices the puck slowing down at all. The coefficient of kinetic friction between a puck and the ice is about 0.03.
   (a) If a player shoots at 80 mph, how fast is the puck moving 200 ft later? Give your answer in miles per hour, and use at least four significant digits, since your answer will still be close to 80 mph!
   (b) What if there were no rink at all, just a large lake of ice? If the player took the same shot, how far would the puck go, before it came to rest? Express your answer in miles. (Good thing rinks have walls!)

2. When the gun went off and OSU sprinter Blake Callahan took off eastward at the Jesse Owens Track Classic, in which direction (up and east, down and west, etc.) did GRF point?

3. Dale Earnhardt Jr.'s car can do 0–60 mph in 2.9 s. Earnhardt's weight is listed at 165 lb. What force does he feel as he takes off from the line? In which direction (forward or backward) does it point?

4. The coefficients of friction between a 250-lb football sled and the grass are $\mu_s = 0.6$ and $\mu_k = 0.4$. For grass and cleats, $\mu_s = 0.9$. Three potential walk-ons are trying to impress Urban Meyer.
   (a) One 280-lb walk-on exerts a horizontal push of 140 lb on the sled. What is the acceleration of the sled?
   (b) The second walk-on is only 220 lb but exerts a horizontal push of 190 lb. What is the sled's acceleration?
   (c) The third walk-on is only 165 lb but as strong as the second player. What is the sled's acceleration when he pushes?

5. The coefficient of static friction for cleats on Astroturf is 1.5. This places a limit on how quickly a player can accelerate without slipping.
   (a) What is the maximum acceleration with which 315-lb offensive lineman Marcus Hall can burst off the line if he accelerates straight ahead?
   (b) How about for 210-lb Braxton Miller? Will his maximum acceleration be more than, less than, or the same as that for Hall?
   (c) Is there anything they can do to increase their maximum acceleration?

6. A 190-lb baseball player slides 4 ft into home plate. The force of friction on him is 120 lb. Pay attention to units in this problem.
   (a) Approximately what is the magnitude of the total force of the ground on this player?

(b) At approximately what angle, relative to the horizontal, is the force of the ground on the player?

7. The coefficient of kinetic friction between a curling stone and ice is 0.05. The stone is typically launched with a velocity of 2 m/s. It reaches this speed after a player pushes the stone from the hashline to the hogline, a total of 33 ft.

    (a) Assuming constant acceleration over this push, what is the force the player puts on the stone? Express your answer in pounds.

    (b) If the player exerts twice the force on the stone, will he generate twice the acceleration? Why or why not?

8. What is the force that a 0.33-lb baseball exerts on the 2-lb bat if the ball is accelerated straight up at 15 m/s^2?

9. When 200-lb (90.8-kg) lacrosse defender Mark Crawford collides with a 150-lb (68.1-kg) opponent, the opponent is accelerated at 15 m/s^2.

    (a) In pounds, what is the force Crawford exerts on this poor guy?

    (b) What is the magnitude of the force that the opponent exerts *on* Crawford?

    (c) What is Crawford's acceleration due to this collision (if any)?

10. In section 3.4.1, we found that Dwight Howard could move horizontally with an acceleration of 21.6 ft/s^2. However, he would be unable to do this in slippery socks.

    (a) What minimum coefficient of static friction must there be between his shoe and the court, so that he does not slip? (Shoe manufacturers pay close attention to exactly these sort of requirements.)

    (b) If Howard weighed 180 lb, rather than 265 lb, would the magnitude of the static friction force required to accelerate him at 21.6 ft/s^2 be reduced? Would your answer to part (a) of this problem change?

11. In section 3.4.2, we saw that, due to friction, the weight of the sled Jerome could push was limited to 350 lb. This was true if he pushed purely horizontally. However, if he directs his push slightly upward, he can do better. If he directs his push 15° above the horizontal, what is the maximum weight of sled he can push, still assuming that friction is the limiting factor?

12. A 42-in. vertical jump consists of two parts. The first stage is the quick acceleration of the center of mass from a crouching to a standing position. The center of mass rises by 20 cm in this stage. In the second stage, the player's feet leave the floor and the center of mass rises 42 in. under the influence of gravity alone.

    (a) What is the player's velocity just as his feet leave the floor?

    (b) What is the average acceleration during the first phase of the jump?

    (c) What is the average acceleration during the second phase of the jump?

    (d) If the player is 185 lb, what force do his legs need to exert, to execute this jump?

13. A shot-putter can exert 80 lb of force on a 16-lb shot. If he directs this force at 50° to the horizontal, what is the direction (it won't be 50° to horizontal!) and magnitude of its acceleration?

14. A 190-lb amateur basketball wannabe (okay, it's me) can accelerate from rest to 8 mph in 1 s. If I accelerate my center of mass only horizontally (i.e., I don't jump up while pushing forward),

    (a) what is the magnitude and direction of the force of the floor on me?

    (b) what is the minimum coefficient of static friction needed so I won't slip?

15. In the rope-climbing competitions discussed in section 2.3, all of the lifting is done by the arms; legs can only dangle. Imagine 160-lb Garvin Smith in training, hanging

stationary from the rope, his legs off the floor. When prompted by his coach, he rapidly lifts his center of mass by 4 ft in 0.75 s. Let's find out how much force his arms are exerting, taking it in steps:

(a) What is his acceleration?

(b) How big is the *net* force ("force imbalance") producing this acceleration?

(c) In what direction does this net force point?

(d) Which forces are acting *on* Smith. (Remember, his arms are part of himself, so they cannot exert a force on him.)

(e) With what magnitude force is Smith pulling down on the rope?

16. Example 3.5 showed how to quantify the horizontal component of the tension in Stephanie Brown Trafton's arm.

    (a) If she maintains the discus's motion in a level circle (i.e., she is not accelerating it up or down), what angle will her arm make, relative to the horizontal? (You will need to account for the weight of the discus, to figure this out.)

    (b) At the beginning of her windup, she is spinning more slowly than she is at the end. At the beginning of the windup, will the angle of her arm relative to the horizontal be different from the angle in part (a)? If so, will it be larger or smaller? Is this consistent with your observations of the event or what seems intuitively clear to you?

    (c) The men's discus is twice as massive as the women's. If a men's discus were spun at the same speed and same radius as Brown Trafton's throw in example 3.5, what would be the horizontal component of the tension in the athlete's arm? What fraction of the discus's weight is that?

17. In the sport of curling, a player slides (or "throws") a 40-lb granite stone along a sheet of ice. The thrower releases the stone at the "hogline," which is 93 ft away from the center of the intended target. The coefficient of kinetic friction between the stone and ice is $\mu_K = 0.020$.

    (a) Draw a free-body diagram of the stone as it passes the midway point on the sheet (long after the thrower has released it).

    (b) What is the net force on the stone (both magnitude and direction) at the midway point?

    (c) How fast (in mph) should the thrower release the stone at the hogline, in order to get it to stop in the center of the target on the other end of the sheet?

    (d) In competitive events, two "sweepers" rapidly brush the ice ahead of the stone, in order to tweak the friction coefficient slightly. If they make just a 5% mistake—i.e., they create ice with $\mu_K = 0.021$ instead of 0.020—how far away from the center of the target will the stone stop, using the throw speed you found in part (c)?

18. The act of bowling a ball can be rather involved, but in its simplest form the bowler swings her arm in a roughly circular arc, releasing the ball just a moment after it passes through the bottom of the arc. By the time the ball is at the bottom of the arc, its speed is not changing anymore. A decent bowler can make the ball travel the length of the 60-ft alley in about 2.1 s. In this problem, we'll ignore the (small) friction between the ball and the lane, so the ball travels at constant speed, once released.

    (a) At what speed is the ball moving when the bowler releases it?

    (b) Sketch the situation and draw a free-body diagram of the ball when it's at the bottom of the swing and the bowler has not yet released it.

    (c) In what direction is the ball aclerating, at that point?

    (d) If our bowler has 2-ft-long arms, estimate the magnitude of the ball's acceleration.

    (e) From this, estimate the **total** force on the 14-lb ball.

    (f) Finally, estimate the tension in the bowler's arm.

19. I'm a scientist, so you can take it from me: People who compete in the Olympic event of skeleton are insane. With their chin often 3 in. from the ice, they rush down the track at speeds of 75 mph on a nearly weightless sled.

(a) In what direction is the total force on the *rider* in the middle panel of figure 3.24? Is she being pushed into her sled (to the right), away from it (to the left), or not at all?

(b) One turn in the Sochi Olympics had a radius of 42 m. At a speed of 75 mph, what is the total force on this 145-lb rider?

20. Fans at the Tour de France can be exceedingly annoying, often impeding the cyclists and sometimes causing wrecks. (I've been there and, trust me, it's a wild scene.) In 1994 Laurent Jalabert crashed into a gendarme (policeman) and was seriously hurt. Sometimes riders can stop to avoid a collision, sometimes not. It depends on friction and reaction time. The 70-kg Jalabert was speeding at 35 mph along the level street on a wet day. The coefficient of kinetic friction between his wheels and the road was 0.15. He looked up to see the gendarme about 40 m (about half of a football field) ahead of him.

(a) If he has typical human reaction time (0.15 s), how close will he be to the gendarme when he *begins* to apply the brakes?

(b) Applying the brakes in something of a panic, he immediately locks them up and begins skidding toward the gendarme. Draw a free-body diagram of Jalabert and his bike (treating them as one composite object).

(c) At what angle is the force of the ground on the Jalabert+bike object?

(d) As Jalabert skids, what is his acceleration?

(e) How fast was Jalabert going, when he hit the gendarme?

21. In a good hit, a baseball and bat are in contact with each other for about 1 millisecond. In the first game of the 2012 World Series, Tigers pitcher Justin Verlander threw a 90-mph pitch to the Giants' Pablo Sandoval. Sandoval's hit returned the ball directly back at Verlander at 70 mph. What was the average force of the bat on the ball, during the contact?

22. Good tennis players have exquisite knowledge and control of the magnitude and direction of a ball's acceleration. In a soft shot, the head of a player's racquet is vertical, so that it exerts a horizontal force on the ball of 4 N.

(a) At what angle, relative to horizontal, will the ball accelerate? (Be sure to consider all forces.)

(b) What is the magnitude of its acceleration?

4 Punts, the Fosbury Flop, and Other Projectile Motions

If we ignore air, the math describing the motion of the center of mass is both simple and the same for all objects and athletes. Because it is so ubiquitous, a solid grasp of this "projectile motion" is crucial to understanding sports. In this chapter, we'll see how athletes use the projectile motion of balls to their strategic advantage. We'll also study the reasons and physics behind the acrobatics of flying athletes pushing the limits of performance.

Figure 4.1. Nate Adams pulling a Cordova backflip during the 2006 Dew Action Sports Tour. No single piece of Adams or his motorbike follows a simple trajectory, but the center of mass traces a perfect parabola. **Top** and **Bottom:** (© *Bo Bridges/Corbis*).

*What goes up, must come down... but what matters in the game is **where** it comes down.*

A basketball arcs its way to the hoop; a kicked football sails over the wall of defenders toward the goalposts; a baseball is swatted deep into right field. From the shot-put to tennis, projectile motion is ubiquitous in sports. And it is not only balls flying through the air. The path of human projectiles—from Olympic long jumpers to the Chicago Bulls' Michael Jordan—is predetermined by the same immutable laws of physics.

As a matter of fact, we already know these laws; they are just Newton's laws and the kinematic equations we're becoming increasingly familiar with. Nevertheless, even though projectile motion involves no "new" physics, it is so central to athletics that we devote an entire chapter to it. In fact, we will focus on the especially simple cases where gravity is the only important force in the game. This is the case for dense bodies at low speeds such as the shot-put and jumping humans.

In chapter 5, we'll delve into the fluid forces mentioned in table 3.1 (air resistance, etc.). These play an important role in flying basketballs, baseballs, and other projectiles. Even in these cases, gravity usually remains the dominant force, and a thorough understanding of the "air-free" approximation will prove valuable. We'll ignore air for the duration of this chapter.

4.1 The Math: Simpler Than You Think

We've discussed motion that is purely horizontal, such as a swim meet, and purely vertical, such as a rope climb. How do we describe the motion of something that is moving both horizontally and vertically at the same time?

Here's the "secret," which initially confuses many students (don't be one of them), perhaps due to its very simplicity:

> The horizontal motion is completely driven by horizontal forces, and the vertical motion is completely driven by vertical forces.

Once we know the forces, the horizontal motion doesn't "care" about what's happening in the vertical direction, and vice versa.

Gravity is the only force for now, so the acceleration in the vertical direction is $a_y = -g$. In the x-direction, there is no acceleration ($a_x = 0$), so the horizontal component of the shot's velocity never changes. A putted shot will rise, peak, and fall, but none of this affects the horizontal motion: the shot moves horizontally at a constant rate.

Mathematically, these facts are expressed by equations we've seen before, but we write them again here:

$$\Delta x = v_x t \qquad (4.1)$$

$$\Delta y = v_{y,0} t - \tfrac{1}{2} g t^2 \qquad (4.2)$$

A high-speed bullet (or football) shot horizontally hits the ground at the same time as a bullet dropped from the same height.

EXAMPLE 4.1 *Range of a rifle*

These simple equations might surprise you. Imagine target shooting with a rifle using 0.22 bullets; they typically leave the barrel at about 720 mph, just below the speed of sound in air. You shoulder the rifle at a height of 5 ft 6 in. and fire it horizontally. At the precise moment the bullet leaves the barrel (thus becoming subject to only the force of gravity), a friend drops a bullet from the same height of 5 ft 6 in. Which bullet hits the ground first?

Well, the dropped bullet started at rest, so $v_{x,0} = 0$ and $v_{y,0} = 0$. It's easy to find the time, using equation 4.2.

$$\Delta y = 0 - \tfrac{1}{2} g t^2 \rightarrow t = \sqrt{-\frac{2 \Delta y}{g}} = \sqrt{-\frac{2(-5.5 \text{ ft})}{32 \text{ ft/s}^2}} = 0.586 \text{ s}.$$

And the one fired from the gun? Since $v_{y,0} = 0$, the same equation holds, and it *also* takes precisely 0.586 s to hit the ground! The vertical motion is not at all affected by the horizontal.

It is easy to figure out how far the bullet travels before it hits the ground, using equation 4.1:

$$\Delta x = \left(720 \frac{\cancel{\text{mi}}}{\cancel{\text{hr}}} \times \frac{1760 \text{ yd}}{1 \cancel{\text{mi}}} \times \frac{\cancel{\text{hr}}}{3600 \text{ s}}\right)(0.586 \text{ s})$$

$$= 352 \frac{\text{yd}}{\cancel{\text{s}}} \times 0.586 \cancel{\text{s}}$$

$$= 206 \text{ yd}$$

When you use an M-16 rifle, with muzzle velocity of 2180 mph (1066 yd/s), the bullet still hits the ground after 0.586 s. In this time, however, it has traveled much farther: $\Delta x = (1066 \text{ yd/s})(0.576 \text{ s}) = 624 \text{ yd}$.

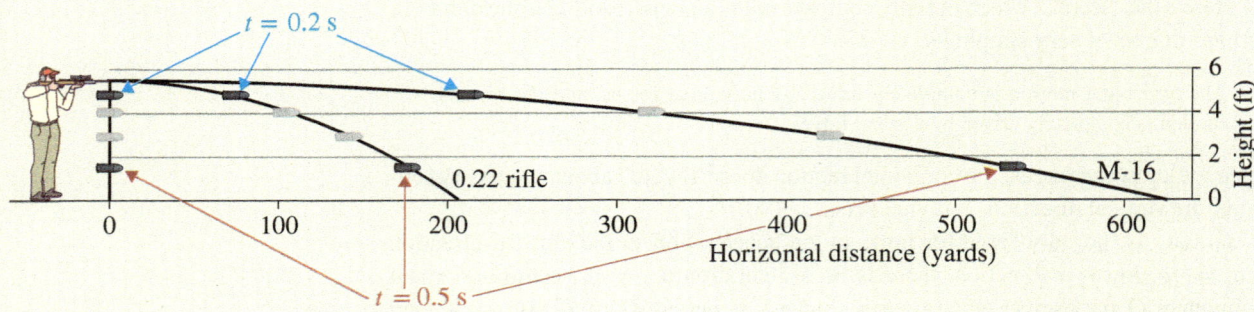

The figure shows the trajectories of a bullet that is dropped, one that is fired from a 0.22 rifle, and one fired from an M-16. The bullet is drawn every 0.1 s after being fired (or dropped). For a given time, the height of the bullet is the same in all three cases.

EXAMPLE 4.2 *Downfield pass*

(© *Royalty-Free/Corbis*)

Especially on a quick post play, a football receiver has to know precisely what to expect from the quarterback when he turns around to catch the pass. On November 18, 2006, number 1-ranked Ohio State and number 2 Michigan, both undefeated as they entered their final game of the season, met in Columbus for what became an epic game. In a key play, OSU's Troy Smith fired off a quick pass from the 14-yd line, to wide receiver Anthony Gonzales; Gonzales caught the pass at the 3-yd line and scored a touchdown.

In photographs and video, it is seen that Smith throws the ball essentially horizontally, similar to the photograph here, from a height of about 6 ft. Video analysis reveals that the ball was in the air for about one-third of 1 s (0.33 s). Here's what the receiver needs to know: How high will the ball be when it reaches Gonzales, and at what angle (relative to the ground) will it be traveling?

Equation 4.2 tells us the height:

$$y = y_0 + v_{y,0}t - \tfrac{1}{2}gt^2 = 6 \text{ ft} + 0 - \tfrac{1}{2}32 \frac{\text{ft}}{\text{s}^2}(0.33 \text{ s})^2 = 4.25 \text{ ft} = 4 \text{ ft } 3 \text{ in.},$$

right on the numbers.

Now, to figure out the angle of the ball's motion, we need both velocity components. The x-component is easy, since it never changes:

$$v_x = \frac{\Delta x}{\Delta t} = \frac{3 \text{ yd} - 14 \text{ yd}}{0.33 \text{ s}} = -33.3 \frac{\text{yd}}{\text{s}} = -100 \text{ ft/s},$$

a little more than 65 mph. (As usual, the negative sign just indicates direction. On a football field, the goal line is "$x = 0$.")

The y-component we get from equation 2.8:

$$v_y = v_{y,0} + a_y t = 0 + (32 \text{ ft/s}^2)(0.33 \text{ s})^2 = -3.48 \text{ ft/s}.$$

Finally, we use equation 3.14 to find the angle of the velocity:

$$\theta = \tan^{-1}\left(\frac{-3.48 \text{ ft/s}}{-100 \text{ ft/s}}\right) = \tan^{-1}(0.0348) = 2°.$$

Notice that when you take an arctangent ($\tan^{-1}$), the argument inside the parentheses must have no units. If you do a problem like this, and end up with "meters" or something inside the parentheses, you've made a mistake somewhere.

The velocity vector is drawn in the figure. The vertical scale is expanded, so the 2° angle can be seen.

Smith hit Gonzales right on the numbers with a 65-mph bullet for a crucial touchdown in a great game.

$v_x = -100$ ft/s
$\theta = 2°$
$v_y = -3.48$ ft/s
(Vertical scale expanded for clarity)

4.2 Football Punt: Range, Hang Time, and Compromise

Often we are interested in projectiles whose paths begin and end at the same height; think of a fly ball in baseball, a golf drive, a football punt. Three quantities are of interest: (1) the **range** R of the projectile, which is the horizontal distance traveled before landing; (2) the **hang time** T_{hang}, which is the total time in the air; and (3) the **peak height** h. We've already discussed the latter two in section 2.3. There, the motion was purely vertical, but since vertical motion doesn't "care" about horizontal motion, equations 2.14 and 2.15 don't change at all.

Here are the relationships most useful to understanding projectiles in sports:

$$T_{\text{hang}} = 2\frac{v_{y,0}}{g} = 2\frac{v_0 \sin\theta}{g} = 2\sqrt{\frac{2h}{g}} \qquad (4.3)$$

$$R = v_{x,0} T_{\text{hang}} = \frac{v_0^2 \sin 2\theta}{g} \qquad (4.4)$$

Figure 4.2 shows some of the quantities used in these equations.

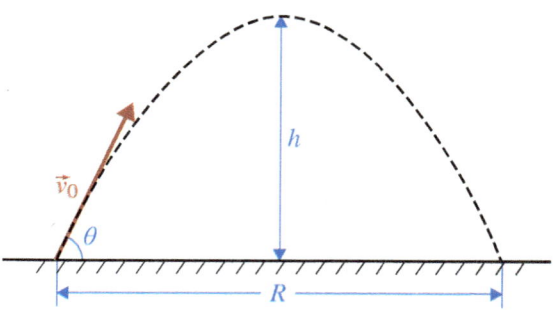

Figure 4.2. The launch angle, launch speed, height, and range of a projectile are related by equations 4.3 and 4.4.

4.2.1 A quick example

Let's take a look at a football punt and how these equations come into play. If a punter kicks a ball straight up at 80 ft/s, equation 4.3 tells us how long the ball is in the air

$$T_{\text{hang}} = 2\frac{v_{y,0}}{g} = 2\frac{80 \text{ ft/s}}{32 \text{ ft/s}^2} = 5 \text{ s}$$

as well as how high the ball goes:

$$2\frac{v_{y,0}}{g} = 2\sqrt{\frac{2h}{g}} \quad \rightarrow \quad h = \frac{v_{y,0}^2}{2g} = \frac{(80 \text{ ft/s})^2}{2 \times 32 \text{ ft/s}^2} = 100 \text{ ft.}$$

A vertical punt has $v_x = 0$, but of course a good punt needs to have some forward motion as well! How do the hang time and height change if, in addition to the 80 ft/s vertical velocity component, the ball moves horizontally with $v_x = 10$ ft/s? The answer: not at all. While it may seem surprising, *any ball launched with $v_{y,0} = 80$ ft/s will rise to a height of 100 ft and remain in the air for 5 s*.

Likewise, *any ball that has a hang time of 5 s will have a peak height of 100 ft*. It does not matter what v_x is. This is the simple "secret" we mentioned at the beginning of this chapter: the vertical motion is independent of the horizontal motion.

Now, how far down the field did this second ball get? That's easy: each second, the ball moves down the field 10 ft and it is in flight for 5 s, so the ball travels 50 ft before hitting the ground.

The hang time of a projectile is determined solely by its peak height, independent of the horizontal velocity component.

4.2.2 The optimum angle

Sometimes the goal is to attain the maximum possible range when launching a projectile (a ball, a human). The athlete can control two things: the launch speed $|\vec{v}_0|$ and the launch angle θ. You won't be shocked to hear that, all other things being equal, the larger the launch speed, the better.

But what about the launch angle? If the athlete is throwing or kicking the ball at the greatest speed he can muster, at what angle should he launch it to achieve maximum range? Clearly, we want the ball to have a lot of horizontal velocity (v_x), but if we put *all* the velocity in the horizontal direction ($\theta = 0°$), then the ball never lifts off the ground—its hang time is zero—and the range is zero.

So that's no good; we need some vertical velocity, too, to give the ball time to travel. We get maximum hang time when the launch angle is $\theta = 90°$, that is, a strictly vertical punt as discussed above. But in this case, too, the ball eventually lands exactly at its launch point, and again the range is zero.

You're probably not surprised to learn that the optimum launch angle is midway between these two extremes: $\theta_{optimal} = 45°$.

Figure 4.3 shows the paths a ball (or a person's center of mass) takes under various conditions.

For a given launch speed, maximum range is achieved for a 45° launch angle.

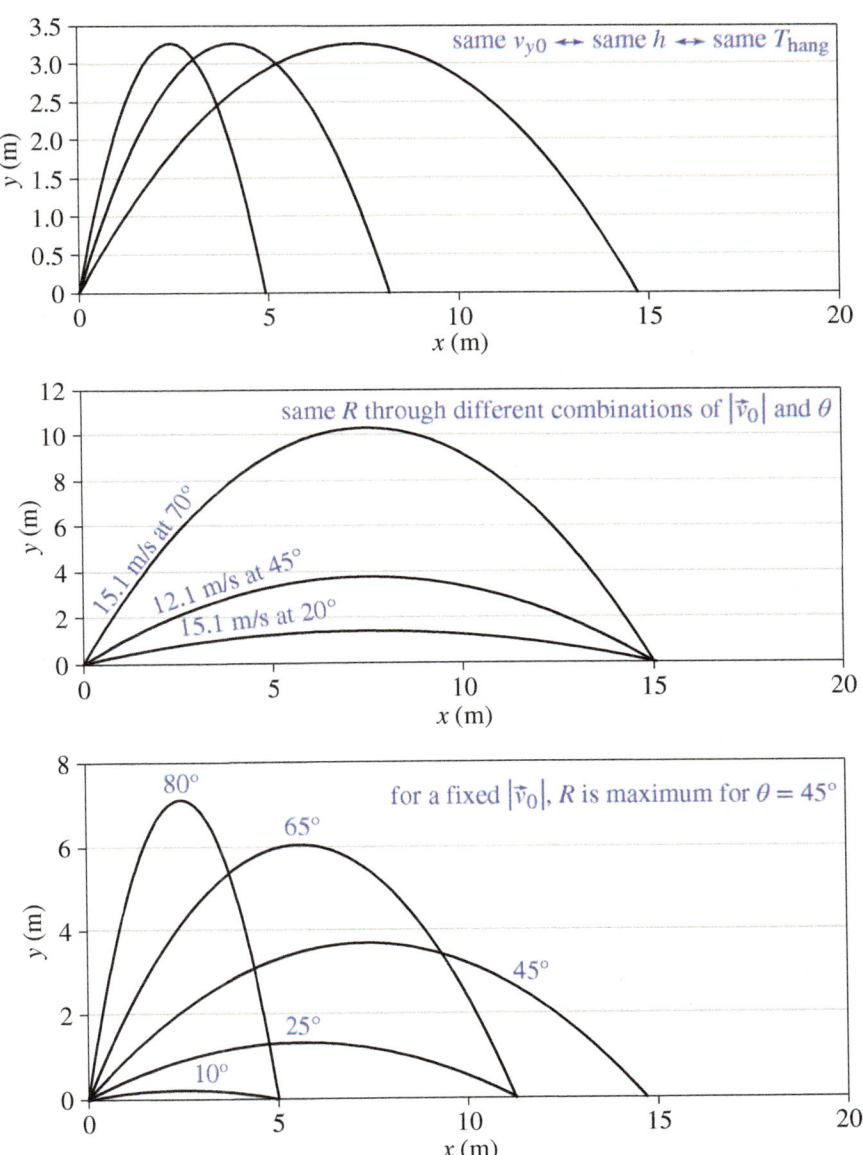

Figure 4.3. The trajectories of objects due to gravity only. *Top:* The hang times of the three projectiles are the same, since they have the same peak height. *Middle:* The range of an object in flight is affected by launch angle *and* launch speed, so an athlete can achieve the same range through a variety of trajectories. *Bottom:* For a given launch speed $|\vec{v}_0|$, the maximum range is achieved when the launch angle is $\theta = 45°$.

> **Question**
>
> The only things an athlete can control are a ball's launch angle and launch speed. Once it leaves the player's hand (or foot, or bat, or racquet), a ball's trajectory is unchangeable. Let's analyze figure 4.3 to see how these affect the ball's path.
>
> The three trajectories in the top panel all have the same hang time, but different ranges. The launch angles are obviously different, but what about the launch speeds? If they are not the same, which one has the lowest launch speed?
>
> In the middle panel, the balls have the same range, but different launch angles. Do they have the same hang times? Are they launched at the same speed?

4.2.3 Punting strategy and compromises

Let's look at an example where an athlete puts these considerations to work on the field, keeping in mind that our conclusions will change somewhat when we consider the effects of air in chapter 5.

A good college or professional kicker can typically launch a football at 55 mph ≈ 81 ft/s. On kickoff, the goal may be to attain maximum range, in which case of course the ball should be launched at 45°. Using equation 4.4, we find that the ball would land $(81 \text{ ft/s})^2 \sin(2 \times 45°) / (32 \text{ ft/s}^2) \approx 68$ yd downfield.

If the chances of attaining a first down or field goal are bleak on fourth down, then it's time to punt. The goal here is to end the down with the ball as close to your goal line as possible. This may or may *not* involve kicking the ball as far as possible.

For example, if the punt is from the defense's 40-yd line, then the 68-yd punt will sail into (and over) the endzone, meaning the receiving team starts play at the 20-yd line. It is much better if the punter adjusts his launch angle so that the ball comes down at the 5-yd line, right near the sidelines—the so-called "coffin corner." With luck, the ball will bounce out of bounds and play will start at the 5-yd line.

The situation is a little less clear when a long expanse of turf lies ahead, say, when the line of scrimmage is at the punting team's 30-yd line, 70 yd from their goal. Say you are the special teams coordinator and your team is punting to Devin Hunter of the Chicago Bears. You have some choices to make. What do you tell your kicker?

As we said before, you don't want the ball launched close to 90° (straight up) or 0°. Perhaps you should instruct the punter to go for maximum range, $\theta = 45° \rightarrow R = 68$ yd? Let's analyze the situation a little bit, and we'll see a potential problem with such a strategy.

For the 45° kick, equation 4.3 tells us how long the ball is in the air:

$$T_{\text{hang}} = 2\frac{v_0 \sin\theta}{g} = 2\frac{81 \text{ ft/s} \times \sin(45°)}{32 \text{ ft/s}^2} = 3.6 \text{ s}.$$

Given that your rushers typically run 9 yd/s, this means they cover only (9 yd/s) (3.6 s) = 32.4 yd before Hunter catches the ball 68 yd downfield. This gives Hunter and the Bears special teams way too much time to set up a coordinated return play that may well penetrate your defenses. Such a punt is shown in the top panel of figure 4.4.

A better strategy might be to sacrifice a little on the range of the punt to increase the hang time, so that your rushers reach Hunter just as the ball comes down, forcing

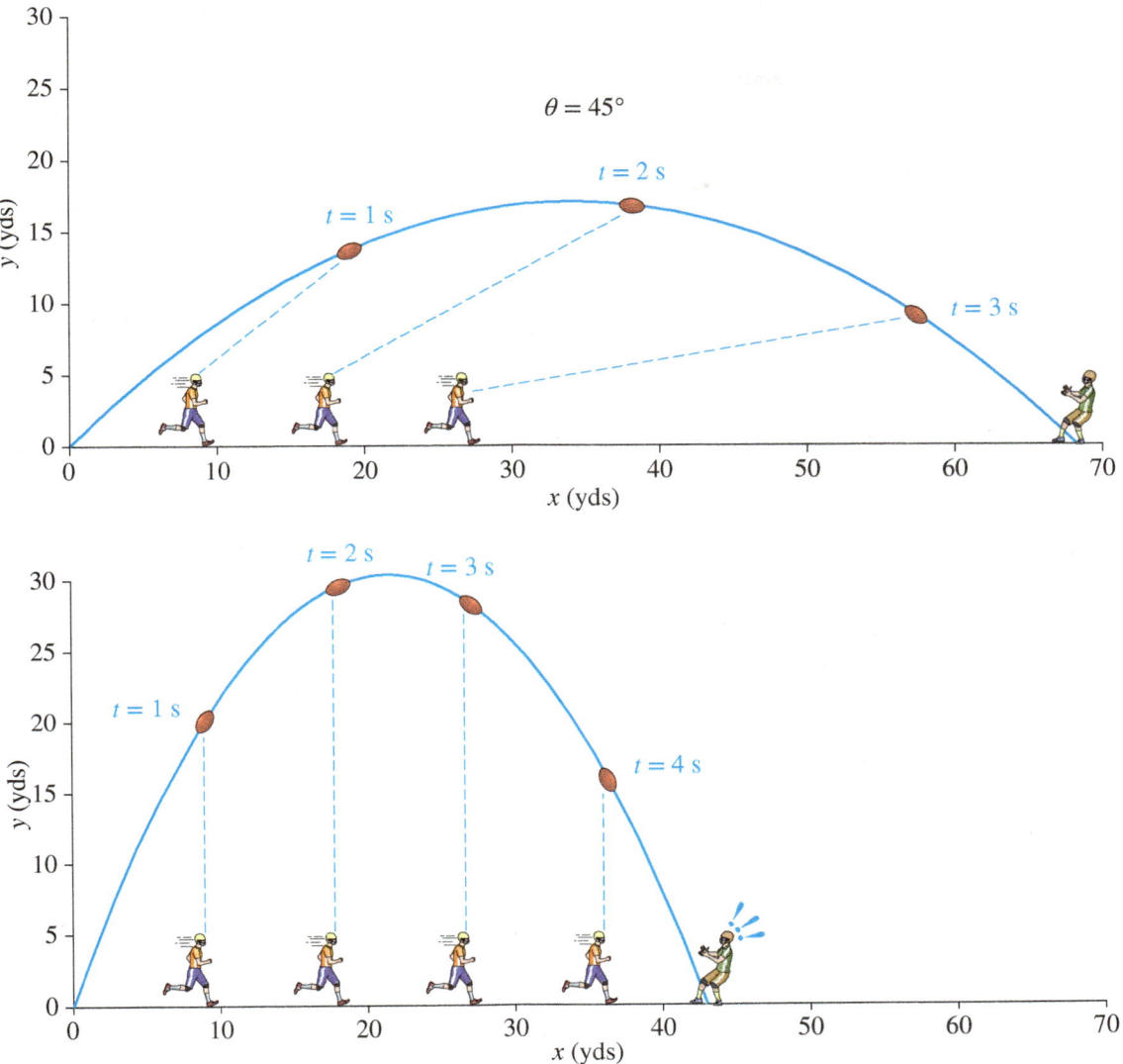

Figure 4.4. Punts launched at 81 ft/s at two launch angles. The ball and the rusher are shown at 1-s intervals. In the upper panel, the receiver catches the ball well before the rusher reaches him. In the lower panel, the ball and the rusher arrive simultaneously, and the receiver is well advised to call a fair catch.

a fair catch, as shown in the bottom panel. The numerical details are worked out in example 4.3.

Now we understand in detail the compromises that enter into different punting strategies. An experienced punter automatically estimates such compromises, factoring in additional effects such as wind speed and direction.

> **Question**
>
> In our discussion of the punt, we talked about the effect of changing the launch angle, all while keeping the launch speed constant at 55 mph. Even though most kickers can't consistently launch the ball much faster than that, is there any reason a kicker might want to launch it at a *lower* speed than the maximum possible?

EXAMPLE 4.3 *Forcing a fair catch*

In the bottom panel of figure 4.4, the ball comes to the receiver at the very moment that the rushers arrive to take him down. If he has any sense, he'll call a fair catch! How can we find the the angle for this optimum punt?

The key is our requirement that the ball travel horizontally down the field at the same rate as the rusher; that is, the rusher remains always directly beneath the ball. This means that the x-component of the ball's velocity is 9 yd/s. Since the punt is launched with a total speed of 81 ft/s = 27 yd/s, we can use equation 3.11 to find θ.

$$v_{0,x} = |\vec{v}_0| \cos \theta \rightarrow \theta = \cos^{-1}\left(\frac{v_{0,x}}{|\vec{v}_0|}\right) = \cos^{-1}\left(\frac{9 \text{ yd/s}}{27 \text{ yd/s}}\right) = 70.5°$$

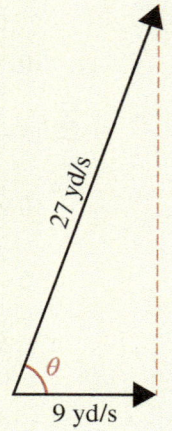

Question

In example 4.3, the optimal launch angle for a forced fair catch was found to be 70.5°. What if the punter gave the ball a larger launch angle? Would it still force a fair catch? If so, why is an angle of 70.5° preferred over a larger one?

A high school punter might only be able to launch the ball at 30 mph, rather than the 55 mph we've been using. In this case, should he kick the ball at an angle larger or smaller than 70.5°, in order to have the ball land just as the rushers arrive?

4.3 Shot-Put

Sometimes equation 4.4 is not appropriate for figuring out the range of a projectile, because the starting and ending heights are not the same. A case in point was the horizontally aimed rifle in example 4.1. Since $\theta = 0$ in that case, equation 4.4 would tell us that the bullet's range is zero, regardless of its muzzle velocity. Example 4.2 discussed a more complicated case where the initial and final heights were not the same.

The simple range equation 4.4 only applies if the launch and landing heights are the same.

Even though the range equation can't be used, the horizontally aimed rifle was easy enough since $v_{y,0} = 0$. But what about the case of the shot-putter? As shown in figure 4.5, he launches the shot with at some angle, and the start and final heights are not the same.[1] In these cases, the range formula is a bit more complicated:

$$R = \frac{v_0^2 \sin \theta \cos \theta + v_0 \cos \theta \sqrt{(v_0 \sin \theta)^2 - 2g \Delta y}}{g} \qquad (4.5)$$

For the shot-put, where $y_{\text{final}} = 0$, Δy is a *negative* number $\Delta y = 0 - y_{\text{initial}}$, where y_{initial} is the release height.

---

[1] Actually, the same is true for a football punt, but the difference between the starting and final heights is so small relative to the typical heights of the ball that it can be ignored.

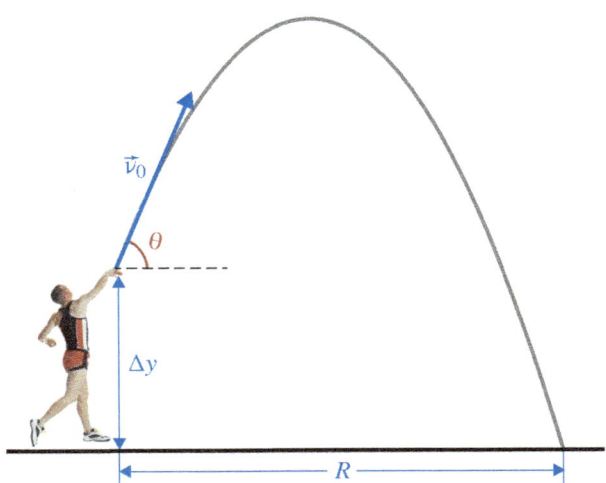

Figure 4.5. The starting and ending heights of the projectile in the shot-put are not the same, so equation 4.5 must be used instead of the simpler range equation 4.4. (© *Rubberball/Getty Images RF*).

4.3.1 Optimum shot-put launch angle

For $\Delta y = 0$ projectiles, the largest range is obtained for a launch angle of $\theta = 45°$, as we've said. For the shot-put, it is not. Through hours of practice, the athlete estimates the best launch angle through trial and error. Mathematically, finding the optimum angle from equation 4.5 takes calculus, but some typical values are given in table 4.1. Example 4.4 shows clearly that, of the several variables under the athlete's control, weight training to increase launch speed even by a small amount is more important than increasing launch height or throw angle.

The table includes a lot of numbers, but a quick study tells us several things. First, the most obvious one: The range (R) numbers are lowest in the upper left corner of the table and largest in the lower right corner. So you want to release the shot high and with as much speed as possible.

We also see that the optimum launch angle is always less than 45° for the shot-put; for typical speeds and release heights, it is right around 41°.

Table 4.1. For a given release height and launch speed, the optimum launch angle and resulting shot distance are given. Inspired by a similar table in *The Biomechanics of Sports Techniques* by James Hay.

Release height	Launch speed				
	22 mph	24 mph	26 mph	28 mph	30 mph
5 ft 6 in. (1.68 m)	$\theta_{opt} = 40.8°$ $R = 11.42$ m	$\theta_{opt} = 41.4°$ $R = 13.32$ m	$\theta_{opt} = 41.9°$ $R = 15.37$ m	$\theta_{opt} = 42.3°$ $R = 17.58$ m	$\theta_{opt} = 42.6°$ $R = 19.96$ m
6 ft (1.68 m)	$\theta_{opt} = 40.6°$ $R = 11.55$ m	$\theta_{opt} = 41.1°$ $R = 13.45$ m	$\theta_{opt} = 41.6°$ $R = 15.50$ m	$\theta_{opt} = 42.1°$ $R = 17.72$ m	$\theta_{opt} = 42.4°$ $R = 20.10$ m
6 ft 6 in. (1.98 m)	$\theta_{opt} = 40.2°$ $R = 11.68$ m	$\theta_{opt} = 40.8°$ $R = 13.58$ m	$\theta_{opt} = 41.4°$ $R = 15.64$ m	$\theta_{opt} = 41.8°$ $R = 17.86$ m	$\theta_{opt} = 42.2°$ $R = 20.23$ m
7 ft (2.13 m)	$\theta_{opt} = 39.9°$ $R = 11.81$ m	$\theta_{opt} = 40.6°$ $R = 13.71$ m	$\theta_{opt} = 41.1°$ $R = 15.77$ m	$\theta_{opt} = 41.6°$ $R = 17.99$ m	$\theta_{opt} = 42.0°$ $R = 20.37$ m
7 ft 6 in. (2.29 m)	$\theta_{opt} = 39.6°$ $R = 11.94$ m	$\theta_{opt} = 40.3°$ $R = 13.84$ m	$\theta_{opt} = 40.9°$ $R = 15.91$ m	$\theta_{opt} = 41.4°$ $R = 18.13$ m	$\theta_{opt} = 41.8°$ $R = 20.51$ m

EXAMPLE 4.4 What to work on?

Equation 4.5 shows that three things—the release height, the launch speed, and the launch angle—determine how far the shot will go. Say you are a shot-putter right in the middle of table 4.1, releasing the shot at 26 mph (11.62 m/s) from a height of 6 ft 6 in. Which of these elements is the most important to focus on? With persistent practice, you might be able to change any quantity by, say, 5%.

Let's start with the angle. Rather than launching at the optimum angle of 41.4°, what if you threw sometimes 5% less, that is, 39.3°? Then, all else being unchanged, your throw would be

$$R = \frac{(11.62 \text{ m/s})^2 \sin 39.3° \cos 39.3°}{9.8 \text{ m/s}^2}$$
$$+ \frac{(11.62 \text{ m/s}) \cos 39.3° \sqrt{((11.62 \text{ m/s}) \sin 39.3°)^2 - 2(9.8 \text{ m/s}^2)(-1.98 \text{ m})}}{9.8 \text{ m/s}^2}$$
$$= 15.61 \text{ m}.$$

This is 3 cm less than the range of 15.64 m obtained at the optimum angle. It seems that, so long as you don't let the angle get too far from optimum, you don't need to ultra-fine-tune it.

How about the height? Rather than launch it at 1.98 m, you might rework your technique to release it 5% higher, at 2.08 m. (This is an increase in release height of 4 in. It is probably doable, with work.) The range is then

$$R = \frac{(11.62 \text{ m/s})^2 \sin 41.4° \cos 41.4°}{9.8 \text{ m/s}^2}$$
$$+ \frac{(11.62 \text{ m/s}) \cos 41.4° \sqrt{((11.62 \text{ m/s}) \sin 41.4°)^2 - 2(9.8 \text{ m/s}^2)(-2.08 \text{ m})}}{9.8 \text{ m/s}^2}$$
$$= 15.75 \text{ m}.$$

This is an 11-cm improvement, perhaps worth a shot (pun intended), if the change doesn't screw up your form in other ways.

Okay, how about increasing the launch speed? Rather than launching at 11.62 m/s, you hit the weights and grow strong enough to launch it at 12.2 m/s. (Approximately speaking, you increase your launch speed by 5% by getting 5% stronger.) Now your range is

$$R = \frac{(12.2 \text{ m/s})^2 \sin 41.4° \cos 41.4°}{9.8 \text{ m/s}^2}$$
$$+ \frac{(12.2 \text{ m/s}) \cos 41.4° \sqrt{((12.2 \text{ m/s}) \sin 41.4°)^2 - 2(9.8 \text{ m/s}^2)(-1.98 \text{ m})}}{9.8 \text{ m/s}^2}$$
$$= 17.06 \text{ m}.$$

Whoa, there it is! That's a 1.42-m increase in range—almost 10%.

Clearly, rather than focusing on hitting precisely the right launch angle or stretching to the highest possible release height, a shot-putter is well advised to work on maximizing launch velocity.

> **Question**
>
> Within a given row (that is, for a fixed release height) in table 4.1, the optimum launch angle gets closer to 45° as the launch velocity increases. Besides simply saying that this is what comes out of equation 4.5, can you give a simple explanation for this?

> **Question**
>
> Within a given column (that is, for a given launch speed) in table 4.1, the optimum launch angle gets closer to 45° as the release height decreases. Besides simply saying that this is what comes out of equation 4.5, can you give a simple explanation for this?

4.4 Human Projectiles

As you know, the laws of physics demand that the center of mass of *any* object in flight follow the parabolic trajectories shown in figures 4.2–4.4 and described by the equations in this chapter. However, as vividly seen in figure 4.1, not all *parts* of an object must follow such a trajectory.

Once an athlete's feet leave the floor, nothing he does can change the parabolic trajectory of his center of mass.

The human body is a special "object" whose shape can change according to the will and abilities of the athlete. Once his feet leave the floor with some velocity $\vec{v}_0$, the motion of an athlete's center of mass is completely determined—there is not one thing he can do about it until he hits the floor (or something else!). However, while his center of mass must rigidly follow equations 4.1–4.4, there is no rule that his foot or hand must obey them, and it's usually the hand or foot—not the center of mass—that matters in sports.

The best athletes contort their bodies with exquisite precision and timing to squeeze the maximum advantage from the stingy laws of nature. Here, we'll describe a few classic cases.

4.4.1 The Blake Griffin ballet

Basketball provides some of the most amazing examples of humans in flight. When the Bulls' Michael Jordan or the Clippers' Blake Griffin takes off at the foul line and soar toward the hoop, it's easy to imagine that they are at least *bending* the laws of physics.

Fans often swore that Jordan had a 2-s hang time, but we know from figure 2.12 that this could happen only if his feet went above the backboard, and even Jordan wasn't *that* amazing. But what about the fact that Jordan is moving horizontally at the same time as he's moving vertically—doesn't that change things? As we discussed in section 4.1 and example 4.1, no, it does not.

However, physics explains a major reason behind the *impression* that something "special" is going on, and it's something ballet dancers instinctively know. These dancers are as tethered by gravity as the rest of us, but the best appear "lighter"—they seem to float just a bit more than they should. How?

Our primate brains have been trained by millions of years of evolution to recognize the parabolic trajectory of an object in flight such as a rock or a spear. The center

Figure 4.6. A dancer performs a grand jeté. Physics demands that her center of mass follow a parabolic trajectory, but she manipulates her body to give the illusion of defying gravity.

Figure 4.7
As graceful as Blake. (© *Comstock/JupiterImages RF*).

High jumpers use the laws of physics to pass their bodies over the bar while their center of mass travels *under* it.

of mass of these dancers follows the same parabolic path, but their head, arms, and feet need not. We tend to watch the face or head of a human performer. Figures 4.6 and 4.7 show a grand jeté in ballet. The dancer raises her arms during the first half of the move; her center of mass rises (as it must) while her head does not. As her center of mass peaks and starts to fall, she lowers her arms, and again her head maintains a level path.

As our primate brain observes the dancer's head, it instinctively knows that something "special" is happening—such horizontal motion seems to defy gravity. Such movements add to the grace of the ballet.

Blake Griffin regularly performs his own grand jeté. On one of his awesome dunks, he tends to leave the floor with his arms and ball low. As he rises, he brings up his arms and the ball (and often his legs, as seen in figure 4.8), slamming them down rapidly just after peaking. It has all the grace of the ballet.

4.4.2 Dick Fosbury's Flop

In the 1968 Mexico City Olympics, Richard Fosbury changed the event of high jumping forever. More than 40 years later, we are accustomed to high jumpers sailing over the bar face up, their arms, legs, and back forming a backward arc, as seen by one of the best, Blanka Vlašić of Croatia, in figure 4.9. The so-called Fosbury Flop is now ubiquitous in the event.

Before 1968, two techniques were in use: the hurdle (or its cousin the scissors kick) and the straddle or roll. These are shown in figure 4.10.

The rules of the event only demand that the jumper's body clear the bar; they make no stipulation about her center of mass. With this in mind, one advantage of the Flop is immediately clear. To avoid the jumper's body touching the bar, the center of mass must clear about 12 cm above the bar, if the scissors kick is used. In the roll, the center of mass must clear the bar by about 2–3 cm. In a well-executed Flop, however, the center of mass can sail more than 15 cm *below* the bar, while the jumper's body clears it from above.

What does this mean? Well, remember the mechanics of a leap from section 3.2.3: the jumper should obviously have strong legs, but not too much mass, and should swing her arms upon launch. There are other mechanics, too, concerning a launch

Figure 4.8. As his center of mass rises, Blake Griffin raises the ball and his arms and feet, later slamming them down reminiscent of a rather violent grand jeté. Los Angeles Clippers's Blake Griffin dunks against San Antonio Spurs during the first half of a preseason NBA basketball game in Mexico City, Tuesday, October 12, 2010. (© *Claudio Cruz/AP Photo*).

Figure 4.9. Croatia's Blanka Vlašić executes a perfect Fosbury Flop in the women's High Jump final at the Athletics World Indoor Championships at Ergo Arena in Sopot, Poland, March 8, 2014. (© *Radek Pietruszka/epa/Corbis*).

from one leg (as the rules for the event stipulate) and a running start. But at the end of the day, optimizing her leap, a jumper is strong enough to leave the ground with some vertical velocity, say, 10 mph (4.47 m/s). Equation 2.14 tells us she can raise her center of mass

$$\Delta y = \frac{v_{y,0}^2}{2g} = \frac{(4.47 \text{ m/s})^2}{2 \times 9.8 \text{ m/s}^2} = 1 \text{ m}.$$

Figure 4.10. Two high-jump techniques commonly used before the invention of the Fosbury Flop in 1968. The center of mass of the jumper at his highest point is indicated by the red X.

Assuming that the initial height of her center of mass is about 1 m, it is 2 m high at the peak of her jump. All other things being equal, then, our jumper could clear 1.88 m using the scissors kick, 1.98 m with the roll, or 2.15 m with the Flop.

As usual, things are complicated, and all other things are usually *not* equal, because of the different run-up mechanics associated with the three techniques. Nevertheless, the huge advantage gained by the lowered center of mass with the Flop more than compensates for the slightly greater effort required by a more awkward run-up.

Hmmm... I wonder why nobody invented the Fosbury Flop *before* 1968? Well, it might have something to do with the fact that, before the 1960s, jumpers landed in unraised sand or sawdust pits. This would not preclude a flop, of course, but the broken neck you'd get would preclude a *second* jump.

4.4.3 Bob Beamon's long jump

In the very same 1968 Olympic games in which Fosbury revolutionized his event, fellow American Bob Beamon obliterated all previous records in his. Before the 21-year-old Beamon took to the runway in Mexico City, the world record for the long jump stood at 27 ft $4\frac{3}{4}$ in. The record had been slowly increasing by fractions of an inch over the decade. Nobody could imagine breaking the 28-ft barrier. In a jump that astounded the world, Beamon sailed through the 28-ft mark in the air, on his way to smashing past the 29-ft mark!

Olympic champion Lynn Davies immediately recognized the magnitude of this feat. He told Beamon solemnly, "You have destroyed this event." In the classic grand style of the Soviets of the time, Olympic medalist Igor Ter-Ovanesyan put it this way: "Compared to this jump, we are as children."

Beamon's jump of 29 ft $2\frac{3}{8}$ in. (8.9 m) stood for 23 years; Mike Powell passed it by 2 in. in 1991, a record that stands today.

Figure 4.11. Bob Beamon in the 1968 Mexico City Olympics in the incredible first jump to smash the 28-ft barrier, returning to Earth again only after smashing the 29-ft barrier as well. (© *Bettmann/Corbis*).

Footage of the event is good enough for video analysis, and it allows us to estimate that Beamon took off with a speed of 10.02 m/s (22.4 mph) at an angle of 23.7°, as indicated in figure 4.12. Details of how to arrive at these numbers are discussed in example 4.5.

The launch angle

But wait! This is the *long* jump. The whole point is to maximize the range, so why wouldn't a long jumper always take off at 45°, the so-called optimal angle? Well, the primary answer is simply that the legs are not strong enough to generate enough vertical velocity to launch at 45°. Look back at figure 3.13. A 45° launch angle would mean that the horizontal and vertical components of the velocity are the same. In 1968, Beamon's horizontal velocity was $v_x = (10.02 \text{ m/s}) \cos 23.7° = 9.175$ m/s, or 20.5 mph. That's fast—comparable to top sprinters.

If the *vertical* component of his velocity were the same as his horizontal, then we'd have a 45° launch angle, but we'd also have a superhuman leap! Equation 2.14 tells us how high Beamon would be at his peak:

$$h = \frac{v_{y,0}^2}{2g} = \frac{(9.175 \text{ m/s})^2}{2 \times 9.8 \text{ m/s}^2} = 4.29 \text{ m} \approx 14.1 \text{ ft}.$$

Beamon's feet would be higher than a basketball backboard. He'd be able to destroy Fosbury's high-jump record (as a hurdle!) in the same leap in which he set the long-jump record!

So we see the point. Beamon's horizontal velocity of 20.5 mph is comparable to that of a sprinter. His vertical velocity of 9 mph (4.03 m/s), which the table in figure 2.12 reminds us, is comparable to a professional basketball player with a 40 in. vertical. (It's hard to imagine Beamon jumping any higher than is seen in figure 4.11!). The only way he could achieve a 45° launch would be to purposely slow down horizontally, which would obviously decrease his range.

Video analysis of long jumpers reveals that a 20° to 30° launch angle is typical.

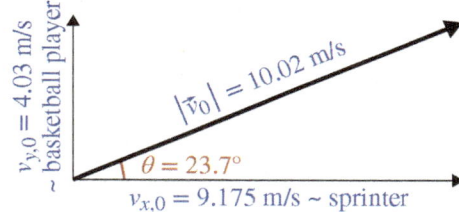

Figure 4.12
The vertical component of Beamon's launch velocity is comparable to that of a basketball player, and the horizontal component comparable to that of a top sprinter. Together, they combine to produce a 23.7° launch angle.

 The only way for Beamon to achieve the "optimum" 45° launch angle would be to slow his run-up considerably.

Maximizing the range

Now, using equation 4.4, we should be able to calculate the range of Beamon's jump:

$$R = \frac{v_0^2 \sin 2\theta}{g} = \frac{(10.02 \text{ m/s})^2 \sin(2 \times 23.7°)}{9.8 \text{ m/s}^2} = 7.54 \text{ m} \approx 24.74 \text{ ft.}$$

This is a nice jump, but hardly Olympic quality and far from the record distance Beamon actually reached. What's wrong? Nothing, really. We just need to remember again that equation 4.4 tells us the horizontal distance the *center of mass moves before it returns to the same height as its launching point*. Once again, the athlete's ability to manipulate his center of mass makes the difference between a decent high-school showing and a breathtaking Olympic world record.

There are several important aspects to this motion, so let's take a look at a typical jump, shown in figure 4.13. Between images 1 and 2, the athlete is taking off. In addition to extending her leg to thrust into the ground, she is rapidly accelerating her arms up. (When we come to angular momentum, we'll see why she chooses to swing one arm clockwise and the other counterclockwise; for now, it is the fact that both swing upward that matters.) As we discussed in gory detail in section 3.2.3, both of these actions increase the ground reaction force (GRF), maximizing the vertical acceleration—and hence the vertical velocity—of her center of mass.

Once her foot leaves the ground shortly after image 2, there is not a single thing she can do about the motion of her center of mass; it will follow a simple parabola until something (the sand pit) exerts a force on the athlete. Since her arms and one leg are raised upon takeoff, the athlete's center of mass is located somewhere above her navel. Her center of mass does not reach this height again until somewhere between images 5 and 6. Here's the important point: the range formula of equation 4.4 tells us the horizontal distance between image 2 and just before image 6. The athlete can eke out greater distance by forcing her center of mass toward the lowest part of her body.

In a good long jump, the center of mass lands at a lower point than it was launched. Extra distance is gained for the same reason that a ball thrown off the top of a parking

Figure 4.13. A stroboscopic photoseries of a long jump. A 5-ft reference is estimated from the athlete's height with bent knee in position 1. The parabola traces out the trajectory of her center of mass, and the horizontal line indicates its height upon takeoff. (© *technotr/Getty Images RF*).

garage has a longer range than the same ball thrown from the ground. Basically, the athlete (or ball) spends more time moving horizontally in the air; even a small amount adds a lot of distance, as discussed in example 4.6. The proper equation to use is the shot-put range equation 4.5.

Bob Beamon's extra distance

Now it's time for some numbers: Just how much extra distance was Bob Beamon able to gain by manipulating his center of mass? Beamon's height is listed at 6 ft 3 in. (1.91 m). In a position similar to image 1 in figure 4.13, we'll estimate the height of his center of mass to be 1.2 m. Beamon landed in a crouch much more extreme than the athlete shown in figure 4.13, and we'll estimate the height of his center of mass upon landing at just 50 cm. His trajectory is shown in figure 4.14.

The change in Beamon's center-of-mass position in the vertical direction is

$$\Delta y = y_{\text{final}} - y_{\text{initial}} = 0.5 \text{ m} - 1.2 \text{ m} = -0.7 \text{ m}.$$

We can use this in the shot-put range equation to find his jump:

$$R = \frac{v_0^2 \sin\theta \cos\theta}{g} + \frac{v_0 \cos\theta \sqrt{(v_0 \sin\theta)^2 - 2g\Delta y}}{g}$$

$$= \frac{(10.02 \text{ m/s})^2 \sin 23.7° \cos 23.7°}{9.8 \text{ m/s}^2}$$

$$+ \frac{(10.02 \text{ m/s}) \cos 23.7° \sqrt{((10.02 \text{ m/s}) \sin 23.7°)^2 - 2(9.8 \text{ m/s}^2)(-0.7 \text{ m})}}{9.8 \text{ m/s}^2}$$

$$= 8.9 \text{ m}.$$

Beamon's center of mass returned to its original height after having traveled 24 ft 9 in. horizontally. However, he continued to "fall" another 70 cm (about 2 ft 4 in.) as he sailed over the 25-, 26-, 27-, 28-, and 29-ft marks!

By precisely controlling his body, Beamon exploited the laws of physics to his advantage to cheat another 1.36 m (4.42 ft) from the tyranny of parabolic motion. That's the stuff of an Olympic champion.

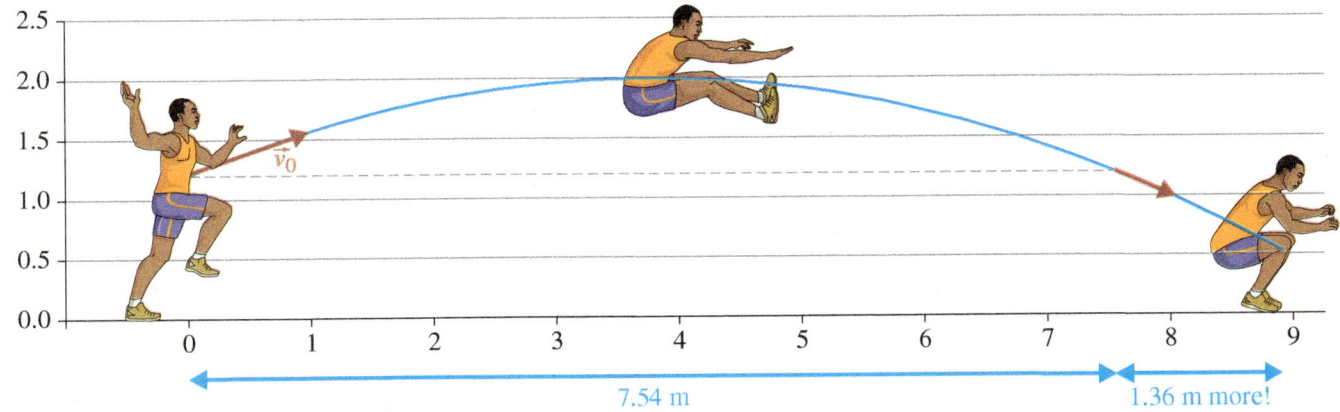

Figure 4.14. The calculated trajectory for Bob Beamon's extraordinary 1968 long jump, using equations discussed in this chapter. The relatively small effects of air drag are ignored. Beamon's center of mass is 1.2 m high at takeoff and 0.5 m at landing. Red arrows indicate his velocity at the two points when his center of mass is 1.2 m high.

EXAMPLE 4.5 *Estimating the launch velocity*

In section 1.2, we discussed how to measure important kinematic quantities from video footage of sports. Here we discuss how to extract the launch velocity, both magnitude and direction, from the footage of Bob Beamon's jump.

The grainy 1968 footage available has far too poor quality for most precise measurements, but frame-by-frame analysis clearly reveals that Beamon was in the air for about 0.97 s. Of course, the horizontal length of his jump need not be estimated from the film; that remarkable number—8.9 m—is quite well documented. Therefore, the horizontal component of Beamon's velocity is easy to calculate:

$$v_x = \frac{\Delta x}{\Delta t} = \frac{8.9 \text{ m}}{0.97 \text{ s}} = 9.175 \text{ m/s}.$$

For the vertical velocity component, we need to know something about height. The video and other photographic evidence indicate that Beamon took off in a stance resembling image 1 of figure 4.13. As mentioned earlier, given his height, we estimate the initial height of his center of mass to be $y_0 = 1.2$ m.

Beamon's landing was extraordinary. He landed in a tight crouch, his bottom entering the sand almost at the same position as his feet did just the instant before. A decent estimate of the final height of his center of mass is $y = 0.5$ m, or about 1 ft 8 in.

Given this information, equation 4.2 tells us his vertical velocity component:

$$\Delta y = v_{y,0} t - \tfrac{1}{2} g t^2 \quad \rightarrow \quad v_{y,0} = \frac{\Delta y}{t} + \tfrac{1}{2} g t$$

$$= \frac{0.5 \text{ m} - 1.2 \text{ m}}{0.97 \text{ s}} + \tfrac{1}{2} \left(9.8 \text{ m/s}^2\right)(0.97 \text{ s})$$

$$= 4.03 \text{ m/s}.$$

Now that we have the horizontal and vertical velocity components, we are done. We've found the velocity vector. If we want to, we can express this vector in terms of its magnitude and angle, using equations 3.13 and 3.14:

$$|\vec{v}_0| = \sqrt{v_{x,0}^2 + v_{y,0}^2} = \sqrt{(9.175 \text{ m/s})^2 + (4.03 \text{ m/s})^2} = 10.02 \text{ m/s}$$

$$\theta = \tan^{-1} \frac{v_{y,0}}{v_{x,0}} = \frac{4.03 \text{ m/s}}{9.175 \text{ m/s}} = 23.7°.$$

EXAMPLE 4.6 *How much extra time is Beamon in the air?*

Long jumpers gain critical extra feet and inches by lowering their center of mass to remain airborne as long as possible. How much extra time did Bob Beamon gain by landing with his center of mass 70 cm below its height (1.2 m) at takeoff?

This is very easy to figure out, since Beamon's horizontal velocity component does not change and we know the extra distance he traveled.

$$v_x = \frac{\Delta x}{\Delta t} \rightarrow \Delta t = \frac{\Delta x}{v_x} = \frac{1.36 \text{ m}}{9.175 \text{ m/s}} = 0.148 \text{ s}.$$

This is almost precisely the same as Usain Bolt's reaction time in the 100-m dash (see section 2.2) and about the time for the blink of your eyes. It made all the difference.

> **Question**
>
> In the bottom panel of figure 4.1 none of the paths is purely parabolic, but which is most nearly parabolic? Which is least parabolic? Why?

Collected Equations

$$\Delta x = v_x t \qquad (4.1)$$

$$\Delta y = v_{y,0} t - \tfrac{1}{2} g t^2 \qquad (4.2)$$

$$T_{\text{hang}} = 2 \frac{v_{y,0}}{g} = 2 \frac{v_0 \sin \theta}{g} = 2 \sqrt{\frac{2h}{g}} \qquad (4.3)$$

$$R = v_{x,0} T_{\text{hang}} = \frac{v_0^2 \sin 2\theta}{g} \qquad (4.4)$$

$$R = \frac{v_0^2 \sin \theta \cos \theta + v_0 \cos \theta \sqrt{(v_0 \sin \theta)^2 - 2g \Delta y}}{g} \qquad (4.5)$$

Problems

1. A baseball is hit so that it travels straight upward after being struck by the bat. A fan observes that it takes 3 s for the ball to reach its maximum height, and her friend asks casually, "Wow, how high do you think that ball went?" Little does her friend know that she read this book, and can answer.

   (a) What was the maximum height?

   (b) A little later, another struck ball still reaches its maximum height in 3 s, but rather than being hit straight up, the ball was a long fly and reached its peak 200 ft in front of the batter. In this case, what was the maximum height?

2. In a preseason game on August 21, 2009, A. J. Trapasso punted a ball that struck the newly installed high-definition JumboTron at Cowboys Stadium. The 600-ton JumboTron was located only 90 ft above the field. The ball traveled 15 yd horizontally before striking the JumboTron. Here, we will find the minimum speed at which the ball left Trapasso's foot and the launch angle.

   (The Raiders's Ray Guy did the same thing in 1976 in the Superdome. The height of the scoreboard then? Also 90 ft.)

   (a) Assume the ball barely makes it up 90 ft (we look for the minimum launch speed, after all) and find $v_{y,0}$.

   (b) Find the total time of the ball's journey.

   (c) Find $v_{x,0}$.

   (d) What are the launch speed and angle?

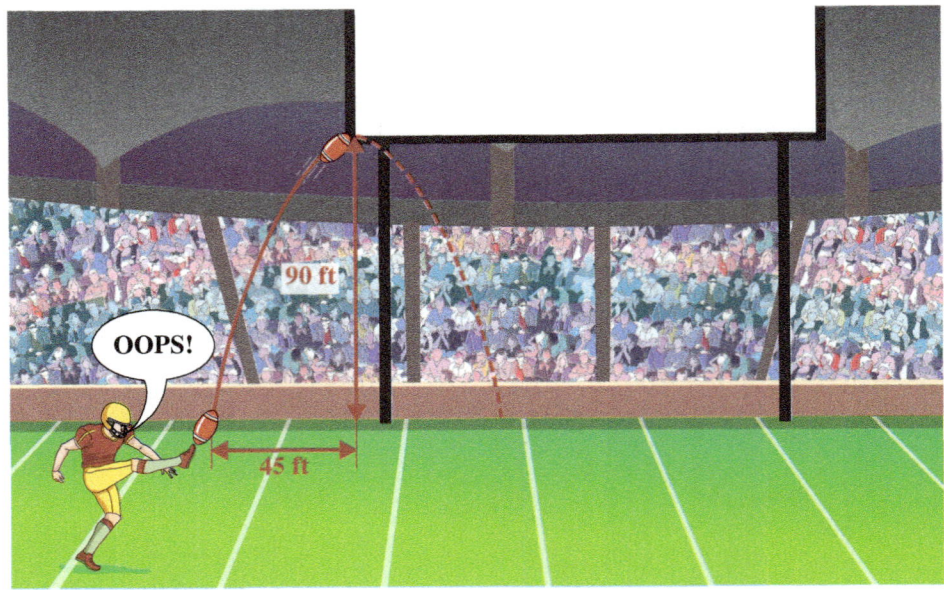

3. A punter launches the ball at 25 yd/s at a launch angle of 50°.

   (a) At what constant speed would a player have to run so that he always remains directly under the ball?

   (b) Can the typical college gunner (the player on the punting team who first races down the field) maintain this speed?

4. An athlete has an intuitive knowledge, gained by experience, of the changing nature of a ball in flight. Ignoring the effects of air as we've done so far, consider the *change* in each of the following quantities from moment to moment. Does it change by a fixed amount every second? Does it change every second, but by a varying amount? Or does it not change at all?

   (a) The x-component of the velocity

   (b) The y-component of the velocity

   (c) The height

   (d) The acceleration

5. A softball is batted at 65 mph at an angle of 50° to the horizontal.

   (a) For how long is it in the air?

   (b) In the horizontal direction, how long does it take the ball to travel 25 yd toward the outfield?

6. Irina Slutskaya was amazing at leaps in her skating routine. In a particular maneuver, she keeps her left leg pointing straight down and lifts her right leg out in front of her while in the air. Does the act of lifting her right leg cause her left to hit the ice earlier, or later, or does it have no effect?

7. Ohio State quarterback Braxton Miller is a master of escaping defenders while still maintaining concentration on receivers downfield.

(a) In the 2012 OSU–Wisconsin game, Miller had plenty of time thanks to an outstanding offensive line, and from a standing position he threw a short horizontal pass at 15 yd/s at an angle of 45° downfield (that is, it is thrown partially toward the sideline and partially downfield). What were the components of the initial velocity in the downfield direction and the sidelines direction?

(b) Earlier in the season, Miller threw almost exactly the same pass (that is, his *arm* made the same motion), but this time he was also scrambling at 5 yd/s toward the backfield. What was the resultant speed of the football? At what angle, relative to the downfield direction, was it thrown?

8. A small difference in pitch speed can make a big difference in the time and position at which the ball crosses the plate. Consider two horizontally thrown pitches.

    (a) How much earlier does a 96-mph pitch reach the plate, compared to a 93-mph pitch?

    (b) What is the difference in height, if any, between the two pitches as they cross the plate?

9. You can tell a lot just by watching the trajectory of a ball. You and a friend are sitting at Huntington Park in downtown Columbus, Ohio, and a long fly just reaches the center field fence, 400 ft from home plate. You figure it took about 6 s to get there. Your friend whistles and wonders aloud, "How fast do you think that ball came off the bat? And how high did it go? And will this guy get moved to the majors?"

    You can help him out on at least some of these questions.

    (a) How high did the ball go?

    (b) How fast did it come off of the bat?

    (c) At what angle did it leave the bat?

10. Consider two baseball hits, one reaching a higher maximum than the other. Is it possible for the lower hit to spend more time in the air than the higher one?

11. Ideally, a basketball shot hits nothing but net. Hit any part of the rim, and you've got a good chance to bounce out. Geometry says that a shot should have a high arc. That way, the ball passes through the hoop with plenty of "wiggle room" on either side. The figure P4.11 shows four shots with different angles of approach. A shot with an approach angle of 33° just barely has a chance of making it; anything less must miss. One reason that players don't use more of an arc is simply the increased strength and effort needed. In the jump shots shown, the ball is released from a height of 10 ft, the same height as the rim, just at the 3-point line (23 ft 9 in.).

    (a) What is the launch speed required to achieve a 70° approach angle?

    (b) What is the hang time for this shot?

    (c) In his book *The Physics of Basketball*, John Fontanella finds that the best players don't usually opt for the largest arc, which helps them geometrically, but rather for the "softest shot," or the lowest approach velocity. Since we are ignoring air, this occurs for a 45° launch. What is the launch speed for this "softest shot," shooting from the 3-point line?

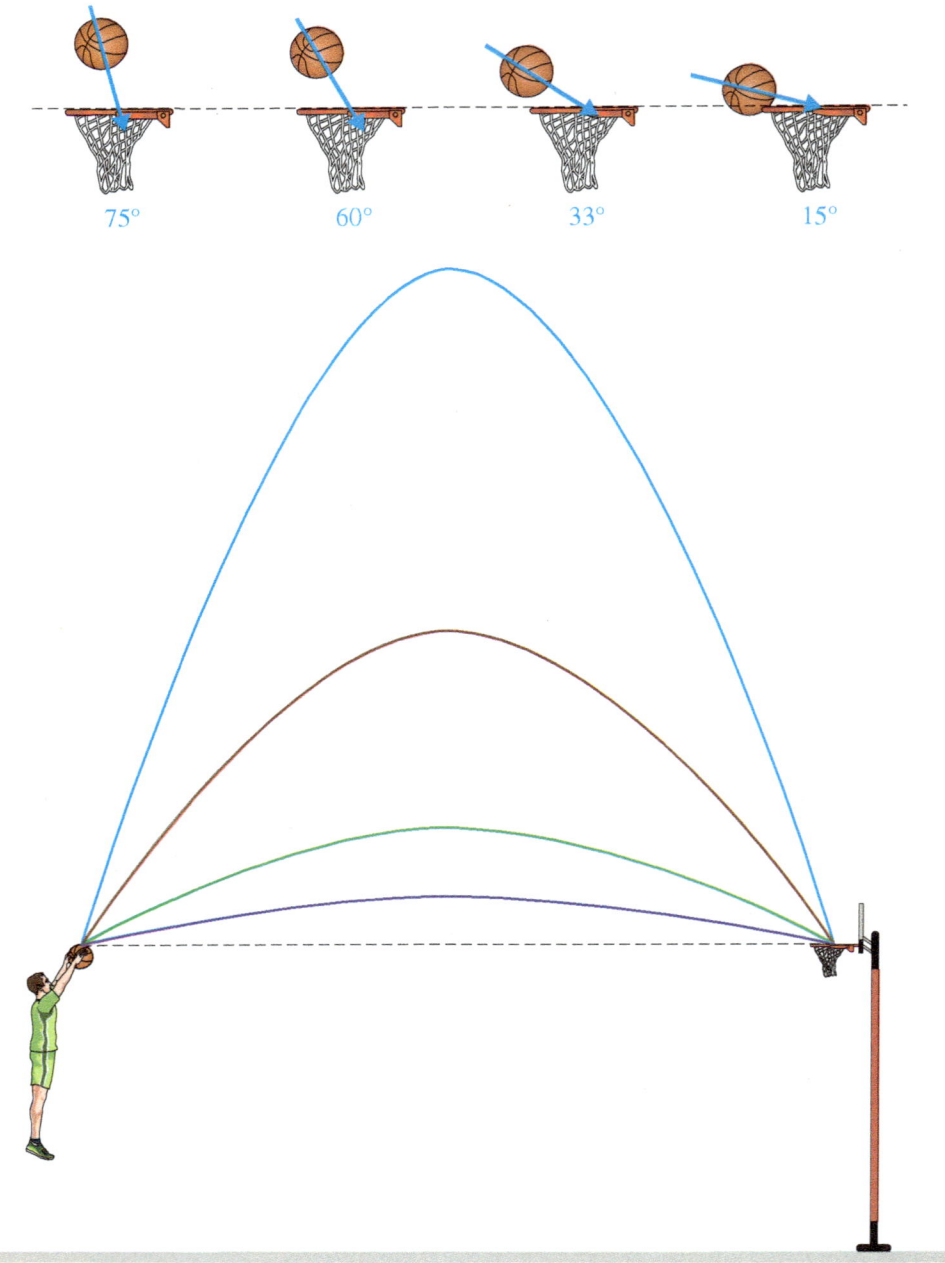

Figure P4.11

12. The left panel of the figure P4.12 shows two possible throws of one player to another, with path B reaching a larger peak height. Answer A, B, "they are the same" or "cannot tell" to the following questions.

    (a) Which throw trajectory requires more time for the ball to make its journey?

    (b) Which throw has the larger launch speed?

    Now look at the panel on the right, where paths A and B reach the same height:

    (c) Which throw trajectory requires more time for the ball to make its journey?

    (d) Which throw has the larger launch speed?

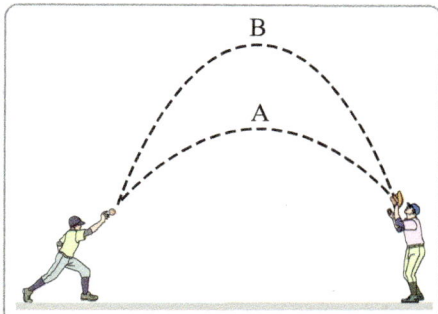
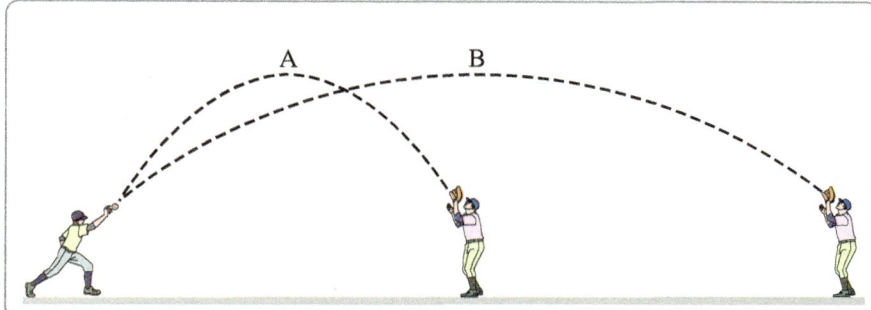

Figure P4.12

13. When Cleveland Cavaliers's Tristan Thompson jumps straight up for a dunk, he usually keeps his legs more or less extended straight down. Would the following reach a greater height, rise to the same height, or not reach as high, if instead he bend his knees once in the air?

    (a) The top of his head
    (b) His center of mass
    (c) His shoes
    (d) The ball in his outstretched hands

14. In many events, coordination and perfect timing among several motions of the body are required for peak performance. In his phenomenal 1968 long jump, Beamon left the board with his hands above shoulder height and forward knee driven up. If he'd swung his arms and thigh only a fraction of a second later, they would have been slightly lower, lowering his center of mass by perhaps 3 in.

    (a) All else being equal, by how much would this have shortened his jump? Give your answer in inches.
    (b) This jump still would have broken the world record. But would it have surpassed the 29 ft milestone?

15. In the game of horseshoes, a player tosses the horseshoe underhand in a swinging motion toward a post that sits in a sand pit. According to the National Horseshoe Pitchers Association (yes, there is such a body), the post is 40 ft from the pitcher.

    (a) Say a player releases the horseshoe 18 in. above the ground at a speed of 38 ft/s (25.9 mph) and launch angle of 45°. How far from the pitcher will the horseshoe land?
    (b) Program the range formula into your calculator or a spreadsheet program, so that you can easily plug in different angles and calculate the range. If the pitcher prefers to throw the horseshoe at the same speed and release height, what would be the best angle for him to launch the pitch? (There are two possible correct answers.)

16. In a jump serve in volleyball, the server throws the ball into the air and strikes it with his open hand from a position 9 m behind the net. An athletic player may be able to hit it when the bottom of the ball is 3 m above the ground. Here you will show that, without aerodynamic effects, a horizontally hit serve will never work—it will go into the net or go too far and out of bounds.

    (a) How much time does it take for the ball to fall to the height of the top of the net (which is 2.41 m above the floor)?
    (b) Based on your answer to part (a), what is the minimum launch speed for the ball to make it over the net?

(© Rubberball/Getty Images RF)

(c) What is the range of the ball struck horizontally at this speed? Is it less than 18 m, the length of the court?

(d) Look ahead to figure 6.29. Do you think putting topspin on the serve might improve it?

(e) The "sky serve" is intended to land on the opponent's side at a nearly straight-down trajectory. If this serve were a sky serve, launched at the same speed, but at the nearly vertical angle of 85°, would it land in bounds on the opponent's side of the net?

Curveballs, Foul Shots, and Bent Kicks

5

Figure 5.1. A player shooting a basketball has several aerodynamic forces to deal with, in addition to gravity. Thinking about them all would be even more distracting than that shirtless guy who always wiggles his beer belly behind the backboard when opposing players are shooting foul shots. (© *Rubberball/Getty Images RF*).

While the parabolic trajectories we studied in chapter 4 are a good approximation for balls and people flying through the air, many sports rely crucially on the details. We will see that for some objects (fast, rapidly rotating, light) aerodynamic forces dominate even gravity! These lifting, curving, and slowing forces are complicated and change in both magnitude and direction during the course of the trajectory. In this chapter, we will use approximate mathematical treatments to explain the importance of backspin on a basketball and the curve of a curveball. We'll also discuss why bicycle velodrome arenas are overheated and why throwing a discus *into* the wind is better than throwing it with the wind.

The way to catch a knuckleball is to wait until it stops rolling and then pick it up.
—Bob Uecker, MLB catcher

Which could you throw farther—a shot-put, a baseball, or a Wiffle ball? After a moment's thought, you'll probably answer that you could throw a baseball much farther than the other two—probably hundreds of feet farther—and you'd be right.

The reason a baseball goes farther than a shot-put is obvious by intuition and also based on our discussion of forces and throw speed in section 3.4.4. You can give the lighter baseball a greater acceleration and hence a greater throw speed.

But what about the Wiffle ball? That's even lighter than the baseball, so doesn't that mean it has a *greater* throw speed? The answer is yes, so according to our discussion in chapter 4 (especially the range equation 4.4), the Wiffle ball would go farther. The issue, of course, is that in chapter 4 we ignored air. You *do* release the Wiffle ball

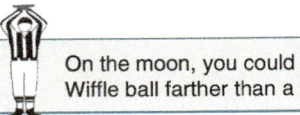

On the moon, you could throw a Wiffle ball farther than a baseball.

at a higher speed, but once you release it, air drag slows it much more than it does the baseball.[1]

Drag is just one of the forces air exerts on balls (and people). In this chapter, we will discuss this and other forces that modify the trajectory of an object flying through the air. The importance of these effects ranges from negligible such as in the shot-put to dominant such as in table tennis. In many cases, what makes an athlete great is his or her ability to manipulate the effects of air. Soccer balls would not bend, pitches would not curve, and golf drives would not slice without air. Bicyclists on the straightaway could easily maintain an 80 mph pace. Indeed, one is hard pressed to imagine a sport unaffected by the forces we'll discuss here; weightlifting may be the only entry on the list.

5.1 Overview

There are four forces that determine the flight of anything—ball or human—as it flies through the air. The basketball player in figure 5.1 is intuitively dealing with them all. We have discussed only one of the forces: gravity. Gravity is surely the simplest force and is often the most important force. The other three are more subtle and obviously related to the air since their formulas contain the density of air ρ_{air}. It is these forces that we will discuss in this chapter. They can be as strong as gravity and change motion dramatically.

These four forces determine the flight of a basketball:

1. Gravity—constant (in magnitude and direction) downward force on any object on Earth.

2. Buoyancy—constant (in magnitude and direction) upward force on any object in air or water.

3. Drag—variable-magnitude backward-directed force on any object *moving* in air or water.

4. Magnus—variable-magnitude sideward-directed force on an object *moving and spinning* in air.

5.1.1 Approximations

The forces due to air can be complicated. Your colleagues taking introductory physics courses for engineers and scientists are routinely told to consider physics in a world without air. The advantage of this approach is that it allows the student to focus on important, but simpler, concepts before considering the additional messy effects due to air. So far, we have followed this path, too, by discussing in chapter 4 the principle of projectile motion for cases in which aerodynamic effects are not dominant.

But now the rubber hits the road, and we have to go beyond what your science and engineering major friends are learning. In addition to gravity, we have other forces to deal with. In principle, this shouldn't be a problem because, as we discussed in chapter 3, we can just add all the forces to find the net force $\sum_i \vec{F}_i$. And once we know the net force, we can find the acceleration by Newton's second law, equation 3.6. And once we know *that*, we can just use our kinematic equations such as 2.5–2.7, right? Wrong.

---

[1] This means that if you throw a Wiffle ball on the moon as hard as you can, it will travel farther than a baseball. And a Ping-Pong ball would go even farther. Even though you "know" this intellectually, doesn't it still seem strange, as if it can't really be true?

The problem with that plan is that all our kinematic equations are true only if the acceleration is constant, which means that the *force* has to be constant. This is true for gravity, but not for aerodynamic forces. For example, the drag force reduces the speed of a ball. However, once the ball has slowed, the drag force changes! Accounting for effects like this requires more advanced math and often a computer.

So, now what? Do we give up and just swallow whatever the computer tells us? Not exactly. Instead, we follow the approach that scientists everywhere take, when they deal with complicated effects in the real world: we will *approximate*. Although the forces at play are not constant, we can still learn a lot by treating them as if they were.

While our calculations won't be perfectly accurate, they will usually be very close to the right answer, as we'll see by checking with a computer. It is important that you appreciate this approach, since essentially *every* real-world problem under study is complicated enough that computer programs are required for a complete understanding. However, nothing can screw up quite as a computer can, and it is reckless in the extreme to trust computer-generated results unthinkingly. In fact, even if the computer gives a completely correct result, it is still crucially important to grasp the main components driving that result by using simple equations. If the computer completely contradicts a simpler analysis that we think should be approximately correct, say, on species extinction, should we just trust the computer and move on? No. Even if the computer is correct, we need to find out what is so wrong in our approximate treatment; only then will we understand the problem of species extinction.

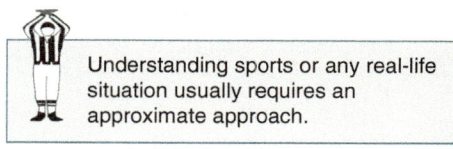

Understanding sports or any real-life situation usually requires an approximate approach.

5.2 Immersion in Fluid: Buoyancy

When you are mostly submerged in the deep end of the swimming pool, why aren't you worried about falling 10 ft to the concrete floor? After all, your weight is pulling you downward as strongly as ever. You'll probably answer that the water is "holding you up," and this is not incorrect. Let's take a look at the source of this force and get a bit more quantitative.

Figure 5.2 shows a swimmer underwater. He doesn't seem worried about falling. Perhaps he is thinking about the man-shaped volume of water that *would* be in his place if he were not present. Think about it: What forces are there on such a volume of water? Certainly gravity pulls it down. But a man-shaped volume of water submerged *in* a pool of water doesn't "fall," so we (or actually Archemides (287–212 B.C.)) conclude that *there must be a force exactly canceling the weight of the man-shaped volume of water*. This is the **buoyant force** $\vec{F}_B$.

22.5 gallons of water displaced by swimmer

Figure 5.2. A swimmer underwater experiences an upward-directed "buoyant force" equal in magnitude to the weight of the water he displaces. (© *Jens Nieth/Corbis RF*).

The water in a pool pushes up on a person with a force equal to the weight of the volume of water that the person's body displaces.

Since $\vec{F}_B$ always opposes gravity, its direction must point up. Its magnitude must equal the weight of the swimmer-shaped volume of water (*not* necessarily the swimmer's weight!).

$$|\vec{F}_B| = |\vec{W}_{\text{displaced H}_2\text{O}}| = (\rho_{\text{H}_2\text{O}})(V_{\text{displaced H}_2\text{O}})g \qquad (5.1)$$

where $\rho_{\text{H}_2\text{O}} = 1000$ kg/m³ is the **mass density of water**; 1 m³ of water has a mass of 1000 kg, or a weight of 2200 lb more than a ton!

The mass density of a typical First World person is a bit less, say 95%, than that of water. Therefore, the water's buoyant force on such a person equals his weight when 95% of his body is submerged. If less of his body is submerged, he sinks farther into the water; if more is submerged, he rises.

Muscle is denser than fat. So a very muscular, fat-free person weighs more than the water displaced by his entire body; the buoyant force will always be smaller than his weight, and he will sink. He'll need all that muscle to swim to the surface.

While the buoyant force on a human or animal body is close to its weight, these forces can differ greatly for objects with higher or lower density, as shown in example 5.1.

Question

Why does it help to expand your chest (often by inhaling fully) when floating on your back in the pool?

EXAMPLE 5.1 *Ball-under foul*

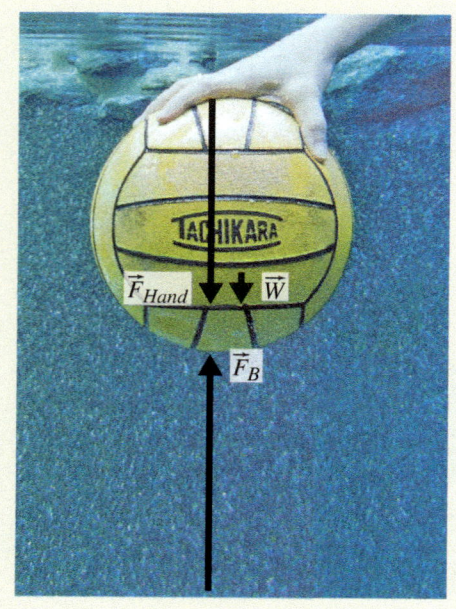
(© McGraw-Hill Education/David Moyer, photographer)

In the intensely physical sport of water polo, if a player pushes the ball completely under water, it is a foul resulting in a turnover. Is this an easy thing to do? How much force does it require to hold the ball underwater?

There are three forces at work here, the downward force of gravity and the upward buoyant force as well as the downward push of the player. The specifications of a water polo ball are listed in table B.3; for a men's ball, the weight is

$$|\vec{W}| = mg = (0.425 \text{ kg})(9.8 \text{ m/s}^2)$$
$$= 4.2 \text{ N} \approx 0.94 \text{ lb}.$$

The buoyant force depends on the volume of water displaced. In this case, the entire ball is submerged, so the volume displaced is the volume of the spherical ball, $V = \frac{4}{3}\pi r^3$. We can use equation 5.1 to find the buoyant force

$$|\vec{F}_B| = (\rho_{\text{H}_2\text{O}})(V_{\text{ball}})g = \left(1000 \frac{\text{kg}}{\text{m}^3}\right)\left(\frac{4}{3}\pi (0.111 \text{ m})^3\right)(9.8 \text{ m/s}^2) = 56.1 \text{ N}.$$

Clearly, then, the forces cancel when the player is pushing down with a force of

$$|\vec{F}_{\text{hand}}| = |\vec{F}_B| - |\vec{W}| = 56.1 \text{ N} - 4.2 \text{ N} = 51.9 \text{ N} \approx 11.7 \text{ lb}.$$

Thus, it takes more than 12 times as much force to push the ball under water as it does to lift it out of the water.

5.2.1 Buoyancy in air

When we step out of the pool, we are immersed in another fluid—air. Since any object will be fully submerged in air, the buoyant force on the object is

$$\text{any object:} \quad |\vec{F}_B| = \rho_{\text{air}} g V \qquad (5.2)$$

$$\text{spherical ball with radius } r: \quad |\vec{F}_B| = \rho_{\text{air}} g \tfrac{4}{3}\pi r^3 \qquad (5.3)$$

With a density of 1.2 kg/m³, air is almost 1000 times less dense than water and so exerts a correspondingly lower buoyant force. Nevertheless, it can play a role where precision counts. One way to estimate its importance to the flight of a ball is by comparing the buoyant force to the ball's weight.

$$\frac{|\vec{F}_B|}{|\vec{W}|} = \frac{\rho_{\text{air}}\, \cancel{g}\, \tfrac{4}{3}\pi r^3}{m\cancel{g}} = \frac{4\pi \rho_{\text{air}}}{3}\frac{r^3}{m} = \begin{cases} (5 \text{ kg/m}^3)\frac{(0.0374\text{ m})^3}{0.15\text{ kg}} = 0.18\% & \text{baseball} \\ (5 \text{ kg/m}^3)\frac{(0.12\text{ m})^3}{0.62\text{ kg}} = 1.4\% & \text{basketball} \end{cases}$$

For baseballs, lacrosse balls, and golf balls, the effects of buoyancy in air are essentially negligible. The same is true for humans, as discussed in example 5.2. For basketballs, footballs, soccer balls, and table tennis balls, the effect is small, 1%–2%, but can be significant.

Rather than a constant downward acceleration of $-g$, the (still constant) acceleration becomes

$$a_y = \left(1 - \frac{|\vec{F}_B|}{|\vec{W}|}\right)(-g).$$

This reduced value replaces $-g$ in kinematic formulas such as 4.3 and 4.4.

The air around us exerts a net upward force on all objects. The buoyant force on a human is about 3 oz.

EXAMPLE 5.2 *Weight-loss effect*

Contrary to the usual assumption, a bathroom scale measures not your weight, which is the force *the Earth* exerts on you, but rather the ground reaction force, or GRF, which is the force that *the scale* exerts on you. (Newton's third law tells us this is the same as the force you exert on the scale.) We've already discussed one aspect of this in chapter 3 and figures 3.5 and 3.6. There, a person was accelerating his center of mass to alter the GRF. But even if the person is standing perfectly at rest, the GRF reading on the scale is a bit less than his weight, due to the buoyant force of the air.

If a 160-lb woman steps onto an accurate scale, what will it read?

To find the buoyant force from air, we need to know the volume of the woman. We've said that the density of the human body is 95% that of water. Turning to table B.5, then

$$\rho_{\text{woman}} = 0.95 \times (1.94 \text{ slug/ft}^3) = 1.82 \text{ slug/ft}^3 \quad (\approx 950 \text{ kg/m}^3),$$

and we can find the woman's volume by turning around the definition of density:

$$\rho = \frac{m}{V} \quad \rightarrow \quad V_{\text{woman}} = \frac{m_{\text{woman}}}{\rho_{\text{woman}}} = \frac{W_{\text{woman}}/g}{\rho_{\text{woman}}} = \frac{160 \text{ lb}/(32 \text{ ft/s}^2)}{1.82 \text{ slug/ft}^3}$$

$$= 2.75 \text{ ft}^3.$$

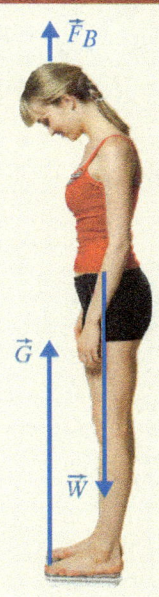
(© Stockbyte/Getty Images RF)

The buoyant force is then

$$|\vec{F}_B| = \rho_{air} g V = (0.0024 \text{ slug/ft}^3)(32 \text{ ft/s}^2)(2.75 \text{ ft}^3)$$
$$= 0.21 \text{ lb} = 3.4 \text{ oz.}$$

A precise scale would therefore read the GRF to be

$$|\vec{G}| = |\vec{W}| - |\vec{F}_B| = 160 \text{ lb} - 0.21 \text{ lb} = 159.79 \text{ lb.}$$

The effect of performing weigh-ins in an air-free room is the same as if our athlete had eaten a 3.4-oz hamburger just before weighing in.

5.3 Moving through Fluid Drag

Considering only his weight and the buoyant force, the net upward force on a typical 180-lb person submerged in water is about 8 lb, the weight of a gallon of milk. Imagine a diver plunging into the water from the 10-m platform and moving at 31 mph straight down, head first. That 8 lb of upward force isn't going to do much for him! He would smash into the concrete bottom of the diving pit 5 m down (the minimum recommended depth) still moving at over 30 mph![2]

Fortunately another, much stronger force comes into play. Whereas there is a buoyant force whenever you are *in* a fluid, a **drag force** is created whenever you are *moving* through a fluid. It is a slowing force, opposing motion as friction does.

5.3.1 A cartoon image of drag

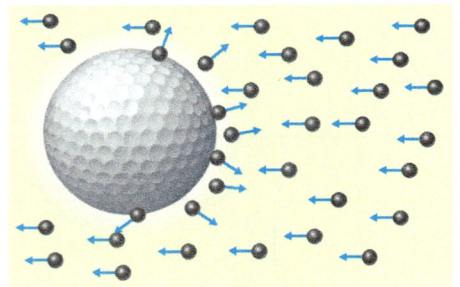

The details of drag or any fluid-dynamic effect are quite complicated, but physicists usually try to understand at least the basics of any phenomenon in the simplest possible terms before moving on to complications.

Drag is caused by many (every second, at least 10^{24}, or a trillion trillion) collisions with air or water molecules as an object crashes forward through the fluid. It's sometimes easier to consider the molecules rushing toward the ball at some velocity $\vec{v}$, rather than the ball rushing toward the molecules—it amounts to the same thing. We can imagine tiny Ping-Pong balls constantly colliding with the much larger ball, as suggested in the figure. Clearly, these collisions tend to slow the motion of the ball through the fluid.

The force on the ball should be essentially the number of collisions times the force each collision produces.

The collision rate will be larger if

- There are more atoms to run into, that is, #collisions $\propto \rho$.

- The cross-sectional area of the ball (the area that "faces forward" as the ball moves) is larger, that is, #collisions $\propto A$.

- More atoms smash into the ball per second, that is, #collisions $\propto v$.

---

[2] Can you prove this assertion?

So,
$$\#\text{collisions} \propto A\rho v.$$

Finally, the force on the ball *per* collision depends on the speed at which the molecules impinge on the ball.

This leads to a reasonable expectation for the drag force formula:

$$|\vec{F}_D| \propto (A\rho v) \cdot v \rightarrow \boxed{|\vec{F}_D| = \frac{C_D}{2} A\rho v^2}. \tag{5.4}$$

Here, C_D is the **drag coefficient** or drag constant. Like the friction coefficients, this constant has to be measured experimentally. Table B.6 in appendix B lists various drag coefficients.

This cartoon picture ignores the complex and frequent interactions among the molecules themselves—the interactions that make air or water a fluid. It certainly can't explain why the dimples on a golf ball reduce drag, or why the drag force is different if the "back half" of the ball is sliced off. Nevertheless, equation 5.4 is a reasonable approximation of the drag force, with details of the ball shape and surface included in the drag constant.

5.3.2 Important complication 1: speed dependence of C_D

For a smooth sphere moving relatively slowly through the air, the drag coefficient is typically $C_D = 0.5$. However, there is much more to the story. When the ball moves slowly, the flow is "laminar" and the flow along the surface of the ball is smooth, only becoming turbulent behind, where a partial vacuum is formed by the ball forcing air from its path. This low-pressure wake is the main component of the drag force. In a real sense, the ball is "dragging along" the air behind it. The wake is generated when the "surface layer" of the airstream on the ball separates from the surface of the ball.

The drag coefficient can depend strongly on ball speed.

The size of this wake is characterized by the drag coefficient. Figure 5.3 shows the wake of a ball moving through air, as measured in a wind tunnel. The left panel shows a soccer ball moving at moderate speed through the air. The surface layer separates from the ball at about its "equator," leaving a large wake characterized by a large drag coefficient C_D.

On the right, the same ball is shown but moving more quickly through the air, and here something interesting is happening. At these higher airspeeds, the surface layer *itself* becomes turbulent, which actually allows airstreams to flow more easily near the ball and the surface layer to separate farther back along its surface. This means that *the drag coefficient of a ball at high speed is smaller than at low speed*, because the size of the wake is smaller. As can be seen in figure 5.4, C_D can drop quickly from 0.5 to lower than 0.2, as the speed of the ball is increased.

It is important to point out that even though the drag *coefficient* is lower at high speeds, the *magnitude of the drag force* is still higher. (It would be strange if this were not so!) This is so because of the factor of v^2 in equation 5.4.

Low speed: Wide wake

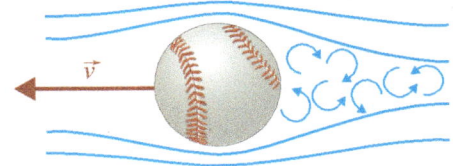
High speed: Narrow wake

Figure 5.3
Top: a ball at moderate speed leaves a large wake as seen in a wind tunnel. **Bottom:** the wake is smaller at high speeds.

Question

Which experiences a stronger drag force, a baseball traveling at 30 mph or a table tennis ball traveling at 30 mph? Why?

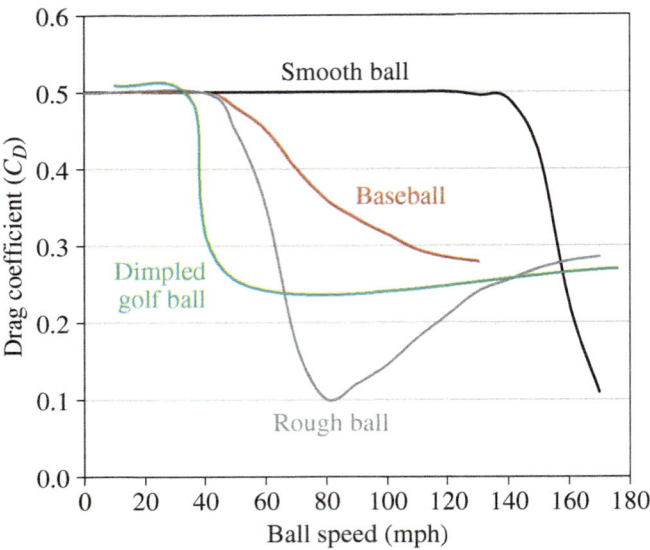

Figure 5.4. The drag coefficient for some typical balls moving through the air, as a function of speed.

5.3.3 Important complication 2: C_D and surface roughness

The discussion above applied to smooth surfaces, where turbulence is generated simply by the nature of fluids at high speeds. When we look at a smaller scale, even at relatively low speeds, small irregularities in the surface of a ball generate their *own* turbulent wakes. This means that each tiny irregularity generates some drag. However, these wakes can combine to make the entire surface layer of the ball turbulent. As we said, a turbulent surface layer leads to a delayed separation of the surface layer, a smaller wake, and a smaller drag coefficient.

The interplay between these effects is complicated and delicate. For example, the fuzz on a tennis ball trips turbulent behavior in the surface layer of the ball, but cumulatively the hairs generate more drag than they eliminate, leading to the tennis ball having *more* drag than a smooth ball. In fact, the drag coefficient for a new tennis ball (with plenty of fuzz) has been measured at $C_D \approx 0.65$, whereas a used one has $C_D \approx 0.5$.

On the other hand, the seams on a baseball trip turbulence at a much lower ball speed than if the ball were entirely smooth. This effect more than compensates for the small extra drag on the seams themselves, and C_D quickly falls below the value for a smooth sphere at speeds as low as 50 mph, as indicated in figure 5.4.

Probably the sport for which surface effects of drag are most intensively studied is golf (see figure 5.5). There are two organizations governing the rules for golf, the R&A in St. Andrews, Scotland, and the United States Golf Association (USGA) domestically. They jointly publish lists of officially acceptable golf balls at www.usga.org. This is serious business—the 45-page list is updated on the first Wednesday of every month!

Prior to 2008, the specifications for a ball involved numerical ranges for the ball's diameter, weight, and color and a general requirement that the ball be spherical. Beginning in 2008, the rules became purposely more descriptive in nature, to ensure that developing new technologies not circumvent the "intent" of the original rules.

The drag coefficient depends on the surface texture in complex ways that are hard to calculate. Surprisingly, smooth balls tend to have the highest drag.

They include the circular-logic rule that "the material and construction of the ball must not be contrary to the purpose and intent of the Rules." Clearly, the governing bodies want flexibility to deal with the huge amount of research going into equipment design, and the balls allowed in competition play have considerably greater drag than balls that can be manufactured.

The effect of any given dimple pattern is extremely complicated and is the focus of serious lab work and intense computer simulations, sometimes involving months of calculations on hundreds of specialized processors in a computer farm. Check out this link for a cool video resulting from a seriously intense calculation: http://hdl.handle.net/1813/11586.

5.3.4 And more complications...

The formula for drag (equation 5.4) is complicated enough. Then it gets more complicated because the drag "constant" is not constant at all—it depends on speed and surface features, as we've just discussed. There are several more complications.

We won't go much further into it, but I thought it would be interesting to show you figure 5.6, based on data from a research journal article on wind tunnel measurements of the drag on a football that faces nose first into the air. The first point I want to make is this: this stuff is serious science and not at all trivial.

Second, the drag force is 10% lower for a football spiraling at 600 rpm (10 rps) than for a nonspiraling football. There is no particular reason that this should be true, but it is. This is yet another reminder that aerodynamics is complicated, and the drag coefficient—like friction coefficients and coefficients of restitution (chapter 6)—need to be measured. They encode complicated microscopic and material physics.

Finally, the elongated shape of a football provides much lower drag than a spherical ball presenting the same area into the wind. When we look at a football, we get the intuitive feeling that it is aerodynamic. Well, it is, and the proof is in the much lower values of C_D seen in figure 5.6, compared to figure 5.4. A well-thrown football slips through the wind beautifully.

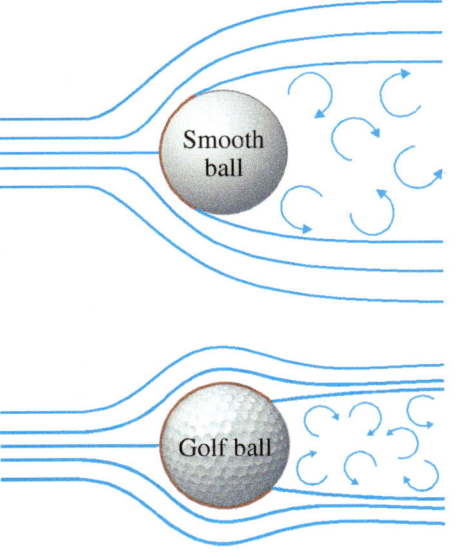

Figure 5.5

Irregularities on the surface of a golf ball trip turbulent behavior in the surface layer, leading to a smaller wake and lower drag force than a smooth ball traveling at the same speed.

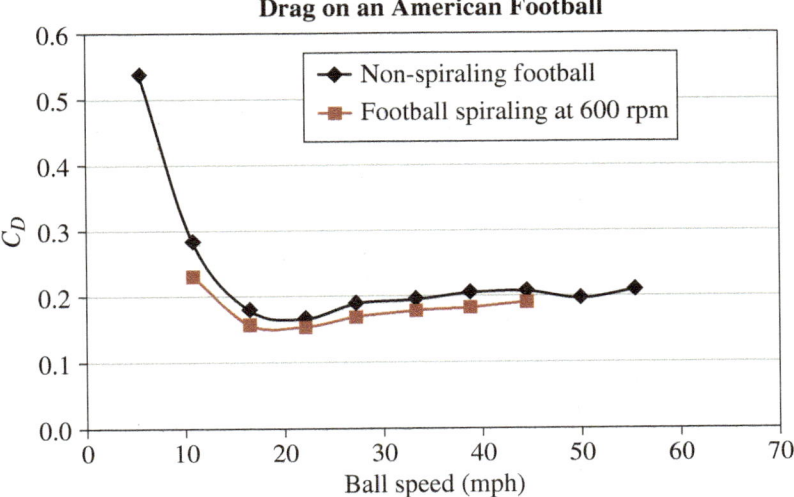

Figure 5.6. Wind tunnel measurements of the drag coefficient for an American football flying nose first into the air. Based on data from R. G. Watts and G. Moore, *American Journal of Physics* 71: 791 (2003).

On the other hand, the drag force on a very rapidly spinning (about 600 rpm) golfball is actually somewhat *higher* at high speeds. Once again, there is no reason to expect this to be true; effects like this simply have to be measured.

5.3.5 Terminal velocity

Is it true that a penny dropped from the top of the Empire State Building would kill a pedestrian or crash through the concrete sidewalk?

Let's consider the tallest building in the world, the Burj Khalifa in United Arab Emirates (UAE), which is an astounding 2722 ft high—more than half a mile! If we threw a penny (in the UAE it would be a dirham coin) from this height, we could use equation 2.13 to find the speed at which it would hit the ground if only gravity were acting:

$$v_{\text{land}} = \sqrt{2gh} = \sqrt{2\left(32\text{ ft/s}^2\right)(2722\text{ ft})} = 417\text{ ft/s} \approx 285\text{ mph}.$$

That's pretty fast. Heck, never mind a penny. How about a baseball, which is 60 times more massive? Now that's *gotta* hurt!

On the other hand, gravity is certainly *not* the only force at work. As the object accelerates downward, its velocity increases and, according to equation 5.4, so does the drag force $\vec{F}_D$ which points upward.

As the ball picks up speed, the weight does not change, but the drag does, and the *net* force decreases. At some point, the ball is moving so fast that the drag force has the same magnitude as the ball's weight. When this happens, the net force is zero, so the acceleration is zero and the velocity stops changing—the ball has reached **terminal velocity**.

An object can move faster than its terminal velocity: v_T is not a speed limit.

The formula for the terminal velocity is found by setting the magnitudes of the drag force and the weight equal:

$$|\vec{F}_D| = |\vec{W}| \rightarrow \frac{C_D}{2}A\rho_{\text{air}}v_T^2 = mg \rightarrow \boxed{v_T = \sqrt{\frac{2mg}{C_D A \rho_{\text{air}}}}} \quad (5.5)$$

The terminal velocity is the velocity at which the drag force has the same magnitude as the weight of an object, even if the object is not in free fall. Most "fast" games such as cricket, baseball, tennis, and hockey are played at or near terminal velocity.

EXAMPLE 5.3 *Baseball dropped from a plane*

So, how fast *would* a baseball hit the ground, dropped from the Burj Khalifa? According to equation 5.5, its terminal velocity is

$$v_T = \sqrt{\frac{2\,(0.15\text{ kg})\,(9.8\text{ m/s}^2)}{0.3\,(\pi\,(0.0374\text{ m})^2)\,(1.2\text{ kg/m}^3)}} = 43\text{ m/s} \approx 96\text{ mph}.$$

The baseball will not hit the ground any faster than this even if dropped from a 747. Now, a baseball traveling 96 mph is nothing to sneeze at—that's good heat even for an MLB pitcher. But it's hardly going to put a hole in concrete.

And pennies? They flutter between falling face down to edge down, but on average they'll hit the ground at about 20–30 mph. That won't put a hole in your skull.

EXAMPLE 5.4 Cooling Verlander's heat

Among the pitches in Detroit Tiger Justin Verlander's repertoire is a fastball that sometimes registers 100 mph on the radar gun. This is some serious heat! But equally important to the batter is the ball's speed as it crosses the plate. How much does air resistance slow down a 100-mph fastball?

We know the distance the ball travels and its initial speed. If the acceleration is constant, then equation 2.6 gives us our answer.

The problem is that the acceleration is *not* constant, because the drag force is not constant—it gets a little weaker as the ball slows on its path. What we'll do is *approximate* this as a constant-force situation and see how well it works.

Figure 5.4 tells us that the drag coefficient on a 100-mph (146.7 ft/s) baseball is about 0.3, and the ball's acceleration is

$$a_x = \frac{F_{D,x}}{m} = -\frac{C_D A \rho v^2}{2m} = -\frac{0.3 \pi (0.125 \text{ ft})^2 (0.0024 \text{ slug/ft}^3)(146.7 \text{ ft/s})^2}{2 \cdot 0.01 \text{ slug}}$$

$$= -38 \text{ ft/s}^2.$$

Now it's easy to find the ball's speed after it has traveled the 56 ft from Verlander's hand to the plate in front of the dazed batter.

$$v_x^2 = v_{x,0}^2 + 2a_x \Delta x \rightarrow v_x = \sqrt{(146.7 \text{ ft/s})^2 + 2(-38 \text{ ft/s}^2)(56 \text{ ft})}$$

$$= 131.4 \text{ ft/s} = 89.6 \text{ mph}.$$

That's a decent approximation, but the drag force we used was for a ball traveling at 100 mph. The average speed of the ball is more like 95 mph; if we use this speed (but keep $C_D = 0.3$), the acceleration is $a_x = -34.3$ ft/s^2 and the final speed is $v_x = 133$ ft/s = 90.6 mph. Actually, since the drag coefficient is slightly higher at 95 mph than at 100 mph, the difference between the two estimates is even smaller.

In fact, doing the exact calculation on a computer yields a speed of 90.4 mph. The point is that the constant-acceleration approximation very often provides an excellent estimate of the right answer, even when air resistance is stronger than the pull of gravity.

> **Question**
>
> Example 5.4 discusses Justin Verlander's 100-mph fastball. But this is more than the a baseball's terminal velocity of 96 mph, according to example 5.3. Is it really possible to throw a ball *faster* than its terminal velocity?

5.4 Sideward Forces from Asymmetries

At sea level, atmospheric pressure is 14.7 pounds per square inch (lb/in^2 or psi). This means that, even as you read this, each square inch of your skin is being pushed inward by a force of 14.7 lb. So don't blame this book for your headache!

A baseball has about 6.8 in.^{2} pointing to one side, so as it flies from the mound toward home, atmospheric pressure exerts on the ball a 100-lb force[3] to the left and

---

[3] $(14.7 \text{ lb/in.}^2) \times (6.8 \text{ in.}^2) = 100$ lb.

an equally strong force to the right. If a pitcher can upset this balance by adding only a single ounce of force to one side, he'll be able to move the ball by more than a foot, on its journey to the plate! That's a serious slider in his arsenal of weapons against the batter.

In this section, we'll discuss the physics behind cricket swings, baseball curves, tennis serve drops, and soccer bends.

Because of the symmetry of a sphere, the drag force on a moving ball points directly opposite to the ball's velocity vector. But this symmetry can be broken by causing the drag force on one side of the ball to be more than that on the other side. There are two sources we'll discuss: asymmetries in the surface roughness and asymmetries in the surface speeds.

5.4.1 The swing of a cricket ball

A cricket ball has several seams that run along the equator of the sphere and is otherwise smooth when new. Bowlers hurl the ball at up to 90 mph at a distance of 22 yd from the striker, almost exactly the distance from mound to plate in baseball.

Often the bowler will hurl the ball in such a way that on one side of the ball, the seam is more toward the front than on the other. The result is shown in figure 5.7. There the ball is moving to the left. The air passing along the top of the ball experiences a smooth surface, so the flow is laminar and separates from the ball about midway between front and back. The air passing along the bottom of the ball, however, runs through irregularities (the seam) early on which "trips" turbulence; the surface layer on that side is turbulent. As we know, this results in a delayed separation of the surface layer; it separates maybe three-fourths of the way to the back of the ball.

The result is an asymmetry in the wake. The air is thrust upward, so by Newton's third law, the ball experiences a force to one side which, of course, causes the ball to accelerate that side. (In the figure, the direction of the force is down, but in a game, this view would be from above and the force would be left or right.) The result is a curve in the ball's path. In cricket, this is known as a ball's "swing."

Over the course of the game, the hard leather surface of the ball will become increasingly scuffed. The bowler will recognize that one side is becoming slightly more scuffed than the other and henceforth will work to preserve and amplify the asymmetry by repeatedly "shining" the smooth side, rubbing it on his shirt, and even spit-shining it. According to the rules of cricket, this is perfectly legal so long as it "wastes no time" in the game. The use of petroleum jelly or other foreign substance is illegal, but there are recent claims that certain mints or candies chewed by the bowler produce a saliva particularly suited to polish a ball. Yuck! Maybe that's why the wicketkeeper (the equivalent of the baseball catcher) is the only member of the fielding team allowed to wear gloves.

A ball is typically hurled hundreds of times before it can be replaced in a test, and toward the end of its use, the asymmetry due to scuffing becomes so dominant that a good swing bowler can choose to swing the ball toward the seams ("conventional swing") or away from them ("reverse swing"), by proper orientation of the ball.

Detailed scientific measurements have been made of the sideward force on a cricket ball, and they find that a force as large as 2 oz (that is, 35% of the ball's weight) can be achieved if the seam is oriented at about 20°, about what is shown in figure 5.8.

Cricket teams often have designated "ball polishers" who chew carefully selected mints.

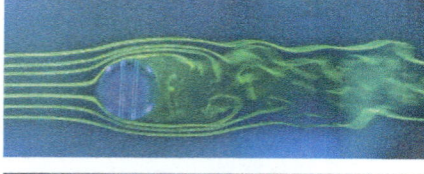
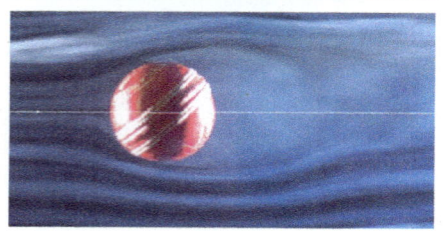

Figure 5.7
Smoke shows the structure of the turbulent wake as air streams to the right past a stationary cricket ball. In the bottom panel, the tilt of the ball's seams with respect to the wind results in an upward deflection of the wake. **Top:** (*Image courtesy of NASA Ames Research Center and Cislunar Aerospace Inc.*); **Bottom:** (*Photo courtesy of Mike Lisa*).

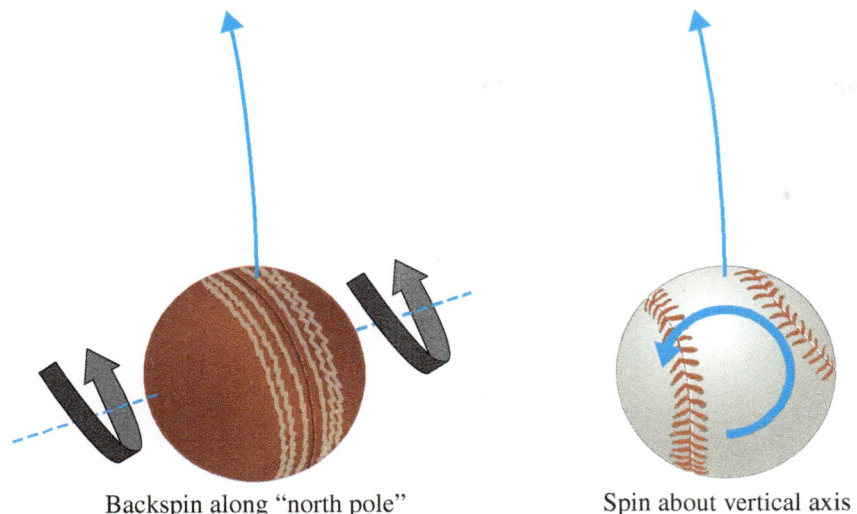

Backspin along "north pole"   Spin about vertical axis

Figure 5.8. The swing of a cricket ball and the curve of a baseball both involve a spinning ball, but the spins serve quite different purposes. In the figure, the balls are seen from above and curve to the left.

To make sure that the ball's orientation doesn't change during its flight, the swing bowlers put a strong backspin about the ball's "north pole" (if the seams are the equator). While pitchers also put a spin on a baseball to make it curve, you should understand that *the purposes of the spins are different in cricket and baseball.* In cricket, spin is used to keep the seams pointed at a constant angle; the asymmetry in the seams then gives the swing. In baseball, tennis, or soccer, it is the spin *itself* that causes the ball to curve. We talk about this next.

5.4.2 Bending a ball's flight

We've seen that if the drag force is stronger on one side of the ball, then there is a sideward push toward the other side of the ball. If we look at equation 5.4, it appears that there are two ways to change the drag on a ball: by changing C_D or changing v. First, we can make the drag coefficient C_D different on the two sides, as discussed in section 5.4.1, by making one side rougher than the other due to seam orientation, dirt, or spit-shining. Second, we can make v, the speed at which air passes over the surface, different on one side versus the other. That is, the second way to spin it.

Figure 5.9 indicates the effect. When there is no spin, the only force is backward, opposite the ball's motion to the right. When the ball spins counterclockwise, however, the bottom of the ball is approaching the air more quickly, causing the surface air layer to separate earlier from the bottom of the ball than the top, and the wake is "forced down." The result is an asymmetry in the interaction with air and a force on the *ball* in the upward direction, perpendicular to the ball's velocity. It is important to point out that there is still a backward-pointing drag force. But what's new is the "sideward" force, called the lift or **Magnus force**.[4]

---

[4] Technically, the Magnus force is only one component to the sideways lift force on a spinning ball. However, it is common to lump all lift forces together and refer to them as Magnus forces.

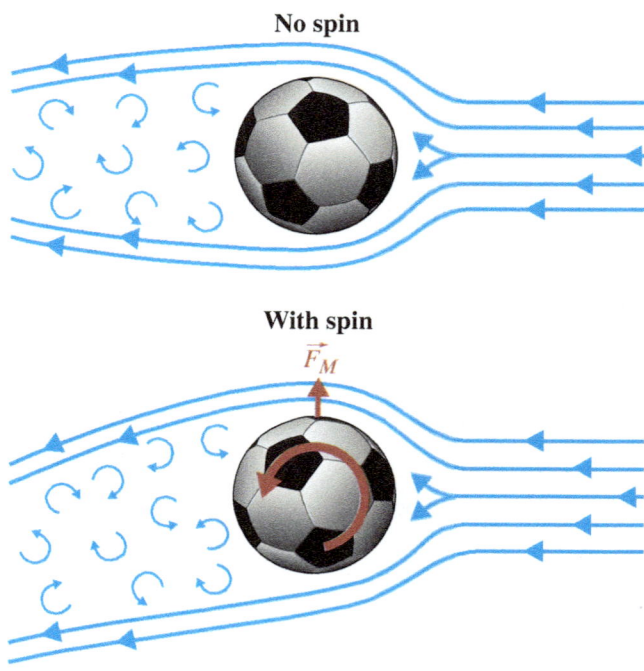

Figure 5.9. The force exerted by air on a nonspinning ball passing through it is directed opposite the ball's velocity vector. If the ball is spinning, the bottom surface of the ball approaches the air at a larger speed than does the top half, and there is an asymmetry in the forces. While there is still a backward drag force in this case, there is also a sideways-pointing Magnus force.

The magnitude of the Magnus force is given by a formula similar to the drag formula:

$$|\vec{F}_M| = \pi C_M \rho_{air} R A v f \quad (5.6)$$

$$|\vec{F}_M| = \pi^2 C_M \rho_{air} R^3 v f \quad \text{for a spherical ball}$$

Again it depends on the density of air, the radius R, and area A that faces the wind as well as the ball's speed v. In this equation f is the **rotational frequency**, measured in rotations per second (rps), and C_M is the **Magnus coefficient** which, like the drag coefficient and coefficients of friction, must be measured. Unless we say otherwise, we'll use $C_M = 1$.

The Magnus force points in the same direction that the front of the ball is spinning.

The Magnus force formula has several terms, but it's easy enough to understand and use. The *direction* of the Magnus force can be a little confusing, so let's develop a rule while looking at figure 5.9. In that figure, the ball is moving to the right, so the right-hand side of the ball is its "front" face. The spin on the ball makes the front of the ball move upward in the figure, and the Magnus force points upward. This is our rule of thumb: The Magnus force points perpendicular to the velocity of the ball, in the direction that the front of the ball is spinning.

We can practice this rule of thumb by looking at figure 5.10, which shows three balls, all moving to the left. The golf ball in the left panel spins clockwise. Looking at the deflected wake, we know that the air has been pushed down. Newton's third law then tells us that the sideward force on the ball is upward—the Magnus force can be seen as a "reaction force" to the force exerted on the air. A clockwise rotation makes the front face of the ball—the left face in this case—move upward, in agreement with our rule of thumb.

Please avoid this misconception!

Aerodynamic forces on flying objects are complex. In this book, we are discussing the most important effects in a somewhat simplified (that doesn't mean "wrong"!) way. Most of the many details are simply not important to sports, so our discussion is appropriate.

Because of the complexity, there is often huge confusion about what is going on. Especially online (sports forums, physics homework forums, YouTube comment threads), misconceptions arise and are endlessly echoed by "answers" sites that simply collect information posted elsewhere, with little quality control beyond spell-checking.

A rampant fallacy is that a smooth ball cannot be curved with a spin, that the seams on a soccer ball or the stitches on a baseball "grab" the air and cause the curved trajectory. You know better. The left-right force imbalance arises because the speed of the ball's surface relative to the wind is different on the left and right sides of the ball because of the spin. Just as you don't need a rough surface to have air drag, you don't need a rough surface to have an *imbalance* of air drag—the Magnus force.

After all, if you've ever played table tennis against an experienced opponent, you've seen a very smooth ball curve wickedly due to spin.

The reason that lacrosse balls don't curve as baseballs do isn't that their surface is smooth. It's because the ball doesn't leave the head of the stick with enough spin.

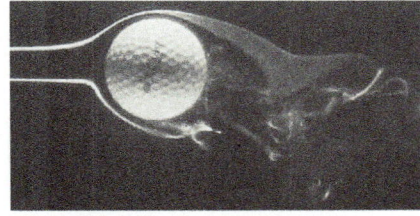
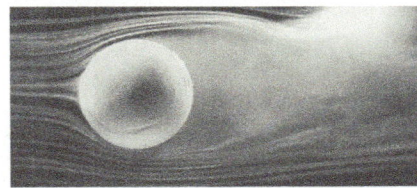

Figure 5.10

Three balls move to the left through air. From top to bottom: a golf ball spins clockwise; a tennis ball spins counterclockwise; a baseball spins clockwise. (Mehta, R.D. and Pallis, J.M., "Sports ball aerodynamics: effects of velocity, spin and surface roughness," in Froes, F.H. and Haake, S.J. *Materials and Science in Sports*, Warrendale, PA, The Minerals, Metals and Materials Society [TMS] 185–197. Copyright © 2001 by The Minerals, Metals & Materials Society. Reprinted with permission).

Question

Way back in section 1.1 we discussed the importance of keeping track of units, and in equations 3.4 we gave the relationships among the units of force, mass, distance, and time.

Look at equation 5.4, which gives the magnitude of the drag force. What are the units of the drag coefficient? That is, is C_D measured in kg, N, $\frac{m}{s}$, or some other unit?

Now look at equation 5.6 and answer the same question: what are the units of C_M?

While we are at it, what are the units of the static and kinetic coefficients of friction, μ_S and μ_K, in respectively, equations 3.9 and 3.10?

 Smooth balls curve as well as or better than textured balls.

5.5 Aerodynamic Forces, One at a Time

A ball flying through the air is subject to many forces simultaneously. In nature, it is usually impossible to switch any of them "off," but it can be helpful to consider the effects of the forces individually. That's what we do in the following sections.

5.5.1 Curveballs and subatomic physics

As scientists attempt to understand and explain sports and the wider world beyond, they are faced with a hugely diverse array of phenomena. They get very excited when they find a deep similarity behind two situations that on the surface look very different. As practitioners of the most mathematical of sciences, physicists love to identify very unrelated phenomena that share a similar mathematical description.

As a subatomic physicist, I'd like to take a moment to show you how to find one such similarity between curveballs and physics used in my area of research. Don't worry: it will become relevant in a moment.

The Magnus force always points sideways, at right angles, to the velocity of a ball in flight. In the situations we've analyzed so far, the force changed the direction of the ball only a little bit, so we could approximate it as having a constant direction. This is fine for most sports situations, since the ball is in the air for less than a second before it hits the ground, or a baseball glove or whatever.

However, just for fun (this is how physicists have fun), let's see what happens to a slider pitch if we "turn off" gravity so it doesn't hit the ground and also remove all the trees and cars and things that it might crash into; let it fly. In this case, the direction of the ball's velocity would really change, and we'd have to account for the fact that $\vec{F}_M$ is perpendicular to $\vec{v}$ in order to describe its motion.

But hey, this is no big deal; we've seen this before. In section 3.5.3, we said that when the force is perpendicular to the velocity, we're talking about a centripetal force, so the baseball will follow a circular path. A pitcher could throw the curve ball, then turn around and catch the ball he'd just thrown!

We can even find the radius of the circular path, which we call r:

$$\left. \begin{array}{l} \text{equation 3.17:} \quad |\vec{F}_M| = m\frac{v^2}{r} \\ \text{equation 5.6:} \quad |\vec{F}_M| = \pi C_M \rho_{air} RAvf \end{array} \right\} \rightarrow r = \left(\frac{m}{\pi C_M \rho_{air} ARf} \right) \cdot v \quad (5.7)$$

For a typical pitch, the diameter of the circle is over 1000 ft, as shown in figure 5.11. (Problem 12 asks you to find the radius.) That figure makes it obvious why we could pretend that a flight path that is actually circular could be approximated by an almost straight path over the short distance between the pitcher's mound and home plate.

I find this interesting because I know of only one other force in nature that is (1) perpendicular to an object's velocity and (2) directly proportional to its speed. This is the force of a magnetic field on an electrically charged particle. The particle also follows a circular path, and the math is just the same as for the curveball:

$$\left. \begin{array}{l} \text{equation 3.17:} \quad |\vec{F}_B| = m\frac{v^2}{r} \\ \text{electromagnetism:} \quad |\vec{F}_B| = eBv \end{array} \right\} \rightarrow r = \left(\frac{m}{eB} \right) \cdot v.$$

We won't go into the electromagnetic physics (well, e is the electron's charge and B is the magnetic field strength, if you must know). But what is striking is that both the baseball and the electron follow a circular path whose radius is equal to

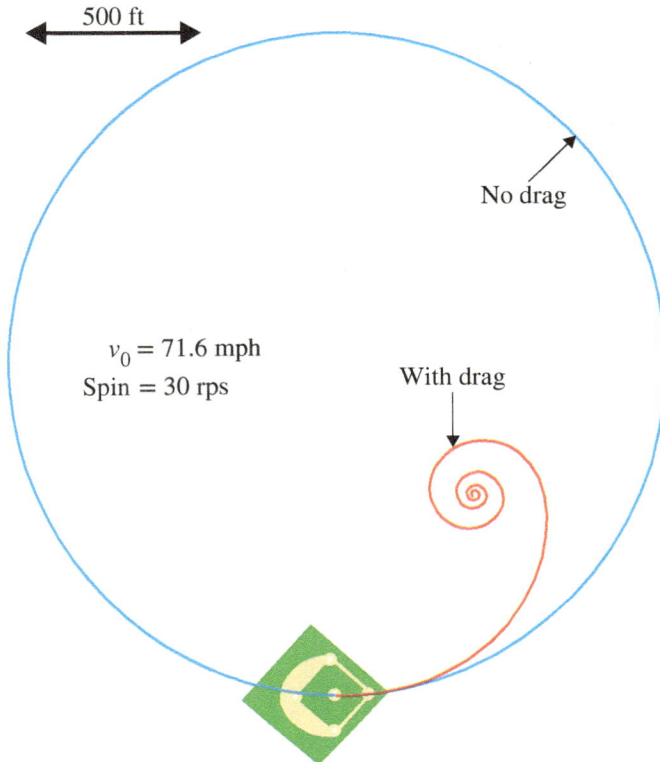

Figure 5.11. The path of a curveball if there were no gravity or drag force is shown by the large circle. Accounting for drag, the path becomes a spiral. A baseball diamond is included for scale.

the velocity times some constant number (given in the parentheses of both equations above).

In fact, physicists use this information to determine how fast a subatomic particle is moving when they measure it. Figure 5.12 shows actual tracks of charged particles as measured on photographic film; you may have seen such pictures before. If a particle makes a big circle, it is moving fast, usually at close to the speed of light. A small circle indicates a slower particle.

A spiraling path

What you can see right away, however, is that the subatomic particles *don't* follow a circular path. What's happening is that, as they pass through the photographic film, they are slowed down through collisions with atoms in the film. (It sounds like air drag, right?) This reduces $|\vec{v}|$ and so reduces the radius of their path r. As they continue to slow, the radius of their path becomes smaller and smaller and the particle traces out a spiral.

In fact, the same thing occurs in baseball! In our discussion in this section, we "turned off" gravity and the drag force. If we get a little more realistic and account for drag, figure 5.11 shows that its slowing effect changes the circular path into a spiraling one, just as for the subatomic particles. The spiraling effect of drag would be dramatic over the long haul, but is not so huge over the smaller part of the pitch that we see between the mound and the plate.

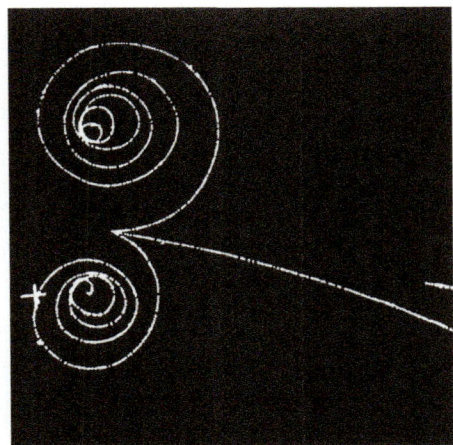

Figure 5.12
A bubble chamber detects three subatomic particles. Two of the particles are relatively slow and curl into spirals of ever-decreasing radius. The third is faster, and the radius of its circle is so much larger that it almost appears to travel in a straight line. (© 2010 The Regents of the University of California, Lawrence Berkeley National Laboratory. Photo courtesy of Berkeley Lab).

Never let anybody tell you that the physics of sports doesn't relate to cutting-edge science!

5.5.2 Do we need a computer?

A computer was used to calculate the spiraling trajectory of the curveball in figure 5.11. However, it looks from the figure as if the simpler circle trajectory is a good description of the pitch for the part we really care about, from mound to plate.

As it turns out, an even simpler approach works almost as well. So far in this book, we've learned how to apply constant-acceleration (that is, constant-force) ideas to mathematically understand the behavior of players and balls. Is a constant-force approach valid here? Well, strictly speaking, obviously not! A constant force toward first base (upward, in the figures) would cause the ball to accelerate faster and faster toward first base, never turning around. This obviously does not describe reality at all.

However, over the relatively short distance between the mound and plate, the ball's path doesn't deviate *too* much from a straight line, so the Magnus force points more or less to the left the entire time. So the constant-force approximation should give an approximately correct prediction. Figure 5.13 shows the result: the constant-force approximation, while completely wrong on a large scale, does a good job on the smaller scale more relevant to baseball.

Please remember this:

This is a very important lesson I hope you take away from these discussions, and it goes well beyond sports. In the scientific approach to real-world problems, we are often unable to calculate the behavior of a complicated system in an absolutely, strictly correct way. Scientists know this at the outset, but they can usually find approximate approaches that work well for a limited range of conditions. They can then build on this solid progress to refine their approach even further, and so on. The fact that a situation or problem is too complicated to deal with 100% correctly does not mean our current tools are impotent.

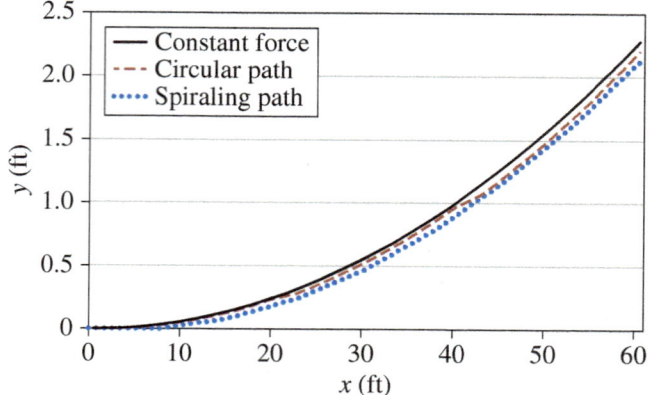

Figure 5.13. The same pitch as drawn in figure 5.11, shown only for the distance between the mound and plate. A constant-force treatment, though not strictly correct, works very well to predict the pitch's deflection over this distance.

Section 5.5 Aerodynamic Forces, One at a Time 123

EXAMPLE 5.5 *Deflection of a curveball*

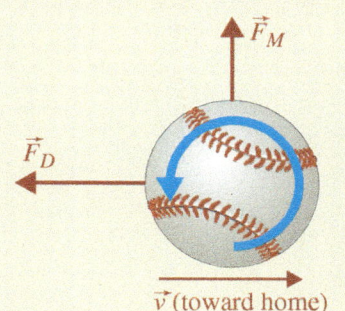

Let's see how to use our constant-force approach to do an approximate analysis of the curveball shown in figure 5.11. The strategy will be to figure out the time it takes to reach the plate, then see how much the ball deflects in that time. The pitch in our example was thrown at 71.6 mph (105 ft/s). Since we know it slows by about 10% on its trip, we'll use a somewhat lower number, say 67 mph (98.3 ft/s), when we calculate forces in our equations. (If we used the *maximum* value of 71.6 mph, we'd overestimate the forces.)

The drag force produces an acceleration in the x-direction:

$$a_x = \frac{\sum F_x}{m} = -\frac{F_D}{m} = -\frac{C_D A \rho_{air} v^2/2}{m}$$

$$= -\frac{0.3\,(0.0022\text{ slug/ft}^3)\,\pi\,(0.123\text{ ft})^2\,(98.3\text{ ft/s})^2/2}{1.03 \times 10^{-2}\text{ slug}} = -14.7\text{ ft/s}^2.$$

We can use equation 2.6 to find the ball's forward velocity as it crosses the plate:

$$v_x^2 = v_{x,0}^2 + 2a_x \Delta x = (105\text{ ft/s})^2 + 2(-14.7\text{ ft/s}^2)(56\text{ ft}) = 9378\text{ ft}^2/\text{s}^2$$

$$\rightarrow v_x = \sqrt{9378\text{ ft}^2/\text{s}^2} = 96.8\text{ ft/s}.$$

(I used $\Delta x = 56$ ft instead of the mound-to-plate distance of 60.5 ft, to account for the stretch of the pitcher off the rubber.)

Then we turn to equation 2.5 to find the ball's travel time:

$$v_x = v_{x,0} + a_x t \rightarrow t = \frac{v_x - v_{x,0}}{a_x} = \frac{96.8\text{ ft/s} - 105\text{ ft/s}}{-14.7\text{ ft/s}^2} = 0.558\text{ s}.$$

Finally, we'll turn to the sideways (y) direction. The Magnus force produces the acceleration:

$$a_y = \frac{\sum F_y}{m} = \frac{F_M}{m} = \frac{\pi^2 C_M \rho_{air} R^3 v f}{m}$$

$$= \frac{\pi^2 \cdot 1\,(0.0024\text{ slug/ft}^3)\,(0.123\text{ ft})^3\,(98.3\text{ ft/s})\,(30\text{ rps})}{1.03 \times 10^{-2}} = 12.62\text{ ft/s}^2.$$

Using this and the travel time in equation 2.10 gives

$$\Delta y = v_{0,y} t + \tfrac{1}{2} a_y t^2 = 0 + \tfrac{1}{2}(12.62\text{ ft/s}^2)(0.558\text{ s})^2 = 1.96\text{ ft}.$$

This is shown in figure 5.13.

5.5.3 A simple formula for a curveball

The math in example 5.5 would have been a lot easier if the ball didn't slow down in flight. A baseball slows down by only about 10% due to drag, so we can get a simple approximate formula for the sideward deflection of a curveball by just saying the ball moves at 95% of its release speed.

As in example 5.5, we'll use the constant-force approximation.

$$\Delta y = \tfrac{1}{2} a\, \Delta t^2 = \tfrac{1}{2}\frac{F_M}{m}\left(\frac{\Delta x}{0.95v}\right)^2 = \tfrac{1}{2}\frac{\pi^2 C_M \rho_{\text{air}} R^3 (0.95v) f}{m}\left(\frac{\Delta x}{0.95v}\right)^2$$

Putting in all the numbers for the mass of a ball, and so on, we get a super simple formula:

$$\boxed{\text{baseball curveball deflection} = \left(7\text{ ft}^2\right)\frac{f}{v}} \quad (5.8)$$

Of course, the number in parentheses will be different for different balls; you'd have to plug in the appropriate numbers for mass, radius, and so on. Still, it's kind of cool that it's as simple as that: the deflection is a number (that depends on the ball) times the spin divided by the speed.

Any pitcher will tell you that to get good lateral motion, you don't want to throw the ball at top speed, even though equation 5.6 says that the sideways force is *larger* at top speed. The reason is that, while the sideways force (thus, the sideways acceleration) is larger for large v, the ball has less *time to deflect* when the speed is high, and this is the larger effect.

Equation 5.6 tells us that if you throw twice as hard, you'll double the Magnus force, but equation 5.8 says you'll get one-half the deflection.

5.5.4 Roberto Carlos's "impossible" free kick

In a 1997 soccer match between Brazil and France, Roberto Carlos scored a free kick from 35 m out. He was one-third of the length of the field away from the goal, with the four-person "wall" directly in front of him. The video quality on YouTube clips of this shot are not phenomenal, but are sufficient to see that the shot was! Despite the long trajectory of the shot, French goalie Fabian Barthez was left standing still, as the ball curved violently behind him. A ball boy *30 ft away from the goal* ducked, as he thought the ball was heading his way at 80 mph. The shot, described at the time as "impossible" or "the goal that defied physics," inspired serious analysis and study by physicists.

In the previous sections, we showed that a curveball in baseball would follow a spiral trajectory, if it were allowed to fly much longer than the mound-to-plate distance of 60.5 ft (18.4 m). For a much shorter distance, though, it was just as good (and much easier) to use a constant-force approximation to calculate the curve.

However, it's different in soccer, where aerodynamic effects are much more important for two reasons. First, the accelerations are so much larger. A 0.425-kg soccer ball is almost three times as massive as a 0.15-kg baseball, so it requires almost three times the force to achieve the same acceleration. However, at 11.1 cm, the soccer ball's radius is three times that of a baseball (3.74 cm). This factor of three more than compensates for the increased mass of the soccer ball, since according to equation 5.6 the Magnus force depends on the *cube* of the radius: tripling the radius means an increase of $3^3 = 27$!

On the other hand, a soccer ball's spin and speed are usually less than those of a baseball, so other factors are at play. Let's put in some numbers. Based on footage, the authors of one study estimated the launch speed of Carlos's kick at 30 $\tfrac{\text{m}}{\text{s}}$ and its spin at 9 rps. The sideward acceleration is then

$$a_y = \frac{F_M}{m} = \frac{\pi^2 C_M \rho_{\text{air}} R^3 v f}{m}$$

$$= \frac{\pi^2 \left(1.2\text{ kg/m}^3\right)(0.111\text{ m})^3 (30\text{ m/s})(9\text{ rps})}{0.425\text{ kg}} = 10.3\text{ m/s}^2.$$

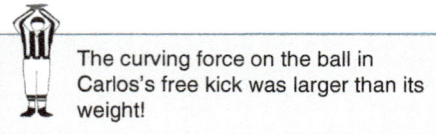

The curving force on the ball in Carlos's free kick was larger than its weight!

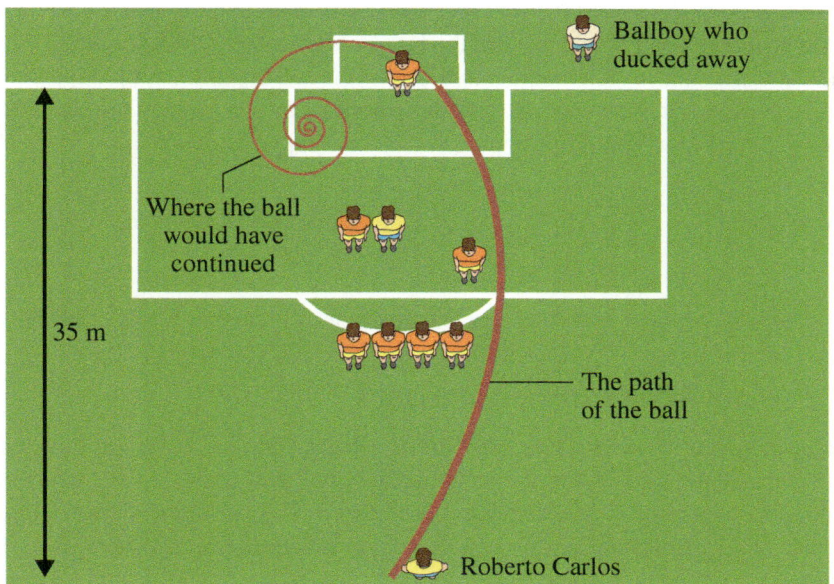

Figure 5.14. The "impossible" free kick of Roberto Carlos in the 1997 Tournoi de France match between Brazil and France. This figure based on a similar one from a 2010 report by the BBC; http://www.bbc.co.uk/news/science-environment-11153466

This is more than 1g and almost triple the Magnus acceleration of the baseball in the previous section!

The second reason Carlos's kick was so wicked and so different from a curveball, is due to the long distance it had to travel to its destination. Carlos took the shot from 35 m out, almost twice as far as the baseball's 18-m journey from mound to plate. This gives the soccer ball more time to deviate from a simple path and develop a spiral. (See figure 5.11 for a reminder; the spiral separates from the simpler circle only after the ball has traveled for a while.)

Figure 5.14 shows the trajectory of Carlos's kick. The spiral was already developing as the ball slipped behind the bedazzled Barthez. This kick was discussed in a detailed study of spiraling trajectories by French scientists in 2010 (New Journal of Physics 12 (2010) 093004).

How do we know when a simple treatment is enough?

Sometimes the complex aerodynamics that determine a ball's trajectory can be handled relatively well with a simple constant-force approach. The baseball's position is predicted to within 3 in. In other situations, a full spiral calculation might be best, as with Carlos's free kick.

As they approach complex situations, scientists have to ask themselves: when is a simple approach likely to be "good enough"? The answer almost always sounds like this: "If quantity X is much bigger than quantity Y, then we can use the simple approach."

For the spirals we're talking about now, the things to compare are (1) the radius of the circle produced by the Magnus force and (2) the distance traveled by the ball. For the baseball, the circle's radius is very large—about 900 ft. This is much bigger than the 60.6 ft that the ball travels from mound to plate.

For Carlos's kick, equation 5.7 tells us the circular radius:

$$r = \frac{mv}{\pi C_M \rho_{\text{air}} ARf} = \frac{(0.425 \text{ kg})(30 \text{ m/s})}{\pi^2 (1.2 \text{ kg/m}^3)(0.111 \text{ m})^3 (9 \text{ rps})} = 87 \text{ m}.$$

This is pretty big, but not much larger than the 35 m that the ball travels from foot to goal. Scientists would say that the two numbers are "of the same order."

A ratio is a handy single-number indicator of the validity of the simple approach. If the ratio of the circular radius to the ball's travel distance is greater than about 5 or so, then a simple constant-force approach is a decent approximation.

$$\text{curveball:} \quad \frac{r}{d} \approx \frac{900 \text{ ft}}{60.5 \text{ ft}} = 15 \quad \rightarrow \quad \text{simple approach probably ok}$$

$$\text{bent kick:} \quad \frac{r}{d} = \frac{87 \text{ m}}{35 \text{ m}} = 2.5 \quad \rightarrow \quad \text{better to use a more precise calculation}$$

5.5.5 John Paxson, master of forces

The legendary 1993 finals series between the Michael Jordan's Chicago Bulls and the Phoenix Suns was among the most exciting and hard-fought in NBA history. Having lost the first two games at home, the Suns battled back to take two of three in Chicago. Back in Phoenix for game 6, they were ahead 98 to 96 with just over 14 s in the fourth quarter. After a quick time-out, the Bulls inbounded and looked for several opportunities to tie the game with a two-pointer, but were repelled. Finally, with just seconds in the game, Horace Grant fed the ball to John Paxson, who was lurking just outside the 3-point arc. Paxson hit nothing but net with less than 4 s in the game, sealing the deal and securing the Bulls's third straight championship.

"I've been playing basketball since I was 8 years old, and I've shot like that in my driveway hundreds of thousands of times," Paxson said later. For practiced athletes, plays like this are intuitive, not intellectual. We'll take the more analytic approach and discuss the interplay among the many forces that determine Paxson's shot.

Paxson's feet were just behind the 3-point line, and the ball was released just above the line, 23 ft 9 in. from the hoop. Video analysis of the play is imperfect, but a decent estimate is that the ball was released at a height of 10 ft (the same height as the hoop), at an angle of 46° and with a speed of 28.5 ft/s (about 19.5 mph).

Figure 5.15
The hoop is surprisingly roomy for a ball sitting at its center... (© *Purestock/SuperStock RF*).

Margin of error

Usually, a basketball player wants the shot to hit nothing but net. If the ball hits the rim, it may take a friendly bounce, but the situation gets a lot more unpredictable.

The radius of a basketball is 4.7 in., about one-half the 9-in. radius of the hoop. The ball sitting in the center of the hoop has a lot of room on each side—4.3 in. to be exact, as shown in figure 5.15. However, if we think about a real shot and still want it to hit nothing but net (see figure 5.16) at any point during its flight, there is only about 1.6 in. of leeway on either side. This is the kind of precision we're interested in, if there's 4 s left in the game and a title on the line.

Gravity alone

A basketball's weight dominates the path it follows, once it's released from the player's hand. Furthermore, the analysis is easy. We can use the range formula (equation 4.4) to find out how far the ball has gone when it returns to its launch height of 10 ft:

$$R = \frac{v_0^2 \sin 2\theta}{g} = \frac{(28.5 \text{ ft/s})^2 \sin(2 \cdot 46°)}{32.17405 \text{ ft/s}^2} = 25.23 \text{ ft} \approx 25 \text{ ft } 2.75 \text{ in.}$$

That's *way* off! In fact, it's 17.75 in.—almost 1.5 ft,—farther than the desired 23-ft 9-in. swish shot Paxson wanted.

Figure 5.16. ...but requiring that the ball never touch the hoop while in flight reduces the margin of error to about 1.6 in. (© *Ryan McVay/Getty Images RF*).

Figure 5.17 shows trajectories with different forces at play. It seems that the gravity-only trajectory may go in the net with a decent rebound off the backboard, so long as Paxson was shooting from the top of the key, straight on toward the basket. However, he was shooting almost from the corner of the court, so this shot wouldn't even have been close. The Suns would have tied the series, the first team to do so after losing the first two games at home.

Gravity and buoyancy

When we ignore all other forces besides gravity, the formulas turn out very simple because the acceleration has a constant direction (down) and a constant magnitude. That magnitude (g) shows up in formulas such as the range equation.

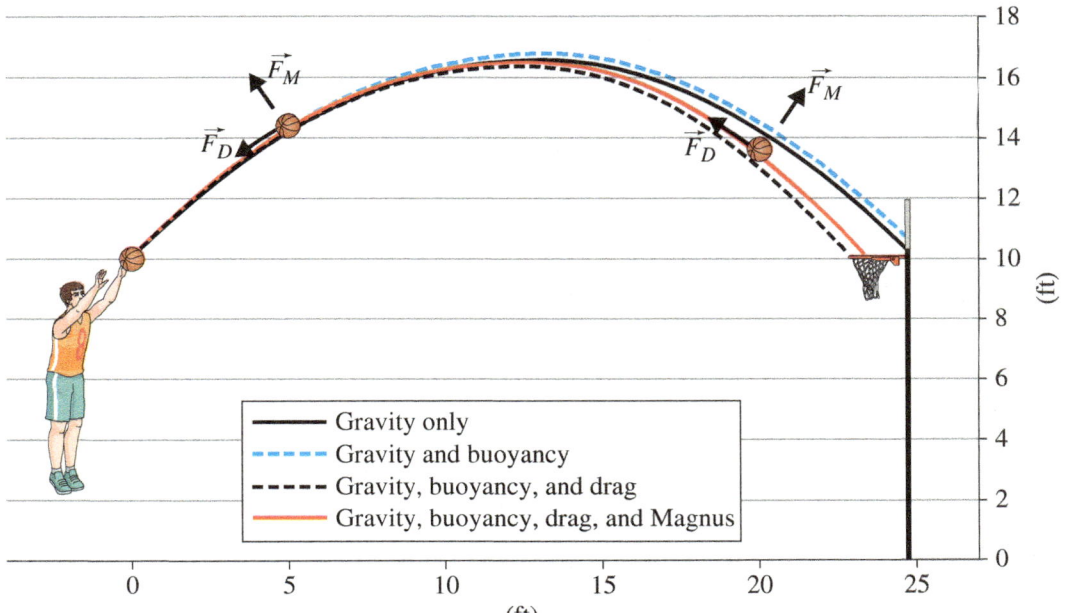

Figure 5.17. The trajectory of John Paxson's 3-point attempt as determined by the interplay of forces. Gravity is the most important force, but if he relies on this alone, he'll overshoot by almost 18 in.; accounting for buoyancy only makes the situation worse. Drag exerts a considerable slowing force, and the ball pounds onto the front of the rim. Paxson's well-practiced backspin generates a lifting Magnus force and just the right touch to put the ball cleanly through the hoop, clinching the win for the Bulls.

Actually, if we include buoyancy in the mix, the acceleration *still* has a constant direction (down) and a constant magnitude. The only difference is that its magnitude is no longer g.

To find out what it is, first we need to know the buoyant force:

$$|\vec{F}_B| = \rho_{\text{air}} g \tfrac{4}{3}\pi r^3 = (0.0024 \text{ slug/ft}^3)(32 \text{ ft/s}^2) \tfrac{4}{3}\pi (0.394 \text{ ft})^3 = 0.0197 \text{ lb},$$

which is about 1.43% of the ball's weight of 1.376 lb.

Now it's easy to use Newton's second law to find the constant downward acceleration of the basketball:

$$a_y = \frac{\sum_i F_{y,i}}{m} = \frac{W_y + F_{B,y}}{m} = \frac{-1.376 \text{ lb} + 0.0197 \text{ lb}}{0.043 \text{ slug}} = -31.54 \text{ ft/s}^2.$$

The constant downward acceleration is 98.57% of what it would be under gravity alone. That makes sense—the downward force is reduced by 1.43%, so the downward acceleration should be reduced by the same amount.

The basketball's path is still a parabola, so we can still use the range equation, but the smaller acceleration replaces g in the formula:

$$R_{\substack{\text{with}\\\text{buoy}}} = \frac{v_0^2 \sin 2\theta}{a_y} = \frac{v_0^2 \sin 2\theta}{0.9857 g} = \frac{1}{0.9857} \underbrace{\frac{v_0^2 \sin 2\theta}{g}}_{25.23 \text{ ft}} = 25.60 \text{ ft} = 25 \text{ ft } 7.2 \text{ in.}$$

So the ball will travel about 4.4 in. farther, due to the buoyant force.

It turns out that the tiny buoyant force actually matters quite a bit because 4.4 in. is quite large, compared to our 1.6-in. margin of error. And for Paxson's shot, it only makes matters worse. Considering only gravity, the shot went about 18 in. too far; when we include buoyancy, the overshoot is 22.4 in., or almost 2 ft!

Taken at almost any position on the court, this shot has no prayer of going in. If it were taken at the 3-point line in the corner of the court (a favorite lurking grounds for long shooters), Paxson wouldn't even hit the backboard, and he would soon hear the taunting chant "aiiiiiiir baaaaall" from the Phoenix crowd.

But Paxson had more forces at his disposal.

Gravity, buoyancy, and drag

The trajectories due to gravity and the buoyant force are easy to calculate, but realistic shots are affected by drag and (if the ball has spin) Magnus forces, and these have major effects. The magnitudes and directions of these forces change throughout the ball's flight, but their effects are easy to understand on a qualitative level.

The drag force points down and leftward on the ball's rise, and up and leftward on its descent, as shown in figure 5.17. This does two things. First, it shortens the range of the shot, as shown in figure 5.17. Second, air drag causes an asymmetric trajectory; that is, the ball comes down more steeply. A glance at figure 5.16 and a little thought should convince you that this is a *good* thing: a steep angle allows greater space for the ball to enter the rim without touching it.

The ball will hit the front of the rim. (Problem 21 asks you to figure out how short the shot is because of to drag.) Whether it would take a lucky bounce and go in anyway is difficult to say exactly. But with a shot this long the ball comes in at high speed, and a "soft" landing is unlikely.

It is time for Paxson to generate one more force, to eliminate the uncertainty of a hard bounce.

The final four: gravity, buoyancy, drag, and Magnus forces

Like drag, the Magnus force changes in strength and direction throughout the ball's flight. A computer was used to plot out the full trajectory shown in figure 5.17, but the qualitative effect of the force is easy to see. Since it always points at right angles to the ball's velocity, it points up and left on the ball's ascent and up and right on the descent. Most players impart a backspin of about 2 revolutions per second (rps). This gave just enough lift for Paxson's shot to pass through the hoop, touching nothing but net, as indicated in figure 5.18.

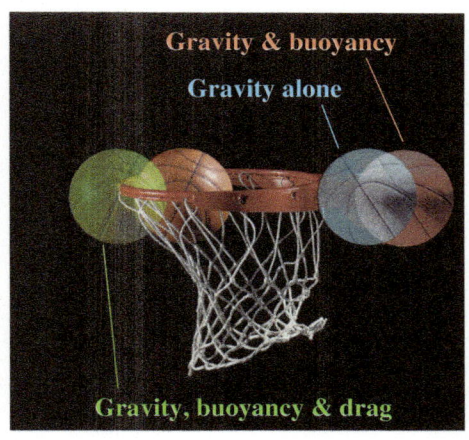

Figure 5.18
The effects of air accumulate as the balls travels to the basket, depending on air density, ball characteristics, speed, and spin. It's a question of inches as a delicate interplay between the various forces puts the ball squarely in the middle of the hoop. (© *Ryan McVay/Getty Images RF*).

5.6 More Complicated Aerodynamics in Sports

The effects of air on flying objects are extremely complicated. The treatments of drag and Magnus forces we've discussed are only approximations (usually good ones) to the much more complex aerodynamics at work. Those approximations allow scientists to calculate air's effects on sports in a straightforward way, as we've done.

In this section, we'll discuss in a more qualitative way some important aerodynamic effects that are difficult to put into simple formulas.

5.6.1 Knuckling

Paxson surely benefits from those hundreds of hours in his driveway, nailing down a reliably consistent touch and spin on the ball. But why put any spin on the ball at all? Why not just use zero spin, and just shoot the ball a little bit harder? Wouldn't that work just as well? Well, no, not really.

Take another look at figure 5.7. The sideways force on the cricket ball comes from the fact that the surface on the top side is smoother than that on the bottom side. The same thing happens with any ball with any structure on its surface.

Figure 5.19 shows sideward forces on a nonspinning baseball as it flies into the wind. Look at the ball in panel (f) of the figure; it has a 2.7-oz force to the right. If it holds its orientation, it will feel a constant lateral force and will curve ever faster to the right as it comes at us.

But what if the ball is spinning *very slowly*, such as less than 1 rps? As the ball works its way through orientations (a) to (f), the lateral force will point first to one side

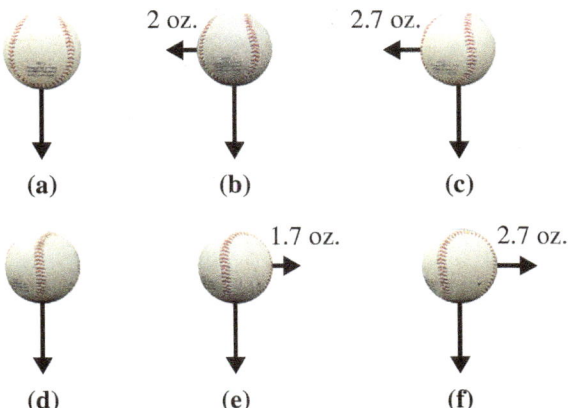

Figure 5.19. The lateral (sideward) force on a ball depends sensitively on the position of the stitches. Here the balls are coming directly toward us at 60 mph. Arrows show the direction of the lateral and weight forces. Based on seminal 1975 measurements of R. Watts and E. Sawyer. (*Photos courtesy of Mike Lisa*).

then to the other. No curveball can ever do this, and the knuckleball is maddeningly confounding to batters (and catchers), as a ball that is obviously accelerating away from the plate suddenly changes its mind and veers back for a strike. This last-minute veering can happen when it is physically too late for the batter's brain and body to coordinate a swing.

Remember that, because of the physical time needed to process information and generate neural signals for a swing, a batter always bases his decision on the first one-third to one-half of the pitch. He must extrapolate the ball's path based on the first part of its flight. To hit the pitch shown on the upper part of figure 5.20, a batter would need to be clairvoyant or lucky enough to "know" that the pitch would change direction, before it ever did.

The figure also shows the importance of a *very* slow rotation on a knuckleball. Increasing the spin to just 0.5 rps (that is, less than a full rotation during its flight!) causes the lateral force to switch direction too quickly for any significant curving to take place. The batter will see the ball flutter a little bit, but it essentially follows a straight line.

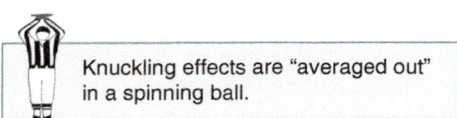
Knuckling effects are "averaged out" in a spinning ball.

This is the physics behind Paxson's instinctive sense to put some spin on a ball. When a ball is spinning rapidly, sideways forces due to surface irregularity switch rapidly back and forth and effectively cancel out. (The force related to the spin *itself*—the Magnus force—is different and is not canceled, of course.)

A final note: the lateral forces on a nonspinning ball are *very* complicated, and the tiniest difference in orientation or stitches on the surface or a thousand other things can lead to very different forces. For this reason, it is not only the batter who is ignorant of the ball's final position on any given pitch. So also are the pitcher and catcher!

Knuckleballs in baseball can be very useful. In basketball, they are definitely a bad idea. Some of the most impressive knuckleballs are seen in soccer shots on goal and volleyball "floater" serves, situations where there is a lot of surface area and a relatively light ball.

5.6.2 Tilting into the wind: discus

Spinning balls flying through air get a sideways force due to the Magnus effect, as we've said, but it turns out that nonspinning flying objects can experience a force

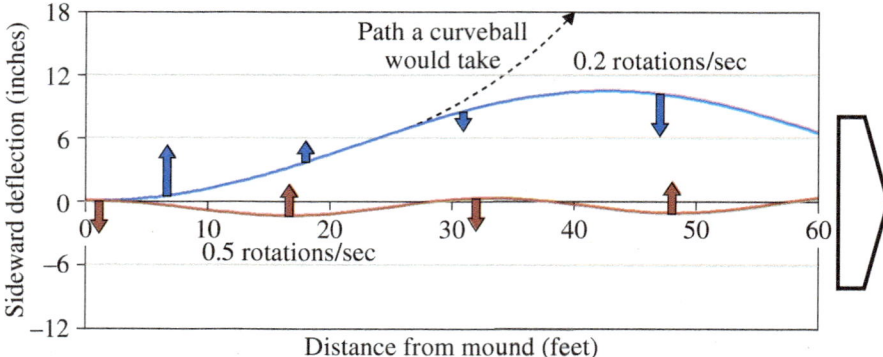

Figure 5.20. The paths of two 60-mph knuckleballs, based on the forces measured by Watts and Sawyer. Arrows show the lateral force on the ball. To create the most confounding pitch, a very slow spin should be given to the ball. To the batter, the slowly rotating knuckleball looks very much like a curveball that has zero chance of crossing the plate . . . until it is too late, and the ball veers in for a strike.

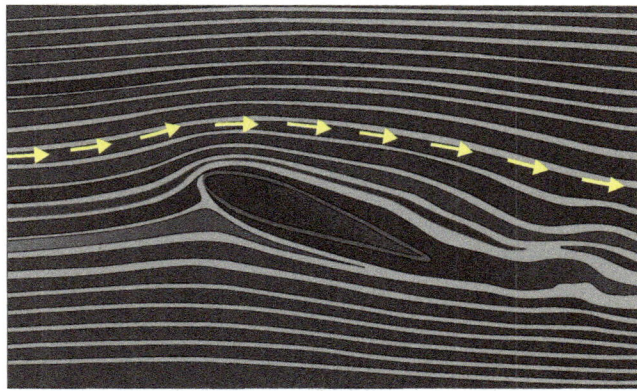

Figure 5.21. The results of a wind tunnel measurement of a wing's aero-dynamics are sketched. Lines of smoke show the path of air molecules at different heights. Arrows on one of them indicate the molecules' velocity. The wing's angle of attack, which is its orientation relative to the air's initial velocity, is about 25°.

perpendicular to their motion. This is a good thing for air travelers, too, since they rely on an upward force to oppose gravity, and airplane wings don't spin!

The aerodynamics of lift[5] on a plane's wing is complicated, and the majority of explanations (even many in introductory physics textbooks) that try to give a simple explanation are at least partly incorrect in important ways.[6] This results in misconceptions, such as the one that says a wing must be curved on top in order to generate lift. While many large-aircraft wings are more curved on top than on the bottom, symmetric wings (same curvature on top and bottom) work quite well and are used frequently on military and stunt aircraft.

Figure 5.21 shows a symmetric wing in a wind tunnel. The air molecules come in horizontally from the far left ($v_{y,0} = 0$). They follow some complicated path, but at the end of the day, they leave with a significant downward velocity component ($v_y < 0$). Well, we know what that means: the molecules have accelerated downward. And we know what *that* means: they experienced a downward force, due to the wing (there's nothing else around to give them a force). Finally, we know that Newton's third law tells us that the wing experiences an oppositely directed (that is, upward) "reaction" force. Voila lift!

Without entering into a long discussion of why the air is forced down, we'll be content with the observation that it is. Furthermore, the story is obviously all about the angle of attack, which is the angle between the velocity of the air relative to the wing and the wing's orientation. The left panel of figure 5.3 shows a similar wing with zero attack angle; the symmetric smoke patterns indicate that there is no lift in that case.

Usually, wing designers are looking to minimize air drag and maximize lift. This occurs at a certain attack angle, depending on the wing's shape. You can get a good feel for this by sticking your arm out of the window of a moving car and shaping your hand into a flat wing. You'll easily find the right angle (probably about 20°) to give a very noticeable vertical lift force with relatively little drag. Increase the attack

---

[5] Sometimes the Magnus force is referred to as a "lift" force, too. Here, we're talking about non-Magnus lift.
[6] A very clear and easy-to-read discussion with instructional applets has been produced by NASA at http://www.grc.nasa.gov/WWW/K-12/airplane/right2.html.

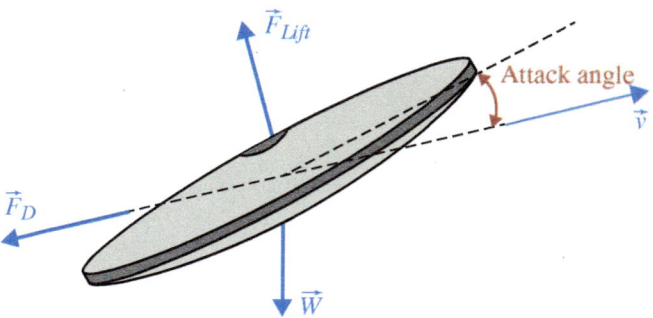

Figure 5.22. The forces on a flying discus. The discus's angle relative to the horizontal is about 45°, while the angle of attack is much smaller, about 10°.

angle much more (so that your thumb points vertically), and the lift will vanish; all you'll feel is a strong drag. You'll also find that a negative attack angle (angling your hand down) gives a downward "lift" force.

A discus is a symmetrical wing, too, and these lift forces play a major role in its flight. Figure 5.22 shows the forces on a flying discus. Notice that in this case, the angle of attack is not the same as the discus's angle relative to the ground, since it's not flying purely horizontally. This also means that the lift force is not purely vertical.

The angle of a discus relative to horizontal stays approximately constant during its flight, but the velocity (obviously) does not. As the attack angle of a discus changes in flight, so does the direction of lift, as shown in figure 5.23. The optimum throwing angle depends on the launch speed and angle, but detailed measurements show that the lift-to-drag ratio is maximized if the discus is oriented about 25° relative to horizontal. When it is thrown at this angle, the lift at the top of the trajectory can be quite high, and the trajectory becomes significantly flatter than the normal parabola we'd have in the vacuum. At the end of the trajectory, the attack angle is too high, and drag takes over. This is like having your hand out the car window oriented perpendicular to the wind.

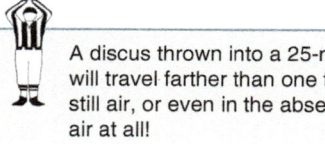
A discus thrown into a 25-mph wind will travel farther than one thrown in still air, or even in the absence of any air at all!

Overall, lift is much more important than drag for the discus. How important? I'm glad you asked. It's so important that a discus thrown *into* the wind will travel farther than one thrown *with* the wind! (This is within reasonable limits, of course. Throw a discus into a hurricane, and it'll come right back at you!) In fact, a discus thrown into a 25-mph wind will travel farther than it would in the absence of any air at all.

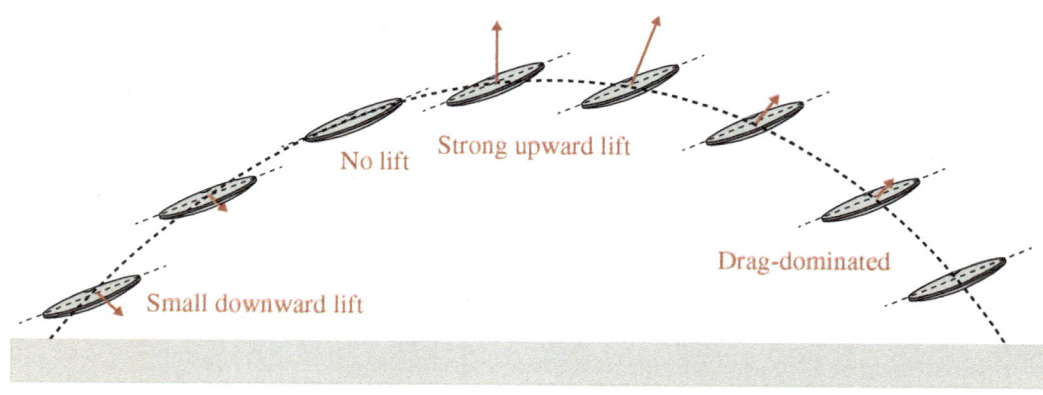

Figure 5.23. The lift force on a discus changes in magnitude and direction during its flight. Based on a figure from Broncazio's *Sport Science*.

5.6.3 Human wings: ski jumps

In chapter 4, we talked about projectile motion in the absence of air. We saw that, while the same rules applied for humans as well as inanimate objects, the fact that humans can change their shape in flight was hugely important for sports. The same is true for flight with lift. Since the discus is stuck in one orientation, it cannot optimize its attack angle from moment to moment and even experiences negative lift during some intervals.

Figure 5.24 shows a flying skier in the iconic ski jumper's stance. His body and skis act as wings with the optimum attack angle to minimize drag and maximize lift. The lift is hugely important. Video analysis of a 170-ft waterski jump by Ben Van Treese, a student in the author's physics of sports class, gave evidence for a lifting force on the order of 20 lb.

By looking again at figure 5.23, it is clear that the discus would do well to maintain its optimum attack angle by reducing the angle it makes relative to the horizontal. While this is impossible for the inanimate discus, just such changes are made instinctively by the best ski jumpers. As their velocity begins pointing downward, they reduce the angle of the skis relative to the horizontal. It is nice that adjusting the skis in this way also brings them more parallel to the hill, easing landing.

5.6.4 Making the world safer for javelin spectators

Figure 5.25 shows the average of the top 25 men's javelin throws in each of the past 125 years. Ignoring the clear of effects of young men going overseas to do more important things in the 1940s, there is steady improvement, as the field sports gained prominence and increased participation and national Olympic teams organized.

Distances improved steadily until the best throws were regularly 90 m or more. In fact, in 1984, Uwe Hohn threw a javelin 104.8 m, which is about the length of an American football field, including the end zones! These throws were approaching the size of sports stadiums and endangering fans, especially given the fact that javelins thrown that far tended to land "flat" and skip up. In response, the International Association of Athletics Federations (IAAF) implemented a rule in 1986 that moved the center of mass of the javelin more toward its front tip. This reduced lift and tipped the nose down so that it stuck in the ground more often. As a result, the best throwers found their distances reduced by 10 m.

But, as often happens, this rule change led to further innovations in equipment design. Manufacturers started adding dimples or rough patches to reduce air drag

Figure 5.24. A ski jumper optimizes the lift force by maximizing the surface area, catching the air, and orienting himself as a discus or wing. (© *Sean Thompson/Photodisc/Getty Images RF*).

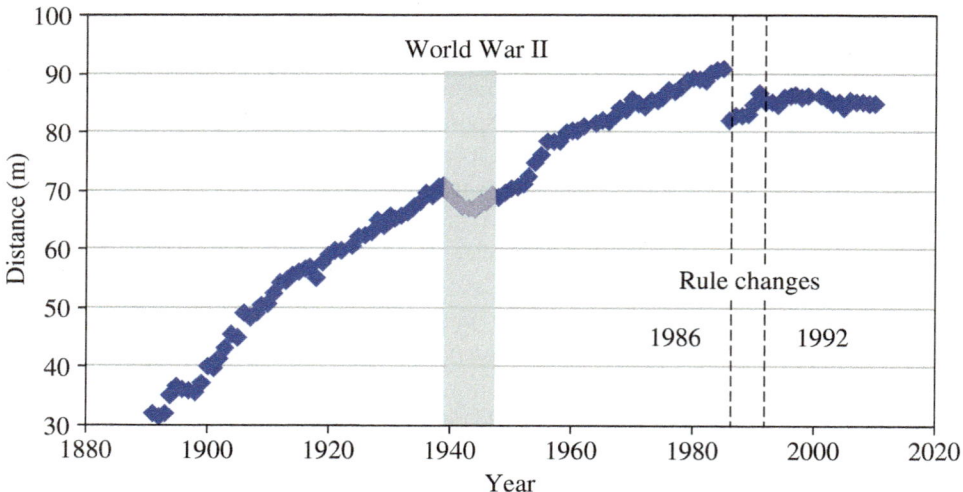

Figure 5.25. The average of the best 25 javelin throws in a given year for each year since 1891. Data collected in an article by Steven Haake in the June 2012 issue of *Physics World*.

and increase flight distance, until the IAAF changed the rules again to prohibit these innovations in 1992.

Of course, modern throwers will almost certainly never reach 100-m distances again, since the equipment that made such throws possible is now illegal. Therefore, performances after 1986 compete in a separate "new rules" record category.

5.6.5 Not so fast! Polyurethane swimsuits

A similar "arms race" between equipment manufacturers and regulating bodies occurred recently in swimming. Figure 5.26 shows that performance in the women's 100-m freestyle improved quickly for the thirty years between 1948 and 1978, and then improved more gradually for the next thirty years. Then, around 2008, there was a sudden drop in times. What was going on around then? Innovation in the form of the polyurethane full-body swimsuits.

These suits are all about drag.

Equation 5.4 says that drag depends on the drag coefficient C_D, the cross-sectional area A, fluid density ρ, and speed v. These suits changed them all! The so-called "hydrophobic" nature of the surface at the microscopic level allowed the water to slip over the surface with much less adhesion than on textile suits or skin; this significantly reduced C_D. They were tight (in fact, it takes about a half hour to put one on!) and squeezed the swimmer's body so that the area A in equation 5.4 was reduced.

Finally, the suits trapped tiny pockets of air, increasing the buoyant force $\vec{F}_B$. Although $\vec{F}_B$ is perpendicular to the drag force $\vec{F}_D$, it has an important effect: it lifts the swimmer slightly higher in the water. Hence, more of her body moves through air. Keeping in mind that the density of air (1.2 kg/m^3) is more than 800 times lower than the density of water (1000 kg/m^3), this significantly changes the effective ρ in the drag formula.

In fact, the only thing in the drag formula that does not decrease is the speed v of the swimmer herself!

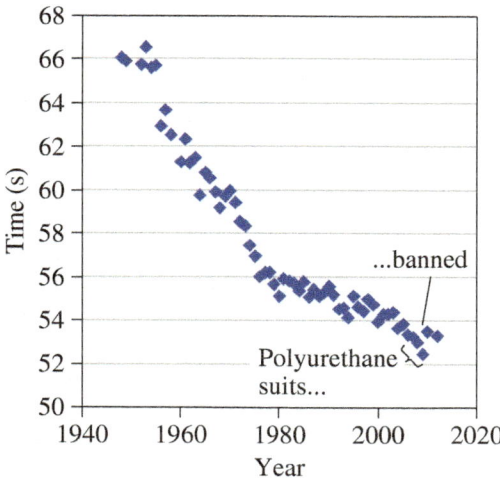

Figure 5.26. The average of top three performances in the women's 100-m freestyle since 1948. Data collected in an article by Steven Haake in the June 2012 issue of *Physics World*.

In 2010, FINA (the French acronym for the International Swimming Federation) made the new polyurethane swimsuits illegal, and times immediately jumped, as seen in the last data points in figure 5.26. As the figure shows, although the effect of the suits was to reduce times by "only" 1.8 s, it took 20 years (1980–2000) for the previous drop of that magnitude. It will likely be quite some time before records set with the polyurethane suits are broken.

While FINA did not create a "new rules" world record category, records set by swimmers with polyurethane suits are often marked with an asterisk in record lists.

5.7 Not All Air Is Created Equal

The formulas for the three fluid forces—buoyancy (equation 5.2), drag (equation 5.4), and Magnus (equation 5.6)—all say that the force is proportional to the density of the fluid. That makes sense, and it's why such forces from water are so much stronger than from air; a 93-mph fastball thrown under water in a swimming pool slows down quickly.

On a 70°F day at sea level, with dry air and normal atmospheric pressure (no storm), air has a density of 1.2 kg/m^3. However, changes in elevation, temperature, humidity, and weather can significantly change the air's density and its effect on balls flying through it. This has measurable effects on sports.

5.7.1 Rocky Mountain (Natural) High

If you've ever gone skiing in the mountains or if you are from a low-elevation area (such as New York City) visiting a higher elevation (such as Denver), you may have experienced the headaches and slight fatigue associated with the first few days in air thinner than you're accustomed to. The density of air drops by 1% for every 275-ft increase in elevation.

The drop is particularly noticeable when looking at air up to where jetliners cruise and above. Figure 5.27 shows a column like this, as well as a blowup of the region

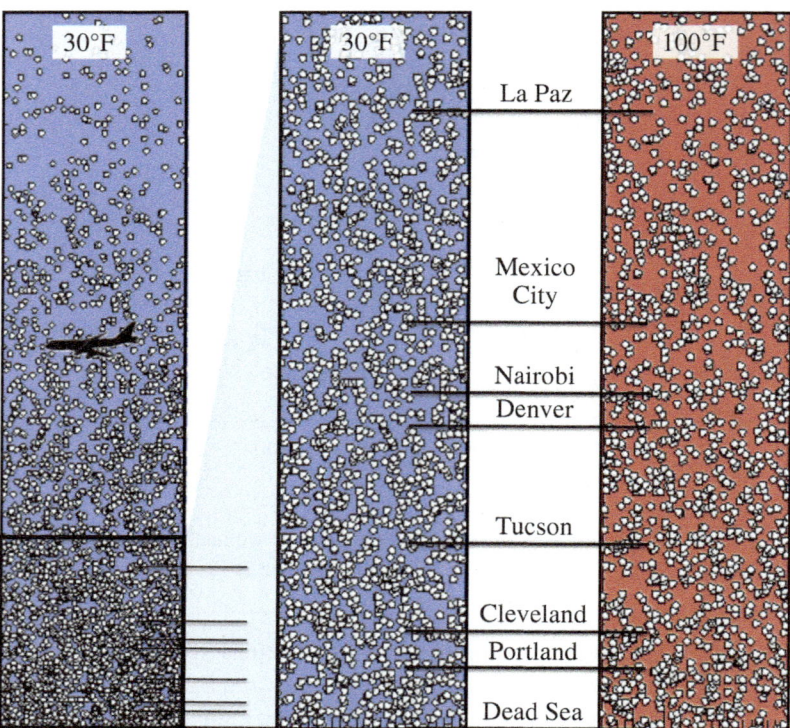

Figure 5.27. The density of air molecules decreases with increasing elevation. This is particularly obvious for a column of air 20 km tall, shown on the left, but still quite noticeable in the range of elevations where people live and play sports. The density also decreases as the temperature increases. There are 13% more molecules in the volume shown on the 30°F day than the 100°F day. Also who knew that Nairobi was so high? No wonder there are so many excellent Kenyan long-distance runners.

Table 5.1. The density of air at various points on Earth.

	Elevation		ρ_{air} (kg/m^3)	
			$T = 40°F$	$T = 70°F$
Dead Sea	−1300 ft	(−396 m)	1.333	1.258
Sea level	0 ft	(0 m)	1.272	1.201
Portland	50 ft	(15 m)	1.270	1.198
Cleveland	791 ft	(241 m)	1.236	1.167
Tucson	2,641 ft	(805 m)	1.155	1.090
Denver	5,280 ft	(1,609 m)	1.047	0.988
Nairobi	5,889 ft	(1,795 m)	1.024	0.966
Mexico City	7,530 ft	(2,296 m)	0.962	0.908
La Paz	11,942 ft	(3,641 m)	0.811	0.765

where we live. In this region, the effect is more subtle, but it is still important. The density of air is almost 20% lower in Denver than it is at sea level (see Table 5.1). This means that all of the forces due to air—buoyancy, drag, and Magnus—are 20% lower there.

For pitchers, this is both good and bad. The retarding force on their fastballs is significantly reduced, so batters have less reaction time. On the other hand, the aerodynamic forces at their disposal are significantly reduced.

EXAMPLE 5.6 *Think fast! Reduced reaction time in Denver*

Let's take Justin Verlander and his 100-mph (146.7 ft/s) pitch on the road. Verlander's pitch also showed up in example 5.4. First stop: outer space. How long does it take the ball to travel the 56 ft from the release to the plate?

This is easy enough, since the velocity of the ball never changes:

$$\Delta t = \frac{\Delta x}{v_x} = \frac{56 \text{ ft}}{146.7 \text{ ft/s}} = 0.382 \text{ s}$$

In New York City, which is essentially at sea level, the initial acceleration due to drag is

$$a_x = -F_D/m = \frac{C_D}{2} A \rho_{\text{air}} v_x^2 / m$$

$$= -\frac{0.35}{2} \left(\pi (0.123 \text{ ft})^2\right) \left(0.0024 \text{ slug/ft}^3\right) (146.7 \text{ ft/s})^2 / (0.01 \text{ slug})$$

$$= -43 \text{ ft/s}^2.$$

In that equation, I used the air density in Imperial units, listed in table B.5.

The ball crosses the plate at

$$v_x = \sqrt{v_{x,0}^2 + 2a_x \Delta x}$$

$$= \sqrt{(146.7 \text{ ft/s})^2 + 2\left(-43 \text{ ft/s}^2\right)(56 \text{ ft})}$$

$$= 129 \text{ ft/s} = 88 \text{ mph}.$$

To find the time it takes, there is one more step:

$$v_x = v_{x,0} + a_x \Delta t \rightarrow \Delta t = \frac{v_x - v_{x,0}}{a_x}$$

$$= \frac{129 \text{ ft/s} - 146.7 \text{ ft/s}}{-43 \text{ ft/s}^2} = 0.412 \text{ s}$$

Final stop: Coors Stadium in Denver. Here the density is 18% lower than in New York, so the equation above shows that the slowing acceleration ("deceleration") will be 18% lower:

$$a_{x,\text{Denver}} = 0.82 a_{x,\text{New York}} = 0.82 \left(43 \text{ ft/s}^2\right) = 35.3 \text{ ft/s}^2.$$

If we use this acceleration in the equations above, we get

$$v_x = 132.5 \text{ ft/s} = 90 \text{ mph}.$$

$$\Delta t = 0.402 \text{ s}.$$

So the ball gets there a *little* faster: 10 ms earlier. In this time, a 100-mph pitch travels $(146.7 \text{ ft/s})(0.01 \text{ s}) = 1.47 \text{ ft} = 17\frac{1}{2}$ in., almost precisely the size of home plate.

Though it is a small effect, example 5.6 shows that a baseball pitch goes slightly faster in Denver than in New York. So, are pitchers happy? No.

As we've discussed, a good pitcher plays with the air to fool the batter, and high up in the Rockies, there are just fewer air molecules between the pitcher and home plate for the ball to interact with. In fact, look back quickly at section 5.5.3. The math says that the deflection of a curveball is directly proportional to the air density. Since

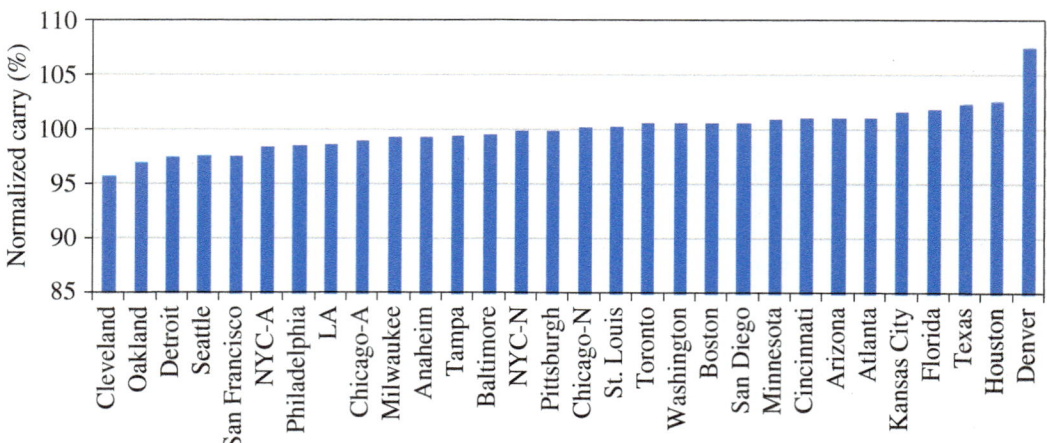

Figure 5.28. The normalized carry is a measure developed by Dr. Alan Nathan to characterize the range of a struck ball in a given stadium, compared to others. The average value of the normalized carry is 100% by design.

the air density is 18% lower in Denver, curveballs will curve—and knuckleballs will knuckle—18% less. So, rather than a 23-in. deflection, the pitcher gets only 18.9 in. This is significant, and batters find it easier to connect in Coors Field Stadium.

At this point, the situation becomes even worse for the pitcher. Once the ball is hit, because of the low drag forces, the ball carries farther at Coors Field than at any other ballpark.[7] Figure 5.28 shows data compiled by Dr. Alan Nathan for the first six weeks of the 2009 MLB season on the "normalized carry." This number quantifies how far a ball travels, compared to the group average. Denver is head and shoulders above its peers. A nuclear physicist by training, Dr. Nathan is among the foremost baseball scientists alive today; he publishes his research frequently and maintains a fascinating website at http://baseball.physics.illinois.edu

Other sports

Looking around online, in sports magazines, and in books, you'll see that the effects of reduced aerodynamic forces (especially drag) at high altitude in other sports are discussed frequently. In his book on the physics of football (cleverly titled *The Physics of Football*), Dr. Timothy Gay documents that the away team puts kickoffs into the end zone just 5% of the time in sea-level stadiums, while the same kickers put it into the end zone 30% of the time in Denver.

There are effects on moving people, too. Several world records in track and field were set in the Mexico City Olympics of 1968, including throws, jumps, and all the men's running events shorter than 400 m. (Endurance runners relying on aerobic energy, on the other hand, were hindered by the reduced oxygen at this altitude.) In problem 23 at the end of this chapter, you'll estimate the effect on Bob Beamon's amazing long jump that we first discussed in section 4.4.3.

5.7.2 Hot days are (not) a drag

If you look closely at figure 5.27, you'll notice that the density of air on a hot day is slightly lower than it is on a cold day. In fact, at a given elevation, the rule of thumb is that the density of air drops 1% for each 5°F increase in temperature. On a 100°F

---

[7] Actually, the low air density also reduces the upward-directed Magnus force on a struck ball, which usually has a large backspin as it comes off the bat. This tends to *reduce* the ball's range. However, the beneficial effect of reduced drag overcomes the detrimental effect of reduced lift, and balls fly farther in the thin air in Denver.

day, the air density is 13% lower than on a 30% day, adding about 25 ft to a 350-ft baseball hit. That can be the difference between a home run and being caught out.

Some sports, such as velodrome bicycle racing, involve a constant battle against air drag. In the London 2012 Olympics, any velodrome spectators sitting at track level might have been a bit uncomfortable, as this part of the arena was heated to a toasty 28°C (82.4°F), specifically to reduce air drag (and foster record-setting performances).

Racing in a 68°F environment might be more comfortable, but would increase drag by 3%. As we discuss in chapter 9, the sacrifice is worth it.

Temperatures are kept high in velodromes to reduce the density of air and hence the drag.

5.7.3 It's not just the heat—it's the humidity

Hey, at least those velodrome racers don't have to deal with *humid* hot air, right? Well, it turns out that this would actually help more! Surprisingly, humid air is slightly *less dense* than dry air.

That doesn't seem right, does it? I mean, water is denser than air, right? And air feels yucky and "heavy" when the air is hot and humid.

Well, yes, *liquid* water is denser than air, but in gaseous form, water molecules—which contain two atoms of the lightest of all elements, hydrogen—are lighter than most air molecules (typically made up of two nitrogen atoms) and take up more space. At a given temperature, water vapor is lighter than air.

Dry air is denser than humid air.

As for the heavy, yucky feeling we get in hot humid air, that is really just due to the particular mechanism our bodies use to get rid of excess heat. Evaporation of sweat cools our skin, and dry air absorbs moisture (that is, allows evaporation) much more quickly than air that already has a lot of moisture.

The density-reducing effect of humidity is small and temperature-dependent but noticeable for long flight trajectories.

A warning: if you read about the effects of humidity on ball trajectories, you may sometimes see that a ball's range is *reduced* due to humidity. This is due not to an increase in air density, but often to reduced bounciness of the ball itself, under humid conditions. We'll discuss this in greater detail in chapter 6.

5.7.4 Storm fronts

In many parts of the United States, spring can be a season of rapid and sometimes violent meteorological events. Just before a major storm, the atmospheric pressure can drop suddenly as the temperature rises (or falls) over the course of an hour. In these cases, the air density can change rapidly.

There are anecdotal reports of athletes noticing these effects in real time. Golfers are notoriously fanatic about their time on the course. If they have reserved a tee time, they will play until absolutely forced off the course by man or nature. Some golfers report that in the hour or so before a big storm, their balls won't fly as far. This is due to changes in the drag and—very important to golf because of the huge backspin on a drive—Magnus forces.

An approaching storm front can rapidly reduce air density—and aerodynamic effects.

Collected Equations

$$|\vec{F}_B| = |\vec{W}_{\text{displaced H}_2\text{O}}| = (\rho_{\text{H}_2\text{O}})(V_{\text{displaced H}_2\text{O}})g \qquad (5.1)$$

$$\text{any object: } |\vec{F}_B| = \rho_{\text{air}} g V \qquad (5.2)$$

spherical ball with radius r: $|\vec{F}_B| = \rho_{air} g \frac{4}{3}\pi r^3$ (5.3)

$$|\vec{F}_D| = \frac{C_D}{2} A \rho v^2$$ (5.4)

$$v_T = \sqrt{\frac{2mg}{C_D A \rho_{air}}}$$ (5.5)

$$|\vec{F}_M| = \pi C_M \rho_{air} R A v f$$
$$|\vec{F}_M| = \pi^2 C_M \rho_{air} R^3 v f \quad \text{for a spherical ball}$$ (5.6)

$$r = \left(\frac{m}{\pi C_M \rho_{air} A R f}\right) \cdot v$$ (5.7)

baseball curveball deflection $= (7 \text{ ft}^2)\dfrac{f}{v}$ (5.8)

Problems

1. Here we will justify the statement made at the beginning of section 5.3. The density of a human body depends on the fraction of fat composing the body, since fat is less dense than muscle. (Human fat tissue has a density of about 0.9 kg/liter (900 kg/m³), so fat floats. Meanwhile the density of fat-free tissue is about 1.1 kg/liter (1100 kg/m³), so this tissue sinks.)

   Consider two males, both 180 lb. One is a typical American physics professor, who shall go unnamed. With 18% body fat, the density of his body is 1.81 slug/ft³.
   (a) What is the volume of his body?
   (b) What is the buoyant force on the professor under water?

   If he jumps off a 10-m platform, he hits the water at 45.7 ft/s (31.2 mph).
   (c) If you consider buoyancy and gravity alone (that is, ignoring drag), will the water eventually bring him to rest?
   (d) If so, how far will he travel in the water before coming to rest?

   The other 180-lb male is Michael Phelps, with a reported 6% body fat, the minimum "essential" fat without being considered medically ill. The mass density of Michael's body is 1.98 slug/ft³.
   (e) What is Michael's body volume?
   (f) What is the buoyant force on Michael under water?
   (g) If you consider buoyancy and gravity alone (that is, ignoring drag), will the water eventually bring him to rest?
   (h) If so, how far will he travel in the water before coming to rest?

2. Footnote 2 of this chapter asked whether you could prove the assertion that a 180-lb diver plunging into the water from a 10-m-high platform would hit the bottom of the pool at 30 mph, if he relied solely on buoyancy to slow him down. Please do so.

3. In problem 1 of chapter 3, you found that a hockey puck shot at 80 mph would be moving at 78.88 mph when it reached the other end of the rink 200 ft (60 m) away.

There we included friction, but ignored air drag; here we'll see whether this is a big deal or not.

The drag coefficient for a thick disk coming end-on into the wind is 0.46. That's not very aerodynamic. You also need to know the official NHL specifications on a puck:

> The puck shall be made of vulcanized rubber, or other approved material, one inch (1 in.) thick and three inches (3 in.) in diameter and shall weigh between five and one-half ounces (5.5 oz.) and six ounces (6 oz.).

(a) What is the value for A in the drag formula (equation 5.4)? Hint: it's not πR^2!

(b) What is the drag force on the puck just after it is shot?

(c) Accounting for *both* kinetic friction *and* drag, estimate how fast the puck will be moving after it has slid 200 ft.

4. (a) Keeping in mind that at terminal velocity the drag force has the same magnitude a ball's weight, what is the drag force on a baseball traveling at twice its terminal velocity?

   (b) A baseball is falling down at twice its terminal velocity. What are the magnitude and direction of its acceleration?

5. This is a multipart problem, where you will be combining many of the concepts we've been discussing. I think you'll find you know more than you think you do. You are going to *estimate* the spin that the player imparted to the ball. We'll consider gravity, buoyancy, and Magnus forces, ignoring drag for now. Part of our estimating technique is to assume that all forces are constant. In the problem below, we'll assume that the Magnus force is constant and always pointing toward the baseline; this is not a bad assumption.

   The Magnus force is what allows soccer players to "bend" the ball around a wall of defenders or away from the keeper. There are a tremendous number of video entries on YouTube, showing bent kicks. Take a look at this one: http://www.youtube.com/watch?v=VMkP3y_QW3s where a young player scores using a corner kick, clearly surprising other players, the crowd, and probably himself. There is no wind to speak of, so without the Magnus force such a kick would be impossible.

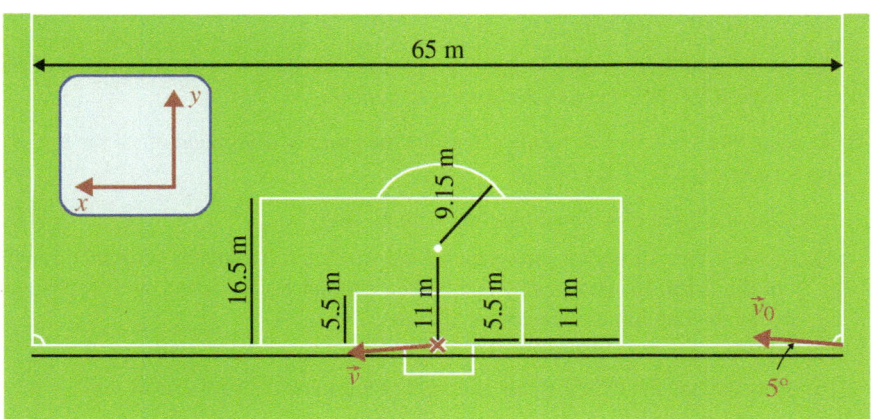

(a) Estimating the flight time with the little scroll bar on YouTube is too crude, so use the fact that the ball reaches a maximum height of about 5 m ("two-and-a-half person's worth") to estimate the flight time. (This can be relatively well estimated, since the ball stays near its peak for a long time, as we've discussed.)

(b) Does the x-component of the ball's velocity change in this problem? Does the y-component change? (Recall we ignore drag.)

(c) In the *xy* plane, will the ball's path be a parabola (like the projectile motion we discussed in chapter 4), a straight line, or some other path?

Lukas kicked the ball precisely from the corner of the field. The field is 65 m wide, and it looks from the video as if the ball fell right in the middle of the goal, marked in the figure by a red X. So it traveled (65 m)/2 = 32.5 m in the *x* direction.

(d) What is the *x*-component of the ball's initial velocity?

(e) What is the *y*-component of the ball's initial velocity? (The tangent function will be useful here.)

If you think about it for a bit, you might realize that because of the symmetry of this situation, the ball will enter the net with the *y*-component of the ball's *final* velocity that is just the negative of the *y*-component of its *initial* velocity.

(f) Now that you have realized that, what is the ball's acceleration in the *y*-direction?

The radius of a soccer ball is 0.111 m, and its mass is 0.425 kg.

(g) Based on your answer to (f), what is the magnitude of the Magnus force?

(h) Based on your answer to (g), how much spin (in rev/s) did Lukas put on the ball?

6. Does a nonspinning fly ball spend more time, less time, or the same amount of time going up compared to coming back down?

7. Can the drag and Magnus forces ever cancel each other?

8. (a) What percentage, by volume, of a water polo ball lies under the waterline as it floats in still water?

(b) The Dead Sea in Israel is the lowest point on Earth. Because of its extreme salinity and mineral content, the water there is extremely dense, 1240 kg/m^3. If the typical human body has a density 95% that of ordinary water, what percentage (by volume) of a person lies under the waterline, as he floats in the Dead Sea?

9. A basketball player makes a 3-point shot from 25 ft from the hoop. The ball is in the air for 1 s, and the player imparts the typical 2 rev/s backspin to the ball. Ignore drag.

(a) Approximately what is the horizontal component of the ball's velocity at the peak of its trajectory?

(b) What is the direction of the Magnus force on the way up? at the peak of the trajectory? on the way back down?

(c) At the peak of the ball's trajectory, approximately what is the magnitude of the Magnus force? Express your answer in ounces.

10. At what speed will a table tennis ball strike the ground, if dropped off of the top of a skyscraper on a 70°F day?

11. Consider the flight of a batted baseball with no spin.

(a) Ignore effects of air. For each quantity, tell whether it (i) remains constant throughout the flight, (ii) changes by some fixed amount every second, or (iii) changes by an amount that changes every second:

x, y, v_x, v_y, a_x, a_y

(b) Now include the effects of air and answer the same questions as in (a).

12. In section 5.5.1, we discussed how, if there were only the Magnus force, a curveball would follow a circular path.

(a) For a typical pitch, what would be the radius of its circular path?

(b) In this unrealistic but interesting case, the pitcher could pitch the ball and then turn around and catch the ball he'd just thrown! How much time would pass between his pitch and his catch?

Top: (© 1998 Copyright IMS Communications Ltd./Capstone Design RF); **Bottom:** (© McGraw-Hill Education/David Moyer, photographer)

13. Ignore drag and buoyancy. You are viewing a baseball pitch from above (such as from a blimp). The curve of the pitch is exaggerated in the figure, for clarity. The ball ($m = 0.15$ kg, radius $= 0.037$ m) is thrown horizontally and straight at home plate at 95 mph. The Magnus force from the spin is 0.44 N.
    (a) Which way is the ball spinning?
    (b) In 0.4 s, how far has the pitch fallen (vertically) in feet?
    (c) What spin has the pitcher put on the ball, in rpm?
    (d) Would this ball be a called strike, if the batter didn't swing? (Home plate is 17 in. across.)
    (e) In 0.2 s (the final moment before the batter must start his swing), approximately how far (horizontally) has the ball deviated from a straight line?

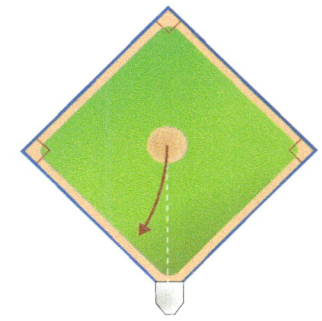

14. Pitchers in baseball and bowlers in cricket have been known to purposely roughen (or "scuff") one side of the ball. (This is legal in cricket but not baseball.)
    If you are a pitcher and you'd like the ball to curve to the left (from your point of view) *without* spinning it, then in what direction should the rough side point (top, left, bottom, etc.)?

15. What must be the spin on a tennis ball moving horizontally at 50 mph, such that it begins rising because of the Magnus force?

16. Ignoring air, you would accelerate downward at 32 ft/s^2 in free fall.
    (a) Accounting now for the buoyant force due to air, do you accelerate downward more, or less, than 32 ft/s^2?
    (b) What is your downward acceleration in free fall, considering only gravity and buoyancy? (Use 5 digits to the right of the decimal.)

17. (a) Ignoring air, how long is a tennis ball in the air if it is hit horizontally 3 ft above the ground?
    (b) Now include buoyancy but not drag, and answer (a) again. For parts (a) and (b), keep 4 digits to the right of the decimal place in your answer.
    (c) If we then include drag, would your answer get larger or smaller? (You don't need to calculate a new number.)

18. Do large hailstones have a larger terminal velocity than small ones? They are larger, so there is greater drag. On the other hand, they are heavier, so it *takes* greater drag to slow them . . .
    To get an idea, we'll consider spherical hailstones that have the density of water.
    (a) In miles per hour, what is the terminal velocity of a 3-in.-radius hailstone?
    (b) How about a hailstone with a 1-in. radius?

19. (a) A basketball is falling *down*ward at 24 mph, without spinning. Accounting for all forces, what is its acceleration? (Don't forget to include its direction.)
    (b) Answer (a) if the ball is moving *up*ward at 24 mph.
    (Use a drag coefficient of 0.4.)
    (c) In which case (ball moving up or ball moving down) is the magnitude of the acceleration larger? Does that surprise you?

20. A lacrosse ball is shot horizontally at its terminal velocity with no spin. Ignoring the tiny buoyant force, what are the magnitude and direction (give an angle relative to horizontal) of its acceleration? (Hint: draw a free-body diagram and think a bit before jumping into formulas.)

21. In section 5.5.5, we discussed the effects of various forces on John Paxson's game-winning 3-point shot. Each had a significant role, and it was crucial that they interplayed exactly right.

(a) Using a constant-force approximation, estimate the drag force on the ball, in ounces. Use $C_D = 0.45$.

(b) Assuming the horizontal component of the drag force is always the same as its initial value, and ignoring the vertical component, estimate how short the shot would be, ignoring the Magnus (lift) force caused by backspin.

(c) Without calculating anything, think about the answer you got for (b). In reality, would Paxton miss by *more* than this amount, or less? (We are still ignoring the ball's spin.) Why do you say that?

22. In section 5.5.4, we talked about Roberto Carlos's "impossible" free kick.

    (a) While it's a cruder approximation than a full spiral, use the simpler curveball treatment of section 5.5.3 to estimate the amazing deflection Carlos achieves after the ball travels 35 m forward. Ignore drag.

    (b) If the velocity vector of this kick were initially directed straight at the goalie, would it enter the goal? Soccer goalposts are 7.32 m apart.

    (c) If drag were accounted for, would the ball's deflection be greater or less than your answer in part (a)? (You don't need to calculate a number.)

23. There are numerous claims that the rarified air in Mexico City led to the many Olympic records set there in 1968. Let's get a rough feeling for whether that could possibly be true, by looking at Bob Beamon's long jump, which we covered in section 4.4.3 without considering air. For this problem, you'll need to know that it was a 70°F day and that Beamon was lean—154 lb (4.8 slug or 70 kg).

    Since the jumper presents less area to the wind during the jump than he would while standing directly upright, use $A = 0.85$ m² in this problem.

    There are two effects that may have helped Beamon, and they both involve drag.

    (a) First, the air is less dense. Estimate the drag force on Beamon as he leaves the launch pad, both at sea level and in Mexico City. The drag constant for a human facing the wind is usually estimated as 0.9.

    (b) By how much does the effect you estimated in part (a) increase Beamon's range in Mexico City as compared to sea level? (Use the crude but simple approximation that the drag force is constant and points directly horizontally backward.)

    (c) Second, there was a 2 m/s tailwind, the maximum allowed to be considered a valid record. This reduces the v in the drag formula (his velocity relative to the air) without changing his velocity relative to the ground. How much lower is the drag force due to this tailwind?

    (d) By how much does the effect you estimated in part (c) further increase Beamon's range?

24. As indicated in figure 5.2, an average man displaces about 22.5 gal of water. That is, the volume of his body is 22.5 gal.

    (a) Consider two swimmers under water, one muscular and one fatty, each having the same 22.5-gal body volume. Is the buoyant force on the muscular swimmer greater than, less than, or equal to the buoyant force on the fatty one?

    (b) Given that 1 gal of water weighs 8 lb, how much would one of the swimmers have to weigh in order to stay stationary under water, neither rising nor sinking?

    (c) Keeping the body volume constant, is there any way the swimmer in part (b) could keep some part of his body out of the water while floating stationary? If so, what would it be?

Game Changers: Collisions in Sports

The sudden, hard interaction between two bodies—ball on bat, player on player, foot on ball—is often a key turning point in a game. Understanding the game means understanding **collisions**, and it involves an entirely new set of concepts including momentum, impulse, and energy. In this chapter, we'll study three types of collisions and how we can predict the effect of violent forces that occur in only milliseconds. Pay attention—this is where "it" happens!

Figure 6.1. Game changers. In many sports, collisions—from tackles to batted balls—drive the flow of competition. A game that lasts for an hour or more can depend crucially on events that occur in a fraction of a second. The forces involved can be huge: a good slug might inflict more than a ton of force on the ball in less than a millisecond. **Left:** (© *Corbis RF*); **Right:** (*Reproduced by permission of The News-Gazette, Inc. Permission does not imply endorsement*).

Football isn't a contact sport, it's a collision sport. Dancing is a contact sport.
—Duffy Daugherty, Michigan State football coach

By now, we've covered several concepts needed to understand a baseball play. We can figure out the time a batter has to make a decision and compare it to his reaction time. We know how to determine how far the pitch falls, how it slows in the air, and even how it curves on its way to the batter. Once the ball is hit, we can figure out the time the fielder has to get under the hit (the hang time) and even how the air temperature affects its range. As the runner heads toward first base, we can analyze the friction on his shoes, the acceleration of his center of mass, and the time it takes to cover the 90 ft to the bag.

But the most important element in the play is the hit, a classic example of a **collision**. Collisions are game changers in essentially every sport involving a ball—the smashing tennis return that clinches a set, the perfectly placed spike in volleyball, even the repeated bounces of a basketball as a player dribbles down the court. And collisions need not involve balls, as hockey and football players can tell you only too well. In this chapter, we'll discuss the physics behind the various types of collisions in sports.

6.1 What a Collision Is and How to Think About It

When two objects or people collide, their motion changes, often violently. These changes in velocity (that is, accelerations) are caused by forces, just as we've been discussing in previous chapters. However, whereas our $F = ma$ treatment is perfect for, say, a ballplayer sliding into third base, a different set of ideas is usually used for collisions.

To start, let's get straight on what we mean when we say "collision."

- A collision always involves a **normal** (pushing) force between the colliding partners, for example, between a ball and a bat, or between colliding football players; interactions through gravity or tension forces are not collisions. (There can be friction during the collision too, changing the ball's spin; we'll come to this later.)
- That normal force is relatively **strong**—often thousands of pounds—as in the baseball hit in figure 6.1, so that other forces also at play (for example, the 5-oz weight of the ball or 3-oz air drag during the hit) can be ignored during the brief collision.
- The force **changes rapidly** with time, so our usual approach of assuming a constant force doesn't work.
- The contact time between colliding objects is **very brief**, about 1–20 ms for collisions involving balls, up to ~200 ms for human-on-human collisions; typical collision times are given in table B.1.

Because of the very short collision times, we usually don't care about the details of an object's motion *during* the collision; the $\frac{1}{4}$-in. displacement of the ball while it touches the bat in figure 6.1 is negligible compared to the several hundred feet it will fly afterward. Usually the most important thing is to figure out the *change* that the collision makes between the situation before and after the event.

In the baseball hit, the obvious change generated by the collision is in the ball's velocity. The bat's velocity changes, too, though less so. So we could try to formulate collision physics in terms of velocities.

But, for reasons we'll mention in a little bit, Isaac Newton (yes, him again) showed that to understand collisions, it's easiest to think about changes in the ball's and bat's momentum and kinetic energy. Any moving object has some momentum and kinetic energy which it can transfer to another object. Understanding hits in sports is all about following this transfer.

Momentum and Impulse

Before we get into formulas, vectors, and mathematics, let's think about what we already know about different collisions—what they share in common and what distinguishes one from another.

In the baseball hit in figure 6.1, the bat and ball approach each other and then separate, both moving in the same direction (toward the outfield) at different speeds. During the brief contact interval, the bat gets a violent "push" to the left that slows it, while the ball gets a big push to the right. Newton called this violent push an **impulse**.

Likewise, the football player with the ball in figure 6.1 got an unwelcome rightward impulse courtesy of his friend on the defending team. In turn, video analysis of this sort of tackle makes clear that the defender endured a leftward impulse in the collision.

With Newton's third law (equation 3.7) in mind, it makes sense to wonder: Are the impulses experienced by the two participants in the collisions equal and opposite? Are the changes in velocity equal and opposite?

Let's look at a couple more collisions.

The left side of figure 6.2 shows before-and-after shots of a play where two football players stick together after the tackle, moving with a common velocity. Once again it appears that the pushes (impulses) on the two players, or perhaps the changes in their velocity, might be equal and opposite.

Finally, the right panel of figure 6.2 shows a bird's-eye view of an off-center bowling collision. After the collision, the pin flies off at high speed forward (up in the figure) and to the right. Since it started from rest, obviously the push (impulse) felt by the pin is in the same direction—forward and to the right. Qualitatively, we see the same thing in this collision as in the previous two: the ball experiences an impulse backward (that is, toward the bowler) and to the left.

However, here there seems to be something different in the bowling collision: the ball's leftward velocity is much less than the pin's rightward one. And the change in the ball's forward velocity (that is, the amount it slows) is much less than the change in the pin's forward velocity.

So much for equal and opposite changes in velocity. That's obviously wrong.

Newton told us that it makes a lot more sense to keep track of the **momentum** of the balls and players, rather than their velocity. A person's momentum is the product of his mass and velocity:[1]

$$\vec{p} = m\vec{v}. \tag{6.1}$$

---

[1] If something is going in a circle or spinning, it also has "angular momentum," which we'll get to in chapter 10. In cases where there could be confusion, we'll get specific and call $\vec{p}$ the "linear momentum."

Section 6.2   The Physics of a Football Tackle   **147**

As always, the arrow above the symbol indicates that momentum is a vector. It points in the same direction as the velocity.

Collisions *always* change the momenta of both colliding partners. In fact, the impulse we were discussing above is simply *defined* as the change in an object's momentum:

$$\vec{J} = \vec{p}_a - \vec{p}_b. \tag{6.2}$$

In this formula, *b* and *a* refer to the momentum before and after the collision. As we've said, with collisions it's all about the difference between before and after. Like momentum, impulse is a vector with the strange unit of kg · m/s or slug · ft/s.

Now that we have some idea of what we mean by a collision and what to look for, we'll divide collisions in sports into three categories: tackles, bowling, and dribbles. As we go along, the reason for this division will become clear.

6.2 The Physics of a Football Tackle

In figure 6.2, 200-lb (91 kg) Chris Johnson streaks toward Chandler Williams at 7 m/s, so his momentum is

$$p_{\text{Johnson},x,b} = (91 \text{ kg})(-7 \text{ m/s}) = -637 \text{ kg} \cdot \text{m/s}.$$

As usual, the negative sign means that Johnson's momentum points to the left.

You'll notice (try to suppress the groan, please) the proliferation of subscripts in in $p_{\text{Johnson},x,b}$. That's the nature of the beast—we need all three of them! First, momentum is a vector, so it has components *x*, *y*, and so on. Second, in any collision there are two participants, each with his own momentum; we have to use the labels "Johnson" and "Williams" (or some shorthand such as 1 and 2) to keep track of them. Finally, since we are interested in the *change* in momentum, we need a label to denote the momentum before the collision (*b*) and after (*a*).

During the collision, forces change wildly. The collision begins upon contact. The normal forces exerted by the players are small but quickly grow. The force on Williams is to the left, and the corresponding (reaction) force on Johnson is to the right, per Newton's third law. These forces grow and then diminish until the two players reach the same velocity and they exert no net force on each other. (This is so because if one *did* exert a net force on the other, he'd accelerate away, and the players wouldn't stick together, right? Make sure you understand this.) At this point, even though they are still touching and moving together, we consider the collision over.

After the collision, video analysis reveals that Johnson and Williams move together to the left at 1.8 m/s. So Johnson's momentum after the collision is

$$p_{\text{Johnson},x,a} = (91 \text{ kg})(-1.8 \text{ m/s}) = -163.8 \text{ kg} \cdot \text{m/s},$$

meaning he's experienced an impulse of

$$J_{\text{Johnson},x} = (-163.8 \text{ kg m/s}) - (-637 \text{ kg m/s}) = +473.2 \text{ kg} \cdot \text{m/s}.$$

Pay attention to the signs. Johnson's momentum still points leftward, but the impulse he experiences—the *change* of his momentum—points to the right.

The miracle

So this is all very interesting. (Right?) Newton tells us to pay attention to changes in momentum, instead of changes in velocity. But why?

Figure 6.2
Before-and-after snapshots of two types of collisions in sports. The arrows show the participants' velocities. In the upper panels, Raiders cornerback Chris Johnson gives a clean hit to the Chiefs's Chandler Williams. After the collision, the players "stick together," and so they have the same velocity vector.

 A collision is finished as soon as the momenta of the participants stop changing, even if they remain stuck together.

Here's why. It turns out that while the changes in *velocity* for participants in a collision aren't always equal and opposite, their changes in *momentum* are. In formula form

$$\vec{J}_1 = -\vec{J}_2 \tag{6.3}$$

Two participants in a collision experience equal and opposite impulses.

This tells us immediately that since Johnson's momentum changed by 473.2 kg · m/s to the right, Williams's momentum changed by 473.2 kg · m/s to the *left*.

Let's just pause for a moment to think about it: if the horizontal component of Johnson's momentum *increased* by 473.2 kg · m/s, and the horizontal component of Williams's momentum *decreased* by 473.2 kg · m/s, then the *sum* of the two didn't change at all. This is the crucial principle of the **conservation of momentum** declared by Sir Isaac in 1687:

The sum of the momentum before a collision is the same as the sum of the momentum after the collision.

$$\underbrace{\vec{p}_{1,b} + \vec{p}_{2,b}}_{\vec{p}_{\text{before}}} = \underbrace{\vec{p}_{1,a} + \vec{p}_{2,a}}_{\vec{p}_{\text{after}}} \tag{6.4}$$

Remember: if "outside forces" are important, it's not a collision.

Almost everything changes violently in sports collisions, but the key to understanding these game changers is that the total momentum never changes.

This is really remarkable. Collisions in sports are violent (just ask Chandler Williams!), complicated, messy events in which so much can happen and change—punts deflected, volleyballs spiked, arms broken, heat generated, helmets dented, balls set spinning. Despite all that, momentum is *always* conserved.

Equation 6.4 is true for collisions (such as tackles) where the participants stick together, for collisions (such as in baseball) where they move apart but in the same direction, and for collisions (such as in bowling) where the participants can move away in totally different directions.

It's almost miraculous and extremely powerful to understand that something as simple as the total momentum *never* changes.

Momentum is *transferred* from one participant in a collision to the other. Some of the most game-changing moments in sports depend crucially on that transfer, and our understanding of the underlying physics will allow us to understand these special moments at a deeper level. Example 6.1 applies these concepts to analyze Williams's motion just before he was hit.

EXAMPLE 6.1 *Williams's speed before the hit*

In figure 6.2, Chandler Williams nabs the ball even though he clearly saw, out of the corner of his eye, Johnson closing in on him. Still he maintained concentration enough to grab it, secure it, and turn to face Johnson in the 0.3 s before he was hit.

Turning was important, as this allowed Williams to maintain his forward velocity before the hit. How fast was he moving at this point? It is actually amazing that we can answer this question without knowing any details of the hit.

We start with the momentum of 176-lb (80-kg) Williams. Equation 6.3 tells us that his impulse is equal and opposite Johnson's: $J_{x,\text{Williams}} = -473.2$ kg · m/s, so

$$J_{\text{Williams},x} = p_{\text{Williams},x,a} - p_{\text{Williams},x,b}$$

$$\rightarrow p_{\text{Williams},x,b} = p_{\text{Williams},x,a} - J_{\text{Williams},x}$$

$$= (80 \text{ kg})(-1.8 \text{ m/s}) - \underbrace{(-473.2 \text{ kg} \cdot \text{m/s})}_{\text{with Johnson}} = +329.2 \text{ m/s}.$$

Now it's easy to find his velocity before he was hit:

$$p_{\text{Williams},x,b} = (m_{\text{Williams}}) v_{\text{Williams},x,b}$$

$$\rightarrow v_{\text{Williams},x,b} = \frac{p_{\text{Williams},x,b}}{m_{\text{Williams}}} = \frac{329.2 \text{ m/s}}{80 \text{ kg}} = +4.1 \text{ m/s}$$

> **Question**
>
> In the brief time he had before he was hit, Williams was unable to accelerate to the 7 m/s that Johnson had. If he *had* been able to achieve a rightward velocity of 7 m/s before the hit, would he still have been turned around, or would the two players essentially come to a dead stop?

A handy formula for tackles

In a head-on tackle, the conservation of momentum produces a handy formula to find the velocity of the stuck-together players after the tackle.

$$\boxed{v_{\text{together},x,a} = \frac{m_1 v_{1,x,b} + m_2 v_{2,x,b}}{m_1 + m_2}} \tag{6.5}$$

6.2.1 The energy of a crunch

Even without a formula, you already know that it requires energy to accelerate from a dead stop to a running speed. Likewise, it takes energy to lift a barbell or squeeze one of those jelly-filled squeezy balls that are supposed to relieve stress. Energy is one of the most important concepts in science, and as with everything in physics, it must be quantified mathematically. There are many types of energy, and we'll discuss the ones relevant for sports later. To understand collisions, the most important type of energy is the energy of motion, the **kinetic energy**. The formula for kinetic energy is[2]

$$\boxed{KE = \tfrac{1}{2}mv^2.} \tag{6.6}$$

An object's kinetic energy is *not* a vector. It is just a positive number and does not depend on the direction in which the object is moving.

All types of energy, including kinetic energy, are measured in a new unit:

$$\boxed{1 \text{ joule} = 1 \text{ J} = 1 \text{ kg} \cdot \text{m}^2/\text{s}^2 \quad \text{also} \quad 1 \text{ J} = 1 \text{ N} \cdot \text{m}.} \tag{6.7}$$

---

[2] This formula gives the kinetic energy for an object whose center of mass is *moving* with velocity v. If the object is also *spinning*, then it has "rotational kinetic energy," which we'll discuss in chapter 9. In cases where there could be confusion, we will be specific and call $\tfrac{1}{2}mv^2$ the "translational" kinetic energy.

Obviously, this is a metric unit. Historically, there have been many, many units for energy, and several are still in use today. But the joule is really the only one to learn. If we are using Imperial units for mass and velocity in equation 6.6, we'll need to convert to joules at the end, using

$$1 \text{ slug} \cdot \text{ft}^2/\text{s}^2 = 1 \text{ lb} \cdot \text{ft} = 1.356 \text{ J} \tag{6.8}$$

Let's look again at Chandler and Johnson. Before the collision, their energies are

$$KE_{\text{Williams},b} = \tfrac{1}{2}(80 \text{ kg})(4.1 \text{ m/s})^2 = 672.4 \text{ J}$$

$$KE_{\text{Johnson},b} = \tfrac{1}{2}(91 \text{ kg})(-7 \text{ m/s})^2 = 2230 \text{ J}$$

Johnson's energy is much higher than Williams's. His larger mass is part of the reason, but mostly it is his higher speed and the fact that KE depends on the *square* of v.

After the collision, they share a common velocity and their energies are more similar.

$$KE_{\text{Williams},a} = \tfrac{1}{2}(80 \text{ kg})(-1.8 \text{ m/s})^2 = 129.6 \text{ J}$$

$$KE_{\text{Johnson},a} = \tfrac{1}{2}(91 \text{ kg})(-1.8 \text{ m/s})^2 = 147.4 \text{ J}$$

Something is obvious right away: both Williams and Johnson lost kinetic energy in the collision, so obviously the sum of their KE went down in the collision. There is no "miracle" analog to the conservation of momentum (equation 6.4) for kinetic energy.

However, it turns out that all the kinetic energy "lost" is converted to other forms of energy in the collision. This is important enough to put into a box:

> In most collisions, kinetic energy is converted into heat, and damage. It is never converted back.

> The total kinetic energy after a collision is the total kinetic energy before the collision, reduced by the energy that went into heat, sound, or permanent structural damage (to bodies, balls, or equipment):
>
> $$\underbrace{KE_{1,b} + KE_{2,b}}_{\text{total } KE_{\text{before}}} = \underbrace{KE_{1,a} + KE_{2,a}}_{\text{total } KE_{\text{after}}} + \underbrace{E_{\text{sound}} + E_{\text{heat}} + E_{\text{damage}}}_{\text{converted to other energy forms}} \tag{6.9}$$
>
> (All energies listed are positive numbers.)

The kinetic energy lost in the Williams-Johnson collision is

$$KE_{\text{before}} - KE_{\text{after}} = (672.4 \text{ J} + 2230 \text{ J}) - (129.6 \text{ J} + 147.4 \text{ J}) = 2625.4 \text{ J},$$

or about 90% of the initial energy. Leaving the details of such comparisons until later, I'll just tell you that 2625.4 J is the amount of energy to power a 60-W lightbulb for 45 s, or the energy to lift about 300 lb a distance of 6 ft off the ground (which turns out to be the same energy as 300 lb *dropped* 6 ft!). It's a lot of energy.

There is no easy way to figure out how the converted energy was distributed among sound, heat,[3] dented equipment, and bodily injury. Clearly, depending on how

---

[3] We will use "heat" in the colloquial sense: the energy an object has, making it hotter than its environment. This is perfectly fine for our purposes, but it sometimes annoys physicists who (correctly) point out that there is a more subtle and correct definition.

it is delivered (more on that in the next section), the energy associated with 300 lb dropped 6 ft is more than enough to break several bones. Equipment manufacturers try to make the equipment—pads and helmet—take most of the blow. As the plastic, foam, and leather components flex/stretch, compress/decompress, and rub against one another, most of the kinetic energy eventually winds up as heat.[4]

Finally, some of the "lost" kinetic energy goes into vibrating the air molecules around the collision. It turns out that this is a relatively small fraction of the energy, but it's an important one: it produces the satisfying crunching sound of a good football hit.

6.2.2 Helmet design

Collision analysis mostly deals with the change of each participant's momentum and energy caused by the collision. But what happens *during* the collision may be crucially important if you happen to a participant, as figure 6.3 makes obvious. Both Mr. Coyote and the champion divers have considerable downward momentum when they reach the top of the pool. And both will soon have zero momentum. So the impulse (change in momentum) is the same in the two situations, but the divers' experience is certainly not!

As we've said, even in a collision, it is still forces that do the acceleration. We don't usually use $F = ma$ because the forces change rapidly and violently, but it turns out that there is an important relationship between the impulse and the *average* force in the collision:

$$\boxed{\vec{F}_{\text{ave}} = \frac{\vec{J}}{\Delta t}.} \tag{6.10}$$

Here Δt is the duration of the collision. (See table B.1 for typical values.)

This connection between force and time is crucially important for the design of sports equipment, which is very often designed to change momentum rapidly.

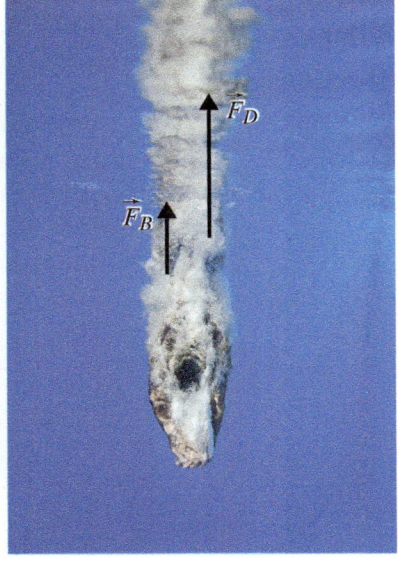

Figure 6.3. There are two ways to end a 10-m dive. Either way, the impulse is the same! **Left:** The unwise "diver" who leaps into an empty pool will experience a rapid change in momentum due to violent normal forces with the concrete. (© *McGraw-Hill Education/Mark Dierker, photographer*). **Right:** When water is present, the athlete experiences the same change in momentum, but over a longer time, because of buoyant and drag forces. (© *microgen/Getty Images RF*).

---

[4] The thermal energy needed to increase the temperature of 1 cup of water by 4.5°F is 2625.4 J.

EXAMPLE 6.2 *Pole vault pit*

Similar to the swimmers in figure 6.3, pole vaulters implicitly trust that the 1-m landing pad will safely bring them to rest; this trust lets them focus on each precious centimeter of the vault. How large is the force experienced by a 80-kg pole vaulter when he hits the pad after a 5-m jump?

The vaulter's momentum after falling 5 m is

$$p_{y,b} = m v_{y,b} = m \underbrace{\left(-\sqrt{2gh}\right)}_{\text{from equation 2.13}} = -(80 \text{ kg}) \sqrt{2 \cdot 9.8 \text{ m/s}^2 \cdot 5 \text{ m}} = -792 \text{ kg} \cdot \text{m/s}.$$

I use the label *b* here, because this is his momentum before he collides with... something. That something could be either a landing pad or the ground. Either way the momentum after the collision is zero, so the impulse is $J_y = 792$ kg · m/s.

Video analysis reveals that a landing pad brings the vaulter to rest in about 150 ms, so the average force on him is

$$|\vec{F}_{\text{ave}}| = \frac{|\vec{J}|}{\Delta t} = \frac{792 \text{ kg} \cdot \text{m/s}}{0.15 \text{ s}} = 5280 \text{ N} \left(\times \frac{0.225 \text{ lb}}{1 \text{ N}} = 1190 \text{ lb}\right).$$

This is more than 0.5 ton, but if distributed over the entire body for a short time, this is not dangerous, so long as he takes care to land on his back.

Example 6.2 shows that, by spreading out the impulse over time, a pole vaulting pit delivers a nonlethal force of about 1000 lb to a vaulter. And Mr. Coyote? What about the force he's going to feel, in figure 6.3? Well, estimating Δt when he strikes hard ground is more difficult, as it is essentially the time it takes his bones and inner organs to squash. However, we can take the contact time of a basketball dribble as an estimate. According to table B.1, this is 10 ms, 15 times shorter than the landing pad. Thus, the average force will be 15 times higher, or over 7 tons. This could present a problem.

It's actually even worse. Equation 6.10 gives the *average* force during the collision, but as we've said, the force varies rapidly. The left side of figure 6.4 plots the force exerted on a helmeted head and an unhelmeted head as it collides with a helmet. The shapes of those force-vs.-time curves are typical of most collisions in sports: once contact is made, the force rises quickly and roughly linearly with time, reaches a peak value, and then falls quickly. We'll call collisions of this type "triangle

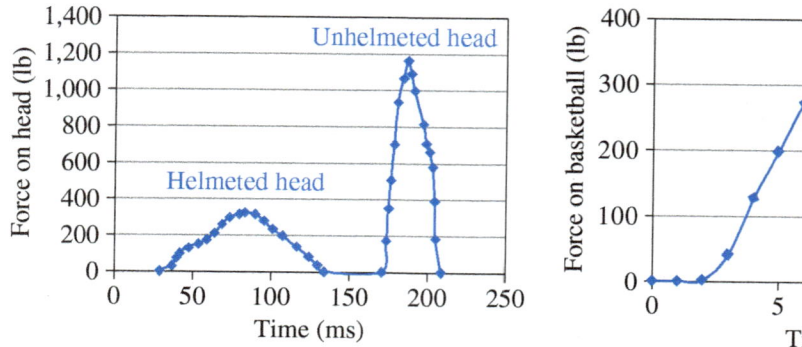
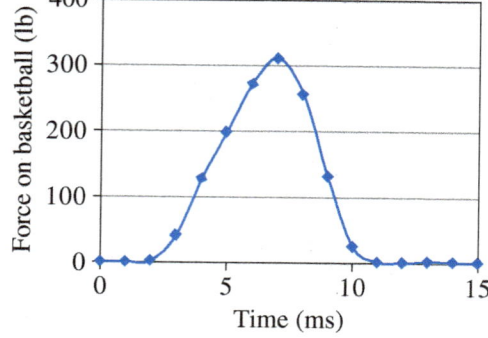

Figure 6.4. Left: the force experienced by a helmeted head, and by an unhelmeted head, when it collides with a helmet. Data from *The Physics of Football*, by Tim Gay. **Right:** the magnitude of the upward force exerted by the court on a bouncing basketball after it's been dropped 6 ft.

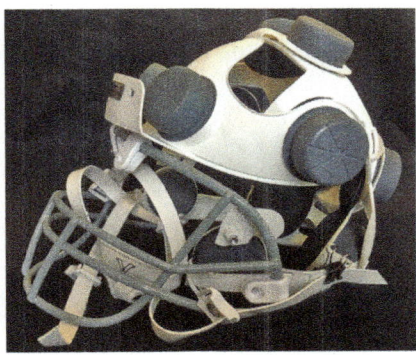

Figure 6.5. The Xenith X2 helmet is in wide use in the NFL. Depicted is the X2 youth model. The impact is spread out *in space* by the hard outer shell. It is spread out *in time* by individual shock absorbers (visible when the shell is removed, on the right) located strategically underneath. (*Photos courtesy of Mike Lisa*).

impacts," inspired by the shape of the curve. A dribbled basketball also experiences a triangle impact, as seen on the right side of the figure.

In a triangle impact, the *peak* force is twice the average force:

$$|\vec{F}_{\text{peak}}| = 2|\vec{F}_{\text{ave}}| = 2\frac{|\vec{J}|}{\Delta t}. \tag{6.11}$$

It is usually F_{peak} that determines the potential for injury, not necessarily the energy. Example 6.3 makes clear that this peak force depends on both participants in a collision.

Visually, it is clear that one function of a football (or other) helmet is to spread out the force of a collision in *space*. Being beaned on a batter's helmet by a fastball isn't fun, but at least the impact is spread over your entire head; it is better than having the ball impact a small area. However, more important is to spread out the collision in *time*. Double the time over which a collision takes place, and you've halved the peak force. This is the purpose of the padding in the vaulter's landing pit, the baseball catcher's chest protector, and the ever-evolving technology of shock absorption in football helmets. Figure 6.5 shows one example.

 A helmet protects by spreading the force both in space via its hard shell and in time via its padding.

EXAMPLE 6.3 *Ray Lewis couldn't hurt a fly*

When I was watching a Ravens game, my ears perked up when an announcer alluded to the physics of linebacker Ray Lewis.

He said that, at top speed, Lewis "has generated almost a ton of force." Hmm... intriguing, but you and I know that this makes no sense. You don't "build up force." Force is the push or pull exerted when Ray "interacts" (it sounds so clinical, almost pleasant, when scientific terms are used, doesn't it?) with another player.

Let's see what kind of forces are involved in a Ray Lewis tackle. At 260 lb (8.13 slug), Lewis regularly reaches 12 mph (17.6 ft/s) as he races toward an opponent. First we'll have Lewis tackle an innocent quarterback, then he'll tackle a fly.

Lewis versus Roethlisberger
Lewis has had some big hits on 241-lb (7.53-slug) quarterback Ben Roethlisberger. When Lewis tackles a stationary Roethlisberger, Big Ben is not stationary for long! According to equation 6.5, both he and Lewis recoil with a speed

$$v_{x,a} = \frac{m_L v_{L,x,b} + m_R v_{R,x,b}}{m_L + m_R} = \frac{(8.13 \text{ slug})(17.6 \text{ ft/s}) + 0}{8.13 \text{ slug} + 7.53 \text{ slug}} = 9.139 \text{ ft/s}.$$

The impulse is the change in Lewis's momentum

$$J_{L,x} = p_{L,x,a} - p_{L,x,b} = m_L \left(v_{L,x,a} - v_{L,x,b} \right)$$
$$= (8.13 \text{ slug})(17.6 \text{ ft/s} - 9.139 \text{ ft/s}) = 68.82 \text{ slug} \cdot \text{ft/s},$$

and equation 6.11 gives us the peak force:

$$|\vec{F}_{L,\text{peak}}| = 2\frac{|\vec{J}_L|}{\Delta t} = 2\frac{68.82 \text{ slug} \cdot \text{ft/s}}{0.15 \text{ s}} = 917 \text{ lb}.$$

Well, it's not a ton of force, but it is almost a half ton. So that's probably what the sportscaster meant. Nevertheless, his mistake is important. Lewis didn't "build up force." He built up momentum. To show the difference, let's have Lewis tackle a housefly.

Lewis versus a housefly
A fly's mass is only 10 mg:

$$m_{\text{fly}} = 10 \text{ mg} \times \frac{1 \text{ kg}}{1000 \text{ g}} \times \frac{1 \text{ g}}{1000 \text{ mg}} \times \frac{1 \text{ slug}}{14.5 \text{ kg}} = 6.9 \times 10^{-7} \text{ slug}$$

After Lewis runs into the fly and they stick together, their speed is

$$v_{x,a} = \frac{m_L v_{L,x,b} + m_{\text{fly}} v_{\text{fly},x,b}}{m_L + m_{\text{fly}}} = \frac{(8.13 \text{ slug})(17.6 \text{ ft/s}) + 0}{8.13 \text{ slug} + 0.00000069 \text{ slug}} = 17.599999 \text{ ft/s}.$$

That fly hardly changed Lewis's initial speed of 17.6 ft/s. (Not surprising, right?)
The magnitude of the fly's impulse is the same as Lewis's:

$$|\vec{J}_{\text{fly}}| = m_{\text{fly}} \left(v_{x\text{fly},a} - v_{x\text{fly},b} \right) = (6.9 \times 10^{-7} \text{ slug})(17.599999 \text{ ft/s} - 0)$$
$$= 1.2 \times 10^{-5} \text{ slug} \cdot \text{ft/s},$$

which means that the peak force is

$$|\vec{F}_{\text{peak,fly}}| = 2\frac{|\vec{J}_{\text{fly}}|}{\Delta t} = 2 \cdot \frac{1.2 \times 10^{-5} \text{ slug} \cdot \text{ft/s}}{0.15 \text{ s}} = 8 \times 10^{-5} \text{ lb}.$$

This is $\frac{8}{100,000}$ lb, or 80 *millionths* of a pound, which is 2.5 thousandths of an ounce. That is not nearly enough force to injure a fly.

The only danger to our housefly is that of having a heart attack (do flies have hearts, actually?) when he sees the Ravens number 52 barreling down on him.

6.2.3 Forcing a runner out of bounds

Some of the most exciting plays in football occur when the ball carrier is streaking alone down the sideline, his blockers having spent themselves on rushing defenders. As the end zone draws near, the best bet for a defender is often to push the runner out of bounds, rather than tackle him. What happens next depends on the sizes and

Section 6.2 The Physics of a Football Tackle

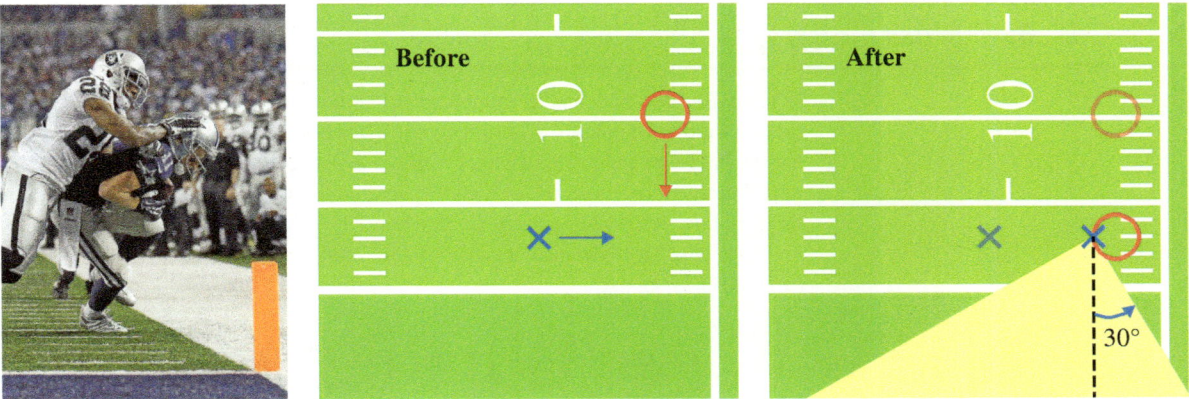

Figure 6.6. Dallas Cowboys wide receiver Cole Beasley (11) is pushed out of bounds by Oakland Raiders cornerback Phillip Adams (28) during a game at AT&T Stadium in Arlington in November 2013. The play is shown schematically on the right. If Beasley had had greater momentum (or if Adams had had less), he would have been in the angular range (shown in yellow) that would have led to a score. (© *Ray Carlin/Icon SMI/Corbis*).

velocities of the two players. Conservation of momentum decides how the whole thing plays out.

In figure 6.6 Raiders cornerback Phillip Adams manages to push Cowboys wide receiver Cole Beasley out of bounds just before he can score. As with the other tackles we've discussed, Adams and Beasley have a common velocity after the collision as they "stick together." If that velocity *vector* points within 30° of forward (see the figure), Dallas scores; if it points outside of that zone, Dallas has to try again or settle for a field goal attempt. The result is in the hands of Isaac Newton and his conservation of momentum.

At 180-lb (5.625-slug) Beasley is fast; he was hauling at about 10 yd/s (30 ft/s) toward the goal line. But cornerbacks are fast, too, and 195-lb (6.094-slug) Adams was moving at about 9 yd/s (27 ft/s) when he hit Beasley. So, in what direction did they fly? That is, in what direction was their velocity pointing after they stuck together?

Well, their posttackle velocity points in the same direction as their combined momentum after the tackle, and this is easy to find: it is just the same as their combined momentum *before* the tackle. If we call the direction toward the goal the x direction and toward the sideline the y direction, the momentum components are

$$p_{both,x,a} = p_{Adams,x,b} + p_{Beasley,x,b} = 0 + (5.625 \text{ slug})(30 \text{ ft/s}) = 168.7 \text{ slug} \cdot \text{ft/s}$$

$$p_{both,y,a} = p_{Adams,y,b} + p_{Beasley,y,b} = (6.094 \text{ slug})(27 \text{ ft/s}) + 0 = 164.5 \text{ slug} \cdot \text{ft/s}$$

Finally, reaching way back to equation 3.14, we can find the deflection angle:

$$\theta = \tan^{-1}\left(\frac{p_{both,y,a}}{p_{both,x,a}}\right) = \tan^{-1}\left(\frac{164.5 \text{ slug} \cdot \text{ft/s}}{168.7 \text{ slug} \cdot \text{ft/s}}\right) = 44.3°$$

The momentum vector and angle are shown in figure 6.7. This is larger than 30°, so the Cowboys have to try again. Because of his speed and mass, Adams had enough momentum going into the tackle to deflect Beasley out of the scoring zone shown by the yellow shaded areas. Example 6.4 examines how a much heavier Beasley could have scored on this play.

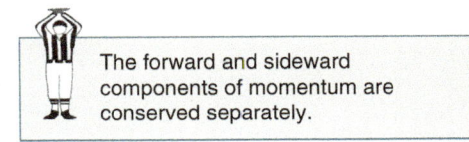

The forward and sideward components of momentum are conserved separately.

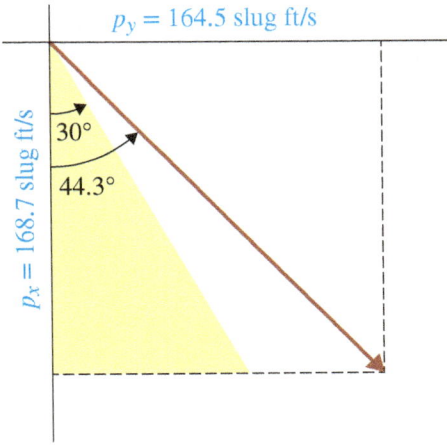

Figure 6.7
Axes rotated to match field orientation. Highlighted triangle shows the 30° scoring zone.

EXAMPLE 6.4 *Mass matters*

If Cole Beasley had had greater forward momentum, the momentum vector after the tackle could have pointed within the yellow zone. Increasing his momentum means either increasing his speed or increasing his mass. A speed of 10 yd/s is about as fast as a player in full gear can be expected to achieve, so let's focus on mass. How much heavier would Beasley have had to be to score on this play?

As is clear from the previous discussion, scoring means $\theta < 30°$. This has direct implications for his mass:

$$\theta = \tan^{-1}\left(\frac{p_{both,y,a}}{p_{both,x,a}}\right) = \tan^{-1}\left(\frac{p_{y,Adams,b}}{p_{x,Beasley,b}}\right) < 30°$$

$$\rightarrow \quad \frac{p_{Adams,y,b}}{p_{Beasley,x,b}} < \tan 30°$$

$$\rightarrow \quad p_{Beasley,x,b} > \frac{p_{Adams,y,b}}{\tan 30°} = \frac{164.5 \text{ slug} \cdot \text{ft/s}}{\tan 30°} = 284.9 \text{ slug} \cdot \text{ft/s}$$

Did you follow that math? If not, go through the three steps again more slowly.

Now we know that his momentum must be larger than 284.9 slug ft/s. Since we really can't ask him to run any faster, now we know what his mass has to be

$$p_{Beasley,x,b} = m_{Beasley} \times (30 \text{ ft/s}) > 284.9 \text{ slug} \cdot \text{ft/s}$$

$$\rightarrow \quad m_{Beasley} > \frac{284.9 \text{ slug} \cdot \text{ft/s}}{30 \text{ ft/s}} = 9.5 \text{ slug}$$

In order for Beasley to have scored in this situation, his mass would have to be larger than 9.5 slug, meaning that instead of 180 lb, he would weigh upward of (9.5 slug) $(32 \text{ ft/s}^2) = 304$ lb! That is the kind of weight you're more likely to find in a lineman. However, this extra mass comes at a cost: linemen are typically significantly slower than wide receivers.

The bottom line is that given Adam's mass and approach speed, there was little chance that *anybody* could have scored on this play.

6.3 Gentler Pursuits: Bowling

We turn now from one of the most dangerous of sports, where participants deal with collisions up close and personal, to one where collisions occur safely 60 ft away from the player.

Together with her husband, Chris, Hall-of-Famer Lynda Barnes heads what's been called the "first family" of bowling. Video analysis of her games reveals that her 14-lb (0.4375-slug) ball is moving at about 25 ft/s, or 17 mph, when it crashes into the pins.

The result is determined by what happens next. At what speed will the struck pins recoil and at what angle? We'll cover this in the next few pages, with a break in the middle to visit a seven-year-old's birthday party a few lanes down.

6.3.1 Beginner's first roll: head-on collision

Bowlers quickly learn that optimum results come from hitting the set of 10 pins at an angle, usually with a curve on the ball. But beginning bowlers often just try to smash the ball head-on into the 1-pin (the pin nearest the bowler). In either case, a lot depends on the speed of the pin's recoil and the ball's deflection after the collision. Let's study the situation in detail.

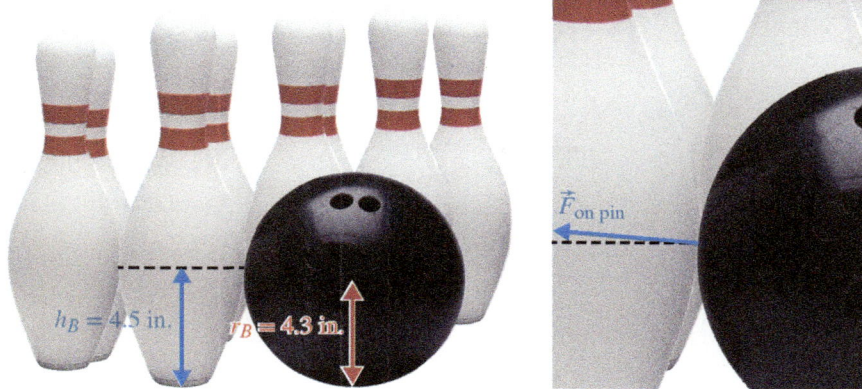

Figure 6.8. The radius of a bowling ball is slightly smaller than the height of a bowling pin's "belly," so the force on the pin is directed very slightly upward, at an angle of about 1.3°. (© *Photodisc/Getty Images RF*).

Since the bowling ball and pin are curved, they come into contact over a very small region—essentially a single point—just below the "belly" of the pin, which is its widest part. As is clear from figure 6.8, because normal (or "pushing") forces always point perpendicular to the surfaces in contact, the ball pushes very slightly up on the pin, as well as forward. The small upward component is important and worth discussing; we'll come to it in a bit. For now, though, we'll talk about the much stronger horizontal push.

To understand the ball–pin collision, we need to know two things. The most important is that the momentum of the collision is conserved, as given in equation 6.4. In this case, the momentum of the pin before it collides with the ball is zero, so the total momentum in the horizontal direction is

$$p_{x,\text{before}} = m_{\text{ball}} v_{\text{ball},x,b} = (0.4375 \text{ slug})(25 \text{ ft/s}) = 10.9 \text{ slug} \cdot \text{ft/s}.$$

This tells us the *sum* of $p_{\text{ball},a,x}$ and $p_{\text{pin},x,a}$, but that's not enough to tell us how fast the pin recoils.

The second thing we need to know is that collisions in bowling are **elastic collisions**, and that in head-on elastic collisions, the approach speed of the participants before the collision is equal to the separation speed after.[5]

 In head-on elastic collisions, the separation speed is equal to the approach speed.

$$\text{elastic collisions:} \quad v_{\text{rel},b} = -v_{\text{rel},a} \qquad (6.12)$$

The important thing here is the relative velocity. Before the collision, the relative velocity (approach velocity) is

$$v_{\text{rel},b} = v_{\text{pin},x,b} - v_{\text{ball},x,b} = 0 - 17 \text{ mph} = -17 \text{ mph}$$

The relative velocity before the collision is always negative, because the distance between the objects about to collide is diminishing with time.

Some math

Now it's time to put these concepts together and figure out the recoil speed. We'll need to go through a bit of math, but hang with it; it's not so much. It just looks cumbersome because we are using the long "ball" and "pin" labels. In fact, keeping

---

[5] This is true only for head-on collisions; we'll talk about off-center collisions later.

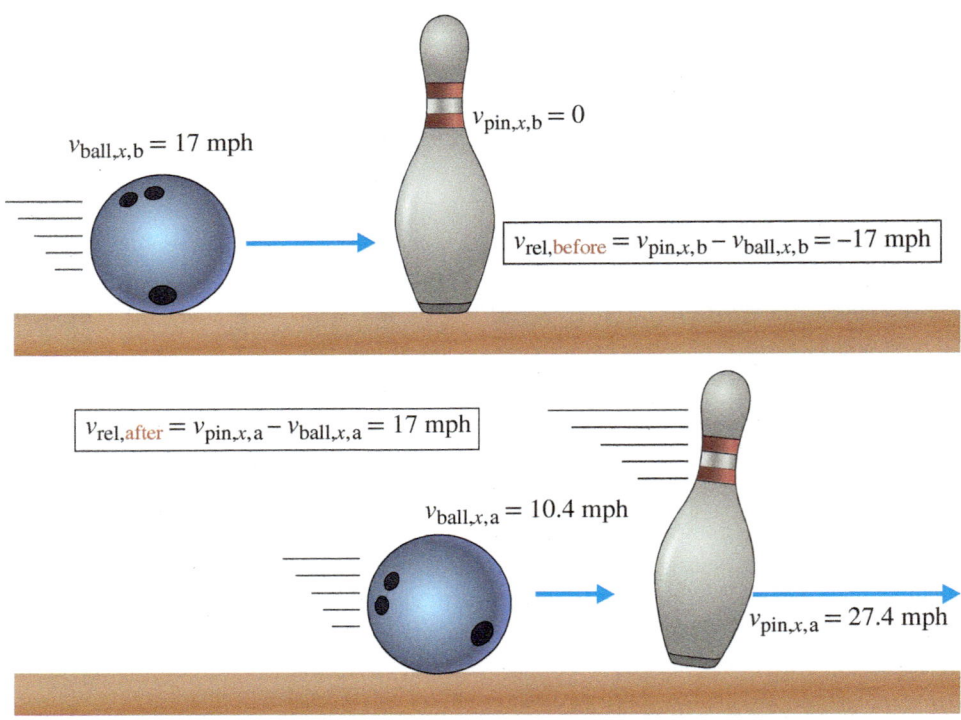

Figure 6.9. Snapshots before and after a ball–pin collision. Because of its much lower mass, the pin recoils backward at a high speed, while the ball slows by about 40%. Because the collision is elastic, the relative speed before and after the collision is the same.

track of the labels (including "before" and "after") is probably the hardest thing about analyzing collision equations (see figure 6.9).

We'll start with the relative velocities, directly from equation 6.12:

$$v_{pin,x,a} - v_{ball,x,a} = -(\underbrace{v_{pin,x,b}}_{\text{0, at rest}} - v_{ball,x,b}) \rightarrow v_{ball,x,a} = v_{pin,x,a} - v_{ball,x,b} \quad (6.13)$$

The only other ingredient to add is that the sum of momenta before the collision is the same as the sum after, as we said in equation 6.4:

$$m_{ball} v_{ball,x,b} + \underbrace{m_{pin} v_{pin,x,b}}_{\substack{\text{zero because} \\ \text{pin at rest}}} = m_{ball} \underbrace{(v_{pin,x,a} - v_{ball,x,b})}_{\substack{\text{this is } v_{ball,x,a} \\ \text{from equation 6.13}}} + m_{pin} v_{pin,x,a} \quad (6.14)$$

Okay, that's pretty much it. Equation 6.14 can be rearranged to give a nice formula for the recoil speed of a pin hit straight on:

$$v_{pin,x,a} = \frac{2 m_{ball}}{m_{ball} + m_{pin}} v_{ball,x,b} \quad (6.15)$$

And this together with equation 6.13 can be rearranged to give the ball's speed after it hits the pin:

$$v_{ball,x,a} = \frac{m_{ball} - m_{pin}}{m_{ball} + m_{pin}} v_{ball,x,b} \quad (6.16)$$

Section 6.3 Gentler Pursuits: Bowling **159**

Looking at the formulas

Now, I promised that this book was not going to be all about deriving formulas and the like. So what's the deal? Why bother deriving equations 6.15 and 6.16? Well, it really is worthwhile to see how the fundamentally important concepts such as momentum conservation (equation 6.4) and the elastic collision condition (equation 6.12) need to be combined to analyze collisions. And come on, it wasn't so bad.

In fact, when they have a formula, physicists like to play with it a little bit; it helps develop intuition. Let's quickly go through some examples to see how this works.

- *Same mass ball and pin.* What if the pin had the same mass as the ball? Then equation 6.16 tells us that the ball would come to a dead stop after crashing into the pin ($v_{ball,x,a} = 0$)! Equation 6.15 says that the pin would fly off at exactly the same speed as the ball initially had ($v_{pin,x,a} = v_{ball,x,b}$). This is very much like a hard head-on shot in billiards, where the cue ball comes to a sudden stop upon collision, and the eight ball recoils, as shown in figure 6.10.[6]

- *Very heavy pin.* Or, what if the pin was very, very heavy, so much heavier than the ball that the ball's mass might as well be zero? In that case, equation 6.15 says that the velocity of the pin after the collision is zero: the super-heavy pin doesn't recoil at all. (But you knew that, right?) And the ball? Equation 6.16 says that the ball recoils directly back with the same speed it came in with ($v_{ball,x,a} = -v_{ball,x,b}$), as shown in figure 6.11. This is like throwing a superball against the side of a building.

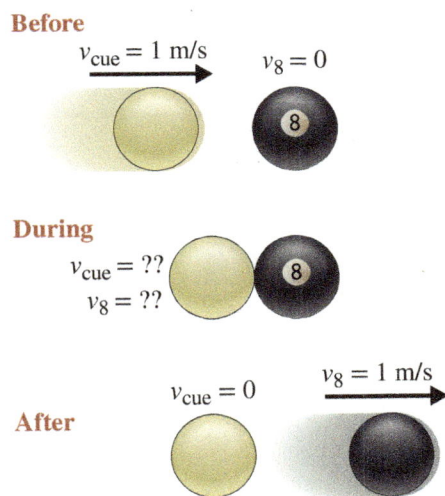

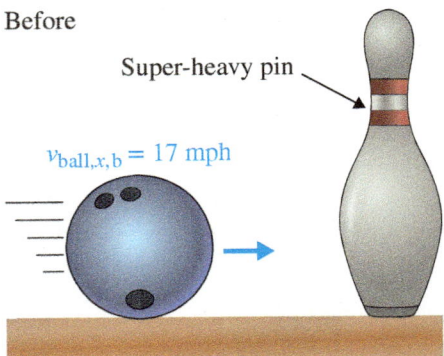

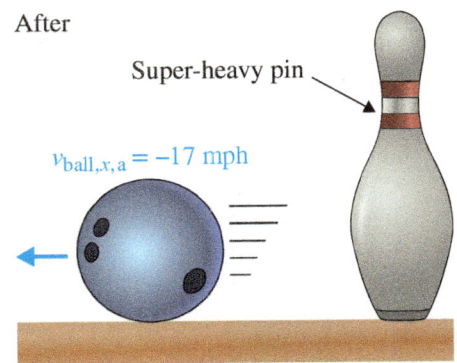

Figure 6.10. In a hard head-on collision between a cue ball and a numbered ball in billiards, the cue ball comes to rest, and the numbered ball recoils at the original speed of the cue ball. During the collision, the speeds change violently and are much harder to analyze.

Figure 6.11
If the pin were made out of uranium, its mass would be huge and it wouldn't recoil at all. The beginner bowler would find the ball coming directly back at him.

---

[6] On some nonprofessional tables, often the pay tables in bars, the cue ball is slightly heavier (6 oz) than the numbered balls (5.5 oz). This makes it easier for the return mechanism on the table to hold the numbered balls until the next customer pays. On these tables, the cue tends to "follow" after it hits a numbered ball. Also, if topspin or backspin is put on the cue, it will tend to follow or pull back after the strike.

Physics sets a limit on the speed at which an object flies away after a collision; this is a key concept for sports such as golf and baseball.

- *Very light pin.* Or, what if the pin was super-light, like a Ping-Pong ball? A physicist will go all the way and say the mass of the pin is zero. Before reading on... what do you *think* will happen in this case?

  ...Done thinking? Okay. Equation 6.16 reveals what you'd suspect: the ball will continue at the same speed as before, 17 mph—it won't slow down at all. You might *not* have guessed what happens with the pin, though Equation 6.15 tells us that it will fly away at twice the ball's speed: 34 mph. The laws of physics don't allow anything to recoil faster than that. In fact, that's a general principle, worth a box.

Collision speed limit: When a stationary object is struck by a moving one, it will *never* bounce away with more than twice the velocity of the initially moving object.

The closest one can come to this speed limit is when a light object (such as a golf ball) is struck by a much heavier one (such as a club head) in an elastic collision.

Real balls, real pins

Okay, let's use the masses of a real ball and pin to return to the original question of the pin's recoil after a ball strikes it at 17 mph. Equation 6.15 says

$$v_{\text{pin},x,a} = \frac{2\,(0.4375 \text{ slug})}{0.4375 \text{ slug} + 0.105 \text{ slug}}\,(17 \text{ mph}) = 27.4 \text{ mph}.$$

As for the ball, equation 6.16 gives us its speed after the collision:

$$v_{\text{ball},x,a} = \frac{(0.4375 \text{ slug}) - (0.105 \text{ slug})}{(0.4375 \text{ slug}) + (0.105 \text{ slug})}\,(17 \text{ mph}) = 10.4 \text{ mph}.$$

And, as promised by equation 6.12, the separation velocity is

$$v_{\text{rel,after}} = v_{\text{pin},x,a} - v_{\text{ball},x,a} = 27.4 \text{ mph} - 10.4 \text{ mph} = 17 \text{ mph}$$
$$= -v_{\text{rel,before}}$$

Example 6.5 weighs the relative advantages of a heavier ball and a faster roll.

EXAMPLE 6.5 *Should you use a heavier ball or should you roll faster?*

The heaviest ball allowed in official bowling tournaments weighs 16 lb, which is 14% more than Barnes's 14-lb ball. To accelerate this heavier ball to the same 17 mph would require the bowler to exert 14% more force (assuming the same Δt).

How much faster would the pin recoil if Barnes used the 16-lb ball?

$$v_{\text{pin},a,x} = \frac{2\,(0.5 \text{ slug})}{0.5 \text{ slug} + 0.105 \text{ slug}}\,(17 \text{ mph}) = 28.1 \text{ mph}.$$

The 14% increase in ball weight leads to only a 2% increase in pin recoil speed and only a 6.7% increase in final ball speed.

On the other hand, if Barnes were 14% stronger, she could forgo the heavier bar and instead use the increased force to accelerate her original 14-lb ball 14% more to $1.14 \times (17 \text{ mph}) = 19.4 \text{ mph}$. If she did that, the pin recoil speed would be

$$v_{\text{pin},a,x} = \frac{2\,(0.4375 \text{ slug})}{0.4375 \text{ slug} + 0.105 \text{ slug}}\,(19.4 \text{ mph}) = 32.9 \text{ mph},$$

a gain of 14% in pin recoil speed.

So, it would make little sense for Barnes to push to the limits to use the heavier ball, especially if it might hinder her amazing consistency. It is best to stick with the comfortable 14-lb ball, and simply roll it a bit faster when she wants more oomph.

How much kinetic energy was "lost"?

In the football collisions of section 6.2.1, a lot of the energy the players had built up while running was converted into heat and damage. We figured out precisely how much, using equation 6.9. Is the same true for the bowling collisions?

Before the collision, only the ball is moving, so

$$\text{KE}_{\text{before}} = \tfrac{1}{2} m_{\text{ball}} v_{\text{ball},b}^2 = \tfrac{1}{2}(6.34 \text{ kg})(7.6 \text{ m/s})^2 = 183 \text{ J}.$$

Here, I decided to use SI units, because as we said, they really are the best when quantifying energy.

After the collision, both ball and pin are moving.

$$\begin{aligned}\text{KE}_{\text{after}} &= \tfrac{1}{2} m_{\text{ball}} v_{\text{ball},a}^2 + \tfrac{1}{2} m_{\text{pin}} v_{\text{pin},a}^2 \\ &= \tfrac{1}{2}(6.34 \text{ kg})(4.65 \text{ m/s})^2 + \tfrac{1}{2}(1.52 \text{ kg})(12.25 \text{ m/s})^2 = 183 \text{ J}.\end{aligned}$$

There is the same amount of kinetic energy after the collision as before.

In fact, in an elastic collision, none of the initial kinetic energy is converted to another form. The kinetic energy is conserved.

$$\boxed{\text{KE}_{1,b} + \text{KE}_{2,b} = \text{KE}_{1,a} + \text{KE}_{2,a}} \tag{6.17}$$

 In elastic collisions, no kinetic energy is converted to heat, damage, or sound.

So, if you hit the cue hard enough to crack the eight-ball in billiards, kinetic energy has been converted to damage, and the collision is no longer elastic. In fact, the fact that you hear a "click" when two billiard balls collide tells you that even low-speed collisions are not perfectly elastic, since some of the kinetic energy went into sound.

6.3.2 Birthday party bowling

One great thing about bowling is that the Barnes family may be practicing in lane 1, the beginner from section 6.3.1 is in lane 16, and happy and loud and oblivious to anyone else, the people in a child's seventh birthday party use lane 22. You know the kind of party: the staff has put bumpers on the sides of the lane, so the ball ricochets from side to side as it lumbers its way toward the pins, rather than falling in the gutter.

One kid rolls the ball with two hands, bent over and swinging it underhand ("Granny style") between his legs. The ball rolls slowly, ka-thunk, ka-thunk, ka-thunk, as the finger holes hit the floor. It's hardly moving at all when it slowly bumps into the head pin and ... stops.

To the delight of the children, an annoyed teenager has to walk down the lane to retrieve the ball. But as he walks, he wonders, What about the conservation of momentum?

After all, it doesn't take any mathematics to realize that before the ball came into contact with the pin, there was some momentum, and after the contact, there was none. Why doesn't Newton's important equation 6.4 work? Well, that's what the qualifier at the bottom of the box for that equation refers to.

It's important to remember that for the collisions we are discussing, the normal force between the colliding partners (the ball and pin) must be overwhelmingly stronger than any other forces. This was true for the adult bowler, but the normal force exerted by the slow-rolling child's ball can't even overcome the static friction of the pin on the wooden alley. As we're using the term, the ball–pin interaction in this case is not a collision.

 Our analysis of a collision is valid only if the only relevant force is the one that the participants exert on one another. In many realistic situations, this will not be the case.

For the record

The physicist in me simply *must* tell you that, in reality, momentum *is* conserved in the interaction discussed above. It always is, in collisions or *any* interaction in the universe. It's a crucial law of nature.

However, the collision is no longer simply between the ball and the pin. Because the pin and the Earth (which is attached to the bowling alley) are coupled by friction, the collision is really between the ball and the pin-plus-Earth system. The pin-plus-Earth system *does* recoil, and so the forward momentum of the colliding pair *is* conserved. It is simply that the recoil velocity is *very* slow, because the mass of the pin plus Earth is *very* large. (Remember, $\vec{p} = m\vec{v}$.)

I simply *had* to tell you that because the scientific principle of momentum conservation is so important in the strictest sense. In most physics courses, the professor would insist upon discussing frictionless alleys, "perfect" pins, zero gravity, and probably no air. But when we look at athletics in the real world, pure "isolated" collisions are rare, and musings on the infinitesimal recoil of the planet is perhaps interesting but ultimately not useful.

More important is the ability to realize when any complicating effects are sufficiently small that an idea such as the conservation of momentum becomes a useful tool to understand the physics of sports. Sometimes it is, and sometimes it isn't. We decide based on the four bullet points in the box at the beginning of this chapter.

6.3.3 Off-center hits: converting a lily

A game of bowling consists of 10 "frames." Each player rolls one or two balls in each frame, the goal being to knock down as many pins as possible. Ideally, all 10 pins are knocked down with the first ball; this is called a "strike," and the frame is over for that player. However, if there are pins still standing after the first ball, the goal of the second ball is to get a "spare" by knocking down the remaining pins.

Returning to the adults' game on lane 16, our teenage worker finds that the beginner's first ball has left a "lily" formation, in which the 5-, 7-, and 10-pins have been left standing; this arrangement is shown in figure 6.12. This formation is essentially never seen in games with expert players, but the beginner who puts little or no spin on the ball will sometimes face it. The teenager shakes his head; the lily is one of the most difficult splits to convert to a spare. The trick to achieving it is known mostly to advanced bowlers, and it's purely driven by physics.

The conversion of any split requires that a pin be hit off-center, so that it recoils to the side as well as backward. You intuitively know two things: (1) the ball is going to recoil partially sideways too and (2) this whole discussion is going to involve vectors and angles. In fact, the algebra here can be rather involved, because of all the variables at play. We'll discuss the main physics point and how it leads to interesting consequences for off-center elastic collisions.

The beginner bowler decides to hit the 5-pin off-center such that it recoils directly backward into the 10-pin. This requires delivering an impulse that points directly from the 5-pin to the 10-pin. The angle of this impulse is pretty easy to find from geometry (even if the shot is difficult to perform in reality). The details are given in the caption to figure 6.13. To keep you on your toes, I decided to use number (1 and 2) subscripts this time, rather than "ball" and "pin." We see that we'd like to hit pin 5 so that it recoils at 60°.

If the beginner rolls his 0.375-slug (14-lb) ball at 25 ft/s and hits the 5-pin at just the right spot, he can indeed get the pin to recoil at 60°. What happens next is determined by the physics we've been discussing. The important thing is that

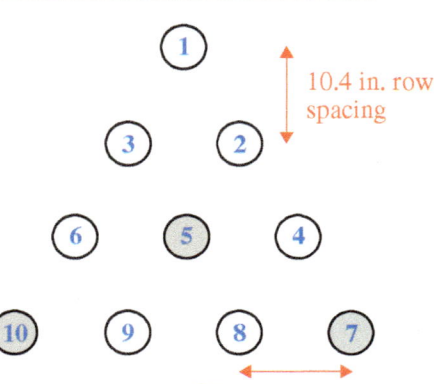

Figure 6.12
Top: a bowling ball approaches a full set of pins. If the lily formation, which is pins 5, 7, and 10, is left standing, the pin spacings make it very difficult to knock them all down with one more roll to convert a spare. (© *Royalty-Free/Corbis*).

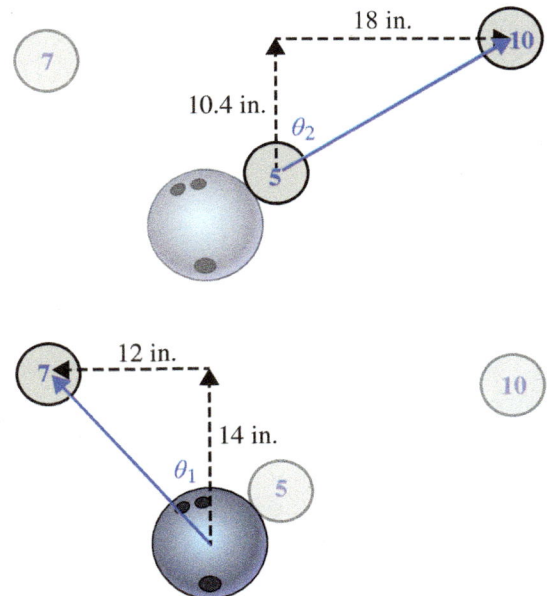

Figure 6.13. Ideally, the momentum of the 5-pin after the collision should point along the hypotenuse of the triangle. Equation 3.14 tells us the angle:

$$\theta_2 = \tan^{-1}\left(\frac{18 \text{ in.}}{10.4 \text{ in.}}\right) = 60°.$$

The bottom panel shows the ideal direction for the ball's momentum:

$$\theta_1 = \tan^{-1}\left(\frac{-12 \text{ in.}}{14 \text{ in.}}\right) = -40.6°.$$

(1) momentum is conserved (equation 6.4) and (2) kinetic energy is conserved (since this is an elastic collision). Using this information and doing some math (which we won't go through in detail), we find that it will crash into the 10-pin with a speed of 20.14 ft/s. The beginner is delighted.

Now, what about the ball? Did it bounce in the other direction to hit the 7-pin, as desired?

To answer this question, we turn back to equation 6.4, which is a *vector* equation. This means that just as the momentum in the forward (x) direction has to be conserved, so, too, the momentum in the sideways (y) direction needs to be conserved. We made this same point in section 6.2.3 when we analyzed a tackler pushing a runner out of bounds. Since the beginner rolled the ball straight down the alley, there was no momentum in the sideways (or y) direction before the collision, so there must be no sideways momentum afterward.

$$\underbrace{0}_{p_{y,\text{before}}} = \underbrace{p_{1,y,a} + p_{2,y,a}}_{p_{y,\text{after}}} \quad \rightarrow \quad p_{1,y,a} = -p_{2,y,a}$$

$$= -(0.105 \text{ slug})(20.14 \text{ ft/s})\sin 60°$$

$$= -1.83 \text{ slug} \cdot \text{ft/s}.$$

As usual, the minus sign means that the ball deflects to the left. The situation is shown in figure 6.14.

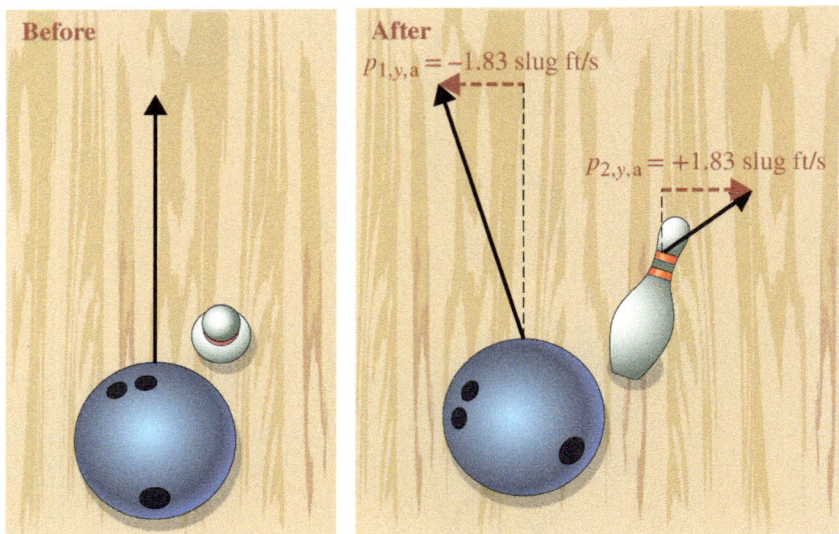

Figure 6.14. Arrows show the momenta of the ball and pin. Even though the ball has a lower velocity than the pin, its momentum is much larger, because of its mass. Since there was no momentum in the sideways (y) direction before the collision, there must be none after: the ball's leftward momentum precisely cancels the pin's rightward momentum.

Likewise, the momentum in the forward (x) direction has to be conserved:

$$\underbrace{p_{1,x,b}}_{p_{x,\text{before}}} = \underbrace{p_{1,x,a} + p_{2,x,a}}_{p_{x,\text{after}}}$$

$$\rightarrow p_{1,x,a} = p_{1,x,b} - p_{2,x,a}$$

$$= (0.4375 \text{ slug})(25 \text{ ft/s}) - (0.105 \text{ slug})(20.14 \text{ ft/s})\cos 60°$$

$$= 9.88 \text{ slug} \cdot \text{ft/s}$$

The ball's forward momentum component is much larger than its sideward component, so it deflects only a small amount. Using equation 3.14, we find the ball's deflection angle to be

$$\theta_1 = \tan^{-1}\left(\frac{p_{1,y,a}}{p_{1,x,a}}\right) = \tan^{-1}\left(\frac{-1.83 \text{ slug} \cdot \text{ft/s}}{9.88 \text{ slug} \cdot \text{ft/s}}\right) = -10.5°.$$

(Again, a negative angle here means deflection to the left.) This is much smaller than the ideal deflection angle of 40.6° shown in figure 6.13. In fact, as shown in the top panel of figure 6.15, even accounting for its relatively large diameter, the ball will completely miss the 7-pin.

It's hopeless. *There is no way to convert this split with a 14-lb ball on a straight shot.* Expert bowlers know the physics secret though: one *must* use a much lighter ball. Because of the conservation of momentum, a lighter ball will be deflected to a greater angle.

How much lighter? To figure this out, the only things we need to know are that kinetic energy is conserved in elastic collisions (equation 6.17) and that the momentum in all directions is conserved in *all* collisions (equation 6.4). After that, there's quite a bit of uninteresting algebra, so we just tabulate the result in table 6.1. We see that the straight-on bowler can achieve the required ball deflection *only* with a ball 6 lb or lighter. In fact, the lightest ball available at most alleys is precisely 6 lb.

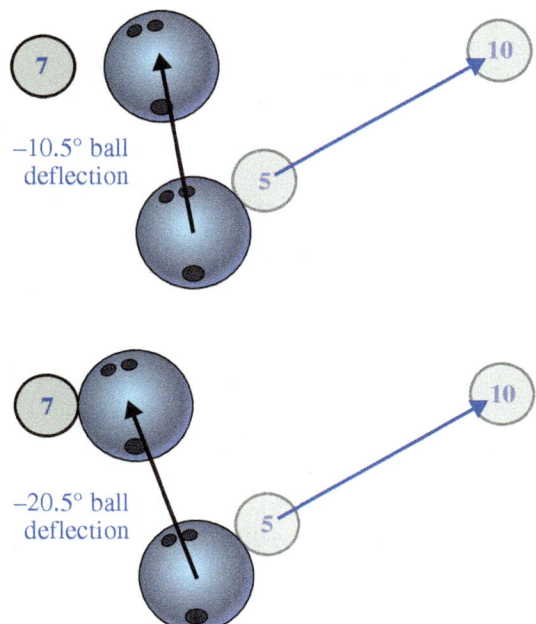

Table 6.1. Ball weight vs. deflection angle after bouncing off a 3-lb 6-oz pin that recoils at 60°.

Ball weight	Deflection angle
16 lb	−9.3°
14 lb	−10.5°
12 lb	−12.0°
8 lb	−16.7°
7 lb	−18.6°
6 lb	−20.7°
3 lb 6 oz	−30°

Figure 6.15. If the 5-pin is deflected into the 10-pin, a 14-lb (0.4375-slug) ball is deflected 10.5° the other way, too little to hit the 7-pin and convert the spare. Using a lighter ball will lead to the same pin deflection while achieving a larger ball deflection.

6.3.4 Off-center billiards shots

In table 6.1, we've included the case when the ball has the same weight (mass) as the bowling pin. In that case, the pin deflects 60° to the right and the ball 30° to the left. That is, they deflect at right angles to each other.

In fact, for off-center elastic collisions, the two objects *always* deflect at right angles if they have the same mass.[7] This fact is not relevant in bowling, since you'll never find a ball as light as a bowling pin. However, it completely dominates the outcome of every collision in billiards, curling, or shuffleboard.

Figure 6.16 shows the "90° rule," well known to experienced players. In a "stun shot" with no topspin or backspin,[8] the cue ball and object ball (that is, numbered ball) leave any off-center collision at right angles to each other if they have the same mass. As we have just discussed, the angle will be smaller if the cue ball is heavier than the object ball, as is often the case in barroom tables.[9]

In an elastic collision, the two balls fly off at right angles to each other.

6.3.5 Beyond two dimensions: The upward hop of the pin

The beginner bowler, unaware that he *must* use a 6-lb ball, fails to convert the lily for a spare. The teenage worker is not surprised, as he knows the "light ball" trick from experience. To augment his intuition, he's also been reading this textbook (there's a lot of downtime, working at a bowling alley), and understands the vector nature of momentum conservation.

---

[7] Actually, this is true only if one of the objects is initially at rest. But this is always the situation in elastic collisions in sports.

[8] The effects of spin and side English on the ball's trajectories can be dramatic and useful; we don't go into them here.

[9] Another caveat: no collision is ever *completely* elastic. Even in billiards, some of the energy goes into the sound of the "click." Therefore, the 90° rule is never perfectly obeyed. In practice, the angle between the recoiling balls is always a little less than 90°.

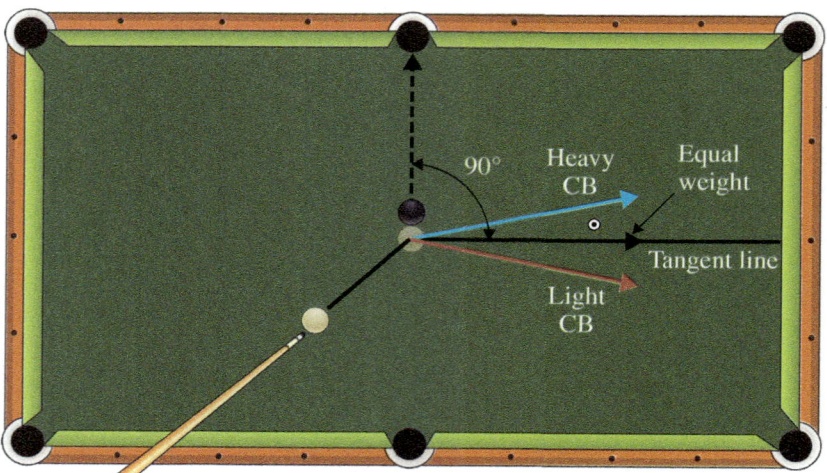

Figure 6.16. An illustration of the 90° rule in billiards when the cue ball (CB) has the same weight as the struck ball. The angle is more or less than 90° if the cue ball is lighter or heavier than the struck ball.

But now his curiosity is piqued. We've been talking about the momentum vector in the two horizontal directions (forward/backward and left/right), but what about the *vertical* direction? In addition to the tumbling motion of a struck bowling pin, if you look closely, you'll notice that the center of mass of the pin generally flies upward. And if you remember the right side of figure 6.8, it is clear why: there is a small upward component of the normal force on the pin.

This all makes sense, but the teenager is right to be puzzled. What about momentum? Obviously *before* the collision, neither the ball nor the pin has any vertical momentum, so the conservation of momentum says that *after* the collision the total momentum should be zero as well. But it's not so: the ball has no vertical momentum after the collision, but the pin does—momentum is not conserved in the vertical direction.

Why? Again it's due to the fact that the system is not isolated. When the ball strikes the pin, it experiences a normal force that is slightly directed downward; this is essentially the "reaction force" to the one shown on the right side of figure 6.8. When this happens, the GRF (the normal force from the floor) increases proportionally, so the vertical force on the ball is zero and it doesn't bounce up or accelerate into the ground. In other words, there is an *external* force in the vertical direction in effect during the collision, so vertical momentum will not be conserved. Example 6.6 looks at this quantitatively.

This is an important point for us. The conservation of momentum is hugely important in science and can be very useful in understanding sports. But it works only if a system is isolated—if external forces are absent or are much, much smaller than other forces at play in the collision. In physics textbooks the systems are isolated (no friction, no air, etc.), but in real life they often are not. In fact sometimes, as in the present case, the system is isolated in some directions (for example, horizontal) but not in others (for example, vertical), so momentum conservation is a useful concept in only some directions. We need to use our common sense when deciding when to apply scientific concepts to any given situation.

> **Question**
>
> In section 6.3.2, we talked about another system that was not isolated, since friction was not small relative to other forces. At the end of that section, we discussed that *total* momentum—if the Earth is included—*is* conserved. Is the same true here? If so, how?

EXAMPLE 6.6 *Vertical impulse*

Even though momentum is not conserved in the vertical direction, the concept of impulse can still be useful. We can actually use it to figure out the vertical speed at which the pin is thrust by the collision, using simple trigonometry.

In the head-on collision described in section 6.3.1, the 0.105-slug pin recoiled backward at a horizontal speed of $v_{\text{pin},x,a} = 27.4$ mph $= 40.2$ ft/s. The horizontal component of its impulse is then

$$J_{\text{pin},x} = p_{\text{pin},x,a} - p_{\text{pin},x,b} = (0.105 \text{ slug})(40.2 \text{ ft/s}) - 0 = 4.22 \text{ slug} \cdot \text{ft/s}.$$

(Remember b and a mean before and after.)

Since we know that the angle of the normal force $\vec{F}_{\text{on pin}}$ is directed $1.3°$ upward, the impulse $\vec{J}_{\text{pin}}$ is as well. Reaching all the way back to what we learned about vectors in chapter 3, we can use equation 3.14 to find $J_{\text{pin},y}$:

$$\theta = \tan^{-1}\left(\frac{J_{\text{pin},y}}{J_{\text{pin},x}}\right) \rightarrow \frac{J_{\text{pin},y}}{J_{\text{pin},x}} = \tan\theta$$

$$\rightarrow J_{\text{pin},y} = J_{\text{pin},x}\tan\theta = (4.22 \text{ slug} \cdot \text{ft/s})(\tan 1.3°) = 0.096 \text{ slug} \cdot \text{ft/s}$$

Now we have the impulse in the vertical direction, which is the same as the change in momentum in the vertical direction. So we can answer the question.

$$J_{\text{pin},y} = p_{\text{pin},y,a} - \underbrace{p_{\text{pin},y,b}}_{0} \rightarrow m_{\text{pin}}v_{\text{pin},y,a} = J_{\text{pin},y}$$

$$\rightarrow v_{\text{pin},y,a} = \frac{J_{\text{pin},y}}{m_{\text{pin}}} = \frac{0.096 \text{ slug} \cdot \text{ft/s}}{0.105 \text{ slug}} = 0.91 \text{ ft/s}.$$

This nice little upward hop of the pin as it flies backward is important. It causes the struck pin to recoil more toward the top of the other pins, making it easier to topple them over.

6.4 A Happy Medium: Dribbling and Driving

We began this chapter by discussing a dangerous game punctuated by the most violent collisions and moved on to safe games where the biggest dangers come from dropping a bowling ball on your foot or getting caught cheating at billiards against a 300-lb biker in a bar. Now we come somewhere in the middle and talk about basketball, baseball, and tennis. It's more athletic than bowling or billiards but a bit more civilized than football.

The collisions in these sports are between the two extremes also in a more important sense. In tackles, the collision participants stick together after the collision: $v_{\text{rel,after}} = 0$. In bowling, the participants fly apart at the same speed as they initially approached one another: $v_{\text{rel,after}} = -v_{\text{rel,before}}$. In baseball, basketball, and tennis, the two objects (for example, ball and racquet) separate after the collision, but more slowly than their approach velocity.

Physicists and sports equipment manufacturers use the **coefficient of restitution (COR)** e to classify collisions. It is simply the ratio of the separation speed to the approach speed:

$$e = -\frac{v_{\text{rel,after}}}{v_{\text{rel,before}}} \tag{6.18}$$

As we said with equation 6.12, the minus sign is there simply because $v_{\text{rel,before}}$ is a negative number.

Every type of collision—bowling strikes, football tackles, baseball hits—has its own COR that tells how bouncy it is. The three categories of sports collisions are summarized in table 6.2.

You can find extensive regulations and specifications on the COR for sports equipment, from baseballs to golf club heads to high-altitude tennis balls. The COR quantifies the **inelasticity** of the collision, which is related to the fraction of the initial kinetic energy converted to heat, sound, or damage.

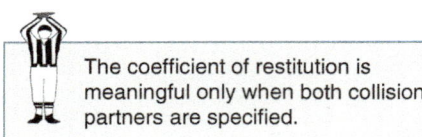
The coefficient of restitution is meaningful only when both collision partners are specified.

An important note: Manufacturers or articles will often talk about the coefficient of restitution of a golfball or of a bat. This is actually misleading and can lead to wrong expectations of the equipment's performance. The coefficient of restitution is defined only between *two* objects, such as a bat and a baseball, or a golfball and a clubhead. Obviously, the coefficient of restitution between a baseball bat and a lump of clay is a lot less than that between the bat and a ball!

Table 6.2. Collisions in sports can be separated into three categories, depending on the fraction of kinetic energy lost to heat, sound, and damage.

		Conserved?		
Type of collision	Comments	KE?	$\vec{p}$?	COR
Elastic (billiards, bowling, curling)	No generation of heat, sound, and so on.[a]	Yes	Yes	$e = 1$
Inelastic (baseball, basketball dribble, hockey, golf)	Some energy converted to heat, sound, breakage, and so on.	No	Yes	$0 < e < 1$
Completely inelastic (football tackle, baseball in mitt)	Participants stick together after the collision. Maximum possible amount of kinetic energy converted to heat, and so on.	No	Yes	$e = 0$

[a] Here is an example where the idealized concepts of the theoretical physicist collide with reality. In sports, there is always *some* kinetic energy lost in the collision, as evidenced by the sound produced when a ball strikes a bowling pin. It turns out that the fraction of energy converted in billiards-type collisions is so small as to be negligible, so we ignore it. This is a common approach in science.

6.4.1 The sad, short life of the NBA's synthetic ball

In 2006, NBA Commissioner David Stern announced that the league would begin using a new synthetic-surface ball, rather than the leather balls used since its inception in the 1940s. Besides a new ribbing pattern, the new ball differed significantly from the old.

Players' initial reaction to the ball during preseason training was almost universally negative, and the complaints only got worse as the early season progressed. Shaquille O'Neal, Dirk Nowitzki, Steve Nash, and Dwyane Wade were among the most vocal critics. Complaints included increased hand cuts and abrasions and a general complaint that the ball didn't "feel" right or "play" right off the boards. Spaulding and Stern insisted that the ball had undergone rigorous testing and had been proved "better" and more consistent than the leather balls.

Finally, Dallas Mavericks owner Mark Cuban commissioned two physics professors at the University of Texas at Arlington to make a scientific comparison of the balls. One of the most important findings they reported was that the coefficient of restitution was significantly lower for the synthetic-ball-on-floor collision than for the leather-ball-on-floor collision. Example 6.7 examines how even this small change can affect a player's natural dribbling rhythm.

To determine e, the scientists did not need high-speed video cameras watching the ball approach and leave the floor, as would seem to be required from equation 6.18. It's actually much easier. From equation 2.13, we already know that the relative velocity between the ball and the floor is related directly to the height it is dropped from:

$$v_{\text{rel,before}} = -\sqrt{2gh_i}.$$

Also, from equation 2.14 the height that the ball reaches after the bounce is totally determined by its speed after leaving the floor:

$$h_f = \frac{v_{\text{rel,after}}^2}{2g} \rightarrow v_{\text{rel,after}} = \sqrt{2gh_f}.$$

Putting these together, we come to a useful equation for finding the coefficient of restitution *when something is bouncing vertically off a nonrebounding floor*:

$$\text{bouncing vertically on floor:} \quad e = \sqrt{\frac{h_f}{h_i}} \quad (6.19)$$

Figure 6.17 shows the situation.

Dropping the ball from 1.3 m, researchers found that the leather ball bounced back to a height of 0.853 m, and the synthetic ball to a height of 0.797 m, so

$$e_{\text{leather}} = \sqrt{\frac{0.853 \text{ m}}{1.3 \text{ m}}} = 0.81$$

$$e_{\text{synthetic}} = \sqrt{\frac{0.797 \text{ m}}{1.3 \text{ m}}} = 0.78.$$

Regulating bodies for many sports use equation 6.19 to specify allowed ranges for the coefficient of restitution; see example 6.8 for tennis.

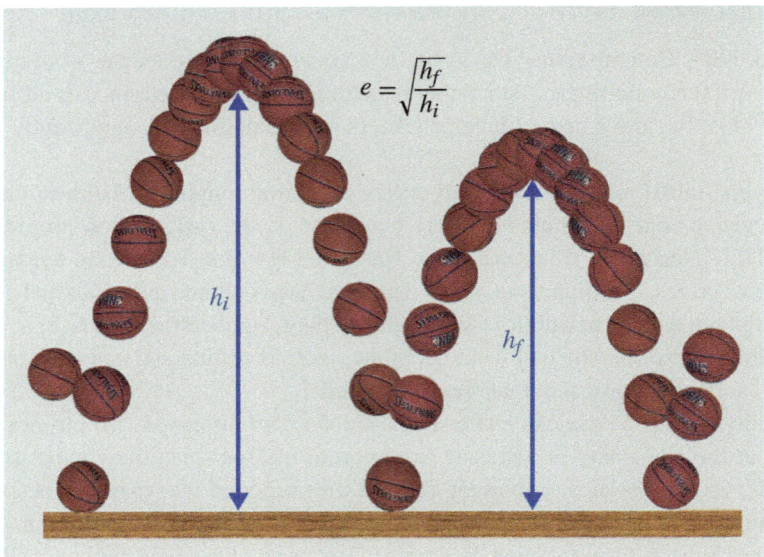

Figure 6.17. The coefficient of restitution for an object bouncing off a stationary floor is related to the heights of successive bounces. (*Photo courtesy of Mike Lisa*).

EXAMPLE 6.7 Chris Paul's "automatic" dribble

Mark Cuban noted in his November 10 blog that there were winners and losers when the new ball was introduced. The winners were guys who make their living making long jump shots. Because of the lowered coefficient of restitution, the ball didn't come off the rim as hard, and shots that would have bounced out with the leather ball were more likely to go in with the synthetic one. Several research articles were written about the ball; figure 6.18 from one of them makes Cuban's point.

The losers were the point guards who rely on muscle memory to know precisely where the ball is, as they make passes or a quick move to the basket, as in figure 6.19. Chris Paul is without question one of the quickest and surest ball handlers in the NBA. How much did the new ball throw off his rhythm?

Video analysis shows that the 6-ft Paul typically releases a dribble about 1 m from the court, and the ball hits the floor 0.16 s later. (At various points in the game, Paul often

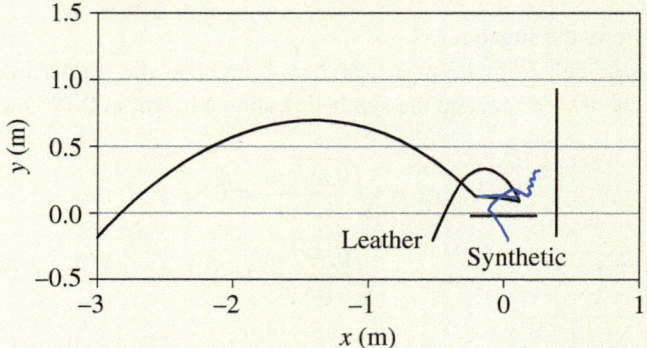

Figure 6.18. The same shot taken with a leather ball and a synthetic ball will have different outcomes. Because of the lower coefficient of restitution, the synthetic ball goes in. From a research article on the topic: Okubo and Hubbard, *Engineering of Sport* 7:705–712 (2009).

dribbles harder and lower, however.) Let's see how long he has to wait for the ball to come back to his hand.

We'll break the dribble into three parts:

1. **Downward journey** Equation 2.10 gives us the speed at which Paul throws the ball down:

$$\Delta y = v_{y,0}\, \Delta t - \tfrac{1}{2} g\, \Delta t^2$$

$$\rightarrow v_{y,0} = \frac{\Delta y + g\, \Delta t^2/2}{\Delta t} = \frac{(-1\text{ m}) + (9.8\text{ m/s}^2)(0.16\text{ s})^2/2}{0.16\text{ s}} = -5.47\text{ m/s}$$

And equation 2.8 tells the ball's speed as it hits the floor:

$$v_y = v_{y,0} - g\, \Delta t = -5.47\text{ m/s} - (9.8\text{ m/s}^2)(0.16\text{ s}) = -7.03\text{ m/s}.$$

2. **Bounce** The coefficient of restitution connects the velocity before the bounce to that after the bounce. We use equation 6.18, with the COR for the leather ball:

$$e = -\frac{v_{\text{rel,after}}}{v_{\text{rel,before}}}$$

$$\rightarrow v_{\text{rebound}} = v_{\text{rel,after}} = -e \times v_{\text{rel,before}} = -0.81\,(-7.03\text{ m/s}) = +5.70\text{ m/s}.$$

3. **Upward journey** Now we just play things in reverse. Equation 2.9 gives us the speed of the ball when it returns to Paul's hand:

$$v_y^2 = v_{y,0}^2 - 2g\, \Delta y$$

$$\rightarrow v_y = \sqrt{v_{y,0}^2 - 2g\, \Delta y} = \sqrt{(5.70\text{ m/s})^2 - 2(9.8\text{ m/s}^2)(1\text{ m})} = 3.235\text{ m/s}.$$

Finally, we can figure out the time it takes to return to his hand, once it's left the court, using equation 2.8:

$$v_y = v_{y,0} - gt \rightarrow t = -\frac{v_y - v_y 0}{g} = -\frac{3.235\text{ m/s} - 5.70\text{ m/s}}{9.8\text{ m/s}^2} = 0.216\text{ s}.$$

Figure 6.19
Basketball is a fast-moving game, and the player must instinctively know when and how hard the ball will return to his hand when dribbled, freeing his mind to deal with the changing situation on the court. (© McGraw-Hill Edication/John Flournoy, photographer).

So, Paul's hand "knows" that the leather ball will return to it 216 ms after hitting the floor.

And with the synthetic ball? Using $e = 0.78$ instead of 0.81 in step 2 above, we'd find that his hand has to wait 229 ms. That extra 13 ms—only 6% longer—corresponds to a 2-in. shift in the ball's position. It won't force Paul to lose control, but it can throw off his rhythm, make him less comfortable. This sense of discomfort was one of the major complaints the players had with the new ball.

The reduced coefficient of restitution was not the only problem. In fact, by over-inflating the synthetic ball to 14 PSI rather than the recommended 8.5 PSI, the new ball could be made to bounce as high as the leather one. The problem then was elasticity: even at the recommended pressure, the synthetic ball was harder to compress (hence, grab); the difference became even greater when it was overinflated.

Friction was the other major issue. The coefficient of friction between skin and a dry leather ball is $\mu_s = 1.69$, whereas the coefficient using a synthetic ball was a

(© Royalty-Free/Corbis)

whopping 3.2! This led directly to cuts, abrasions, and bandages on players' hands. When the skin was moistened somewhat with sweat (actually, Visine was used by the physicists), the situation reversed: the synthetic ball was suddenly much more slippery than the leather one.[10]

Amid the outcry from star players, lab results from the Texas physicists, and grievances brought on by the players' union, the NBA took the remarkable step of returning to the leather ball midseason, after only 2 months of play with the synthetic one.

> **Question**
>
> The International Basketball Federation (FIBA) rule 5 for ball testing states: "The internal pressure of the balls will be increased until the rebound height from a drop height of 1800 mm is at least 1300 mm ± 100 mm. The test is performed on a plane surface with a mass of more than 1 ton." Why the requirement of a mass of 1 ton?

6.4.2 Pádraig Harrington's drive and swinging harder

Golfers are more familiar with the coefficient of restitution than are other athletes, since the elasticity of a ball–club collision is critical in determining the ball's launch speed off the tee. Let's see what goes into 2012 PGA Grand Slam winner Pádraig Harrington's drive.

Three ingredients determine the launch velocity of a golfball: (1) the energy and momentum of the clubhead just before the collision; (2) the transfer of that momentum to the ball; and (3) the loft angle. We'll take these in reverse order.

Loft and tension in the shaft

Figure 6.20 shows Harrington just before and just after he makes contact with the ball. Just before the clubhead strikes the ball, it is moving horizontally—all

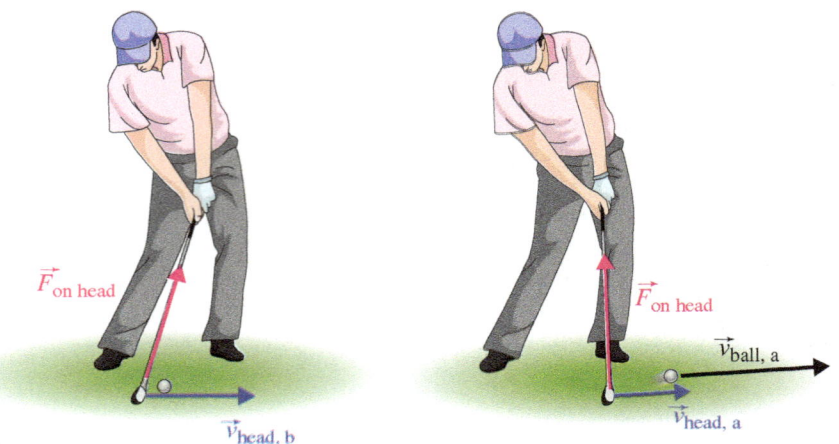

Figure 6.20. Pádraig Harrington drives one. Based on high-speed video footage, the right image represents the situation only 6 ms later than the left one.

---

[10] One reason is that leather absorbs liquid much more efficiently. During the course of a game, a leather ball may increase in weight by as much as 10%, owing to absorbed sweat.

momentum points to the right. After the ball is struck, both the ball and the head have some upward momentum; vertical momentum is not conserved. It's important to understand why.

A golfer's bag contains clubs with heads of varying mass, composition, and loft; some are shown in figures 6.21 and 6.22. Since the normal force is always perpendicular to the surface of contact, a large loft (for example, the right panel of figure 6.21) means a horizontally moving clubhead accelerates the ball up and to the right. As we discussed in chapter 5, air drag means that maximum range is achieved for angles smaller than 45°. For the very high speeds in a long drive, the optimum angle turns out to be about 11°–15°. For that reason, drivers (often identifiable by silly sock covers in the bag) have the smallest loft angles.

Okay, fine. Now we know why the ball's momentum after the collision has an upward component. But shouldn't that mean that the clubhead feels a *downward* force in the collision, so should be moving *down* and to the right after the collision? Well, yes, that would be true if this were an isolated system. But it's not—there's a net force upward when the collision happens, because Harrington is moving the head on a curved path that can be approximated by a circle. Well, we know what that means: he's exerting a centripetal force toward the center of the circle, which is upward in this case. This force—the tension in the club shaft—is indicated in figure 6.20.

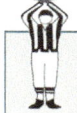
The dominant force exerted by a club shaft is the tension running along the shaft. This centripetal (upward) force means that the vertical component of momentum is not conserved in this collision.

It turns out that the tension in the shaft, which is a force that runs *along* the shaft, is essentially the only large force on the clubhead at the bottom of the swing. This may be surprising to the nongolfer; we might imagine horizontal forces due to the torque exerted by Harrington. But such misconceptions are easily dispelled by stunt golfers who drive a ball using a clubhead on a chain, which can *only* support forces directly along the chain.

Figure 6.21. Golf heads with varying loft are designed for long drives (left) or closer to the green. (*Photos courtesy of Mike Lisa*).

Figure 6.22. The loft angle of a clubhead is the angle between the normal force on the ball and the horizontal velocity of the club upon contact. The normal force is perpendicular to the head's face. Harrington's driver has a 9° loft. (*Photo courtesy of Mike Lisa*).

Therefore, in the horizontal direction, the system is isolated.[11] Any residual horizontal force due to shaft elasticity is negligible compared to the normal forces exceeding 1000 lb. Since Harrington's driver loft is small (9°), we'll ignore the vertical direction altogether until the next section.

Transferring momentum to the ball

Harrington uses a TaylorMade Burner Superfast driver with a 200-g titanium head and 9° loft. His average head speed is about 115 mph (51.4 m/s). What kind of ball speed does that produce?

In section 6.3.1, we talked about the completely elastic (no kinetic energy lost, $e = 1$) collision between a bowling ball and pin. The separation speed after the collision was the same as the approach speed before the collision: $|v_{\text{rel,after}}| = |v_{\text{rel,before}}|$. When some kinetic energy is lost ($e < 1$), as in golf, the separation speed is lower ($|v_{\text{rel,after}}| = e|v_{\text{rel,before}}|$), and we get the following formulas for the velocity of the clubhead and the ball after the collision:

$$v_{\text{ball},x,a} = \frac{1+e}{1+m_{\text{ball}}/m_{\text{head}}} v_{\text{head},x,b} \quad (6.20)$$

$$v_{\text{head},x,a} = \frac{1 - e\, m_{\text{ball}}/m_{\text{head}}}{1+m_{\text{ball}}/m_{\text{head}}} v_{\text{head},x,b} \quad (6.21)$$

The math that produces these formulas is just the same as what gave us the bowling equations 6.15 and 6.16, but with $e \neq 1$.

Using the mass of a ball and Harrington's clubhead, $m_{\text{ball}}/m_{\text{head}} = (45.9 \text{ g})/(200 \text{ g}) = 0.23$, so the denominators in the formula above are 1.23.

What's more interesting is the numerator of equation 6.20. One thing it tells us is that the faster the club is swung, the faster the ball comes off the tee, which is hardly a surprise. There are "long drivers" out there, who can achieve head speeds in excess of 150 mph. But professional player, who live by precision and work on their long *and* short games, are more typically in the range of 110–120 mph.

The only other thing under our control is the coefficient of restitution; equation 6.20 quantifies what we know, which is that a more elastic collision results in higher ball speed. Around 1970, the COR of a golfball with the hard floor was about the same as a golf ball with a club, about 0.7. Since then balls have improved, and there are USGA-compliant balls that have $e = 0.81$ on a hard floor. Drivers have also improved; by making the head act more as a trampoline, the driver–ball COR is larger than the floor–ball COR. Today, there are drivers whose COR with a golfball is $e = 0.9$.

In something of an arms race with equipment manufacturers, the USGA (U.S. Golf Association) tightly regulates clubs to ensure that talent, and not technology, dominates the sport. Current rules forbid drivers having a COR with a golf ball of more than 0.83. Not surprisingly, manufacturers make sure their clubs are right up against that limit.

---

[11] Remember in section 6.3.5 we talked about isolation of the system in one direction but not the other.

So, with all this information, how fast can Harrington launch the ball? Here are some numbers:

$$v_{\text{ball},a} = \frac{1 + 0.70}{1.23} (115 \text{ mph}) = 159 \text{ mph} \quad e = 0.70, 1970\text{'s equipment}$$

$$v_{\text{ball},a} = \frac{1 + 0.83}{1.23} (115 \text{ mph}) = 171 \text{ mph} \quad e = 0.83, \text{current equipment (legal)}$$

$$v_{\text{ball},a} = \frac{1 + 0.90}{1.23} (115 \text{ mph}) = 178 \text{ mph} \quad e = 0.90, \text{current equipment (illegal)}$$

$$v_{\text{ball},a} = \frac{1 + 1.00}{1.23} (115 \text{ mph}) = 187 \text{ mph} \quad e = 1.00, \text{ideal}$$

$$v_{\text{ball},a} = \frac{1 + 1.00}{1.00} (115 \text{ mph}) = 230 \text{ mph} \quad e = 1.00, \text{ideal and super-heavy}$$

The "ideal" case corresponds to the unattainable limit in which no kinetic energy would be lost in the collision, not even to make the "whack!" sound of the drive. The final one listed is for a "super-heavy" clubhead, so that $m_{\text{ball}}/m_{\text{head}} = 0$ in equation 6.20. This would correspond to an ideal (no energy lost) collision between a Mack truck and the golf ball. Clearly, increasing the head mass so much would have a much larger impact than the incremental improvements due to a larger COR. However, accelerating a Mack truck to 115 mph takes, well, a truck engine! As it turns out, 200 g is pretty much ideal, because of human body mechanics.

The diminishing return of a faster swing

Armed with a driver at with the legal-limit COR of 0.83, the golfer who longs for greater range naturally thinks of increasing his swing speed. In fact, we would naively expect that increased club speed should *really* help, because of the details of the range equation. Do you remember those details? Well, here's a reminder:

When we first started talking about projectile motion in chapter 4, equation 4.4 related the range of a projectile with its initial speed:

$$R = \frac{v_0^2 \sin 2\theta}{g}.$$

The fact that the range increases quadratically with v_0 (that is, with the square of v_0) means that the launch velocity matters a *lot*. For example, if we could somehow double the launch velocity, we'd increase the range by a factor of 4! Or, more realistically, if we increased the speed by 20%, say from 100 to 120 mph, the range would increase by 44%. Such an improvement would change a 250-yd drive to a 360-yd one. It would seem worthwhile to push hard for the 20% increased swing speed!

Unfortunately (or, for those of us who cannot achieve high speeds, fortunately), however, the gain in range is much less than these expectations, for two reasons.
1. **COR decreases with $v_{\text{rel},b}$.** According to equation 6.20, doubling the club speed should double the ball speed; this comes directly from equation 6.18, which says that the separation speed is proportional to the approach speed. These statements would be true, if e were a constant number. However, in most collisions, the COR is reduced at higher impact velocities.

Figure 6.23 shows data compiled by the USGA for various clubs and balls. The COR falls essentially linearly with club speed—you get less bang for the buck as

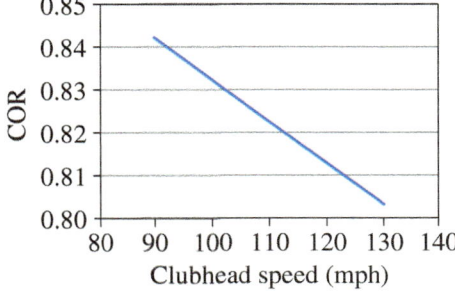

Figure 6.23
USGA data show that the coefficient of restitution decreases at higher club speeds.

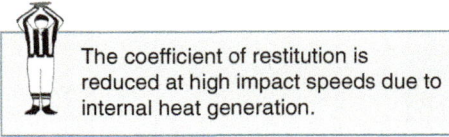

The coefficient of restitution is reduced at high impact speeds due to internal heat generation.

you swing faster! Some golfers believe that players with fast swings get an unfair advantage with modern balls and clubs, but this data refutes the idea.

The reason for this strange effect is shown in figure 6.24, again from the USGA. We can see that the ball compresses significantly for both swing speeds shown, though, not surprisingly, the compression is a bit more for the high swing speed. This is where some golfers worry: they think that the ball is "activated" only at higher compression. However, just the opposite is true. The fact is that *the more a ball compresses, the more internal friction heats it up, so the more kinetic energy is lost*. That is, the collision becomes less elastic and *e* goes down.

The reason that bowling or billiard collisions are elastic is that those balls and pins don't compress (much).

2. Air drag. The second reason for the diminishing return is something you already know: aerodynamics. There are two aerodynamic effects, one good and one bad. The lift (Magnus force) from the ball's backspin helps the ball go farther, but the drag reduces its range. Equation 5.6 tells us that the lift grows linearly with the ball's speed, but equation 5.4 tells us that drag grows quadratically. That is, the "bad" aerodynamic effect counteracts the "good" one increasingly at high speed.

These two reasons together lead to diminishing returns on launch speed. Figure 6.25 shows USGA measurements of range vs. club speed. Far from our naive quadratic dependence on $v_{\text{head},b}$, the range doesn't even grow linearly with it!

To be sure, a faster club speed leads to a ball hit farther. It's just that you get less bang for the buck at higher speeds.

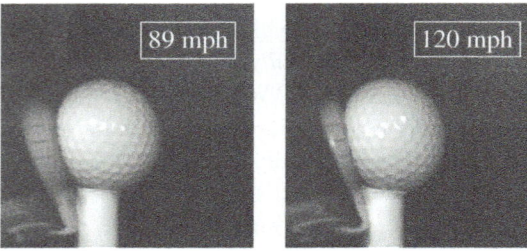

Figure 6.24. Golf balls are compressed significantly at all driving speeds, though a bit more as the club speed increases. (*Images courtesy of US Golfing Association*).

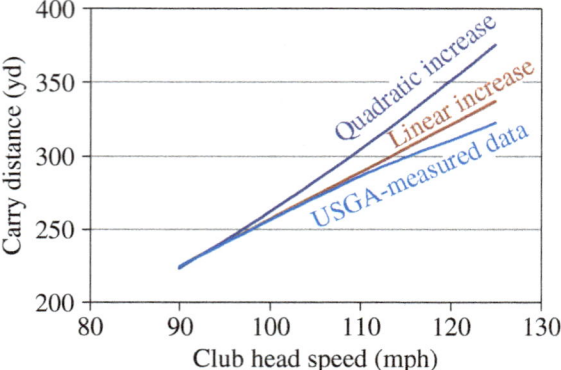

Figure 6.25. USGA data on driving range vs. club swing speed. The range increases as club speed increases, but less than linearly and much less than the naive quadratic expectation.

EXAMPLE 6.8 *COR of tennis balls*

The International Tennis Federation (ITF) regulates equipment for the sport and specifies that type 1 (fast) balls must rebound between 53 and 58 in. "when dropped from 254 cm (100 in.) onto a smooth, rigid and horizontal surface." What is the range of COR for legal tennis balls colliding with a hard ground?

This is easily found with equation 6.19.

$$e_{minimum} = \sqrt{\frac{53 \text{ in.}}{100 \text{ in.}}} = 0.728 \qquad e_{maximum} = \sqrt{\frac{58 \text{ in.}}{100 \text{ in.}}} = 0.762$$

This range of e corresponds to dropping a ball from a height of just 4 ft, so $v_{rel,b} \approx$ 14 mph. Just as with golf balls, however, e is much lower for harder collisions. At 45 mph, $e \approx 0.59$ and at 100 mph, $e \approx 0.42$.

> **Question**
>
> The ITF sets a different specification for high-altitude tennis balls. When dropped from 100 in., they must rebound between 48 and 53 in. Do they have a higher or a lower COR, compared to regular balls? Why do you think the specifications on these balls are different?

6.5 Off-Center Hits: Spinning the Ball

All collisions involve strong normal forces, which can change the momentum of a ball violently. Often, however, the **spin** of the ball is as important as its final velocity.

Now, whenever things start spinning in physics, the math can get complicated, fast. Fortunately for us, however, we can stick to the basics and things won't get too hairy. There are just three things we need to know to get started:

1. A nonspinning object will remain nonspinning until a **torque** is applied to it.

2. Assuming it doesn't change shape, a spinning object will remain spinning at the same rate until a **torque** is applied to it.

3. If it spins at all, a free object will spin about an axis that passes through its center of mass.

Okay, the first two items are actually two sides of a coin, right? And they look a lot like Newton's first law: if there's not a *force*, an object moves at constant *velocity*. With spin, we just replace "force" with "torque" and "velocity" with "spin rate." Any force can make a torque if its "line of action" doesn't pass through the center of mass. The line of action is easier to see than to define with words. In figure 6.26, the line of action force from the player's finger goes directly through the center of the basketball (where the center of mass is), so it exerts no torque and won't slow the ball. (As usual, however, friction and air drag eventually do slow the ball.)

Okay, "torque" is a pretty impressive word, but that's about as specific as we'll get for now. We'll come back to it in chapter 9.

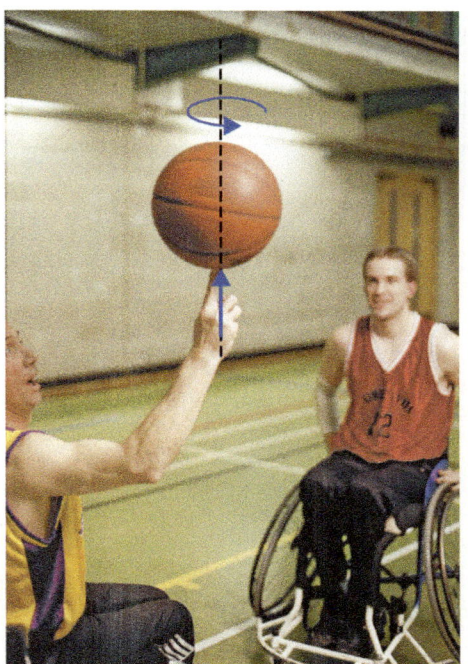

Figure 6.26
When he spins the ball on his finger, the "line of action" of the normal force from a player's finger passes through the ball's center of mass, so it exerts no torque. This means it doesn't increase or decrease the ball's spin rate. (© Image100/Alamy RF).

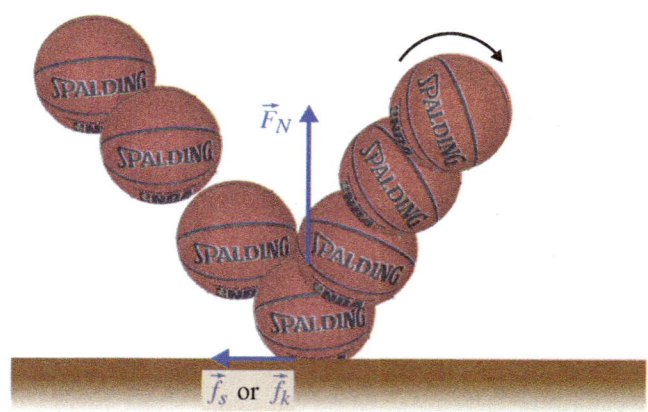

Figure 6.27. When a basketball bounces, its velocity and spin change. (*Photo courtesy of Mike Lisa*).

6.5.1 Bounce pass

A common example is shown in figure 6.27. There are two forces exerted by the floor. The first is the familiar normal force, which has to point straight up. The line of action of $\vec{F}_N$ points through the ball's center of mass, so doesn't exert a torque so doesn't cause a spin.

In addition, however, there is a friction force. Friction always opposes sliding, and the bottom surface of the ball wants to slide forward on the floor during the ball's contact with it. Therefore, friction points backward. If $\mu_s|\vec{F}_N|$ is large enough, static friction will act and totally stop the sliding. Otherwise, the ball will slip and it will be kinetic friction. Either way, we have a backward-pointing force.

The friction force *does* exert a torque, since its line of action does not pass through the ball's center. The frictional torque generates a topspin. How *fast* the ball ends up spinning depends on several factors, including the frictional coefficients, contact time, normal force, and ball speed. We won't get into that here.

The main points are that during the bounce, three things have changed:

1. $v_{x,\text{after}} < v_{x,\text{before}}$ due to the friction: forward motion slowed.
   Important: a force that exerts a torque still follows Newton's second law, just as always.

2. $|v_{y,\text{after}}| < |v_{y,\text{before}}|$ because of inelasticity ($e < 1$): vertical motion slowed.

3. No spin before the collision; a topspin after, due to torque.

> **Question**
>
> The arrow in figure 6.27 indicates that the friction could be either kinetic or static. If the ball never stops moving, how could there be *static* friction?

Figure 6.28
Spin doctor: Point guard Steve Nash executes another perfect bounce pass through traffic. The ball flies fast and low between Lakers Kobe Bryant and Pau Gasol, but will bounce up more slowly and steeply due to backspin. (© *NBAE/ Getty Images*).

Now think about a dry court, where the friction on the bounce pass is static. If the ball *initially* had a backspin, then the static friction force would be much larger. In that case, the x-component of the velocity comes *way* down as it hits the floor. After the bounce, a hard-thrown ball would be much slower (and easier to catch) and come up at a more vertical angle. This is the perfect bounce pass: fast and low through traffic, bouncing up and slow just before the intended recipient of the pass. If you want to see the master at work, watch a game with Steve Nash (figure 6.28).

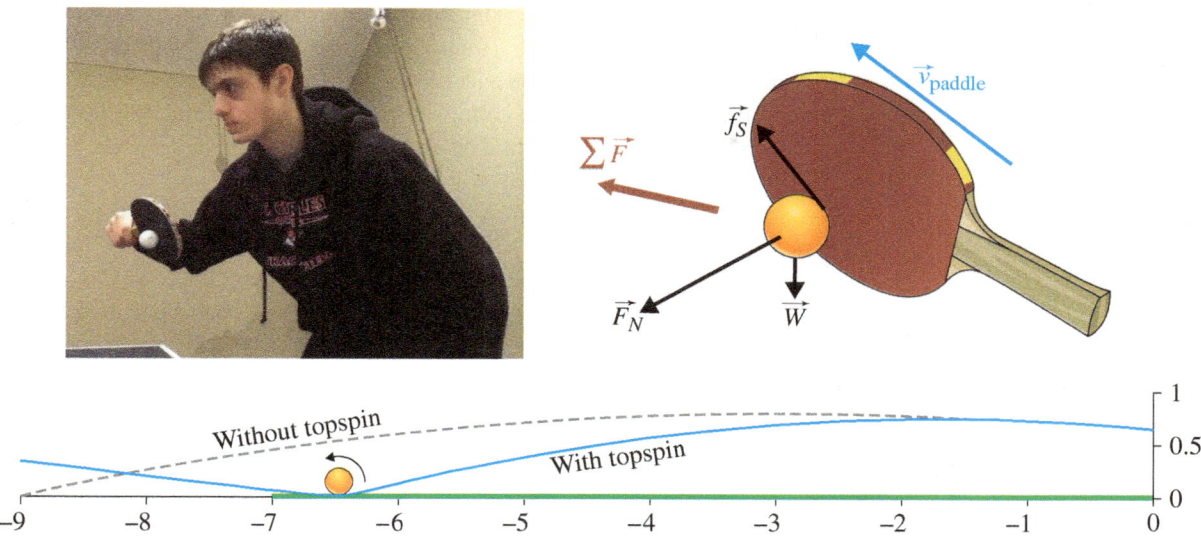

Figure 6.29. The best table tennis players can impart tremendous topspin by rapidly flicking their rackets perpendicular to the normal force. The bottom panel shows the same 22-mph shot with a 1800-rpm topspin (solid curve) and with no topspin (dashed). Distances are shown in feet. (*Photo courtesy of Mike Lisa*).

6.5.2 Diving shot

The entire game of table tennis is determined by two things: reaction time and spin. Players put spin on the ball by quickly moving the racket along the surface direction, as shown in figure 6.29.

The International Table Tennis Federation (ITTF) tightly regulates balls and rackets to ensure that skill, rather than technology, dominates performance. Surface dimple size, friction coefficient, and thickness of sponge padding on the racket face (which allows the ball to "sink in" to the surface, increasing friction and contact time) are all specified in ITTF laws.

According to our rule of thumb from section 5.4.2, the Magnus force on a ball with topspin is directed downward. Therefore, the hit shown in the top panel will lead to a diving shot. Our brains have developed for eons to deal with the flight of rocks and spears, and so on. Rocks follow nice parabolic paths. So, our hardwired prehistoric instinct tells us to expect the long shot shown in gray in the figure. This shot would go off the end of the table, scoring a point for us.

However, with topspin, the ball "breaks" and dives, hitting the back of the table. And it's worse! Having hit the table, the ball is still spinning, and we get the bounce-pass effect of figure 6.28, but in reverse! Upon contact, the bottom surface of the ball is moving *backward* relative to the table (even though the center of the ball is moving forward), so friction *accelerates* the ball. The ball tends to scoot toward us, low and fast. This didn't happen to our ancestors; here again, table tennis challenges our caveman brains.

6.5.3 Backspin on a golf shot

It's a fair bet that more technical, scientific, and pseudo-scientific writing has been done on golf than on any other sport. The amount of money, speculation, and analysis that goes into the game is astounding. This is all in pursuit of perfection in "a game whose aim is to hit a very small ball into an even smaller hole, with weapons singularly ill-designed for the purpose," according to the great statesman and occasional golfer Winston Churchill.

Even if a golfer manages to handle the ill-designed weapons perfectly to achieve the launch he desires, there is a tremendous amount of strategy involved in the endless variations of weather, course conditions, and hole layout. We won't discuss strategy, but rather the physics at the golfer's disposal, when he strategizes. So far we've discussed launch speed and a little bit about launch angle. Now let's talk about the other important parameter for a ball coming off a club—spin.

As discussed in detail in chapter 5, the backspin on a golf ball (often thousands of revolutions per minute) gives it tremendous lift. With the same launch velocity and angle, a ball hit with backspin will remain airborne 2–3 s longer and can travel 60–100 yd farther than a ball without backspin. So, if backspin is one key to greater range, why do so many manufacturers advertise low-backspin clubs and balls, and why are online chat boards filled with golfers looking to reduce their backspin? To answer that, let's first see what causes backspin.

The physics behind a backspin

The physics behind a golf backspin is a little harder to envision than the spin on the basketball or table tennis ball, so let's break it up. Figure 6.30 shows the stages of a collision between a ball and clubhead. For now, we're going to think of the clubhead approaching the ball horizontally. Initial contact is made in the first stage, and the force is perpendicular to the clubhead surface—it is only the normal force. The ball begins to accelerate along the direction of the normal force, which is the head's loft angle (remember figure 6.21).

Here's the point: if the club suddenly stopped there, and delivered its force to the ball, then the ball would accelerate along the direction of the normal force and it would launch at the loft angle. But the club *doesn't* stop there; it keeps moving at about 100 mph. Because of its mass, the ball doesn't accelerate away infinitely quickly ($\vec{a} = \vec{F}/m$); it doesn't have time to get out of the way. The bottom of the club pushes under the ball, and the ball begins to slide up.

As the normal force gets stronger, friction gets huge ($|\vec{f}| = \mu|\vec{F}_N|$) and the ball begins to roll up the face, as in panel 2 of figure 6.30. There's your backspin.

Ball launch angle

Okay, that wasn't so bad. But here's a question that is the source of so much confusion among golfers:[12] **Why is the golf ball's launch angle always lower than the loft angle** for a horizontal swing?

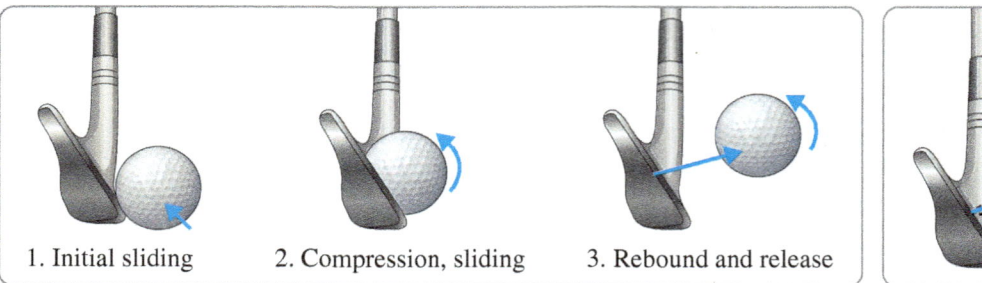
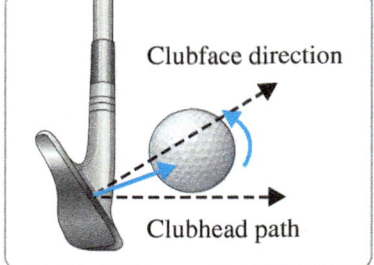

1. Initial sliding 2. Compression, sliding 3. Rebound and release

Clubface direction
Clubhead path

Figure 6.30. The three stages of a club–ball collision are shown on the left. On the right, the launch angle always lies between the clubface direction (which is the same as the loft angle here) and the club velocity.

---

[12] It's also the source of income for gurus who confuse golfers even further.

Here's the reason, and it's part of the "reduced backspin" mania of many golf enthusiasts: the very strong friction force that caused the backspin in panel 2 points diagonally *down*. The net force exerted by the club is therefore a bit lower than the loft angle.[13]

For low-loft clubs such as drivers, the launch angle is only a little lower than the loft; for a 30° loft club, it can be 5° lower.

If the club or ball gets wet with grass or water, the friction coefficient is reduced, leading to reduced spin and a little higher launch angle. This may or may not be desirable for the golfer; he might prefer a large backspin near the green, so the ball "bites" upon landing to prevent rolling.

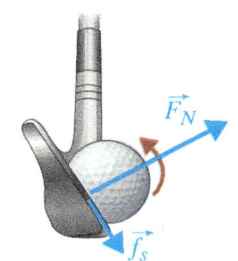

So why is backspin bad? And how can we fix it?

We still haven't answered the question: if backspin provides a lift force that keeps the ball airborne longer, why do golfers want to reduce it, especially on long drives? Well, it is true that at high speeds the drag on a ball with high spin is a little larger than one with no spin; nevertheless, this is not enough to "cancel" the beneficial lift effects of spin.

One valid reason to limit backspin is to allow the ball to roll once it's landed, especially if the fairway is dry and firm. Similar to Steve Nash's bounce pass, a large backspin reduces the ball's forward velocity. So although the "carry" (distance traveled in the air) is less, the total distance, including rolling, might be greater if the ball doesn't come to a dead stop. That's one easy reason that any golfer will tell you.

But the other reason is that much of the force of the collision is directed downward by the friction. In other words *it usually isn't the backspin that's the problem; it's the friction that causes the backspin that's the problem.*[14]

To achieve a large launch angle, either you can use a large-loft club (which leads to this undesirable downward friction force and large backspin), or you can increase the angle of attack. This is shown in figure 6.31. To achieve the same launch angle, an upward angle of attack requires less loft, so less force is "wasted" in spin-generating friction.

 In golf, a huge range of physical effects are at play. Optimizing performance means striking a delicate balance through equipment and style.

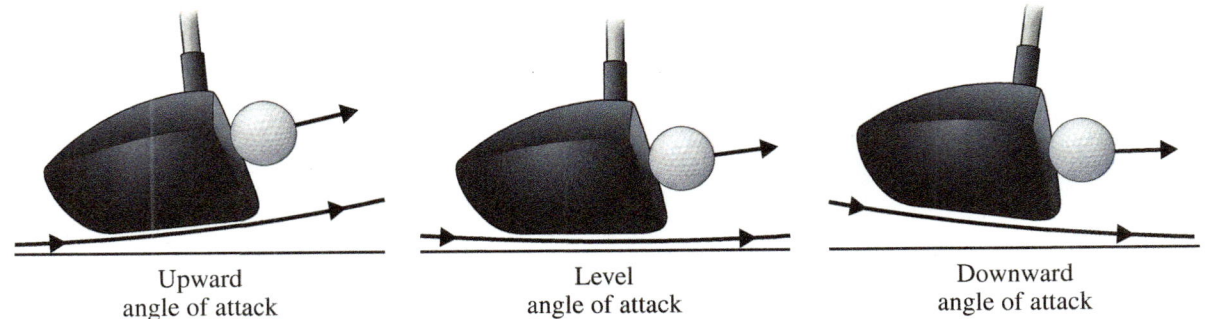

Figure 6.31. The angle between the ground and the velocity of the clubhead at the time of contact is known as the **angle of attack**.

---

[13] Warning: some manufacturers label their clubs with a bit lower loft than the club's true value, partly to compensate for the effect we're talking about, and partly to give the enthusiast the idea that he's using a low-loft club similar to what many of the pros use.

[14] We can mention an entirely different benefit of low spin for the nonprofessional. If the player hits the ball with an "open" or "closed" clubface (that is, not square-on), the ball picks up a sidespin and will "hook" as a curveball does. Low-spin balls obviously reduce hook as well as lift.

Collected Equations

$$\vec{p} = m\vec{v} \tag{6.1}$$

$$\vec{J} = \vec{p}_a - \vec{p}_b \tag{6.2}$$

$$\vec{J}_1 = -\vec{J}_2 \tag{6.3}$$

$$\underbrace{\vec{p}_{1,b} + \vec{p}_{2,b}}_{\vec{p}_{\text{before}}} = \underbrace{\vec{p}_{1,a} + \vec{p}_{2,a}}_{\vec{p}_{\text{after}}} \tag{6.4}$$

$$v_{\text{together},x,a} = \frac{m_1 v_{1,x,b} + m_2 v_{2,x,b}}{m_1 + m_2} \tag{6.5}$$

$$\text{KE} = \tfrac{1}{2}mv^2 \tag{6.6}$$

$$1 \text{ joule} = 1 \text{ J} = 1 \text{ kg} \cdot \text{m}^2/\text{s}^2 \quad \text{also} \quad 1 \text{ J} = 1 \text{ N} \cdot \text{m} \tag{6.7}$$

$$1 \text{ slug} \cdot \text{ft}^2/\text{s}^2 = 1 \text{ lb} \cdot \text{ft} = 1.356 \text{ J} \tag{6.8}$$

$$\underbrace{\text{KE}_{1,b} + \text{KE}_{2,b}}_{\text{total KE}_{\text{before}}} = \underbrace{\text{KE}_{1,a} + \text{KE}_{2,a}}_{\text{total KE}_{\text{after}}} + \underbrace{E_{\text{sound}} + E_{\text{heat}} + E_{\text{damage}}}_{\text{converted to other energy forms}} \tag{6.9}$$

$$\vec{F}_{\text{ave}} = \frac{\vec{J}}{\Delta t} \tag{6.10}$$

$$|\vec{F}_{\text{peak}}| = 2|\vec{F}_{\text{ave}}| = 2\frac{|\vec{J}|}{\Delta t} \tag{6.11}$$

$$\text{elastic collisions: } v_{\text{rel},b} = -v_{\text{rel},a} \tag{6.12}$$

$$v_{\text{pin},x,a} - v_{\text{ball},x,a} = -(\underbrace{v_{\text{pin},x,b}}_{0,\text{ at rest}} - v_{\text{ball},x,b}) \rightarrow v_{\text{ball},x,a} = v_{\text{pin},x,a} - v_{\text{ball},x,b} \tag{6.13}$$

$$m_{\text{ball}} v_{\text{ball},x,b} + \underbrace{m_{\text{pin}} v_{\text{pin},x,b}}_{\substack{\text{zero because} \\ \text{pin at rest}}} = m_{\text{ball}} \underbrace{(v_{\text{pin},x,a} - v_{\text{ball},x,b})}_{\substack{\text{this is } v_{\text{ball},x,a} \\ \text{from equation 6.13}}} + m_{\text{pin}} v_{\text{pin},x,a} \tag{6.14}$$

$$v_{\text{pin},x,a} = \frac{2 m_{\text{ball}}}{m_{\text{ball}} + m_{\text{pin}}} v_{\text{ball},x,b} \tag{6.15}$$

$$v_{ball,x,a} = \frac{m_{ball} - m_{pin}}{m_{ball} + m_{pin}} v_{ball,x,b} \tag{6.16}$$

$$KE_{1,b} + KE_{2,b} = KE_{1,a} + KE_{2,a} \tag{6.17}$$

$$e = -\frac{v_{rel,after}}{v_{rel,before}} \tag{6.18}$$

$$\text{bouncing vertically on floor: } e = \sqrt{\frac{h_f}{h_i}} \tag{6.19}$$

$$v_{ball,x,a} = \frac{1+e}{1 + m_{ball}/m_{head}} v_{head,x,b} \tag{6.20}$$

$$v_{head,x,a} = \frac{1 - e m_{ball}/m_{head}}{1 + m_{ball}/m_{head}} v_{head,x,b} \tag{6.21}$$

Problems

1. Ray Lewis may not be faster than a speeding bullet, but is he more energetic? How fast would a .22-gauge bullet need to fly, to have the same kinetic energy as Ray Lewis running at 12 mph?

2. In his book *The Physics of Football*, Prof. Tim Gay points out that reaction time and the conservation of momentum collaborate to give a built-in advantage to the offense, at the line of scrimmage. The offense knows the snap count and rhythm and can explode off the line the instant the ball is hiked. The defense, on the other hand, must react to movement on the offensive line. As we discussed in chapter 2, the time delay before they start to move is about one-fifth of a second (0.2 s).

   (a) A fit 300-lb offensive lineman can accelerate, immediately after the snap, at 16 ft/s². With what magnitude of force is the ground pushing him forward?

   (b) A 300-lb defensive lineman accelerates just as quickly, but with a 0.2-s delay. They collide 0.52 s after the snap in an inelastic ("tackle-type") collision. How fast, are the linemen moving *just before* they collide?

   (c) How fast and in what direction (toward the quarterback, or away from him) are they moving *immediately after* they collide?

   (d) What is the magnitude of the impulse of their collision?

3. A soccer ball flies at a player at 20 m/s, who "heads" it directly back in the opposite direction at 8 m/s.

   (a) What is the impulse of the collision?

   (b) As we discussed, soccer "heads" are triangle impacts. What is the maximum force *on the player's head* during the collision? Express your answer in pounds.

   (c) What is the maximum force *on the ball* during the collision? Express your answer in pounds.

4. In hockey, bumps and checks are useful to throw an opponent off. Even small, unexpected changes in his velocity can disrupt an opponent's timing. In a game, an 80-kg player skating at 10 m/s overtakes and bumps from behind a 100-kg player moving in the same direction at 8 m/s. As a result, the 100-kg player's speed increases to 9.42 m/s.

(© skynesher/E+/Getty Images RF)

(a) How fast is the 80-kg player moving after the bump?

(b) What is the coefficient of restitution associated with this collision?

5. A 0.35-kg tennis racquet moving to the right at 20 m/s hits a 0.06-kg tennis ball that is moving to the left at 30 m/s. After the collision, the racquet continues to the right, but at the reduced speed of 10 m/s.

   (a) What is the total momentum of the ball–racquet system before the collision?

   (b) What is the ball's speed after the collision?

   (c) What is the coefficient of restitution?

   (d) Using a contact time of 6 ms, estimate the peak force on the ball in this collision. Give your answer in pounds.

6. Tell whether each of the following increases, decreases, or leaves unchanged the speed at which a golf ball leaves the tee: COR; clubhead speed before the shot; mass of clubhead, mass of ball.

7. The coefficient of restitution between a hockey puck (made of vulcanized rubber) and hard ice is 0.5 if the puck is at room temperature.

   (a) If a room-temperature puck is dropped onto the ice from a height of 36 in., how high will it bounce back?

   (b) A cold puck (near 32°F), however, has a COR of 0.35. How high will the cold puck bounce?

   (c) If we treat them as simple collisions between puck and stick, would slapshots typically be faster or slower with the cold puck? (In reality, a slapshot is more complicated than a simple collision, as the stick springs from a flexed position and can strike the puck several times during the shot.)

   (Pucks tend to warm up when they are struck, even though they spend time on the ice. In the NHL, buckets of hockey pucks are kept refrigerated during the game, to keep the puck's "liveliness" constant.)

8. In a slapshot, a 0.17-kg puck is accelerated from rest to 80 mph while in contact with the stick for 2.5 ft. The weight of the shooter plus his stick is 180 lb.

   (a) What is the puck's acceleration?

   (b) What is the contact time between stick and puck?

   (c) What force was exerted on the stick by the puck during the shot?

   (d) If the shot is taken from the blue line (64 in. in front of the goal), how long does the goalie have to react?

   (e) If the coefficient of kinetic friction between skates and ice is $\mu_k = 0.02$, how far does the shooter slide after the shot, before coming to rest?

9. In the tackle discussed in section 6.2, how fast would Chandler Williams have to be moving before his collision with Chris Johnson such that they would both come to a dead stop in the collision?

10. Getting tackled by the 250-lb linebacker Ray Lewis moving at 9 yd/s is no fun. Imagine you are a 200-lb quarterback. Would you rather get hit while running directly at Lewis at 9 yd/s or while standing stock still?
    In each case

    (a) What is your velocity immediately after the tackle?

    (b) What is the peak force on you, assuming that the duration of a tackle is 200 ms (see table B.1)?

    (c) How much kinetic energy got converted to other forms (for example, heat, sound, broken bones)?

11. Using information we've already covered, how high will a tennis ball, dropped from a helicopter, bounce after it hits the ground? ($C_D = 0.5$ and $e = 0.74$.)

12. In section 6.2.3, we discussed in detail the hit that knocked Cole Beasley out of bounds. In that case, Phillip Adams was running directly toward the sidelines when he hit Beasley. But often the defender hits the runner at an angle. What if Adams had been running at 9 yd/s, but hit Beasley partially from behind, at 70°, rather than at 90°, as indicated in the figure. Would Beasley have scored in this case?

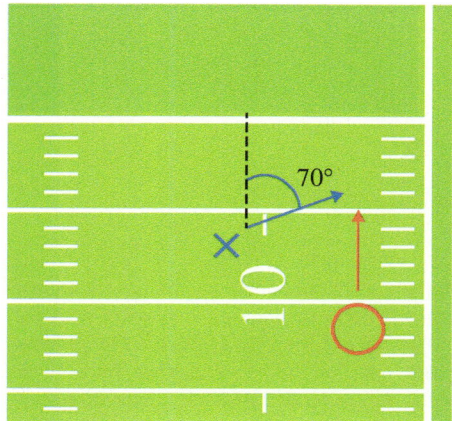

13. In a hard pool shot, the cue may hit the numbered ball at 18 mph head-on, coming to a dead stop. Estimate the peak force between the balls during the collision. (Some relevant information may be found in the appendices.)

14. A basketball is dropped on the court from a height of 4 ft. It experiences a "triangle impact," being in contact with the court for 15 ms and experiencing a peak GRF of 135 lb.
    (a) How high will it bounce back up?
    (b) What is the coefficient of restitution for this collision?
    (c) How much kinetic energy was lost in this collision?

15. A trash-talking 260-lb football player on an ESPN Sport Science show runs at 12 mph directly toward a dummy approaching him at 14 mph. They collide for 160 ms and come to a dead stop, stuck together.
    (a) How heavy is the dummy?
    (b) Assuming a triangle impact, what is the peak force exerted on the dummy?
    (c) What is the coefficient of restitution?

16. Two figure skaters stand stationary and facing each other on (frictionless) ice, waiting for the music to start. The woman weighs 130 lb and her partner weighs 195 lb. As the music starts, she pushes off him, gliding backward at 4 ft/s.
    (a) How fast does her partner recoil?
    (b) If it had been *he* who had done the pushing instead of she, but she'd still glided off at 4 ft/s, how fast would he have recoiled?

17. A 195-lb safety is running at 10 yd/s, chasing down a 180-lb wide receiver running at 9.1 yd/s in the same direction. When he catches him, the safety wraps him up and they go down together.
    (Keep track of four digits past the decimal point on your calculator.)
    (a) What is the horizontal component of their velocity just after the tackle?
    (b) What is the coefficient of restitution of this collision?
    (c) How much kinetic energy was lost in this collision?
    (d) Do you think the initial collision (that is, before they hit the ground) was painful? Why or why not?

7 Energy in Sports: Bursts of Power

Sports is all about controlling the flow of energy from one form to another, beginning with the food we eat and the air we breathe. This chapter covers the five forms of energy relevant to sports, with a focus on quick bursts of energy conversion. Along the way, we'll discuss the human engine and the concepts of efficiency and power.

(© *Eyewire/Getty Images RF*).

We've seen that many scientific concepts are useful to understand the actions of an athlete—the force he exerts, his momentum, changes in body shape, and so on. The ability to do these things, of course, ultimately comes from the food eaten by the athlete.

What is it *in* the food that gives us the ability to perform athletically? Does food contain force? momentum? impulse? No. Food (properly mixed with oxygen) contains energy.

It turns out that the best way to understand certain aspects of sports is to follow the energy. We can already learn a lot from a further discussion of the humble bounce of a basketball, so let's start there.

7.1 Bouncing Basketball: The Whole Process

7.1.1 Heat in basketball: not just for Miami

In the analyses of the bouncing basketball from chapter 6, some kinetic energy was lost in collision. But that energy didn't disappear. As equation 6.9 says, it was converted to other forms. Since we hear each bounce, we know that some energy went into sound energy. (Each sound emitted from a collision means the ball will rise a little less high after the bounce.) However, most of the energy goes into heat.

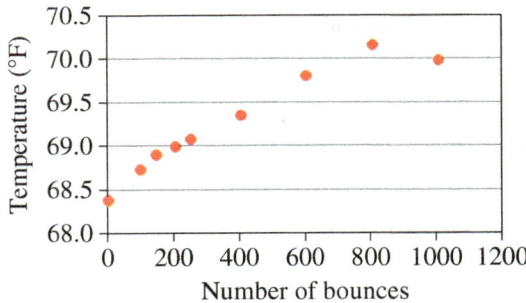

Figure 7.1. The temperature of a basketball increases with each bounce, as some of the ball's kinetic energy is converted to heat. From *The Physics of Basketball* by Fontanella. Fontanella dribbled the ball to keep it bouncing.

Figure 7.1 shows the temperature of a basketball versus the number of bounces as it is dribbled; it increases from about 68°F to 70°F, certainly noticeable.[1] Something similar happens in the football tackle: some of the initial kinetic energy is converted to heat (and possibly broken bones).

We're going to use the umbrella term **"thermal energy"** to cover heat, damage, and sound energy. When kinetic energy is converted to thermal energy, it cannot be converted back.[2] For our purposes, it is "lost." The golf club and ball making contact loses energy to sound, but that sound energy won't ever be converted to a form to move a golf ball. It's a one-way street.

Kinetic energy is converted to heat in a golf ball or basketball via internal friction of the rubber layers as they stretch and rub against one another; any flexible material will heat up as it stretches and pulls. Generally, the greater and faster the compression, the more energy is converted to thermal energy. (Remember figure 6.24 and the accompanying discussion.)

 Energy that is converted to heat, damage, or sound will never be converted back to an energy form useful for sports.

7.1.2 Energy during the bounce

The basketball bounce is different from the football tackle in an important way. In the tackle, the kinetic energy is reduced and never comes back—it's all converted to thermal energy.

Not so with the basketball. On the first half of the bounce, *all* the kinetic energy is converted to some other form, as the ball comes to a complete stop. If the ball were totally uninflated and "dead," it would simply stay on the floor; the kinetic energy would have been converted irrevocably to thermal energy.

A properly inflated ball, however, will start moving upward on the second half of the bounce, because most of the initial kinetic energy was converted not to heat but to **elastic potential energy**. "Potential" energy is just a form of energy into which kinetic energy can be transformed and which can be transformed directly back into kinetic energy. There are lots of examples of elastic potential energy, such as a coiled spring (a physics book favorite) or a hockey stick flexed during a slapshot.

Elastic potential energy, which we'll label PE_{elast}, is measured in joules, just like any other energy. When the ball is squished, PE_{elast} is large; when it's round,

Q: Equation 6.9 doesn't say anything about potential energy. You told us that any converted kinetic energy went into thermal energy. Are you changing the story?

A: The story is not changing and nothing's wrong, because equation 6.9 relates the energy *just before* the collision to the energy *just after* the collision. The elastic potential energy we're talking about here is nonzero only *during* the collision.

---

[1] Once the ball gets significantly hotter than the room temperature of 68.3°F, it begins losing heat to the surrounding air. That's why it stops heating up after about 1000 bounces.
[2] In general this is not entirely true. Steam and internal combustion engines do convert heat back into something useful, but there are other issues involved there, and these are not relevant for sports.

188 *Energy in Sports: Bursts of Power* Chapter 7

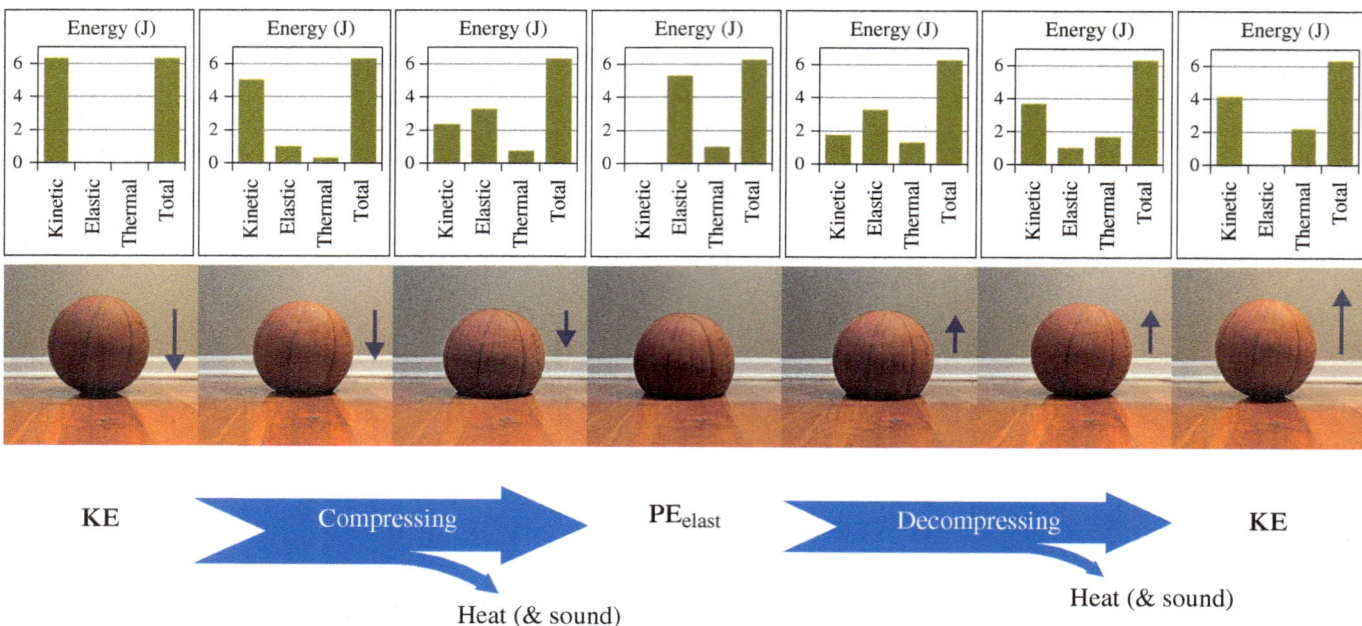

Figure 7.2. Following the energy *during* the bounce: As a basketball bounces on the floor, energy is transformed between kinetic, elastic, and thermal forms. The small changes in gravitational potential energy aren't plotted, to reduce clutter. The ball's changing velocity is shown by the small arrows. (A somewhat deflated ball was used in this image, to emphasize the compression.) (*Photo courtesy of Mike Lisa*).

$PE_{elast} = 0$. Figure 7.2 shows several video frames of a basketball as it comes into contact with the floor, compresses, decompresses, and is just about to take off.

There are many things to learn from figure 7.2:

- As the ball comes to a dead stop, it becomes maximally deformed. In terms of energy, this means KE is minimum when PE_{elast} is maximum.

- Energy is transformed from kinetic to elastic during the compression, then back to kinetic during the decompression.

- However, since energy is lost to thermal forms throughout the entire process, the thermal energy component continually grows, and the kinetic energy at the end of the bounce (last panel, just as the ball is about to leave the floor) is less than the kinetic energy before the bounce.

- The **total energy** never changes; it is the sum of kinetic, potential, and thermal energy.

In our upcoming discussion of archery and slap shots, we treat elastic energy in greater depth.

7.1.3 Details of energy during the rise

In the last three panels of figure 7.2, energy is being converted back from elastic to kinetic energy, as the ball gets an upward velocity. What happens next? Well, it rises and slows under the force of gravity.

But instead of talking about forces, physicists sometimes find it useful to stick with energy and define another form of potential energy, **gravitational potential energy**, which we'll write as PE_{grav} in formulas.

$$PE_{grav} = mgy_{c.m.} = |\vec{W}|y_{c.m.} \qquad (7.1)$$

If elastic potential energy is "the energy of flexing," then gravitational potential energy is "the energy of being high off the ground." What's nice is that if something is 1 m off the ground and its weight is 30 N, then it has (1 m)(30 N) = 30 J of gravitational potential energy. If it's sitting still at 1 m up, then its total energy is 30 J. If it is 1 m up but moving, then it has a total energy of 30 J plus whatever its kinetic energy is. Energies just add. There is no need for vectors or components. Example 7.1 discusses a hypothetical case in which the energy of a Michael Jordan leap is concentrated into a baseball.

The situation is shown in figure 7.3. In terms of energy, during the rise the kinetic energy is transformed into gravitational potential energy plus, of course, a little thermal energy as air friction heats the air. During the fall, the gravitational potential energy

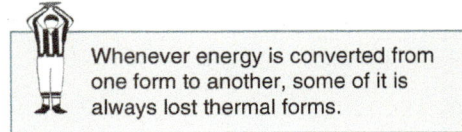
Whenever energy is converted from one form to another, some of it is always lost thermal forms.

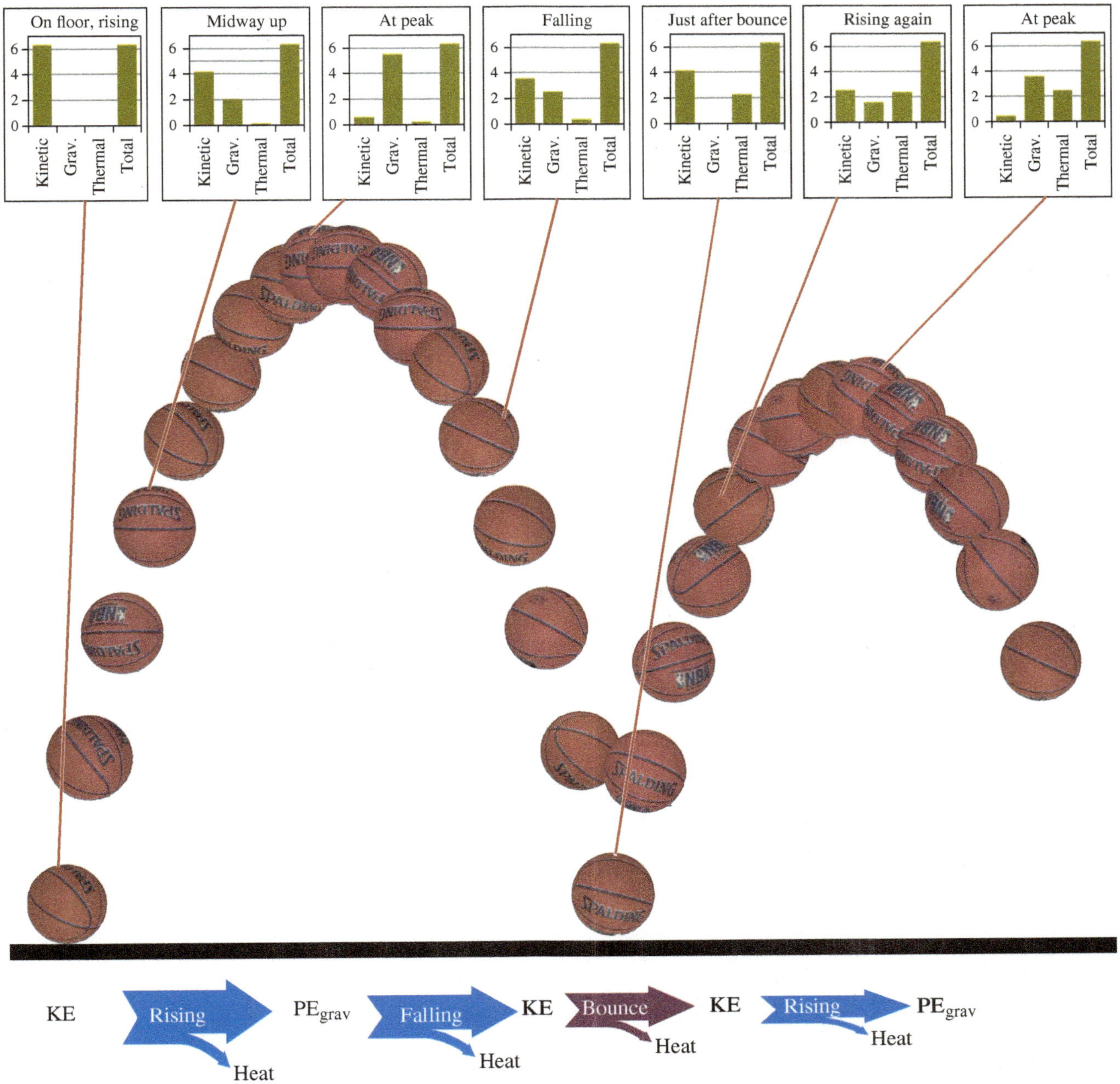

Figure 7.3. As a basketball arcs through the air, energy is transformed between kinetic, gravitational, and thermal forms. During the rapid bounce process (red arrow), energy is transformed in and out of the elastic form; this is shown in detail in figure 7.2. (*Photo courtesy of Mike Lisa*).

is transformed back into kinetic energy plus a little thermal energy, again due to air friction. During the bounce itself, the kinetic energy is transformed into elastic energy (plus some thermal energy), and then the elastic energy transformed back into kinetic (plus some thermal) energy. (Are you sensing a pattern here, regarding the thermal energy?)

The bar graph at the top of figure makes clear several things:

- Energy is transformed between kinetic and potential energy as the ball rises and falls.

- The *total* energy never changes.

- At every step of the way, the thermal energy increases, sometimes quickly (during the bounce) and sometimes slowly (due to air friction). This leaves less and less energy available to raise the ball (increase gravitational energy) or move it (increase kinetic energy).

> **Question**
>
> In figure 7.3, why is the kinetic energy not zero at the peak of the ball's bounce? (See the third bar graph at the top of the figure.)

EXAMPLE 7.1 *Michael Jordan, baseball player*

In 1993, basketball legend Michael Jordan shocked the NBA by announcing his retirement from the game. Shortly thereafter, he began what turned out to be a brief career as a professional baseball player.

As a basketball player, Jordan's weight was listed at 195 lb. There are several somewhat conflicting claims about his vertical jump (without a run-up), but the most reliable reports put it at about 41 in. (3.42 ft).

(a) If Jordan could put the energy of his vertical leap into a baseball, rather than into the motion of his body, what would be the vertical height of the ball? That is, how high would it go, thrown or batted straight up?

First, we need to find the energy of his leap. The flow of the energy is as follows: food energy was converted initially into kinetic energy KE of his body until his feet left the ground. After this, during his rise, the energy was converted from KE to PE_{grav}, until at the peak of his jump, all energy is in the form of gravitational potential energy:

$$PE_{grav} = mgh = |\vec{W}|h = (195 \text{ lb})(3.42 \text{ ft}) = 666.25 \text{ lb} \cdot \text{ft}.$$

This same energy, put into a baseball, leads to a height of

$$PE_{grav} = mgh \rightarrow h = \frac{PE_{grav}}{mg} = \frac{666.25 \text{ lb} \cdot \text{ft}}{(1.03 \times 10^{-2} \text{ slug})(32 \text{ ft/s}^2)} = 2021 \text{ ft}$$

That's more than 0.25 mi! This number is unrealistically high for two reasons. First, we ignored air drag, which is considerable for the high launch speed of this ball; drag would reduce the maximum height to about 500 ft, still incredibly high.

(b) Again if Jordan could put all the energy of his vertical leap into the ball, how far could he hit or throw it?

Equation 4.4 tells us that the range of a ball depends on its launch speed and angle. In section 4.2, we learned that maximum range (ignoring air) will be achieved when the ball is launched at 45°. The only thing we need to know is the launch speed, which is simply given by the initial KE of the ball. At launch, all Jordan's energy has been put into the kinetic energy of the ball, so

$$KE = \tfrac{1}{2}mv_0^2 \rightarrow v_0 = \sqrt{\frac{2KE}{m}} = \sqrt{\frac{2\,(666.25\text{ lb}\cdot\text{ft})}{1.03\times 10^{-2}\text{ slug}}} = 360\text{ ft/s}$$

which is about 250 mph!

The range of the ball is

$$R = \frac{v_0^2 \sin 2\theta}{g} = \frac{(360\text{ ft/s})^2 \sin(2\cdot 45°)}{32\text{ ft/s}^2} = 4050\text{ ft}$$

which is almost 10 times the distance from home plate to the center field fence in a major league park. Again, drag is huge and reduces the maximum range of a ball hit at 360 ft/s to "only" about 700 ft, well beyond the 500-ft range only achieved a few times in the history of baseball.

However, neither Jordan nor anybody else can impart 666.25 lb · ft of energy to a baseball, since the process of hitting or throwing a ball is not as efficient as the process of a vertical jump. Efficiency is what we discuss next.

7.2 Efficiency

The concept of "efficiency" comes up a lot when we talk about energy conversion, for both people and machines. It's a natural concept, but we should keep in mind that it can be defined in various ways and sometimes it doesn't make sense to discuss efficiency at all.

Generally, we want to talk about the energy conversion efficiency of a process. Of course any process (such as a chemical reaction or a tennis serve) conserves the *total* energy, but we are usually interested in just the "useful" energy (such as the kinetic energy of the tennis ball) that results from the process. We'll use the symbol η, the Greek letter eta, for energy conversion efficiency.

The term "efficiency" can take on different meanings, depending on context.

$$\eta = \frac{\text{useful energy output from the process}}{\text{total energy input to the process}} \qquad (7.2)$$

That seems natural enough and can be very useful in the analysis of sports, as long as you are clear about what is "useful energy" and "the process." Let's look at some examples.

7.2.1 The efficiency of a basketball bounce

A vertical bounce of a basketball off the court is a pretty simple process in which the direction of the ball is reversed. The energy flow involves kinetic, elastic, and thermal forms, as we covered in detail in section 7.1.2.

Here it seems pretty obvious that the "input" energy is the ball's kinetic energy just as it hits the floor, and the "useful output" is its kinetic energy as it leaves the floor.

$$\eta_{\text{basketball bounce}} = \frac{\text{KE}_{\text{after}}}{\text{KE}_{\text{before}}} = \frac{\frac{1}{2}mv_{\text{after}}^2}{\frac{1}{2}mv_{\text{before}}^2} = \left(\frac{v_{\text{after}}}{v_{\text{before}}}\right)^2.$$

The nice thing is that we know how the velocities before and after the bounce are related, through the coefficient of restitution from equation 6.18: $v_{\text{after}} = ev_{\text{before}}$, so we can write

$$\eta_{\text{basketball bounce}} = e^2 = \underbrace{0.81^2 = 0.66}_{\substack{\text{NBA leather ball} \\ \text{from section 6.4.1}}}.$$

So, the bounce of a properly inflated NBA ball is 66% efficient. The other 34% of the energy goes into heat and sound and is not considered useful.

7.2.2 The efficiency of a golf drive

Analyzing the basketball bounce gave a nice simple formula for the efficiency of an inelastic collision. Examining a golf drive will show us why this formula might not be the right one to use in all cases.

In section 6.4.2, we discussed the physics of a golf drive in detail. I told you that the maximum coefficient of restitution allowed in USGA tournaments is 0.83. So, does that mean that $0.83^2 = 69\%$ of the energy in a golf swing goes into the ball's kinetic energy? Sadly, no.

On page 177, we did a short calculation for a collision between the 200-g clubhead and the 45.9-g ball. If the clubhead was initially traveling at 115 mph (51.4 m/s), the ball was launched at 171 mph (76.4 m/s). The "input" energy is the kinetic energy of the clubhead:

$$\text{input energy} = \text{KE}_{\text{head,before}} = \tfrac{1}{2}(0.2 \text{ kg})(51.4 \text{ m/s})^2 = 264.2 \text{ J}.$$

Probably everyone would consider the "useful output energy" to be the kinetic energy of the launched ball:

$$\text{useful output energy} = \text{KE}_{\text{ball,after}} = \tfrac{1}{2}(0.0459 \text{ kg})(76.4 \text{ m/s})^2 = 134 \text{ J},$$

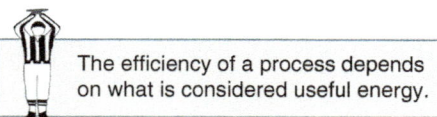
The efficiency of a process depends on what is considered useful energy.

which is almost exactly one-half of the energy that Harrington put into the swing.

Why is the efficiency of a Harrington's drive only 50%, instead of 69%? Well, in addition to excluding thermal energy from the category of useful output energy, we are not including his club's continued kinetic energy *after* the collision. Any energy retained by the clubhead is considered "wasted" energy here.

Using equation 6.20, we can write a formula for the energy efficiency of a golf drive, with this definition:

$$\eta_{\text{golf drive}} = \frac{m_{\text{ball}}}{m_{\text{head}}}\left(\frac{1+e}{1+m_{\text{ball}}/m_{\text{head}}}\right)^2.$$

This equation clearly shows that the mass of the clubhead matters. Figure 7.4 shows this as a graph.

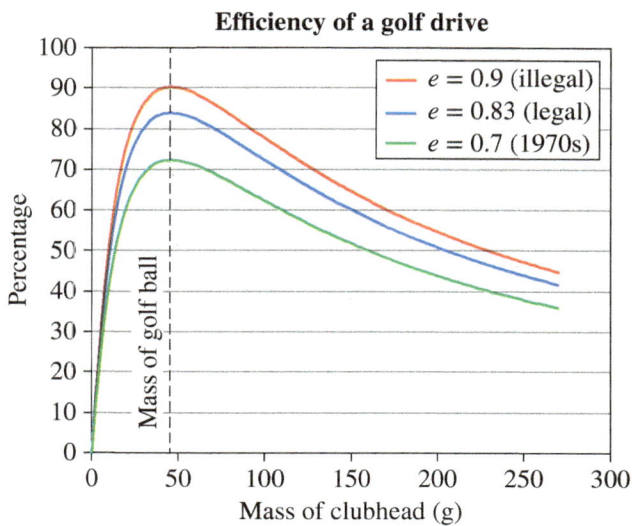

Figure 7.4. The energy efficiency of a golf drive depends on the coefficient of restitution e as well as the mass of the clubhead. The efficiency is always maximum when the clubhead has the same mass as the ball.

Most driver heads are between 195 and 205 g, even though the energy efficiency would be maximized for if the head were the same mass as the ball—45.9 g. Clearly, the choice of clubhead is driven by more than simply the efficiency of energy transferred from head to ball. A main consideration is that, for this type of controlled motion, muscular motion is optimized at lower speeds, so that even though the *fraction* of energy transferred is larger for a light head, the *total amount* of energy transferred is larger for a heavier one.

7.2.3 Heat death

A basketball loses energy to heat during its rise, during its fall, and during its bounce. This is one example of a great and overarching law of nature, the second law of thermodynamics. This law is in fact deep and can be surprisingly subtle, but for our purposes it is fair to say that in *any* process in sports, some energy is *always* lost forever to thermal forms. That lost energy is henceforth useless in terms of the game.

In fact, the laws of thermodynamics seem to imply an eventual "heat death" to the entire universe, when all energy is uniformly distributed as thermal energy. Basketballs will neither bounce nor roll. Life, sports, and any organized activity will then be impossible. The universe will eventually be a uniformly warm, dull bath.

Well, that's a cheerful thought, eh? Did you know that basketball could be so depressing? As a Cleveland Cavaliers fan, I've known the feeling on occasion.[3]

When we discuss sports and many other activities, thermal energy is usually excluded from the useful energy category when we quantify efficiency[4] using equation 7.2. Considering the eventual heat death scenario, this definition would say that the evolution of the universe over its entire lifetime will be a process with zero efficiency.

---

[3] I also know that you can't keep a good team down for long. Go Cavs!

[4] Again, what you consider useful energy may vary. Water heaters, for example, may have energy efficiency ratings of 67%. Naturally, in this case, thermal energy *is* the output useful energy, and the rating says that 67% of the chemical energy in the fuel winds up as thermal energy in your water.

7.3 The Athlete: The Energetic Starting Point

For inanimate objects, there are kinetic, gravitational, and elastic energy forms, and they can be transformed into one another (with a "tax" for each step, paid to thermal energy). Often, we can think of people as objects, such as when they are projectiles in flight, swapping kinetic for gravitational energy and vice versa; that works just fine. It's when we want to talk about athletes injecting energy into the game through their own exertions that things become a little more subtle. But hey, without that exerted effort, there's no game, so let's jump in.

7.3.1 The source of energy and its flow

The energy flow in sports begins with the food we eat and air we breathe.

In the last section, we learned that all energy winds up in the thermal form eventually. Where does it *start* from? In sports, it all comes from the food processed by the athlete's body. Of course, motor sports also use fossil fuels as a source of energy.

In fact, our bodies are in many ways similar to automobile engines. In both cases, **chemical energy** is obtained through chemical reactions between hydrocarbons (gasoline for cars, sugars for us) and oxygen. In both cases, there is chemical "waste" (carbon dioxide (CO_2) from us; carbon dioxide and carbon monoxide (CO) from cars) released into the air. And in both cases, unavoidable energy "waste" in terms of heat is released. The remaining energy is available for conversion into useful mechanical forms; for us, lifting a barbell or throwing a ball; for a car, turning a crankshaft.

In equations, we'll use the symbol E_{chem} for chemical energy. Like all energy, it can be measured in joules. However, when we quantify the amount of chemical energy released when some food is combined with oxygen, it is quite common to use the unit of **Calories**:[5]

$$1 \text{ Cal} = 4184 \text{ J}. \tag{7.3}$$

And with that, we have now introduced all the energy forms relevant to sports. There are many other forms, of course, from nuclear to electromagnetic, but table 7.1 lists the only ones relevant to us.

Figure 7.5 is a sketch of energy flow. One way or another, food as chemical energy is transformed into thermal energy. Athletic performance is, at its essence, the controlled flow of energy from its chemical form into kinetic, gravitational, and elastic forms, and eventually to thermal forms.

Table 7.1. The five types of energy important to sports.

KE	Kinetic energy (energy of motion)
PE_{grav}	Gravitational potential energy
PE_{elast}	Elastic potential energy
E_{therm}	Thermal energy (includes sound, vibrations, damage)
E_{chem}	Chemical (food) energy

---

[5] The unit used to talk about food is the capitalized Calorie, to distinguish it from the "thermodynamical calorie" or just "calorie." So 1 Cal is 1000 calories, also known as 1 kilocalorie, or kcal. It can be confusing. I just wanted you to be aware of this, in case you find yourself puzzled by something you read elsewhere. We will never use the calorie (cal) but always the Calorie (Cal).

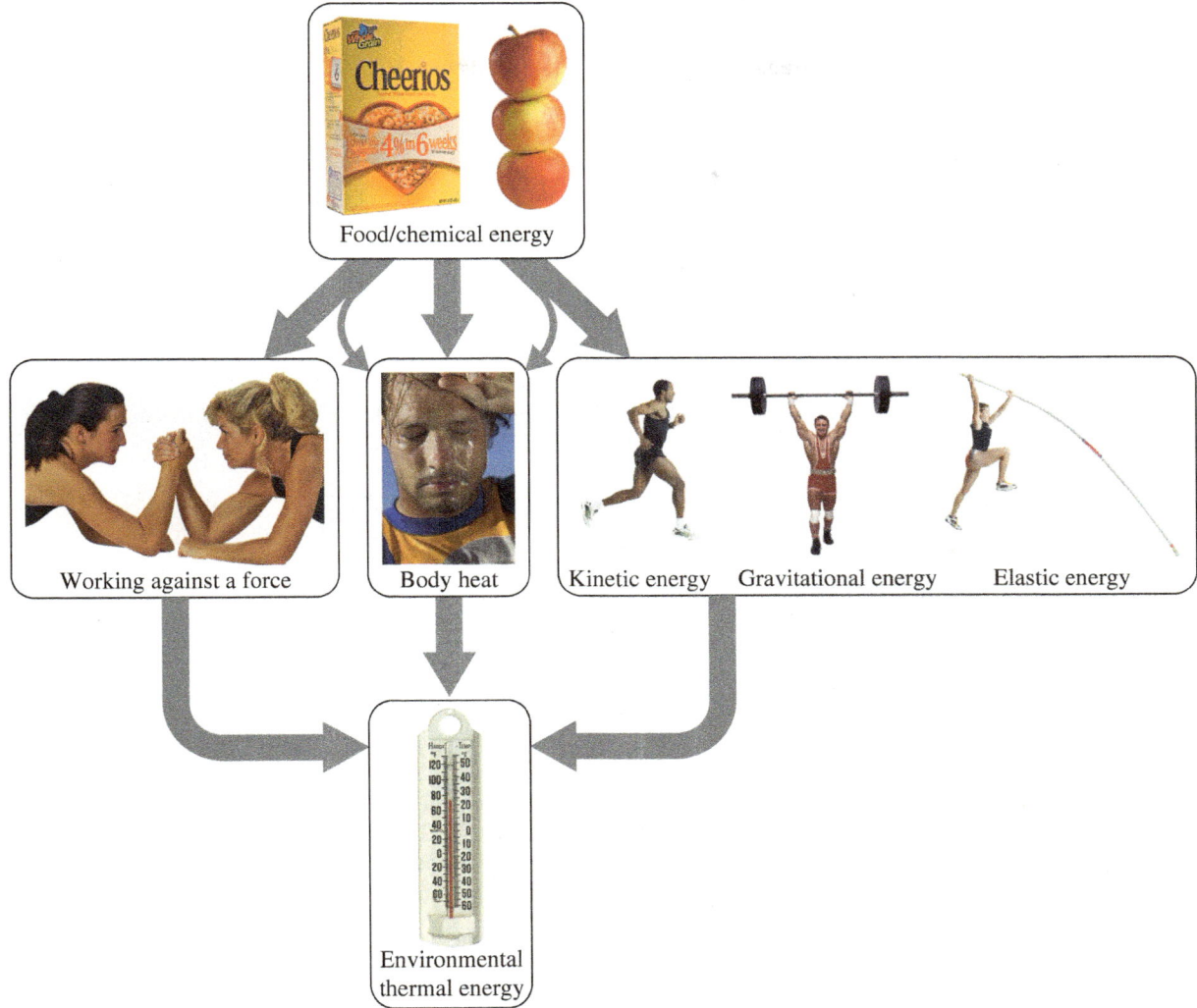

Figure 7.5. The big picture. Sports is determined by how the athlete controls the flow of energy from the food or chemical form to its eventual thermal form. The arrows are one-way streets, but kinetic, gravitational, and elastic energy forms (grouped in the box) can be interchanged among themselves. **Cereal:** (© *The McGraw-Hill Companies, Inc./Jill Braaten, photographer*); **Apples:** (© *Lynx/iconotec.com/Glow Images RF*); **Arm Wrestling:** (© *McGraw-Hill Education/Aaron Roeth, photographer*); **Sweating:** (© *Brand X Pictures/PunchStock RF*); **Runner:** (© *Photodisc/Getty Images RF*); **Weightlifter:** (© *Jack Mann/Photodisc/Getty Images RF*); **Pole Vault:** (© *Rubberball/Getty Images RF*); **Thermometer:** (© *Comstock Images/Getty Images RF*).

EXAMPLE 7.2 *Exactly how much energy is in an energy bar?*

A PowerBar has 240 Cal of food energy. This is 240 Cal × (4184 J/1 Cal) = 10^6 J (1 million J). Let's see what that much energy that corresponds to.

Weight lifting: How much weight can 240 Cal lift 2 m off the ground?
Equation 7.1 gives the energy of a mass lifted off the ground:

$$\text{PE}_{\text{grav}} = |\vec{W}| y_{\text{c.m.}} \rightarrow |\vec{W}| = \frac{\text{PE}_{\text{grav}}}{y_{\text{c.m.}}} = \frac{10^6 \text{ J}}{2 \text{ m}}$$

$$= 5 \times 10^5 \text{ N} = 500{,}000 \text{ N}$$

$$\approx 112{,}000 \text{ lb}.$$

That's 56 tons—much more than the weight of a World War II Sherman tank!

Pitching: How fast can 240 Cal throw a baseball?
Equation 6.6 gives the energy of a moving object:

$$KE = \tfrac{1}{2}mv^2 \rightarrow v = \sqrt{\frac{2KE}{m}} = \sqrt{\frac{2 \times (10^6 \text{ J})}{0.15 \text{ kg}}}$$

$$= 3651 \text{ m/s}$$

$$\approx 8170 \text{ mph.}$$

That's 10 times the speed of sound and fast enough to circle the Earth's equator in just 3 hr.

Track and field: How far can 240 Cal throw a shot-put?
The distance that a put is shot depends on its launch velocity, so again we are interested in the kinetic energy, just as in the example above:

$$v = \sqrt{\frac{2KE}{m}} = \sqrt{\frac{2 \times (10^6 \text{ J})}{7.26 \text{ kg}}} = 525 \text{ m/s}$$

Then we can just use the range equation 4.4 to find the distance put, using the 45° optimum launch angle:

$$R = \frac{v_0^2 \sin 2\theta}{g} = \frac{(525 \text{ m/s})^2 \sin(2 \cdot 90°)}{9.8 \text{ m/s}^2} = 28{,}125 \text{ m} = 28 \text{ km}$$

That's 17.5 mi! That 16-lb shot, thrown from the 2012 London Olympic Stadium, would have sailed over Big Ben, 6 mi away, still rising. It would have just about reached Heathrow Airport on the outskirts of the *other* side of the city.

Is it clear that 240 Cal is a *lot* of energy, if it could all be converted into useful (for sports) energy forms?

7.3.2 The human engine I: energy conversion

Note: The discussion here may go into greater detail than the instructor wishes to cover. The important points are listed in the summary at the end of this subsection.

The conversion of energy between the kinetic, elastic, and gravitational forms is pretty easy to understand through mechanical forces such as gravity and drag. But the conversion from food energy to other forms is less obvious. Are there "forces" involved here, too?

It turns out that the answer is yes, but they are complicated and even involve some quantum mechanics. My editors forced me to remove the 450-page chapter on quantum mechanics and microbiological forces, so we'll just cover the basics. We'll discuss the energy flow in terms of chemistry, instead of microscopic forces,[6] and our treatment will be very broad strokes.

All chemical reactions involve energy as an ingredient. If energy is released in the reaction, it is called **exothermic**. **Endothermic** reactions require energy input to proceed.

---

[6] A physicist always has to point out that chemistry is a sort of shorthand glossing over the more fundamental physics underneath.

The universal fuel: ATP

Our bodies can extract energy from a number of different chemical fuels (glucose, fats, starch, proteins, etc.), but *the only fuel used by muscle cells is adenosine triphosphate* (ATP). Any other fuels are only useful inasmuch as they can regenerate ATP in the cell.

ATP is the only fuel used directly by muscles.

The conversion of ATP into adenosine diphosphate (ADP) in the cell mitochondria is an exothermic reaction, which means that some chemical energy is released. Part of that energy makes muscle fibers contract,[7] providing energy to throw a ball, lift a weight, or draw a bowstring, that is, something athletically useful. But at least one-half goes into useless heat that the body releases as waste into the air. That is, the efficiency is

$$\eta_{\text{ATP}\to\text{mechanical energy}} \lesssim 50\% \qquad (7.4)$$

The efficiency for any given muscular action will vary, but the best it can be is about 50%.

Each muscle cell stores enough ATP for about 2–4 s of maximal effort. It also contains a store of a chemical called phosphocreatine (PCr) that can be converted very quickly to ATP within the cell for another 10 s of hard effort. Additional ATP for continued functioning of the muscle cell must be generated from other fuel sources.

Food to ATP

We don't eat ATP in our food. Instead, ATP is produced in an endothermic reaction. Remember, this means that energy must be input to produce ATP.

The combination of glucose (the crucial caloric ingredient extracted in our digestion of food) with oxygen is exothermic. The energy released from that *exo*thermic reaction is the input to the *endo*thermic reaction that produces ATP. That's why we die if we can't breathe—we very quickly run out of oxygen to power our heart and everything else.[8] The important reaction here is

$$\underbrace{C_6H_{12}O_6}_{\text{eat}} + \underbrace{6O_2}_{\text{inhale}} \to \underbrace{6H_2O}_{\text{sweat}} + \underbrace{6CO_2}_{\text{exhale}} + 686 \text{ Cal.} \qquad (7.5)$$

(This reaction occurs in reverse in plants, in the photosynthesis process; that's where the glucose we eat and the oxygen we breathe comes from in the first place.)

In an equation like 7.5, $C_6H_{12}O_6$ represents 1 **mole**, abbreviated as 1 mol, of glucose molecules, not just one molecule. A mole is about 6×10^{23}, or 600 billion trillion (!) molecules. A mole of glucose molecules is 180 g of glucose. One mole of oxygen is 22.4 liters of pure O_2 gas, so $6O_2$ represents $(6 \text{ mol } O_2) \times (22.4 \text{ liter/mol}) = 134$ liters of O_2.

So, mix 180 g of glucose with 134 liters (about 32 gal) of oxygen[9] and you've released 686 Cal. About one-half of that energy is immediately released as waste heat. The other half is used to produce ATP. That is, the energy efficiency of the glucose-to-ATP conversion is about 50%.

---

[7] Muscles can exert only contracting forces; they cannot stretch themselves. For a muscle to stretch, it relaxes as an "antagonistic" muscle contracts. The triceps are antagonistic to the biceps.

[8] It's no problem to hold your breath for a short time, because your blood holds an oxygen reservoir to keep you going for a while.

[9] I'm excluding many details that would complicate this discussion, but I have to point out that when we breathe, roughly 24 liters of air must pass through our lungs, for each liter of oxygen absorbed. Therefore, the quantity of *air* we must breathe is 24×134 liters = 3200 liters, or 845 gal.

Actually, if the glucose is processed very quickly, the efficiency can be a lot less, so we'll write

$$\eta_{\text{glucose} \to \text{ATP}} \lesssim 50\%. \tag{7.6}$$

Overall efficiency

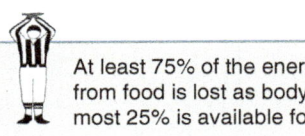

At least 75% of the energy extracted from food is lost as body heat. At most 25% is available for sports.

To summarize, our bodies turn about one-half of the available energy from the food we eat (and air we breathe) into heat; the other half is stored as energy in ATP. And at most one-half of *that* energy can be used in sports; the rest also goes into waste heat. So, if we consider the Calories listed on the food package as the input and the energy we put into sports as the output, the energy efficiency of a muscular process is

$$\eta_{\text{glucose} \to \text{mechanical energy}} = \underbrace{\eta_{\text{glucose} \to \text{ATP}}}_{\lesssim 50\%} \times \underbrace{\eta_{\text{ATP} \to \text{mechanical energy}}}_{\lesssim 50\%} \lesssim 25\% \tag{7.7}$$

Stored chemical energy

Oxygen absorbed from the environment (through breathing) is used almost as quickly as it is acquired. Glucose, on the other hand, need not be. If digested glucose is not immediately needed for muscular (or cerebral, that is, brain) activity, it is stored as **glycogen** or as **fat**. Glycogen is a starch stored in the muscles and liver that can be converted to glucose and ATP when needed. A healthy human typically has about 20 Cal (5 g) of glucose circulating in the bloodstream and a maximum of 1800 Cal (450 g) of glycogen stored in the muscles and liver. (Persons with greater muscle mass have proportionally higher glycogen storage capacity.) This is the **glycogen saturation point**.

If the body tries to absorb glucose when the cells are full of ATP and the glycogen stores are at their saturation point, the excess is stored as **body fat**, by far the largest reservoir of stored energy. Fat is stored throughout the body and is energy-dense stuff. A gram of pure fat, mixed with oxygen, provides 9 Cal of chemical energy, compared to 4 Cal/g of glucose or glycogen. We store more than 50,000 Cal (200,000,000 J) of chemical energy in the form of body fat! It's funny: the fatter I get, the less energetic I feel.

By way of summary, figure 7.6 shows the routes that glucose takes in the body. It includes a fourth pathway we'll mention but won't discuss. Very high levels of glucose in the blood can be very harmful. If the pancreas cannot produce enough insulin to stimulate the conversion of glucose to glycogen or fat quickly enough, some glucose will be eliminated in the urine. Obviously, the potential energy in this glucose is lost to the body altogether.

With all these storage forms flying around, let's be very clear. When I write E_{chem} in formulas, I mean all the energy stored as available ATP, PCr, glucose, glycogen, and fat.

Important points from section 7.3.2

There are three main take-away points of this section:

1. Chemical reactions between glucose in the food we eat and oxygen in the air we breathe release energy.

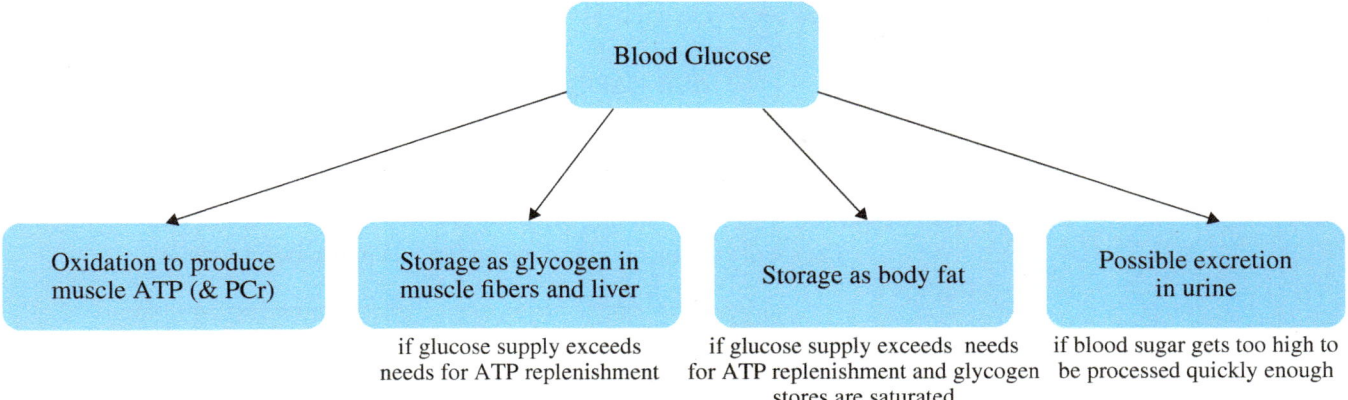

Figure 7.6. The possible paths for blood glucose extracted from food. Pathways on the right in the figure are chosen as pathways to their left are saturated.

2. At least one-half of that energy is converted directly to thermal energy (heat). The remainder is stored in the chemical ATP in the muscles or, if the muscles are full of ATP, as glycogen or fat. Glycogen and fat can be converted later to ATP as needed by the muscles.

3. To flex, muscles release the chemical energy stored in their ATP. At least one-half of that energy is released as heat. The remainder goes into muscular movement, which is the source of all energy in sports.

Taken together, items 2 and 3 say that the athlete can convert at most 25% of the energy in his food to forms useful in sports (see equation 7.7).

7.4 Keeping Score: Energy Accounting in Sports

 Energy flows from one form to another, but the total amount of energy never changes.

Whether it is a ball bouncing on a court, a weight lifter lifting a barbell, or a football player pushing a sled, energy is always just being *converted* from one form to another. In figures 7.2 and 7.3, the little bar graphs show that while the amount of energy in any form might rise or fall, the total amount of energy never changes. This is one of the most fundamental laws in all of science, and it continues to hold true when we add the athlete into the picture. The **law of energy conservation** says that the total energy[10]

$$\text{total energy} = \underbrace{KE + PE_{grav} + PE_{elast}}_{\text{also known as mechanical energy}} + E_{chem} + E_{therm} \quad (7.8)$$

is an unchanging number, for any sports situation. As indicated in the formula, the sum of the kinetic and potential energies is often called the **mechanical energy**. Mechanical energy consists of those forms that can be interchanged back and forth among themselves.

---

[10] If our discussion were not confined to sports, the definition of "total energy" would need to include many other forms, such as nuclear energy.

Another way to write this is to say that the change in the total energy is zero:

$$\Delta KE + \Delta PE_{grav} + \Delta PE_{elast} + \Delta E_{chem} + \Delta E_{therm} = 0$$

As usual, the Δ symbol just means the change in something, over some period of time. (For example, ΔPE_{grav} can be the difference between the gravitational energy now and what it was 30 s ago.)

Let's write the energy conservation formula into a useful form:[11]

$$\underbrace{\Delta KE + \Delta PE_{grav} + \Delta PE_{elast}}_{\text{change in mechanical energy}} = \begin{pmatrix} \text{food} \\ \text{energy} \\ \text{burned} \end{pmatrix} - \begin{pmatrix} \text{thermal} \\ \text{energy} \\ \text{generated} \end{pmatrix} \quad (7.9)$$

The "thermal energy generated" can be from friction, air drag, or body heat.

Equation 7.9 tells the same story told by figure 7.5. The game is determined by the balance between energy input from burning Calories and the energy output to heat. The change in mechanical energy (left side of the equation) is positive if something speeds up ($\Delta KE > 0$), rises higher ($\Delta PE_{grav} > 0$), or gets elastically squished ($\Delta PE_{elast} > 0$). All these things require energy, and the change in mechanical energy is positive only if more energy is put in than is taken out, that is, if

$$\begin{pmatrix} \text{food} \\ \text{energy} \\ \text{burned} \end{pmatrix} > \begin{pmatrix} \text{thermal} \\ \text{energy} \\ \text{generated} \end{pmatrix}.$$

This should make sense to you. Does it? If not, pause and look over it again. It's worthwhile to get it down before you move on.

7.4.1 The water analogy

All the different forms of energy and "one-way streets" can get confusing. Some students find it useful to consider an analogy with water and buckets, just to solidify the concept.

In this analogy, each drop of water represents 1 J of energy. Three buckets, which represent kinetic energy, elastic potential energy, and gravitational potential energy, can hold water. A full KE bucket represents an object moving quickly, whereas a full PE_{grav} bucket represents an object high in the air. A sports process involves pouring water from one container into another.

As a concrete example of this analogy, consider the basketball from figure 7.3 as it travels from its highest point, down to the ground, and back up to its next peak. In the water analogy, this process is represented by water poured from a full PE_{grav} bucket into the KE bucket, then into the PE_{elast} bucket, then back into the KE bucket, and finally back into the PE_{grav} bucket. The total amount of water (representing the total amount of energy) never changes; it just moves from bucket to bucket.

Oh, except for one thing: you *always* spill some water onto the dusty ground, every time you transfer water between buckets. This, of course, represents the loss to thermal energy. That water sinks into the dust and can't be put back into the buckets. That's the one-way street for thermal energy.

The total amount of water stays the same (conservation of energy), but the amount of *useful* water—the amount in the buckets—goes down because of the spillage.

---

[11] In equation 7.9, the "food energy burned" (that is, "Calories burned") is the *decrease* in chemical/food energy, or $-\Delta E_{chem}$.

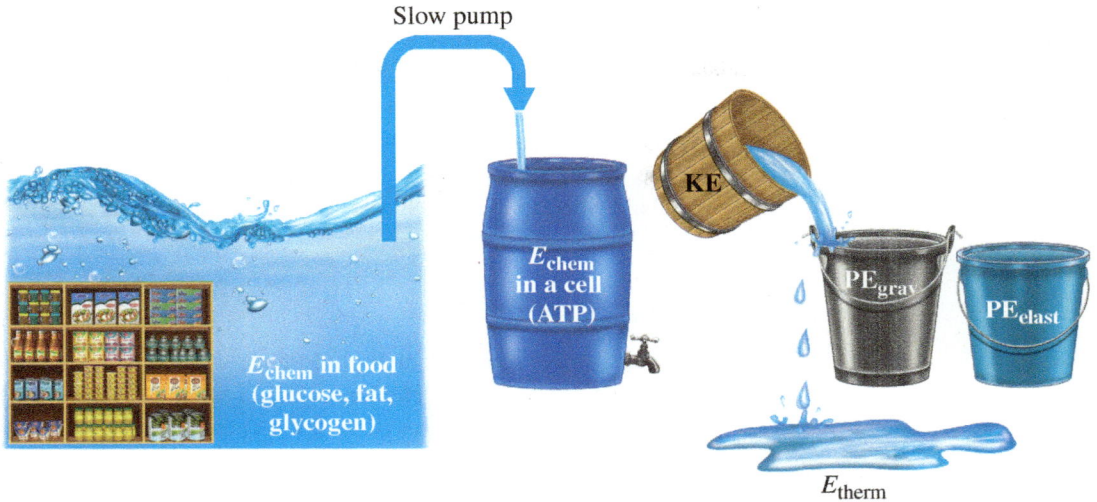

Figure 7.7. The water analogy for energy.

Eventually, you wind up with a wet ground and empty buckets. That's the "heat death" we mentioned in section 7.2.3.

We can refill our buckets from the spigot, which is connected to a barrel. If you've been following the analogy, you've guessed that the water from the spigot represents chemical energy and the barrel holds energy stored as ATP in our cells. The fact that water can flow from a spigot into buckets (or onto the ground), but not from buckets back into a spigot, represents the one-way flow of energy *from* the chemical form to other forms.

Finally, as we work our way to the left in the sketch in figure 7.7, water slowly is being pumped into the barrel from a vast reservoir or lake. The situation sketched is identical to the system used in U.S. cities, in which water is continually pumped at a slow rate from a reservoir into water towers. The water can flow *out* of spigots connected to the tower much more rapidly than it is pumped in.

In our analogy, the lake is the glucose, fat, and other chemical energy stored in our bodies. The barrel is the muscle cell, and the "slow pump" converts glucose energy into ATP energy in the muscle. The pump works one way only: energy stored as ATP in the muscle can never be put back into the glucose "reservoir."

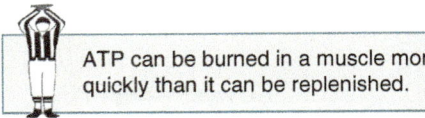

To complete the analogy, the "barrel" can only hold so much water, representing the fact that a muscle cell can only hold so much ATP. The barrel can be emptied much more quickly than it is being filled. In our analogy, this represents the quick burst of energy available by converting ATP into mechanical motion, compared to the relatively slower rate at which glucose can be converted to ATP. We'll discuss this shortly.

7.4.2 How useful is the energy conservation concept really?

The conservation of *momentum* was great for figuring out whether a given tackle would knock a running back out of bounds before he reached the end zone or the speed at which a ball leaves a bat. Will the conservation of energy be as useful?

Well, in introductory physics books, it sure is. In those books, you'll see lots of blocks attached to "ideal" springs (which don't heat up when they stretch or compress), often sliding on "frictionless" surfaces. Body heat generation is *never* included in those introductory physics books, and the burning of food calories is ignored. Air and air drag are almost always neglected.

Well, sure, then it's easy! The right side of equation 7.9 is zero, and an unchanging amount of energy sloshes between the kinetic, gravitational, and elastic forms, a never-ending perpetual-motion machine. That's what the engineer's first physics class covers.

Meanwhile, back in the real world, our athletes eat food, breathe air, and get hot. While the law of energy conservation is absolutely true, it is sometimes of limited practical use when applied to athletes. This point will be clear if we come back to the seemingly simple concept of efficiency.

Efficiency and human motion

We've seen that the term "efficiency" depends on what you consider to be useful energy. To see that sometimes efficiency isn't a useful concept, consider again a basketball that rises and then returns to the floor. There are two possible ways for this to happen, shown in figure 7.8.

On the left, a basketball bounces up from the floor, trading kinetic energy for gravitational energy; on the way back down, the energy flow reverses. The energy can flow back and forth between the kinetic, gravitational, and elastic forms, always losing a bit to thermal energy, of course. The efficiency of the process is never 100%, because of thermal losses, but it's always positive.

What about when a person gets involved? Well, that's when the one-way street of chemical energy flow changes things.

When a player picks it up, he's converting some of his food or chemical energy into the gravitational energy of the ball. The energy flow for that is shown on the upper right of figure 7.8. It looks very much like the upward bounce, except it's E_{chem} that's being converted to PE_{grav}, rather than KE being converted. The useful output is the change in mechanical energy generated by the athlete. The input is the food energy burned. It would make sense to divide the two to find the efficiency of lifting.

What's very different is the energy flow for the ball returning to the floor. The human body can convert chemical energy to other forms of energy, never the other

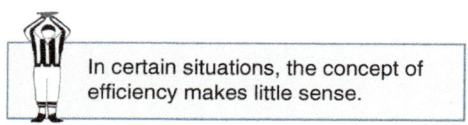
In certain situations, the concept of efficiency makes little sense.

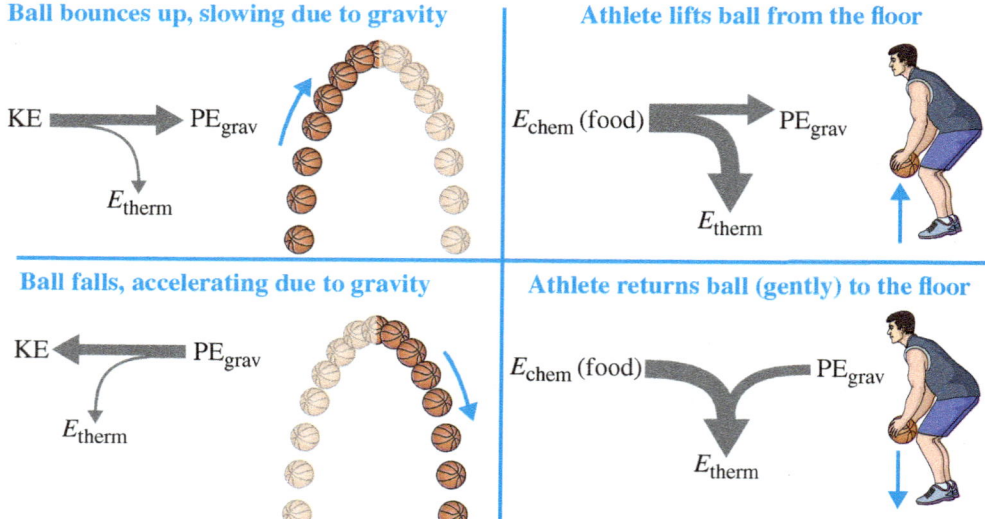

Figure 7.8. Two possibilities for a basketball to go up and then down. **Left side:** energy flows from the kinetic to the gravitational form on the upward journey of a ball, and back into kinetic on the downward journey. For both journeys, some energy is lost to thermal forms due to drag. **Right side:** When a person is involved, the picture is a bit different. Energy never flows back into the chemical form (which would be like magically putting food into your stomach!).

way around. (That's what plants are for.) So, while it obviously takes a positive number of food calories burned to *increase* the energy of the ball, it also takes a person a positive number of food calories burned to put the ball back down, that is, to *decrease* the ball's energy. One way or another, the food energy burned and the ball's gravitational potential energy are turned into thermal energy.

If you lift weights, you know this already: it takes almost as much effort to very slowly *un*curl a dumbbell as it does to curl it. All the energy that you expend while uncurling goes into heat. In this case, the output is a negative change in mechanical energy, so using equation 7.2 would give a "negative efficiency."

Efficiency is a natural concept and can be very useful in analyzing energy flow in athletics. But you have to use common sense when using it, be clear about what energy is useful, and know that sometimes a simple formula for efficiency makes little sense.

7.5 Uncle Rico's Hopes Dashed

In the movie *Napoleon Dynamite*, Napoleon's self-deluded Uncle Rico boasted that he could "throw a football over them mountains" in the distance. Since the steaks he eats every day contain more Calories than the PowerBar from example 7.2, so he should be able to do it!

The problem with Uncle Rico is that he only read up to this point in this book. That was just far enough to prove to himself and the world that, of course, any book about sports must include him. So he never learned about power (which we come to next) and efficiency. He was afraid of the word "biochemistry," and, unlike you, his eyes glazed over in section 7.3. Don't be like Uncle Rico; keep reading.

First, there is efficiency. A small part of Rico's problem is the body's efficiency of energy conversion. According to equation 7.7, only at most 25% of the 240 Cal of energy can be converted to mechanical energy. Still, that's more than 0.25 million J $\left(\frac{1}{4} \times 240 \text{ Cal} \times \frac{4184 \text{ J}}{1 \text{ Cal}} = 251{,}000 \text{ J}\right)$, probably enough to reach the foothills.

Alas, Rico has bigger problems. While he can eat and store sufficient energy to fulfill his fantasy, he has neither the strength nor the power to use that energy. The crucial concepts here—and in much of sports—are **work** and **power**.

7.5.1 Work

Always on the hunt for get-rich-quick schemes, Uncle Rico had a strange conception of the whole "work" thing. For his sake and ours, science demands a precise definition of this word that has so many meanings in everyday speech.

Work quantifies the transfer of energy when an athlete exerts a force. If the force is constant,

$$W = F_x \Delta x + F_y \Delta y = |\vec{F}| \times \left(\text{distance moved parallel to } \vec{F}\right) \quad (7.10)$$

As we know, the force is usually *not* constant in sports; we'll discuss this in section 7.6 on weight lifting.

Doing work *always* involves exerting a force. And it *always* involves moving the object (or person). Work has units of energy, because $\vec{F}$ is in newtons and $\Delta \vec{x}$ is in meters, and $1 \text{ J} = 1 \text{ N} \cdot \text{m}$, as it says in equation 6.7.

In fact, not only does work have the *units* of energy, but also there is a stronger connection between work and energy, and this connection has a name that's near and dear to scientists:

Work-Energy Theorem

The work done by the athlete plus the work done by other forces (except gravity and elastic forces) equals the change in mechanical energy:

$$W_{\text{athlete}} + W_{\text{other forces}} = \Delta KE + \Delta PE_{\text{grav}} + \Delta PE_{\text{elast}} \qquad (7.11)$$

If there is an "other force," it is usually a dissipative force such as drag or kinetic friction, though it can be from anything (say, work done by an opposing athlete). The reason gravity and elastic forces are excluded from being "other forces," is that their effects already appear on the right-hand side of the equation.

Comparing equations 7.11 and 7.9, we see the connection between food energy, heat, and work:

$$\begin{pmatrix} \text{food} \\ \text{energy} \\ \text{burned} \end{pmatrix} - \begin{pmatrix} \text{thermal} \\ \text{energy} \\ \text{generated} \end{pmatrix} = W_{\text{athlete}} + W_{\text{other forces}}$$

The work-energy theorem already helps us understand part of Rico's problem, as illustrated in example 7.3.

EXAMPLE 7.3 *Rico attempts some work*

If Rico wants to give the ball 251,000 J of kinetic energy, how much force does his arm have to generate? Assume his hand (and the ball) travels 1.3 m during the throw before he lets go.

Equation 7.11 tells us that to change the ball's kinetic energy from zero (when Rico's arm is cocked back, ready to throw) to 251,000 J, he must do 251,000 J of work. He does this work by exerting a force on the ball while he moves it 1.3 m. Equation 7.10 lets us find the force he has to exert:

$$W = |\vec{F}_{\text{hand}}| \times |\text{distance hand moves}|$$

$$\rightarrow |\vec{F}_{\text{hand}}| = \frac{W}{|\text{distance hand moves}|} = \frac{251{,}000 \text{ J}}{1.3 \text{ m}}$$

$$= 193{,}000 \text{ N}$$

$$= 43{,}400 \text{ lb} \qquad (21.7 \text{ tons})$$

So while Rico has easily stored enough energy in his body to "throw the football over them mountains," he's certainly not strong enough (that is, he cannot generate enough force) to do it.

Question

The force that Rico had to exert on the ball didn't depend on the ball's mass. (We never used it in the calculation in example 7.3.) Does that make sense? Shouldn't the force be larger if he were throwing a bowling ball rather than a football?

7.5.2 Power

Maybe we're underestimating Rico. I bet he'll tell us that if he keeps working out, he'll be able to exert 22 tons of force, no problem. Alas, even if he could become *strong* enough, he'd never become *powerful* enough to achieve his goal, as we'll see. Once again, we need to precisely define a term that is loosely used in everyday speech.

Some sports columns described the unforgettable Mike Tyson as an incredibly *strong* boxer. However, by far the word most often used to characterize him was *powerful*.

What is the difference between a strong athlete and a powerful one? In everyday speech, these words are used interchangeably, but in a scientific analysis of sports, they mean very different things.[12]

Power is the rate at which energy flows from one form to another.

The rate of energy transfer from one mode to another is called *power*. Whereas "strong" indicates that Tyson can exert large forces, "powerful" indicates that he can rapidly convert chemical energy in his muscles to the kinetic energy of his fist (which, in turn, is usually converted to thermal energy in the form of his opponents' broken bones or torn tissue). Make no mistake, Iron Mike was both strong *and* powerful.

The power of any given process is just the amount energy transferred, divided by the time it takes to transfer the energy:

$$\bar{P} = \frac{\text{energy transferred}}{\Delta t}. \tag{7.12}$$

When we talked about velocity and force earlier, we saw that it was often simplest to talk about their average value for some length of time. The same is true here, so equation 7.12 is actually the **average power** for some duration of time Δt. If we really want to talk about the rate of energy transfer *at* some instant of time, we talk about the **instantaneous power** and use the symbol P instead of $\bar{P}$. Again, we did the same things with velocity, acceleration, and force.

In the language of the water analogy of energy from section 7.4 and figure 7.7, *power is the rate at which water flows from one container to another*. When we talk about the power of a hockey stick unflexing or the release of a bow string, this refers to the rate of water flow from the PE_{elast} bucket to the KE bucket. Very often, however, the "energy transferred" that interests us is the work done on something (such as a football). In that case,

$$\bar{P}_{\text{work output}} = \frac{\text{work done}}{\Delta t} = \frac{W}{\Delta t}. \tag{7.13}$$

Power is measured in **watts**[13] or **horsepower**.

$$1\ W = 1\ \frac{J}{s} \qquad 1\ \text{horsepower} = 1\ \text{h.p.} = 746\ W \tag{7.14}$$

When not talking about sports, we traditionally use watts to quantify power from electrical devices and horsepower for gas-powered engines. But in both cases, it is the same quantity—the conversion from a stored form (electrical or fossil fuel) to another. When discussing sports, you might encounter either unit.

---

[12] We've seen this sort of distinction several times before, such as between "fast" (capable of large speeds) and "quick" (capable of large accelerations).

[13] It's an unfortunate fact of life that W is so often used as the symbol for the unit watt; it can be easily confused with W, the symbol for work. Pay attention to the italics.

EXAMPLE 7.4 Rico's explosive power requirement

Assuming Rico somehow bulked up enough to generate 193,000 N of force (as we found in example 7.3), how much *power* would he have to generate to give the football 251,000 J of kinetic energy?

We know how much work Rico needs to perform; to find the power, we need to know the time he has to do it in. This is a simple matter of force laws and kinematics. According to Newton's second law (equation 3.6),

$$a_{\text{ball},x} = \frac{\sum F_x}{m_{\text{football}}} = \frac{193{,}000 \text{ N}}{0.42 \text{ kg}} = 4.6 \times 10^5 \text{ m/s}^2$$

That's a lot of acceleration—47,000 g's $\left(\frac{4.6 \times 10^5 \text{ m/s}^2}{9.8 \text{ m/s}^2} = 47{,}000\right)$!

Equation 2.7 lets us find the time he has to move the ball 1.3 m during the throw:

$$\Delta x = \underbrace{v_{x,0}}_{\text{zero}} \Delta t + \tfrac{1}{2} a_x (\Delta t)^2$$

$$\rightarrow \Delta t = \sqrt{\frac{2 \Delta x}{a_x}} = \sqrt{2\,(1.3 \text{ m})\,4.6 \times 10^5 \text{ m/s}^2} = 0.002 \text{ s}$$

The power he needs to generate, then, is

$$\bar{P}_{\text{work output}} = \frac{251{,}000 \text{ J}}{0.002 \text{ s}} = 125{,}500{,}000 \text{ W} = 168{,}200 \text{ h.p.}$$

The average American home burns about 1000 W of electricity, so we are asking Rico to generate power enough for 125,000 homes—a city the size of Columbus, Ohio! The 2013 Ford F-150 5.0-liter V8 generates 360 h.p., so Rico needs to generate as much power as 467 F-150's, which is clearly impossible.

Pay attention to an important point. In example 7.4, I said that an average household consumes 1000 W of electricity, not "1000 W of electricity *per day*." The "per day" (or "per second" or whatever) is already included in the definition of power: it is the *energy used* "per day." Power is the rate of energy use. This is a perennial source of confusion among students, so before you move on, be sure you understand this important point.

The human body can convert food energy at only a limited rate.

7.5.3 The human engine II: power

Good grief! These numbers are getting out of hand! Example 7.2 tells us that the energy in a PowerBar can throw a shot-put 17.5 mi. Examples 7.3 and 7.4 tell us that for Rico to do even one-quarter of that work on a football, he needs to exert 22 tons of force and generate power enough to light a medium-sized city (for 2 ms anyway)! Do we really consume this much energy in the food we eat?

Yes, we do. We consume a lot of energy. We store it; we convert it into heat; we convert it into useful work. The point is, however, that the human engine can convert that energy only so quickly. Just as a truck engine can't burn through its entire tank of fuel in an instant, neither can we process a PowerBar in a very short time.

This isn't a book on biochemistry, but the physics of sports is to a large extent a study of energy and power in athletics. Since all athletic energy begins in the body, we'll need to know something about how the body generates power.

Equation 7.13 is the formula for the rate at which the athlete adds mechanical energy to the game—the "output" of his efforts. The *input* power is the rate at which chemical energy is burned:

$$\bar{P}_{\text{chemical input}} = \frac{\text{food energy burned}}{\Delta t} = \frac{-\Delta E_{\text{chem}}}{\Delta t}. \qquad (7.15)$$

That annoying minus sign is there because when you burn energy, the chemical energy in your body goes down. That is, $\Delta E_{chem} < 0$. The power defined in equation 7.15 is always a positive number.

We talked about efficiency in terms of energy. It applies to power as well:

$$\bar{P}_{\text{work output}} = \eta_{\text{chemical}\to\text{mechanical energy}} \times \bar{P}_{\text{chemical input}}. \qquad (7.16)$$

The efficiency, denoted by $\eta_{\text{chemical}\to\text{mechanical energy}}$, is always less than 1, and it depends on the activity done and the chemical reaction involved. Equation 7.7 is for the special case in which the chemical reaction is between glucose and oxygen. In this section, we'll discuss other reactions, too.

Note: The discussion here may go into greater detail than the instructor wishes to cover. The important points are listed in the summary at the end of this subsection.

In section 7.3.2, I said that ATP is the only fuel that our muscles use to generate mechanical forces. ATP generation processes from other fuel can be broken into two categories: anaerobic (not requiring oxygen) and aerobic (requiring oxygen). The glucose oxidation process shown in equation 7.5 is an example of an aerobic process.

Anaerobic processes

There is a small amount of preformed ATP in our muscle cells ready for immediate use; under maximum exertion, it is spent in about 2 s. About twice as much energy is also stored in the muscle cells in the form of PCr, which can be rapidly converted to ATP and depleted in 8–10 s. Since more than one-half of the energy from burning ATP goes into heat, the very limited amount of readily available ATP and PCr is actually a self-preservation mechanism. If an athlete could burn much more ATP at its maximum rate, his body could not emit the waste heat quickly enough, leading to overheating and death.

An athlete can generate maximum power for up to 2 s, by burning all the preformed ATP in his muscles.

Since all the energy is converted in a very short time, consumption of preformed ATP and PCr is used for short bursts of great power. There are two important points to make. First, reducing the duration of the effort (Δt) below 2 s will not increase the power of the burst; therefore, Rico can generate the same power in a 2-*milli*second effort as in a 2-s effort.[14] Second, each muscle cell contains its own store of PCr and preformed ATP, and the cells work cooperatively to produce motion; therefore, the peak power depends on the athlete's muscle mass.

The higher the intensity of an athletic event, the greater the reliance on PCr as a source of energy. After an event requiring maximum power generation for up to 10 s (for example, sprinting or jumping), athletes need to rest a few minutes, as their reservoir of PCr is regenerated.

The second important anaerobic process is ATP production from glycogen.[15] This is called *anaerobic glycolysis*, and it kicks in with a delay of 5 to 10 s after intense physical activity starts. Anaerobic glycolysis can convert three times as much energy as the direct PCr-burning process before resources are depleted, but at only one-half the power. A by-product of this process is *lactic acid*, which causes pain to the athlete and decreased performance if too much of it builds up. Lactic acid can be eliminated

Anaerobic processes can fuel high-power bursts up to about 60 s, which is useful for football, hockey, boxing, and so on.

---

[14] The *amount* of energy consumed will be 1000 times larger in the 2-s effort, but not the *rate* of consumption.

[15] Remember? This was the starch stored in muscle fibers (though not in the cells themselves). Look back at section 7.3.2 for a refresher.

by the body or burned aerobically to produce ATP, but at a slower rate than it is produced by anaerobic glycolysis.

Together, PCr conversion and anaerobic glycolysis can fuel hard exertion for about 90 s. After this, power output necessarily drops, as anaerobic fuel is depleted and aerobic processes do most of the work. An athlete needs a break to recover, "catch his breath," restock PCr supplies, and process built-up lactic acid. A prime challenge is to recover quickly enough for the next bout of activity—breaks in the action of around 1 min are given between boxing rounds, ice hockey shifts,[16] and downs in American football.

It is no coincidence that several high-power-output athletic events are timed to match the energy processes of the participants' bodies. Figure 7.9 lists some events of this type. It is interesting to note that wrestling matches increase from 2 min in high school to 5 min in the Olympic event. The reliance on longer-term aerobic energy processes is much stronger at the higher level.

Aerobic (oxygen-consuming) processes

Partly because the body can absorb oxygen only at some maximum rate (more on this in chapter 9), aerobic processes convert chemical energy more slowly. On the

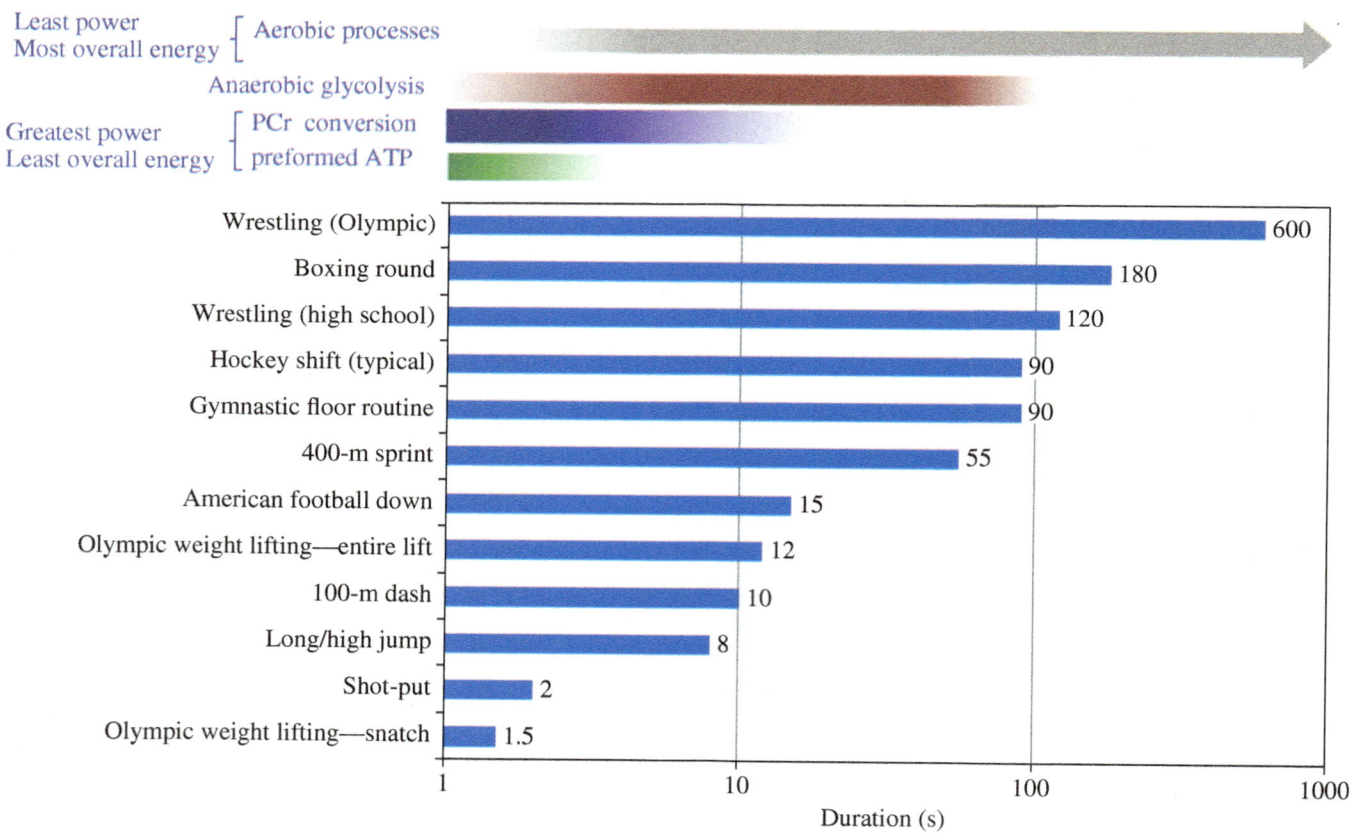

Figure 7.9. The duration of several high-power athletic events has naturally been matched to the body's anaerobic processes, which can produce high energy output but only for a limited time. Notice that the time axis is on a logarithmic scale.

---

[16] Even though a hockey period lasts 20 min, players are continuously being shuttled on and off the ice; the shift of any one player between short rests to recover is rarely much more than 1–2 min. For forwards, 45 s is more typical.

other hand, aerobic processes can utilize the vast reserves of energy stored in the body as glucose, glycogen, fat, and (in extreme cases) some protein.

They are also considerably more efficient. While anaerobic glycolysis generates two or three times as much power as aerobic glycolysis, it uses only about 7% of the fuel's energy; the rest is wasted. Therefore, the anaerobic process burns through glycogen up to 40 times more rapidly, and a glycogen store that could fuel 2 hr of hard aerobic effort can be depleted in just a couple of 1-min-long bursts of anaerobic burning. Tour de France cyclists have to keep this in mind when they consider a sprint to break away from the peloton.

Figure 7.10 shows measurements of human power output for efforts of different durations. The transition from high-power, short-duration to lower-power endurance exertions is clear.

The main aerobic fuels and typical energy stores are listed in table 7.2. This chapter is focused mostly on anaerobic bursts. In chapter 9, we'll discuss endurance events and slower-burn aerobic power production.

Aerobic energy conversion is less powerful than anaerobic conversion, but much more efficient. Aerobic processes also access vastly larger energy stores.

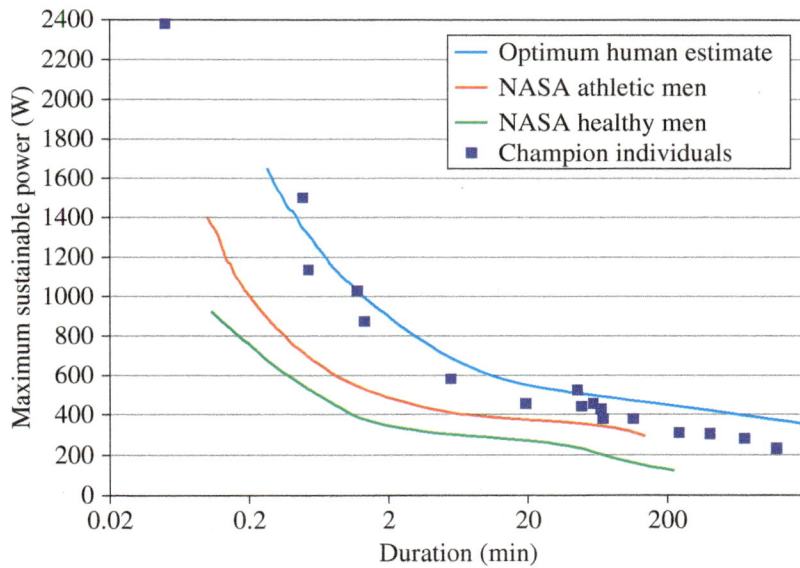

Figure 7.10. The power output by top athletes varies strongly as the duration of the effort is lengthened. The data points represent measurements of record-holding athletes in bicycling or rowing exercises, as collected in *Bicycling Science*, 3rd edition by David Wilson. Curves are from a NASA report from the early days of manned space missions and a theoretical estimate, from Wilson's book, of the maximum power output humanly possible.

Table 7.2. Stores of aerobically available fuel in the average person. The right most column tells how long each fuel source could sustain vigorous effort (approximately running at 8 mph) before being completely depleted. Notice that burning *all* fat is impossible and that fatigue and need for sleep would necessarily break up the exercise into more than one session. Adapted from Gleeson, 2000, Biochemistry of Exercise, in *Nutrition in sport*.

Fuel	Mass (kg)	Energy (Cal)	Exercise duration (min)
Liver glycogen	0.08	306	16
Muscle glycogen	0.40	1,529	80
Blood glucose	0.01 (10 g)	38	2
Fat	10.5 (23 lb)	93,000	4856 (3 days!)

Aerobic and anaerobic processes occur simultaneously

You should keep in mind that both anaerobic and aerobic ATP production are taking place at the same time; it's just the balance between them that varies, depending on the activity. After all, even though Usain Bolt gets most of the energy for his 9.8-s sprint anaerobically, he certainly isn't holding his breath while running! There are three reasons for him to breathe deeply while he sprints. First, while the muscles can use PCr and glycogen, the brain uses mostly glucose for its fuel, and glucose processing is an aerobic process (remember equation 7.5); deprived of oxygen, the brain starts stressing. Second, once the race is over, he's going to need energy too, in order to celebrate and give his signature "archer's" victory pose. Finally, the aerobic process works together with the anaerobic one; even though it is not dominant for his short burst, every bit helps.

During intense effort, the first 10 s or so is almost entirely anaerobically driven. In fact, aerobic processes contribute only about 20% of the energy to the first 30 s. During the next 30 s, that fraction is up to 40%. For the third minute of exercise, the aerobic contribution is 70%. After this, it's all aerobic, and the whole story is determined by how quickly the athlete can absorb oxygen. We'll discuss this in detail in chapter 9 on cycling.

Power to keep the engine idling

When a truck is performing some high-power feat such as dragging boulders up rocky mountains in the tundra in television commercials (while the driver listens to Sirius radio and sips coffee in his heated leather seat), it's pretty easy to see the work done. As defined by equation 7.13, the output power $\bar{P}_{output}$ is large. Now, you already know that the *input* power—the rate at which chemical energy (from the reaction between gasoline and air) is being consumed—is higher than the output rate.

In fact, when it's just sitting there idling, the truck is consuming fuel, too. It is doing no mechanical work on anything outside itself (for example, boulders), but some chemical energy is being converted to internal motion and eventually heat. Even if we're not involved in a game, we have to keep idling too, or else it's game over ... forever. It takes energy just to breathe, keep the heart beating, and read this book.

The rate at which we consume chemical energy when we are at rest is called the **resting metabolic rate** (RMR) or **basal metabolism rate**. For a 180-lb person, it is just about

$$\bar{P}_{RMR} \equiv \bar{P}_{\text{chemical input, while at rest}} \approx 100 \text{ W} \approx 80 \frac{\text{Cal}}{\text{hr}} \quad (7.17)$$

The events we are focusing on in this chapter (weight lifting, slap shot, balls bouncing) are quick, high-power events in which the power generated is much higher than $\bar{P}_{RMR}$. When we discuss endurance events such as cycling in chapter 9, we'll revisit RMR and human power one more time.

Important points from section 7.5.3

Even if you forget words like "glycolysis," remember the main points:

1. The human body stores a tremendous amount of energy and so can perform a lot of work. Limitations on the *rate at which* that work can be performed (that is, the power output by an athlete), rather than the total amount work an athlete can perform, is often what determines athletic performance.

2. In short bursts of a few seconds, an athlete can generate about 1600 W of mechanical power by burning all the ATP stored in muscle cells anaerobically.

3. The athlete can generate several hundred watts for somewhat longer bursts, by similar anaerobic processes.

4. After efforts of type 2 or 3 above, one must rest to allow aerobic processes to replenish the ATP in the muscles. The duration of many "power events" is tuned to the time scales of anaerobic depletion of ATP in muscle cells (see figure 7.9).

5. For long-duration events such as marathons, a fit athlete can output about 200 W of mechanical power. Here ATP is being generated from stored energy sources at the same rate at which it is being converted to mechanical work by the muscles.

6. While short maximum-effort bursts produce much more mechanical power than endurance efforts, they are considerably less efficient. Only about 7% of the food energy consumed shows up as useful mechanical energy in a maximum-effort burst, compared with 25% in an endurance effort.

7. Even with no mechanical power output at all, the human body consumes food energy at a rate of about 100 W, simply to remain alive.

7.6 Behdad Salimikordsiabi's Clean and Jerk

Olympic gold medalist Behdad Salimikordsiabi is a phenomenon of nature and part of a growing line of dominant Iranian weight lifters. When he lifts 543 lb above his head in two quick bursts, we know he is using the highest-power anaerobic chemical processes available, as discussed in section 7.5.3. In this section, we'll quantify the work and power he generates. We'll also analyze some interesting aspects of the lifting techniques used in competition.

7.6.1 Work during a lift

Step 1: Lifting: positive work

When the 370-lb Salimikordasiabi lifted 543 lb (2416 N or 247 kg) above his head in the London 2012 Olympics, it was pretty easy to believe that he was doing work. If he lifts at a constant rate (he doesn't; we'll come to this in a moment), he's exerting an upward force equal to the barbell's weight,[17] and equation 7.10 tells us precisely how much work he does:

$$W = F_x \Delta x + F_y \Delta y = 0 + (+2416 \text{ N})(+2 \text{ m}) = 4833 \text{ J}$$

The signs are important. Since both F_y and Δy are positive, Salimi has done a positive amount of work. A similar situation is shown in figure 7.11.

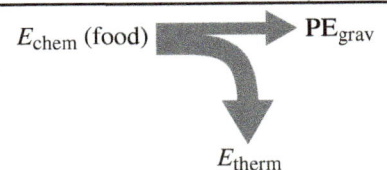

Figure 7.11
This lifter is doing positive work because the force he exerts is in the same direction as the motion. The increase in the barbell's energy equals the work he does: $\Delta PE_{grav} = |\vec{F}| \Delta y$. (© McGraw-Hill Education/Aaron Roeth, photographer).

---

[17] To lift it, doesn't he have to exert a force *greater* than the barbell's weight? Well, yes, to get it going (that is, to accelerate it from rest), he does. But once it's moving up *at a constant rate*, the barbell is not accelerating, so the net force on it is zero. Salimi's force exactly "cancels" the barbell's weight.

In fact, the work he did went directly into changing (increasing) the energy of the barbell:

$$\Delta PE_{grav} = PE_{grav\ after} - PE_{grav\ before} = mgy_{after} - 0 = (247\ kg)(9.8\ m/s^2)(2\ m)$$
$$= 4833\ J$$

which is about 1 Cal.

But you know that weight lifters do *not* lift the barbell at a fixed, steady rate. If they did, then the GRF on their feet would always equal their body's weight plus the weight they are lifting. However, this is far from realistic, as we discuss later in section 7.6.2. Accounting for the changing force mathematically would require a computer or calculus. This is not needed, however, because equation 7.11 tells us that the work done is simply the change in energy of the barbell, 4833 J.

That's an important and nice thing about work: it doesn't matter whether the energy is changed smoothly or in fits and starts. The work done is just the amount of energy added to the system.

 Whether work is done at a constant rate or not, the total work is simply the change in the energy of the system. This can greatly simplify our analysis of an event.

Step 2: Waiting for the judges: zero work

It is not surprising that it takes work to increase the energy of the barbell. But the work-energy theorem tells us that if the energy of the barbell is not changing, then no work is being done. That makes sense if the barbell is sitting on the ground, but consider again Behdad Salimikordasiabi.

He has done his 4833 J of work, and in figure 7.12, he is holding over 500 lb above his head, waiting for the judges' white lights indicating a valid lift. While he waits, he is not changing the barbell's speed, so he is not changing its kinetic energy. Nor is he changing its height, so he is not changing its gravitational energy. He's not

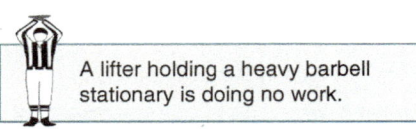 A lifter holding a heavy barbell stationary is doing no work.

E_{chem} (food) ⟶ E_{therm}

Figure 7.12. "Hey Salimi, that ain't workin'!" Behdad Salimikordasiabi at the 2012 Olympics in London. He has lifted 247 kg or 543 lb. The gravitational energy stored in the weights is about 1 Cal. That is enough to light a 75-W bulb for about 1 min. At the bottom is the energy flow diagram while he waits for the judges to declare it a valid lift. (© *Ggrigory Dukor/Reuters/Corbis*).

bending it, so he is not changing its elastic energy. In fact, he is doing no work at all. I wouldn't want to tell him that!

I'm sure Salimi found our definition of "work" just as nonsensical as Michael Phelps found our comments on his zero average velocity in section 2.1. Salimi would correctly say that he's expending a lot of energy while holding up that weight. He is right. But it's not work in the scientific sense. All the food energy that he is converting is going directly into global warming. Well, that and a gold medal.

As we discussed with the velocity-vs.-speed and other concepts in this book, scientific and mathematical terms need to be used with some care; otherwise you annoy Salimi and Phelps with nonsense. The terms and concepts are useful, but only in the correct context.

Step 3: Returning to ground: negative work

After the referee has signaled a good lift, the rules of the International Weightlifting Federation (IWF) specify that the weight lifter cannot simply drop the barbell, but must lower it at least to shoulder height before releasing it. To lower the barbell the approximately 50 cm to shoulder height, Salimi has to exert an *upward* force on the bar while the bar moves *down*. All lifters allow the barbell to accelerate somewhat (but not at -9.8 m/s^2) on the way down, even while gripping it. Salimi exerts an upward force of about 1800 N on the 2415-N barbell as he lowers it. The work he does is then

$$W = F_y \, \Delta y = (+1800 \text{ N})(1 - 0.5 \text{ m}) = -900 \text{ J}$$

Even though it required the expenditure of chemical energy from Salimi's food reserves, the work he has done is negative. This is a quantitative example of the "one-way street" energy flow sketched in the lower right corner of figure 7.8.

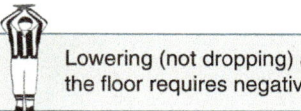

Lowering (not dropping) a weight to the floor requires negative work.

While the barbell's energy has been reduced by Salimi's negative work, it is not zero. After Salimi lets the bar go, it is converted from the gravitational to the kinetic form, as discussed in example 7.5.

EXAMPLE 7.5 *Releasing the barbell*

How fast will the bar hit the floor at Salimi's feet?

When the bar was above Salimi's head, it had 4833 J of energy, in the PE$_{grav}$ form. Since then, he performed -900 J of work on it, so now it has 4833 J $-$ 900 J $=$ 3933 J of energy.

The impressive noise and vibration released by the crash is the sound of less than 1 Cal being released! Think of that the next time you eat a Tic Tac!

Just as the barbell is about to hit the floor, none of the energy is in gravitational form; all of it is kinetic. Then it's easy to find its speed:

$$\text{KE} = \tfrac{1}{2}mv^2 \rightarrow v = \sqrt{\frac{2\text{KE}}{m}} = \sqrt{\frac{2 \times 3933 \text{ J}}{247 \text{ kg}}} = \sqrt{\frac{2 \times 3933 \text{ kg} \cdot \text{m}^2/\text{s}^2}{247 \text{ kg}}}$$

$$= 5.6\sqrt{\text{m}^2/\text{s}^2} = 5.6 \text{ m/s}$$

Notice how I drew that calculation out, to make clear how the units work. The energy unit, joule, is defined in equation 6.7.

7.6.2 The snatch and clean and jerk techniques

There are two Olympic weight-lifting events. In the *snatch*, a barbell is lifted over the athlete's head in a smooth movement, usually starting from the squat position. In

Figure 7.13. Cartoon illustration of the two Olympic weight-lifting events, the snatch and the clean and jerk.

the *clean and jerk* event, the "clean" and the "jerk" motions are clearly distinct, with a pause in between. Medals are awarded based on the sum of a competitor's snatch weight and his clean and jerk weight. In London, Salimi lifted 208 kg in the snatch and 247 kg in the clean and jerk, for a total of 455 kg and the gold.

Salimi could lift almost 86 lb more in the clean and jerk by exploiting its twofold motion and timing his body's energy generation precisely to the event. To understand the physics behind this feat, we need to understand the motions in some detail. Figure 7.13 sketches the steps of the two lifts, and figures 7.15 and 7.14 show each in action.

The snatch versus the clean

The most intense and difficult parts of the lift are those in which the arms and shoulders are doing the work—the snatch, the clean and the jerk. In the "recovery" phases shown in figure 7.13, the lifter is mostly driving upward with his legs and keeping his back straight. This is not to dismiss the recovery stages as easy! They take tremendous strength and very importantly balance and technique.

Focusing on the first burst of power, figure 7.13 indicates that the snatch and the clean motions take approximately the same amount of time, usually from 0.7 to 0.95 s. In that interval, the lifter is generating his maximum power in an anaerobic burst, using the readily available ATP in his cells. Example 7.6 gives a quantitative estimate of Salimi's power in the clean maneuver.

There is a difference in the maximum barbell heights achieved in the snatch and clean maneuvers. Because the lifter must get the barbell above his head in one motion in the snatch, he must lift the barbell higher in the snatch than in the clean. For a male of average height, the bar is raised approximately 1 m in the snatch and 80 cm in the clean.

There's no point flying all the way to London if you're not going to give it your best shot, so presumably any lifter in the Olympics will generate his maximum power ($\bar{P}$) in both events. In a little while, we'll estimate the amazing power generated by Salimi in London 2012, but for this discussion we'll just keep it as a variable. What we're going to see is that we can find the relationship between m_{snatch} and m_{clean}, and it won't depend on exactly what $\bar{P}$ is. We'll do this by combining the concepts of gravitational potential energy, work, and power. This is the power of scientific reasoning: simple concepts are used in combination to understand more and more complex situations.

Hang in there.

Mathematically, equation 7.13 relates power to work; equation 7.13 relates work to energy (here, gravitational energy); equation 7.13 relates gravitational energy to height lifted. We can use them one after the other in a formula that looks like this:

$$\underset{\substack{\uparrow \\ \text{equation} \\ 7.13}}{\bar{P}\,\Delta t} = \underset{\substack{\uparrow \\ \text{equation} \\ 7.11}}{\text{Work}} = \underset{\substack{\uparrow \\ \text{equation} \\ 7.1}}{\Delta PE_{grav}} = mg\,\Delta y$$

Since we said that the lifter would use his maximum power (whatever it is) for both lifts, and the duration of the clean and the snatch maneuvers is the same, $\bar{P}\,\Delta t$ is the same for both. So the above equation implies that $m\,\Delta y$ will be the same for the two maneuvers.

$$m_{snatch}\,\Delta y_{snatch} = m_{jerk}\,\Delta y_{jerk} \rightarrow \frac{m_{snatch}}{m_{jerk}} = \frac{\Delta y_{jerk}}{\Delta y_{snatch}} \approx \frac{0.8\text{ m (typically)}}{1\text{ m (typically)}} = 80\%$$

This explains why snatch weights for world-class lifters are typically about 80% of their clean and jerk weights, as seen in table 7.3.

Table 7.3. Performances of the men's +105-kg weight-lifting competition in the London 2012 Olympic Games. Best lifts in the snatch and clean and jerk are given in kilograms. While the weight lifted varies a lot from one athlete to another, the ratio between m_{snatch} and m_{clean} is always around 80%. This can be understood by considering power and energy, as discussed in the text.

Place		m_{snatch}	m_{clean}	Ratio (%)
1	Behdad Salimikordsiabi (IRI)	208	247	84
2	Sajjad Anoushiravani (IRI)	204	245	83
3	Ruslan Albegov (RUS)	208	240	87
4	Jeon Sang-Guen (KOR)	190	246	77
5	Irakli Turmanidze (GEO)	201	232	87
9	Yauheni Zharnasek (BLR)	196	230	85
10	Chen Shih-chieh (TPE)	182	236	77
11	Péter Nagy (HUN)	191	225	85
15	Christian López (GUA)	172	215	80
16	Damon Kelly (AUS)	165	216	76
17	Frederic Fokejou (CMR)	160	202	79
18	Carl Henriquez (ARU)	122	160	76

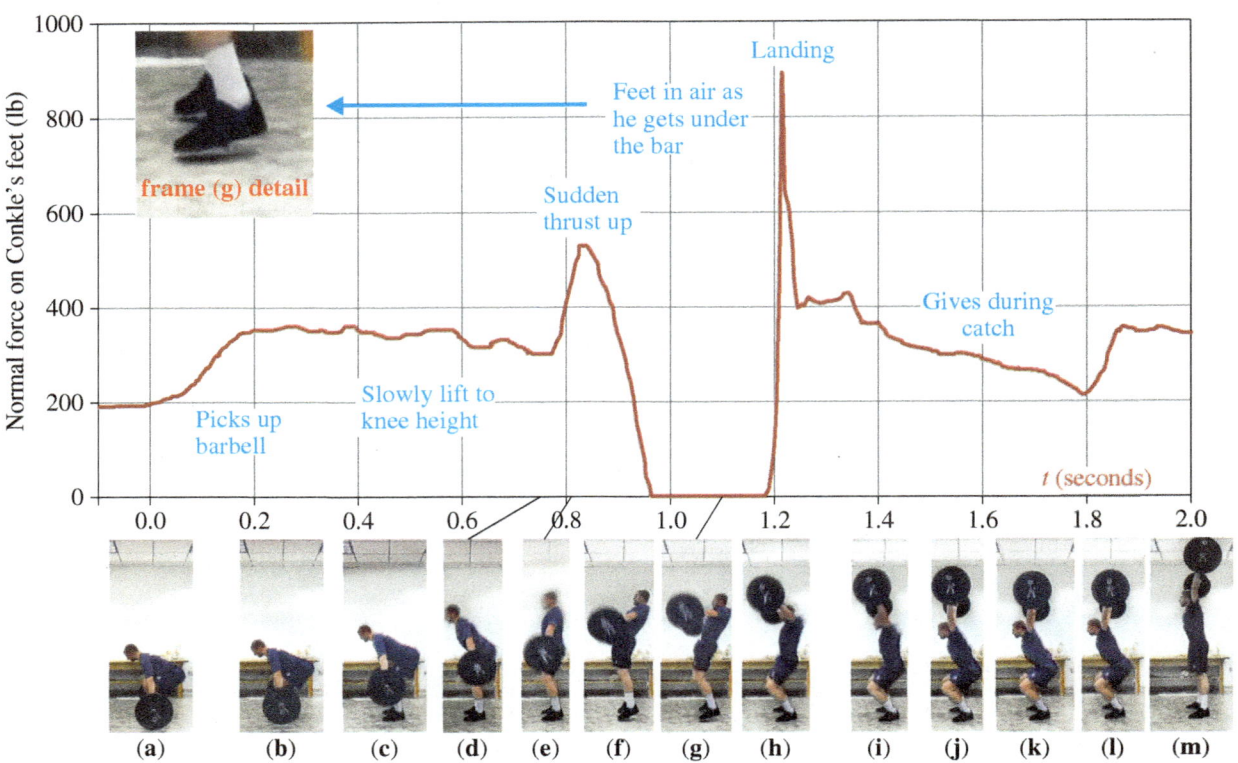

Figure 7.14. Amateur weight lifter Jeffrey Conkle executes a snatch. In panel (a), he prepares to lift, so the normal force on his feet is just equal to his weight, 195 lb. In panels (b) and (c), he lifts the bar slowly, and the normal force increases to compensate for the 160-lb weight. As shown in the blowup of panel (g), after he exerts a huge upward force to accelerate the barbell upward in panels (d)–(f), Conkle's feet leave the ground, as he gets under the bar to catch it around panel (h). Panels (i)–(m) represent the recovery phase. (*Photos courtesy of Mike Lisa*).

Airborne with 500 lb!

Figure 7.14 shows a time sequence of Jeff Conkle's snatch lift, together with the normal force on the lifter's feet. Let's take a look at that force curve. Not surprisingly, the normal force increases when Conkle picks up the barbell.

In panels (d)–(f), Conkle rapidly accelerates the barbell upward. To do this, he exerts a large upward force on it with his hands, and we know that this means the barbell exerts a large downward force on him. Unless he himself is going to accelerate downward, a large upward force from the floor has to counteract this. For that reason, we see a "bump" in the normal force above frame (e).

Okay, that makes sense. (If not, go over it again. It's worth it, and you'll get it.)

But what happens next? The normal force on Conkle's feet goes to zero. That means that he's not even touching the floor for about 0.2 s! In fact, a blown-up image of panel (g) shows clearly that his feet have left the ground. The same thing happens in a clean and jerk lift, as shown in figure 7.15.

Doesn't that seem a bit odd? There you are, lifting the heaviest object that you can possibly manage... and you decide to take a jump. Yeah, that sounds like a good idea.

Actually, figure 7.16 illustrates the physics behind why it *is* a good idea. In figure 7.15, the maximum height of the barbell in the clean phase is seen in panel (f). The job of the lifter now is to get under the thing before it falls again, so ideally his body accelerates downward at a rate higher than the barbell does. As shown in figure 7.16, and as you know, the center of mass of the combined lifter-plus-barbell system will accelerate downward at 9.8 m/s^2 when nothing "external" (such as the ground) is touching it. That's absolutely unavoidable.

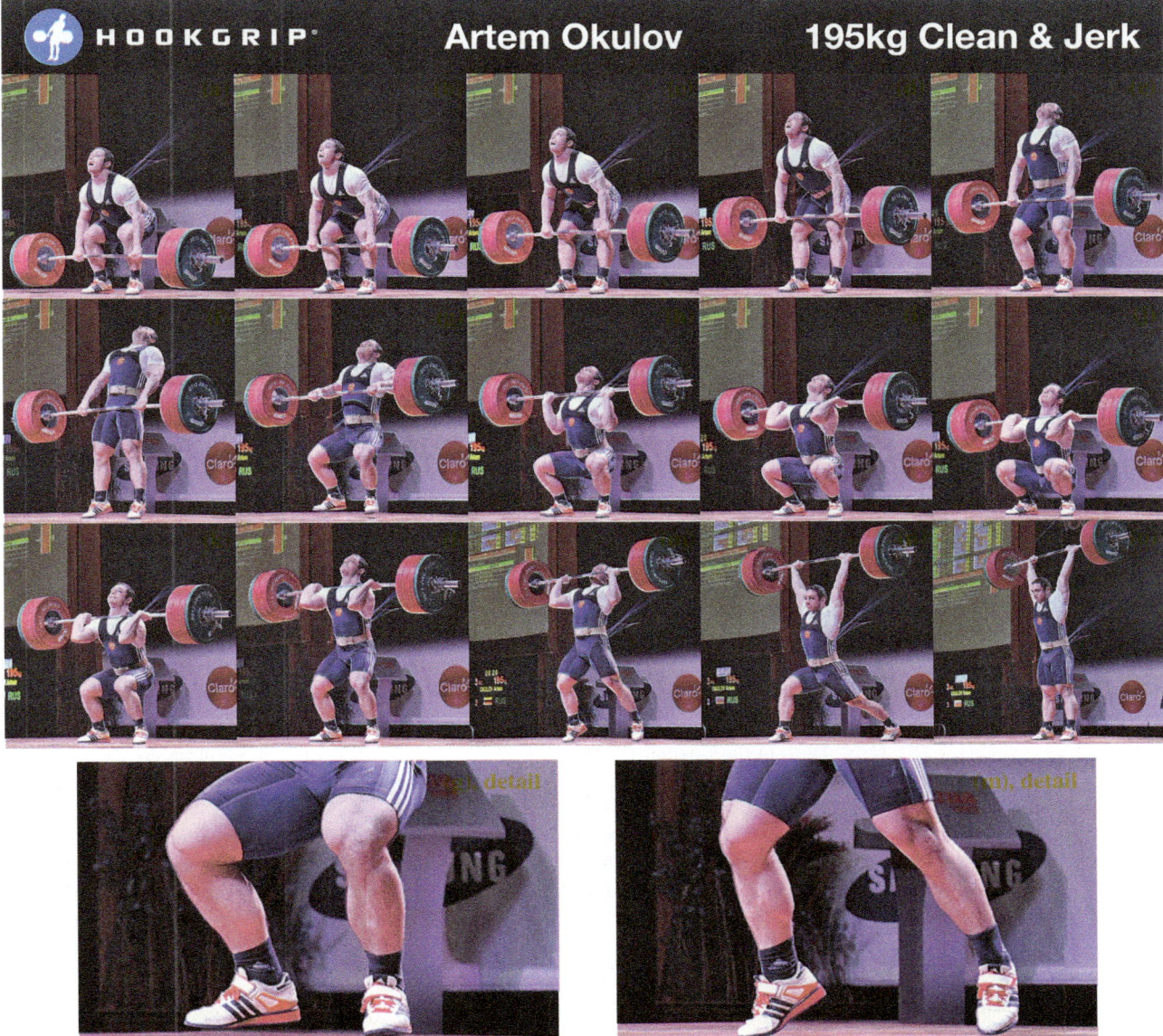

Figure 7.15. Photosequence of a clean and jerk lift by the young Russian lifter Artem Okulov. Panels (a)–(j) are the clean; panels (k) and (l) are the first recovery. The jerk is shown in panels (m) and (n); the second recovery results in the final position in panel (o). The blowups of panels (g) and (m) show clearly that Okulov's feet leave the ground twice: once during the clean and once during the jerk. (© *hookgrip*).

But taken individually, the center of mass of the lifter and the center of mass of the barbell can move at different rates. The middle panel shows the forces on the barbell. By continuing to push up on the barbell while he's airborne, the lifter partially cancels the downward force of the barbell's weight. So even though the barbell is accelerating downward, it is doing so at *less than* 9.8 m/s^2.

Conversely, the "reaction force" that the barbell exerts *on* the lifter points down, just as the lifter's weight does. Therefore, those two forces combine to accelerate the lifter downward at *more than* 9.8 m/s^2.

The fraction-of-a-second burst of anaerobic power, combined with this surprising technique of lifting the feet to "force oneself under the bar," exploits the physics of gravity and of the human body to perform feats like Salimi's.

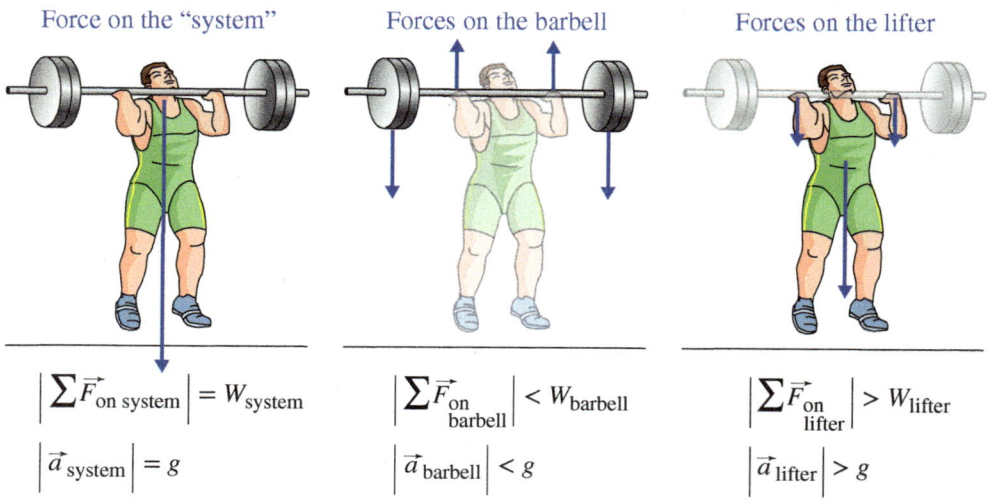

Figure 7.16. A lifter and barbell are briefly airborne. The three panels repeat the same lifter and barbell. In the left panel, the overall (gravitational) force on the lifter-and-barbell system is drawn. The center panel shows the forces on the barbell, from the lifter and from gravity. On the right, the forces on the lifter, from his own weight and from the barbell, both point down.

EXAMPLE 7.6 *Power lift*

Frame-by-frame analysis of video footage of Behdad Salimikordsiabi's clean and jerk lift in the London Olympics reveals that he reached full extension on the clean phase (see panel (f) of figure 7.15, for example) 0.9 s into his lift. How much power did he generate in this first part of his lift?

Here, we'll follow the same logic we did earlier, but with numbers this time. We start with the mechanical energy he's given to the barbell. The barbell began at rest, and at its highest point, it is at rest again (remember from looking at projectile motion?) so the kinetic energy hasn't changed. Rather, all the energy change has been gravitational potential energy:

$$\Delta PE_{grav} = mg\,\Delta y = (247\text{ kg})(9.8\text{ m/s}^2)(0.8\text{ m}) = 1936.5\text{ J}.$$

We already learned about the work-energy theorem (equation 7.11) which tells us that the work Salimi has done is equal to the change in the barbell's energy, $W = \Delta PE_{grav} = 1936.5$ J.

Finally, equation 7.13 lets us find the average power Salimi generates in this part of the lift:

$$\bar{P} = \frac{W}{\Delta t} = \frac{1936.5\text{ J}}{0.9\text{ s}} = 2152\text{ W}.$$

Two kilowatts! That is a tremendous amount of power—enough to power two or three households.

Converting units, $2152\text{ W} \times \frac{1\text{ h.p.}}{746\text{ W}} = 2.9$ h.p., so Salimi produces almost as much power as three horses.

Most estimates of human power suggest that a healthy male can generate 1.2 h.p. for a brief burst; the maximum quoted for a professional athlete is about 2.5 h.p. At 2.9 h.p., Behdad Salimikordsiabi is truly one of the most powerful athletes in the world.

Collected Equations

$$\text{PE}_{\text{grav}} = mgy_{\text{c.m.}} = |\vec{W}|y_{\text{c.m.}} \tag{7.1}$$

$$\eta = \frac{\text{useful energy output from the process}}{\text{total energy input to the process}} \tag{7.2}$$

$$1 \text{ Cal} = 4184 \text{ J} \tag{7.3}$$

$$\eta_{\text{ATP} \to \text{mechanical energy}} \lesssim 50\% \tag{7.4}$$

$$\underbrace{C_6H_{12}O_6}_{\text{eat}} + \underbrace{6O_2}_{\text{inhale}} \to \underbrace{6H_2O}_{\text{sweat}} + \underbrace{6CO_2}_{\text{exhale}} + 686 \text{ Cal} \tag{7.5}$$

$$\eta_{\text{glucose} \to \text{ATP}} \lesssim 50\% \tag{7.6}$$

$$\eta_{\text{glucose} \to \text{mechanical energy}} = \underbrace{\eta_{\text{glucose} \to \text{ATP}}}_{\lesssim 50\%} \times \underbrace{\eta_{\text{ATP} \to \text{mechanical energy}}}_{\lesssim 50\%} \lesssim 25\% \tag{7.7}$$

$$\text{total energy} = \underbrace{\text{KE} + \text{PE}_{\text{grav}} + \text{PE}_{\text{elast}}}_{\text{also known as mechanical energy}} + E_{\text{chem}} + E_{\text{therm}} \tag{7.8}$$

$$\underbrace{\Delta \text{KE} + \Delta \text{PE}_{\text{grav}} + \Delta \text{PE}_{\text{elast}}}_{\text{change in mechanical energy}} = \begin{pmatrix} \text{food} \\ \text{energy} \\ \text{burned} \end{pmatrix} - \begin{pmatrix} \text{thermal} \\ \text{energy} \\ \text{generated} \end{pmatrix} \tag{7.9}$$

$$W = F_x \Delta x + F_y \Delta y = |\vec{F}| \times \left(\text{distance moved parallel to } \vec{F}\right) \tag{7.10}$$

$$W_{\text{athlete}} + W_{\text{other forces}} = \Delta \text{KE} + \Delta \text{PE}_{\text{grav}} + \Delta \text{PE}_{\text{elast}} \tag{7.11}$$

$$\bar{P} = \frac{\text{energy transferred}}{\Delta t} \tag{7.12}$$

$$\bar{P}_{\text{work output}} = \frac{\text{work done}}{\Delta t} = \frac{W}{\Delta t} \tag{7.13}$$

$$1 \text{ W} = 1 \frac{\text{J}}{\text{s}} \qquad 1 \text{ horsepower} = 1 \text{ h.p.} = 746 \text{ W} \tag{7.14}$$

$$\bar{P}_{\text{chemical input}} = \frac{\text{food energy burned}}{\Delta t} = \frac{-\Delta E_{\text{chem}}}{\Delta t} \tag{7.15}$$

$$\bar{P}_{\text{work output}} = \eta_{\text{chemical}\to\text{mechanical energy}} \times \bar{P}_{\text{chemical input}} \qquad (7.16)$$

$$\bar{P}_{\text{RMR}} \equiv \underset{\text{while at rest}}{\bar{P}_{\text{chemical input}}} \approx 100 \text{ W} \approx 80 \frac{\text{Cal}}{\text{hr}} \qquad (7.17)$$

Problems

1. The Top Thrill Dragster at Cedar Point amusement park in Sandusky, Ohio, is one of the world's most impressive roller coasters. Wearing only a lap harness in the open air, riders on a 7.5-ton train are accelerated by hydraulic motors on a level track to 120 mph in just 3.8 s. After this, no more power is supplied to the train—it's on its own. The train shoots straight up (twisting along the way) along the track to a height of 420 ft, before a straight-down descent.
   (a) What is the kinetic energy of the train after the 3.8-s launch?
   (b) How much power is generated by the motors as they launch the train? Give your answer in megawatts (MW) and horsepower (h.p.).
   (c) What is the gravitational potential energy of the train at the top of the ride?
   (d) If you ignore air drag and friction, how fast is the train moving at the top of the ride, in miles per hour?
   (e) In reality, drag and friction do about −1.2 MJ of work on the train before it reaches the top. If you account for this, how fast is the train moving at the top of the ride? (The fact that you are moving so slowly at the top is one of the most terrifying aspects of the ride.)

2. Estimate the power generated by 94-kg Usain Bolt in the first 2 s of his 100-m dash in Berlin. (Refer to section 2.2 for information.) Give your answer in watts and horsepower.

3. In 1985, legendary American power lifter Lamar Grant became the first man to deadlift five times his body weight. All muscle at 132 lb, Grant lifted 661 lb about 2.5 ft off the ground! In the dead lift, a barbell is lifted off the ground to about thigh level, the back is straightened, and then the weight is lowered.
   (a) How much work did Grant do on the bar while lifting it?
   (b) Grant took about 2.5 s to lift the bar. Why is this the optimal time to take for such an activity? Why no slower or faster?
   (c) In horsepower (h.p.), what was the power of the lift?
   (d) As discussed in section 7.5.3, this type of highly anaerobic effort is inefficient in terms of food energy usage. Assuming $\eta_{\text{glucose}\to\text{mechanical energy}} = 7\%$, how many Calories did Grant burn in the lift?

4. In problem 2 in chapter 4, we considered A. J. Trepasso's Jumbotron-striking punt, launched at 53.3 mph at an angle of 75.9° with respect to the horizontal. For now, we'll ignore the effects of air and treat the punt using free-fall kinematic equations. We're also going to pretend that the Jumbotron didn't get in the way. That is, only gravity matters.
   (a) What is the ball's kinetic energy just as it leaves Trepasso's foot?
   (b) What is the ball's speed at the highest point in its flight?
   (c) What is the ball's kinetic energy at the highest point in its flight?
   (d) What is the gravitational energy of the ball at its highest point?
   (e) Find the gravitational and kinetic energy of the ball halfway up to its highest point. [Hint: start with gravitational.]

(f) The bullet from a typical 0.357 Magnum pistol is 8 g and listed muzzle velocities are typically about 1400 ft/s. Compare the kinetic energy of a just-punted football (your number from part (a)) to such a bullet. How many times more or less is it?

(g) Actually, because of air drag the ball was measured to hit the ground at 39 mph. How much thermal energy was generated by the ball's passage through the air?

5. In 1932, a group of five men from the Polish Olympic ski team, in training for the Winter Games in Lake Placid, climbed 362 m from the 5th floor to the 102nd floor of the Empire State Building in New York City; the building had just been completed a year before. Bounding up the stairs two at a time, they reached the top in 21 min. They were surprised at the top by the Czech team, who had taken the elevator, but who then challenged the Poles to a stair race. Building management, tired of all the clomping of athletes in training, forbade the race, however.

   (a) Assuming each man's mass to be 80 kg and a constant rate of climbing, how much power did each man generate?

   (b) Clearly this was an aerobic activity by top athletes, so $\eta_{glucose \rightarrow mechanical\ energy}$ was probably the maximum value of 25%. How many food Calories did each man burn?

6. A 100-kg skydiver falls with terminal velocity of 50 m/s.

   (a) Is air drag doing positive or negative work on the diver?

   (b) How much power is generated by air drag? That is, at what rate is work being done?

   (c) How much work is done by the air drag when the skydiver falls 1 km at terminal velocity?

   (d) For how long would this much energy provide power to a typical household? (The typical household consumes about 1 kW of electrical power.)

7. We've had fun discussing 118-kg Ray Lewis. He is running and has momentum of 590 kg·m/s. How does his kinetic energy compare with the gravitational potential energy of a 80-kg diver standing on the 10-m-high platform?

8. According to Wikipedia, the mass of Mark Cavendish plus his bike is 79 kg. For moderate lengths of time, he can do useful work on a bike at 250 W. Ignoring air drag and rolling friction, how long will it take him to ride 2 mi (3218 m) up the hill shown in the figure? Assume he rides at a fixed speed.

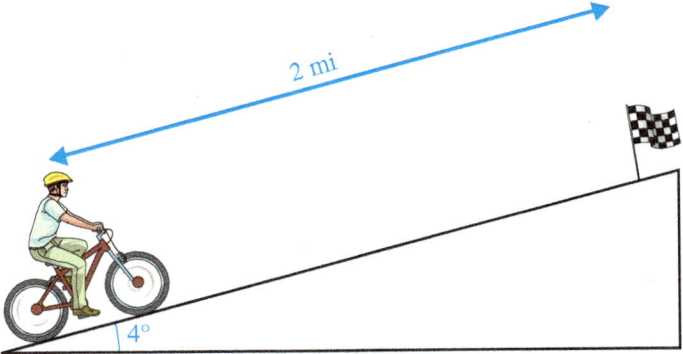

9. For short periods of time, some football players can put out 1.2 h.p. If such a player keeps it up for 3 s, how far could he push a training sled at constant speed while exerting a force of 250 lb?

10. A kilowatt-hour (kWh) is a tremendous amount of energy.

    (a) If this energy is used to lift a 2-ton car off the ground (that is, converted into gravitational potential energy), how high would the car be lifted? Give your answer in miles.

(b) About how long does it take the typical household to consume this much electrical energy? (Imagine a household's consumption is the equivalent of fifteen 60-W lightbulbs burning. This accounts for the refrigerator, computer, etc.)

11. An 80-kg skater is initially moving at 8 m/s, then allows air drag and kinetic friction with the ice to slow him to a stop. How much work was done by the air and friction combined?

12. As we discussed in section 2.3, 170-lb Don Perry set the world record for rope climbing by raising his center of mass by 16 ft in 2.8 s.
    (a) What was his power output?
    (b) Do you think he could he have kept up this pace, for example, climbed 60 ft in 8.4 s? If not, why not?

13. In a good tackle, a 120-kg football player running at 6 m/s hits a 100-kg receiver initially running at 8 m/s. After the hit, they move off together at 4 m/s.
    (a) What was the work done on the tackler?
    (b) What was the work done on the receiver?
    (c) How much energy went into thermal modes (heat, sound, broken bones)?
    (d) Could this have been a head-on collision? (Think about momentum.)

14. When a bat hits a ball, it can do work on the ball. Consider these three cases:
    (1) A 100 ft/s pitch is batted back at 100 ft/s.
    (2) An 80 ft/s pitch is batted back at 100 ft/s.
    (3) A ball is batted off the tee (that is, the ball is initially stationary) at 100 ft/s.
    (a) In which case did the bat do the most work?
    (b) In which case did the bat do the least work?
    (c) Do your answers make sense? Why or why not?

15. The shot-put is a power event. American Olympic medalist Reese Hoffa can attain launch speeds of 33 mph with the 16-lb shot.
    (a) How much work must Hoffa perform on the shot?
    (b) In horsepower (h.p.), how much power would Hoffa have to generate with his one arm if his technique were to stand stationary and throw the shot by thrusting his arm straight out 2.5 ft? (To figure this out, you'll need to figure out the time it takes him to do this; use the constant-acceleration kinematics of chapter 2.)
    (c) Considering that Behdad Salimikordsiabi generates 2.9 h.p. using all of his back and leg muscles, this would be a tremendous amount of power to achieve with one arm—too much for a human, in fact. What do shot-putters do, to achieve such launch speeds?

16. (a) Is it possible for a tackle (in which the players stick together) to *increase* the kinetic energy of one of the players? If not, why not?
    (b) Is it possible for a tackle (in which the players stick together) to *increase* the kinetic energy of *both* players? If not, why not?

17. In sections 6.4.2 and 7.2.2, we discussed the golf drive; you may want to refer to that material for this question.
    When no energy is lost to thermal modes in a collision, the coefficient of restitution is $e = 1$. This represents the ideal golfing equipment. The legal maximum is $e = 0.83$. A 200-g clubhead moving at 100 mph strikes a golf ball.
    (a) What is the total energy of the system just before the collision?
    (b) What is the efficiency of the shot (that is, what fraction of the total energy goes into the ball's kinetic energy) for $e = 1$? For $e = 0.83$?

(c) How much energy winds up as $KE_{clubhead}$, KE_{ball} and E_{therm} for $e = 1$?

(d) How much energy winds up as $KE_{clubhead}$, KE_{ball} and E_{therm} for $e = 0.83$?

(e) Compared to the launch speed attained with the legal equipment, how much higher would be the ball's launch velocity with the ideal equipment?

18. A stick of dynamite contains 1 megajoule (MJ, or 10^6 J) of energy.

   (a) A Snickers bar contains 260 Cal. This is equivalent to how many sticks of dynamite?

   (b) Why don't we explode when we eat a Snickers bar?

19. Everyday lighting takes a lot of energy, and a comparison with sports can be illustrative.

   (a) How much energy is consumed by a 60-W lightbulb if it's turned on for only 1 min?

   (b) If a kicked soccer ball had this much kinetic energy, how fast would it be moving, in miles per hour?

   (c) How long would it take a ball moving at this speed to cross the 100-m soccer field?

   (d) Realistically, a soccer ball might be kicked at 20 m/s. If the kinetic energy of such a ball were used to light the 60-W bulb, how long would the bulb be on?

20. Before the rise of cable television, ABC's *Wide World of Sports* was *the* place to watch sports. The show opened with the now familiar catchphrase "the thrill of victory and the agony of defeat," while film clips appeared on the screen. The clips used for the thrill of victory varied over time, but for decades, "the agony of defeat" line was spoken over a famous clip of a spectacular wipeout of Yugoslavian ski jumper Vinko Bogataj. During terrible weather at a ski-flying world championship in Oberstdorf, West Germany, Bogataj tried to adjust for bad conditions midway down the ramp but completely lost control and tumbled wildly off the side at its bottom into a scattering crowd.[18] A video search for "agony of defeat" on the Internet will instantly bring up the clip.

   (a) The 82-kg Bogataj left the ramp about 45 m lower than his starting position. (That is, the difference in elevation was 45 m.) If you ignore the slowing forces of kinetic friction and wind drag, what was his kinetic energy at that point?

   (b) What was his speed at that point?

   (c) If Bogataj has simply *fallen* straight down 45 m rather than skiing down a ramp, what would be his kinetic energy?

   (d) After hitting the ground, Bogataj crashed wildly and descended down the hill another 20 vertical meters before coming to rest. Here friction cannot be ignored! How much total work was done on Bogataj to bring him to rest?

   (e) In part (a) you were told to ignore friction and air, but based on your knowledge of friction (from chapter 3) and drag (chapter 5), which one would be more important to account for?

(© Ingram Publishing/Fotosearch RF)

---

[18] Millions of Americans saw Bogataj's epic crash every week, with the result that he was much more famous in the United States than in his home country. On his way to an interview with ABC's Terry Gannon years later, Bogataj got into a fender bender. He quipped, "Every time I am on ABC, I crash."

8 Energy and Timing in Elastic Equipment

Elastic equipment plays a prominent role in many sports. But if all energy in sports originates with the conversion of food by the athlete, then what is the benefit of storing and then releasing that energy from the elastic form, especially since one pays an "efficiency penalty" in the process? In this chapter, we'll answer this question.

Left: (© *Image Source/Getty Images RF*); Right: (© *Corbis Super/Alamy RF*).

The energy involved in some sports originates in fossil fuels. An obvious example is motor sports, but the category also includes downhill skiing events, in which coal (or diesel) is burned to produce the electricity that runs the ski lifts. These lifts give the skier gravitational potential energy "for free" (he needn't generate it), which is then converted into kinetic (and heat) energy during the event.

In most sports, however, all the action originates in the conversion of food energy by the athlete's body and its expression through muscular motion. Equipment used in these events is not a *source* of energy, but a tool to more effectively *convert* energy from one form to another, often by storing it temporarily in the elastic form. The equipment often addresses a mismatch in timing; humans may be strong and energetic enough to perform the work required for an event, but not quickly enough without aid. That is, they are not *powerful* enough.

The timing issues and the physics of the flexible equipment used to store and release energy are often surprisingly complicated in the details. In this chapter, we'll cover some of the most important points.

8.1 The Physics of Archery I: Energy Storage and Transfer

With 20/200 vision in one eye and 20/100 in the other, Olympic gold medalist archer Im Dong-Hyun from South Korea is not legally blind (as is sometimes claimed), but is certainly severely visually impaired. The colors on the target are blurry and look like "paint dropped in water," he told the New York Times.

The 48-in. target, seen from the shooting distance of 70 m, is the same angular size as your pinky nail at the end of your extended arm; the yellow bull's-eye is the

(© *McGraw-Hill Education/Aaeron Roeth, photographer*)

size of a sesame seed stuck to that pinky nail.[1] Expert archers say that particularly sharp vision might actually be a *handicap*, as the archer "overaims" and tires his muscles attempting on-the-spot adjustments way too minute to control perfectly. Im, on the other hand, is a master of consistency. Naturally, he needs to know where the target is, but after that, it's all muscle memory.

The science of archery is much richer and more detailed than you might expect. It turns out that more important than eagle vision is the consistent and intelligent use of energy and timing issues.

8.1.1 The arrow's energy

An arrow is typically launched toward an Olympic target at about 150 mph (67 m/s). Its mass[2] may be 26 g (400 gr, about 100 gr of which is in the tip), so its kinetic energy is

$$\text{KE} = \tfrac{1}{2}mv^2 = \tfrac{1}{2}\left(26 \times 10^{-3} \text{ kg}\right)(67 \text{ m/s})^2 = 58.4 \text{ J}.$$

Im can certainly perform this amount of work (roughly equivalent to lifting a 16-lb bowling ball 2 ft off the floor) on the arrow. However, no human could throw an arrow at 150 mph. So Im increases the elastic potential energy of the bow by deforming it. There is no rush in doing this work; he can take a half hour, if he wants. When he releases it, the bow quickly puts that energy into the arrow in kinetic form.

In terms of our water analogy of figure 7.7, this is equivalent to filling the potential energy bucket from the slow-flowing spigot, then suddenly dumping all the water into the kinetic bucket.

8.1.2 The bow's energy

You might think that Im increased the energy of the bow by just 58.4 J (that is, did 58.4 J of work) in the example above, but he input more energy than that. To understand this, we have to consider the bow itself.

The recurve bow (the type used in the Olympics) is similar to a spring; the more it is stretched, the greater is the restoring force it exerts. For many elastic objects (such as springs), the relationship between the distance stretched and the elastic force is linear:

$$\vec{F}_{\text{elastic}} = -k\,\Delta\vec{x} \qquad (8.1)$$

This equation is known as **Hooke's law**. The displacement from the unstretched ("equilibrium") position is $\Delta\vec{x}$. The negative sign in Hooke's law just says that the restoring force is in the opposite direction to the displacement from equilibrium, which makes sense if you think about it. Figure 8.1 shows the elastic restoring force and the displacement vector.

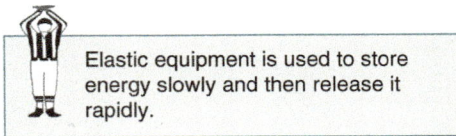

Elastic equipment is used to store energy slowly and then release it rapidly.

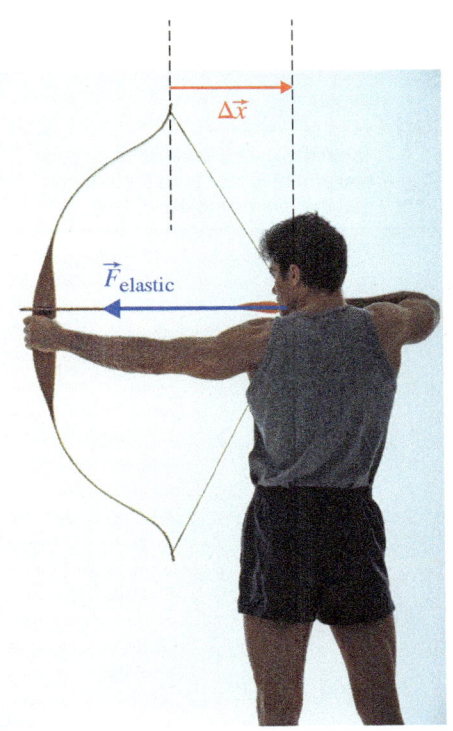

Figure 8.1
The displacement of the bowstring from its equilibrium value is $\Delta\vec{x}$ and points to the right. The elastic restoring force $\vec{F}_{\text{elastic}}$ points in the opposite direction. In this figure, the bow is at full draw, so $|\Delta\vec{x}|$ is the draw length and $\vec{F}_{\text{elastic}}$ is the bow's draw weight. (© *Comstock/JupiterImages RF*).

---

[1] Isn't that amazing? Our brain's ability to focus on an object makes it seem much larger in our mind, especially in some contexts. Check this out for yourself: When the moon is full, hold your hand at arm's length and compare the apparent area of your thumbnail with that of the moon. You'll see that they are nearly identical, regardless of whether the moon is high in the sky or low on the horizon. Someone who never does this thumbnail thing will swear it's visually bigger when on the horizon.

[2] Arrow masses are usually quoted in grains (gr). 1 gr = 64.799 mg.

The constant *k* in equation 8.1 is the **spring constant**. Archers don't refer to the spring constant, but to the **draw weight** of a bow. The draw weight is the bow's restoring force (or the force you have to exert to keep the bow stretched) when $|\Delta \vec{x}|$ is equal to the **draw length**. So for bows that follow Hooke's law,[3] the spring constant is

$$k_{\text{bow}} = \frac{\text{draw weight}}{\text{draw length}}. \tag{8.2}$$

Olympians typically use 50-lb bows and a draw length of about 26 in. (2.17 ft), so a typical spring constant is $k = (50 \text{ lb}) / (2.17 \text{ ft}) = 23 \text{ lb/ft}$.

The more a bow (or any elastic object) is stretched, the more elastic potential energy it has, and this is related to the spring constant as well:

$$\text{PE}_{\text{elast}} = \tfrac{1}{2} k |\Delta \vec{x}|^2. \tag{8.3}$$

The amount of energy that Im put into the bow is then

$$\text{PE}_{\text{elast}} = \tfrac{1}{2} (23 \text{ lb/ft}) (2.17 \text{ ft})^2 = 54.2 \text{ lb} \cdot \text{ft} \times \frac{1.356 \text{ J}}{1 \text{ lb} \cdot \text{ft}} = 73.4 \text{ J}.$$

This is more than the kinetic energy of the arrow eventually fired by the bow.

8.1.3 Bow and arrow efficiency

> Often the useful work done by elastic equipment is less than the energy that the athlete stored in it.

We've just found that Im puts more energy into the bow than the energy that the bow put into the arrow, once released. What gives?

Well, consider that 73.4 J is the amount of energy Im puts into the bow, *even if there is no arrow at all!* In a "dry fire," the bow is released with no arrow, and all the 73.4 J is very rapidly converted into thermal modes, namely, vibrations of the bow string and often damage to the bow itself. The string may snap and injure the archer. One should never dry-fire a professional bow.

In the example we are discussing, the transfer of energy was in fact quite efficient: $\frac{58.4 \text{ J}}{73.4 \text{ J}} = 79.6\%$ of the work performed by Im was eventually converted to the arrow's kinetic energy. Efficiencies of modern bows and arrows are typically in the 50% to 85% range; that is, they give 50%–85% of their stored energy to the arrow.

The driving factor behind the bow and arrow efficiency is the interplay between the magnitude of the force (given by the stiffness of the bow) and the masses involved. Masses, plural. The arrow has inertia—its mass—but so does the bow. The inertia or "effective mass" of the bow is not simply its mass, as it is not moving as a single fixed unit during its release. Nevertheless, massive parts of the bow and string *are* moving, together with the arrow.

The situation is somewhat akin to the "throw speed" discussion of section 3.4.4, where the force needs to be applied to the hand *and* the ball. If the ball is very light, it will be thrown faster, but most of the energy is spent moving the hand. Expert archers know that light arrows will always be launched at higher speed than heavy

---

[3] Olympic recurve bows follow Hooke's law approximately. Compound bows, however, do not. In fact, for most compound bows, the restoring force at draw length is *lower* than it is when the bow is only partially drawn.

ones, but the light arrows may leave too much energy in the bow, potentially a very dangerous and harmful situation. (It's essentially the same as a dry fire, actually.) For this reason, all bows come with specified minimum arrow weights.[4]

The heavier the arrow, the more efficient the energy transfer. A 20-lb (!) arrow will absorb essentially *all* the energy originally in the bow,[5] but it sure won't be moving very fast, and it won't get far.

We learn something from this: While manufacturers and archers often refer to the "energy efficiency of a bow," the efficiency depends on the bow *and arrow*. The same bow will have two different efficiencies with two different arrows.

8.2 The Physics of Archery II: Fire Power

The fact that Im Dong-Hyun could store 73.4 J of energy in the bow is hardly impressive. However, to *throw* the arrow at 150 mph without the bow would require a "powerful" burst.

Video analysis shows that Dong-Hyun draws his bowstring in about 1.5 s. So the average power of his draw is

$$\bar{P}_{\text{draw}} = \frac{73.4 \text{ J}}{1.5 \text{ s}} = 49 \text{ W}.$$

Put another way, Im could have powered a 49-W lightbulb for 1.5 s with his effort.

Once the string is released, this energy is converted to kinetic energy (79.6% of which ends up in the arrow, as we said in section 8.1). Without a high-speed camera, the duration of the release can't be determined by video, but we can figure it out by understanding a little bit about the timing vibrating motion of elastic systems.

Vibrations are very important in physics, and the *timing* of elastic systems is crucial in the physics of sports. Just check out videos of a pole vaulter who has chosen a pole with an incorrect stiffness (that is, spring constant k). If it releases its energy too quickly, he will be "denied" and flung backward into the run-up track; too slowly, and he could end up on his head holding a broken pole.

Some elastic systems react very quickly—they snap back into shape or oscillate at a high rate; these systems have large elastic restoring forces and low mass. Others—the ones with low restoring forces and/or large mass—oscillate slowly. The rate of the oscillation is determined primarily by two things: the spring constant (larger means a higher rate) and the inertia of the system (larger means a lower rate).

The time for one complete oscillation is called the **period** of the oscillation and depends on the square root of the ratio of these quantities:

$$T_{\text{one oscillation}} \propto \sqrt{\frac{m}{k}} \tag{8.4}$$

The characteristic time response of an elastic system depends on its mass and Hooke's constant.

The constant of that proportionality depends on all sorts of details of the object and in most cases is pretty complicated.

---

[4] An entirely different reason not to use the very lightest arrows is their increased susceptibility to small cross-breezes that knock them off their trajectory.
[5] You can imagine this, right? A super-light arrow will zip away, leaving the bowstring quivering violently because of the energy left behind. On the other hand, the bowstring will slowly, slowly release a super-heavy arrow and simply stop once the arrow disconnects. The super-heavy arrow has taken all the energy.

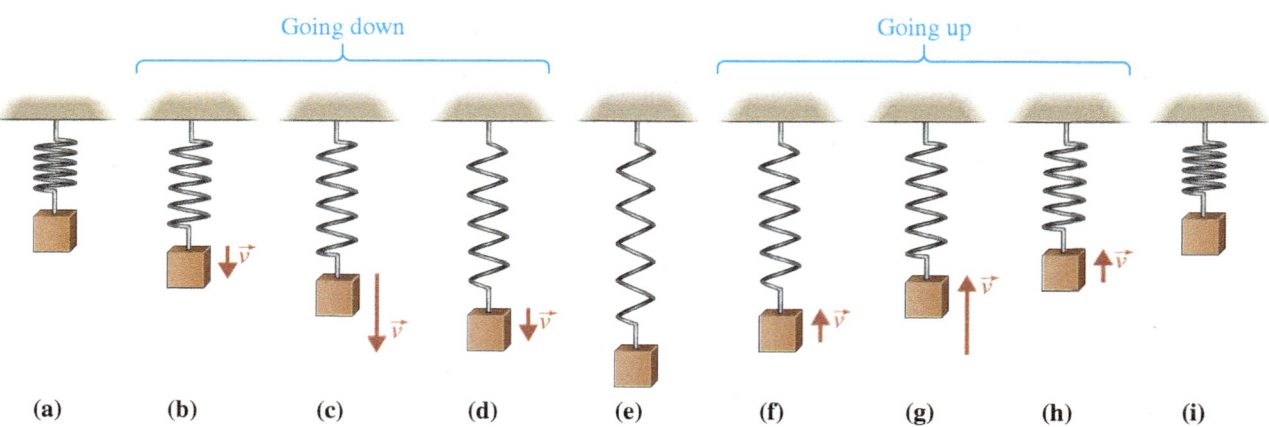

Figure 8.2. The classic "block on spring" system. (Now this is officially a college physics textbook.) Starting from panel (a), the block accelerates downward in panel (b), then through its equilibrium position (no net force) in panel (c). In panel (d) it is moving down but accelerating up, and in panel (e) it is momentarily at rest while it changes direction. In panels (f)–(h), the journey is reversed until the block returns to its original position in panel (i). When the block returns to its starting position, it has completed a full oscillation. The time it takes for this to happen is called the period. Left to itself, the system will keep repeating the cycle over and over.

We won't go into every case, but introductory physics books love to focus on the simplest case of one massive object connected to a spring, where the proportionality constant is a nice, neat 2π:

$$T_{\substack{\text{one oscillation} \\ \text{simple mass-spring}}} = 2\pi\sqrt{\frac{m}{k}}. \quad (8.5)$$

Figure 8.2 shows the full cycle of a block hanging from a spring.[6]

Sometimes when scientists describe a vibrating system, they use another term: "oscillation frequency." Frequency just expresses how often something happens. Mathematically, it is just the inverse of the period:

$$f = \frac{1}{T_{\text{one oscillation}}} \quad (8.6)$$

So if $T_{\text{one oscillation}} = 0.1$ s (one-tenth of a second), then f is "10 per second" or 10 **hertz** (10 Hz).

In the real world of sports, almost nothing is as simple as the mass and spring... except the bow and arrow! The bow is the spring, and the arrow is the mass that it pushes. There is a difference, however. Whereas the little block in figure 8.2 is glued to the spring, the arrow actually detaches from the bowstring after one-quarter of a period, as indicated in figure 8.3.

Numbers

Now we can figure out the time it took for the bow to fire and the firing power.

Equation 8.5 gives us the period of a full oscillation, so the time it takes the bow to convert elastic to kinetic energy is

$$\Delta t = \tfrac{1}{4} T_{\text{one oscillation}} = \tfrac{1}{4} \times 2\pi\sqrt{\frac{m}{k}} = \tfrac{1}{4} \times 2\pi\sqrt{\frac{0.0018 \text{ slug}}{23 \text{ lb/ft}}} = 0.014 \text{ s} = 14 \text{ ms}.$$

---

[6] The Union of Physics Teachers threatened to revoke my teaching license if I didn't include this figure. It is required in any introductory textbook.

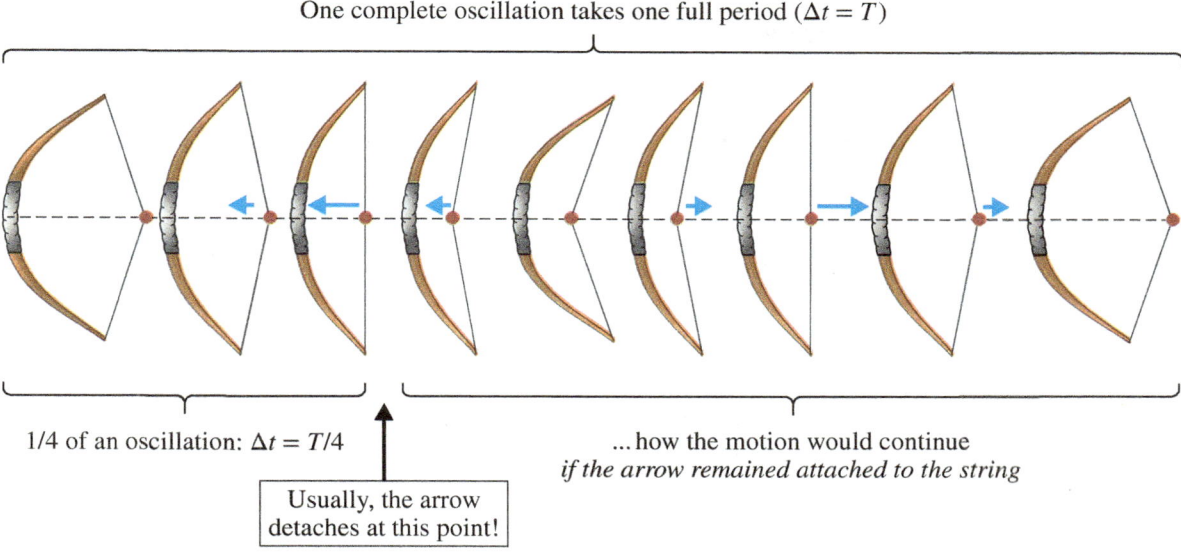

Figure 8.3. An archer's bow and arrow is an oscillatory system. If the arrow stayed connected to the bowstring, it would oscillate back and forth with a period given by equation 8.5. In reality, the arrow detaches from the string after one-quarter oscillation. The small arrows indicate the velocity of the archer's arrow (its rear notch is indicated by a small circle, for clarity). When the system is maximally stretched at the ends, the arrow is momentarily at rest.

(Here, I used the k for Im's bow that we found in the previous section and the m of an arrow.)

The average power during the release, then, is

$$\bar{P}_{\text{release}} = \frac{73.4 \text{ J}}{0.014 \text{ s}} = 5243 \text{ W},$$

which is 7 h.p.!

Here is the important point. The system *output* energy (in the arrow's flight and residual quivering of the bow) is just equal to its *input* energy (by Im pulling back the bowstring). The bow doesn't add any energy. What is different and crucial to the sport is the much greater output *power*.

Very often, elastic sports equipment is designed to change the timing of energy transfer from what humans could achieve on their own.

Question

Doesn't it seem odd that equation 8.5 tells us the time for a full oscillation of the bow (or a block on a spring), without even "knowing" how far back the bow is drawn? If the bowstring is drawn back more, shouldn't it take a longer time to complete an oscillation?

8.3 The Physics of Archery III: Archer's Paradox

Okay, now we come to something really cool, so grab a coffee.

When we think of elasticity in archery, we naturally think of the bow itself. We might imagine the arrow itself to be a rigid rod, bending hardly at all while being accelerated or in flight. Nothing could be further from the truth! Seen in slow motion video, a target arrow in flight almost resembles a fish swimming upstream (see figure 8.4), and one is almost amazed that it can fly straight at all.

Far from being an annoying and undesirable feature, the fact that arrows flex and wiggle is an absolutely critical part of the sport. Any serious archer needs to understand and be able to use this flexibility to advantage. To understand this, let's focus on how a flexible arrow (or any rod) wiggles.

8.3.1 Oscillations of an arrow

When an arrow (or any rod, such as a baseball bat or golf shaft) is distorted, an elastic restoring force arises that tries to restore it to its original straight shape (unless, of course, the arrow is so bent that it breaks, or the golf shaft is permanently damaged). For small deformations of the shaft, the restoring force is proportional to the deformation; that is, the arrow obeys Hooke's law (equation 8.1).

Arrows are rated in terms of their "spine." An arrow's spine is given by its deflection when a 1.94-lb (880-g) weight is hung from its center and it's supported across a 28-in. span. Figure 8.5 shows how it works. The deflection is measured in thousandths of an inch, so that a 520 arrow, such as the Easton ACE 520, will

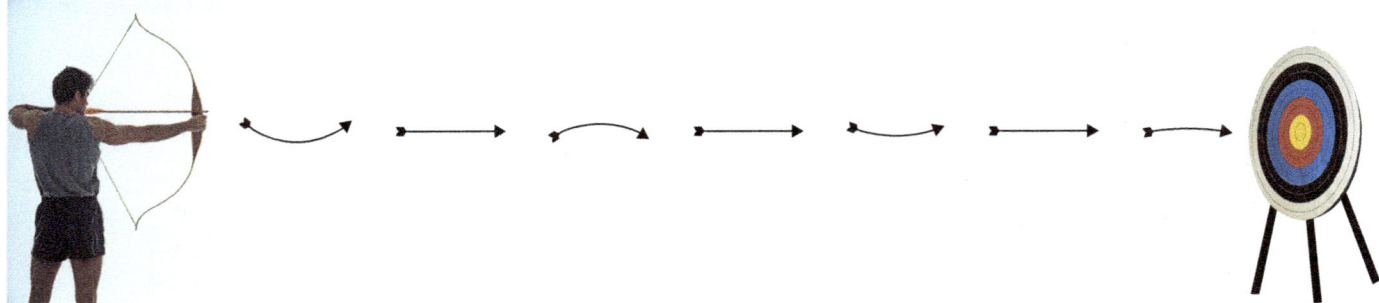

Figure 8.4. If an arrow is deformed and free of other forces, it will oscillate between being bent in one direction and being bent in the other. As time passes, internal friction forces and stabilization by the fletching at the tail damp the amplitude of the oscillations. (Note: while I show an up-and-down oscillation for the purpose of illustration, we are actually more interested in a left-and-right oscillation.) **Left:** (© *Comstock/JupiterImages RF*); **Right:** (© *Comstock/Alamy RF*).

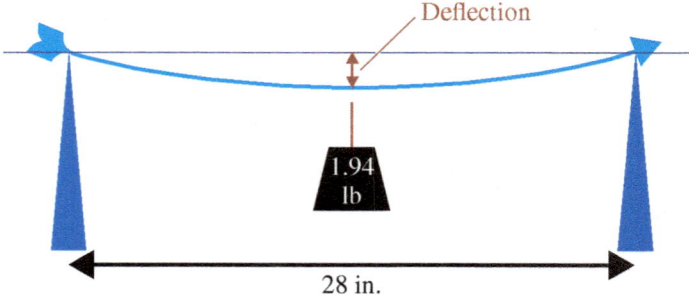

Figure 8.5. The spine of an arrow is the deflection of its center, when an 880-g mass is hung from its center.

deflect 0.520 in. when 880 g is hung from its center. For this kind of deformation,[7] then, the arrow's spring constant is inversely related to its spine deflection:

$$k_{\text{arrow}} = \frac{1.94 \text{ lb}}{\text{spine deflection}} \qquad (8.7)$$

A large spine deflection means low stiffness, or increased flexibility.[8]

If an arrow is bent but free to move (no weights or anything hanging on it), the elastic restoring force will tend to return the arrow to its straight shape. But if there is no internal friction to prevent it, it will overshoot, and the arrow will start bending the *other* way! And back and forth it goes, the arrow oscillating between two bent shapes, as shown schematically in figure 8.4. In the figure, the arrow is wiggling at the same time that it flies between archer and target. Check out some high-speed camera videos on YouTube; sometimes the wiggling action is quite dramatic.

This wiggling is a more complicated example of oscillations in an elastic system, similar to what we saw in figures 8.2 and 8.3. As equation 8.4 says, the time required for one full wiggle ("cycle") is longer for heavier arrows (large m) and for arrows with little spine (low k_{arrow}). Selecting precisely the right combination of spine and mass allows the archer to tune the period of the arrow's oscillation. And that's crucial for the Archer's paradox.

Figure 8.6

When released, the bowstring will accelerate directly toward the riser, that is, in the direction of the target. However, to avoid the riser, the arrow points strongly to the left. Is this archer aiming in the right direction? (© *McGraw-Hill Education/Aaron Roeth, photographer*).

8.3.2 How the archer's paradox works

Okay well, very nice. But I told you to get a coffee and now it's getting cold and you're wondering what is the big deal. All this wiggling around seems to be if not detrimental to the archer, then at most a cute amusement. Not so! It actually provides the solution to one of the most basic problems facing the user of a recurve bow.

The problem is this. The elastic force of the bowstring will pull the string—and the arrow nocked into it—*directly toward* the bow's riser (the "handle part" of the bow). So it would seem logical to make this direction line up with the target, right? The problem is that the arrow can't go *through* the riser because it points off slightly to the left (on a right-handed bow), as shown in figure 8.6. It sure looks as if the arrow is going to deflect off the riser and go wide of the target.

What to do? Should the archer in figure 8.6 rotate slightly to the right, so the *arrow* points at the target, while the force from the bowstring points away from it? Maybe something in between? The surprising answer is that if he has selected the proper arrow, the archer in figure 8.6 will strike the target!

The first step toward understanding this is to realize that the flexible shaft of an arrow will "buckle" if pushed on from both ends, as shown by the hands in the top of figure 8.7. (If this is not clear to you, try to make a plastic ruler buckle in this way.)

The second step is to realize that as the arrow is being accelerated, the shaft is being pushed from the rear by the bowstring, and from the front by the relatively

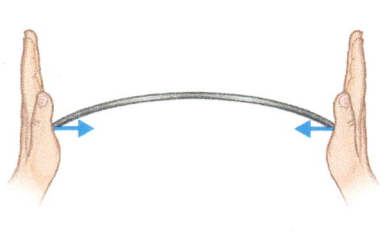
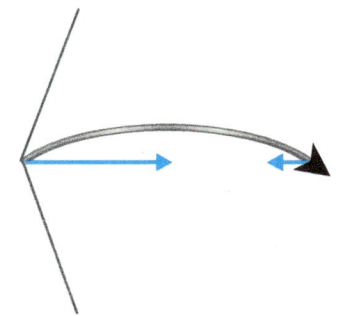

Figure 8.7

Two ways to buckle an arrow shaft: The bottom case is relevant for archery. The rightward force on the shaft comes from the bowstring. The leftward force on the other end is the "reaction force" (a lá Newton's third law) of the force of the shaft on the arrowhead as it accelerates the head to the right.

---

[7] The single bend-at-the-middle deformation we'll talk about here is the simplest and most important type of deformation, but there can be other more complicated ones. These can be important, for example, when considering baseball bats.

[8] Sometimes spine is expressed in pounds, by dividing 26 by the deflection in inches. In this way, a 0.500-in. deflection corresponds to a spine of 52 lb.

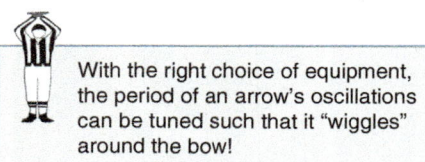

With the right choice of equipment, the period of an arrow's oscillations can be tuned such that it "wiggles" around the bow!

more massive arrow tip. Newton's third law tells us that since the shaft is accelerating (that is, exerting a force on) the tip to the right, the tip must be "pushing back" on the shaft. Just as with the two hands, the combination of these forces will buckle the shaft.

The third and final step is to combine the buckling effect and subsequent oscillations with the forward motion of the arrow. Then if everything is *just* right, you get the perfect archer's paradox effect, as shown in figure 8.8. There the center of mass of the arrow is accelerated in the desired direction, which is the way the *bow* is aligned, as in figure 8.6, deflecting neither left nor right.

The arrow is performing something like a sideways Fosbury flop (section 4.4.2)! In both cases, a perfectly timed and finely tuned combination of motion and bending allows the center of mass of an object to pass through a solid barrier without contact between the object and barrier.

There are a tremendous number of variables to account for when selecting an arrow. The bow draw weight, tip mass, shaft mass, and shaft spine and the bow and arrow efficiency all combine in different ways to determine the amount and timing of buckling effects, launch speed, resistance to air currents, and impact momentum (of importance to hunters). Matching an arrow to a bow to an archer to a particular situation is a highly complex series of balances and compromises.

The amount of materials research that goes into the design and production of athletic equipment in *any* sport is tremendous. Archery is no exception, and arrow manufacturers use a variety of materials and combinations. Figure 8.9 shows just

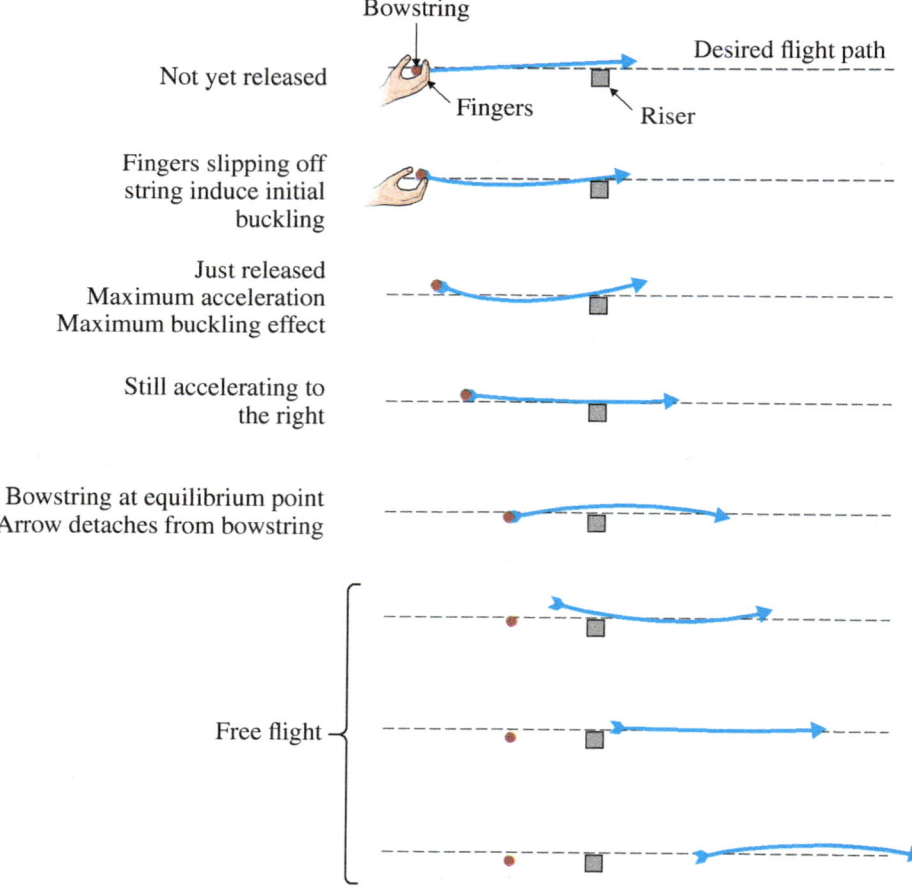

Figure 8.8. By selecting a properly spined arrow for a given bow, arrow weight, and tip, the archer can make the arrow wriggle around the riser of the bow with little or no contact.

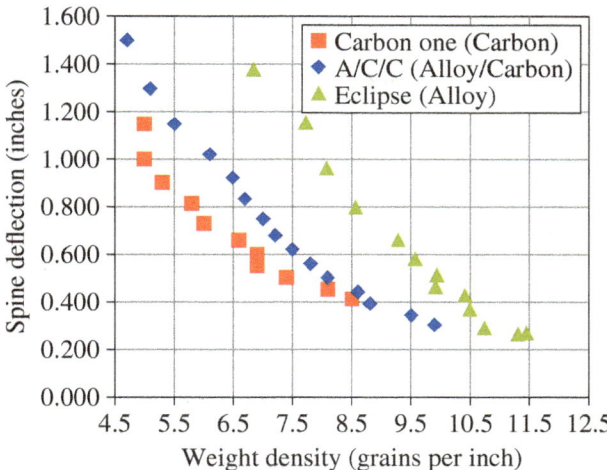

Figure 8.9. Specifications for three arrow shaft designs available from Easton Technical Products. To obtain the shaft weight (in grains), you have to multiply the value on the x-axis by its length in inches.

three models of arrow available from Easton Archery. Typically, a carbon arrow of a given stiffness is significantly lighter (and more expensive) than a model made from metallic alloy. With any material, obtaining greater stiffness (in some sense, a stronger arrow) naturally implies a heavier arrow; this is why the curves on figure 8.9 trend downward.

As in many of our discussions about the "real world" of sports physics, I've only given the broad strokes of the archer's paradox. Experts will correctly point out that the "arrow snakes around the bow" picture I've described here is only approximately true. Nevertheless, it should make clear the importance of precision timing of the elastic elements in this sport and the coordinated choices that must be made. Archers spend an incredible amount of time researching, experimenting, and "dialing in" (that is, fine-tuning) their equipment.

> **Question**
>
> Tips come in various weights. If a slight breeze kicks up during a competition, Im Dong-Hyun might consider reaching for a heavier arrow *tip* than what he's been using. This one simple adjustment, however, leads to many consequences that he'll need to account for.
> These are not super-easy questions; give them some thought.
>
> - How (if at all) will the bow and arrow efficiency change?
> - Will the arrow (once launched) have more or less kinetic energy than it did before?
> - How will the arrow's speed (once launched) change?
> - Will the shaft flex more or less than it did with the lighter tip?
> - Will the arrow's oscillation period increase or decrease?
> - There can be disagreement whether, with the heavier tip, Im should change the spine of his shaft. What might be a reason to choose a stiffer shaft? How about a weaker one?

8.4 Zdeno Chára's Slap Shot: Fast Storage, Faster Release

Every year, the National Hockey League sponsors a "skills competition" in which stars show off their talents. A highlight is always the hardest slap shot competition, where the very best players accelerate a puck from rest to 100 mph in a spectacular, violent shot.

Issues with collisional energy transfer

The 170-g puck is a bit heavier than a 145-g baseball, but the best pitchers could probably adjust their technique to throw a puck at 100 mph if they really wanted to. In either case, the energy transferred to the puck is $\frac{1}{2}mv^2 = \frac{1}{2}(0.17 \text{ kg})(44.7 \text{ m/s})^2 = 170$ J.

However, there is a huge difference between the two methods of transferring this energy. It is generally much more efficient to do work on a projectile in a long, smooth acceleration, such as a baseball pitch, than in a short, violent inelastic collision. The downside is that in a pitch, the projectile moves at the same speed as the hand at release. That's a natural speed limit.

In a collision, however, the "speed limit" is higher—the light projectile (puck) in principle can fly off with a speed twice that of the object (stick, bowling ball, whatever) that strikes it. I put this in a box for you on page 162; there, it says that you hit the speed limit if the striking object is very heavy and the collision is elastic.

So the speed limit is higher in a collision; that's the upside. The downside is that the fraction of energy transferred to the projectile is very small. Remember the 80% energy-transfer efficiency of the bow and arrow[9] from section 8.1? Forget about it in this collision, for two reasons:

1. For a heavy-on-light collision (where the speed limit is approached), the heavy object continues moving forward as a result of momentum conservation. You can see this in the bowling ball collision of figure 6.9. The lighter object moves away quickly, but has taken a small fraction of the energy.

2. The hockey stick–puck collision is far from elastic, due to stick vibrations; high-speed video shows this clearly. So even more energy is lost.

The slap shot

The slap shot addresses these downsides in two ways.

First, since the *fraction* of energy transferred to the puck is going to be small, a solution is to maximize the *total* energy available. This occurs during the windup stage of the slap shot, when chemical energy is converted to kinetic energy. The player rapidly shifts his weight from the rear to the front skate while swinging his shoulders and arms. A large, muscular upper body represents the m in the player's $\frac{1}{2}mv^2$, while a powerful torso and legs get it moving with a large v.[10]

Second, since the high forward momentum of the stick plus player is part of the problem of reducing the fraction of energy transmitted to the puck, the slap shot actually reduces this forward momentum while not greatly reducing the energy. How can this happen? If you reduce the player's momentum, don't you necessarily

---

[9] A bow shooting an arrow is the same sort of smooth energy transfer as a pitcher throwing a baseball.
[10] The fact that the motion is rotational, rather than in a straight line, means that $\frac{1}{2}mv^2$ isn't precisely the correct term for the player's kinetic energy; we'll come to this in chapter 9. However, this is not important for the discussion here.

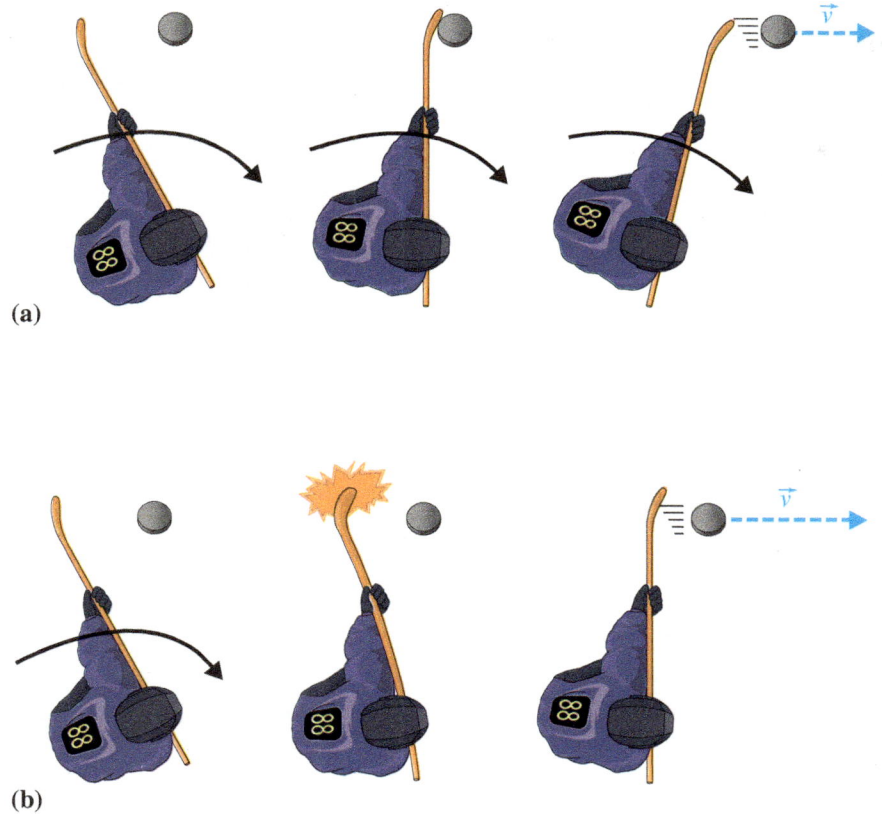

Figure 8.10. A shot viewed from above. (a) The player strikes the puck directly; conservation of momentum keeps him moving after the shot. (b) The player strikes the ice before striking the puck, reducing his momentum while maintaining most of the energy.

reduce his energy? The answer: sure, reducing his momentum means the *kinetic* energy is reduced, but it can be converted to a stored (potential energy) form, to be released later!

Figure 8.10 sketches how this works. Rather than striking the puck directly after the windup, the player strikes the ice about 1 ft behind the puck. This slows him and converts his considerable kinetic energy into elastic potential energy of the flexed stick.

Figure 8.11 shows a dramatic example, courtesy of Boston Bruins team captain Zdeno Chára, whose blistering slap shot was measured at 108.8 mph in the 2012 NFL skills competition. The flex of his stick is dramatic, and we can get a quantitative estimate of the energy stored in it.

Sometimes analysts talk about the *angle* at which the stick is flexed during a slap shot. This is shown in the right panel of figure 8.11. However, this doesn't really make a lot of sense, since it's not as if the stick is a straight line above Chára's left hand and another straight line below his left hand, with some angle between those two lines. Rather, the stick forms a more or less circular arc, and it makes more sense to quantify the flex by the deflection s shown in the middle panel.

In fact, this is how hockey sticks are rated. Each stick comes with a "flex rating," which indicates the amount of force, in pounds, required to deflect the middle of the stick 1 in. While quantified in a different unit (every sport has its own idiosyncrasies), the flex rating is essentially the same as the spine of an arrow we just discussed in section 8.3, and it is measured in the same way.

 Striking the stick on the ice allows the player to convert his body's kinetic energy into elastic energy of the stick before it collides with the puck.

Figure 8.11. Boston Bruins defenseman Zdeno Chára demonstrates the perfect slap shot while warming up before a game against the Washington Capitals in April 2013. He has thrust his weight to his front skate and powerfully swung his 255-lb frame forward. By striking the ice and pushing on the center of the stick, he has reduced his forward momentum and converted his kinetic energy to stored elastic energy. In the photograph, a fraction of this energy is being transferred to the puck, turning it into a fearsome weapon. The right two panels show two possible ways of quantifying the stick's flex. **Left, Middle** and **Right:** (© *Mark Goldman/Icon SMI/Corbis*).

The flex rating of a hockey stick is Hooke's constant for that stick. Hooke's constant for a "100 flex stick" is $k = 100$ lb/in. This is considered a stiff stick, compared to a less stiff "45 flex stick" designed for a child, which has $k = 45$ lb/in.

Typically, experts advise using a stick with a flex rating that is a bit more than one-half the player's weight in pounds, although technique and "feel" preferences play major roles in stick selection. A very low-flex stick can feel "mushy" on hard shots in the hands of a heavy, strong man, whereas a very stiff stick in the hands of a smaller person will not flex enough.

Chára uses an amazingly stiff 155-flex stick, a somewhat higher rating than the factor-of-2 rule of thumb for the 255-lb defenseman. However, the 6-ft 9-in. Chára is brutally strong and attains significant flex, as seen in figure 8.11. Using the estimate of $s \approx 6$ in. from the center panel, we can use equation 8.1 to estimate the force he's exerting on the stick:

$$|\vec{F}| = k|\Delta \vec{x}| = \left(155 \, \frac{\text{lb}}{\text{in.}}\right)(6 \text{ in.}) = 930 \text{ lb!}$$

Such a large force is possible for two reasons: First, it is partially a "static" force of Zdeno slowing himself down; we can often exert temporarily large forces to slow ourselves from crashing into a wall or floor. Second, Zdeno's strong.

We can also figure out the amount of energy that Chára has stored in the stick, using equation 8.3:

$$\text{PE}_{\text{elast}} = \tfrac{1}{2}k|\Delta\vec{x}|^2 = \tfrac{1}{2}\left(155 \, \frac{\text{lb}}{\text{in.}}\right)(6 \text{ in.})^2 = 2690 \text{ lb} \cdot \text{in.}$$

$$= 2690 \text{ lb} \cdot \text{in.} \times \frac{1 \text{ ft}}{12 \text{ in.}} = 232.5 \text{ lb} \cdot \text{ft}$$

$$= 232.5 \text{ lb} \cdot \text{ft} \times \underbrace{\frac{1.356 \text{ J}}{1 \text{ lb} \cdot \text{ft}}}_{\text{from equation 6.8}} = 315 \text{ J}.$$

Energy transfer revisited

So how did we do in terms of energy transfer efficiency? Assuming 315 J of stored energy in the stick results in a 170-J, 100-mph shot, we find that more than 50% of the energy was transferred to the puck. That's way more than we'd expect from a simple collision between heavy Chára plus the stick with the light puck.

We've just gone through the mechanics of the slap shot, but it's worthwhile to look at it another way, to see how this great increase in efficiency occurred. A direct stick-on-puck shot (panel (a) of figure 8.10) is a collision between the heavy, slow Chára-plus-stick object with the light puck. The energy transfer efficiency would then be low because of momentum conservation, as we discussed.

In the slap shot, while most of the mass stops (or nearly so), the blade itself whips much faster, and the collision is between a lighter but faster blade and the light puck. This is much closer to the billiard ball collision of figure 6.10, where fully 100% of the kinetic energy of the cue ball is transferred to the 8-ball.

The amazing effectiveness of the slap shot relies crucially on the ability to convert energy easily from one form to another and on the selection of materials suited to a given player and situation.

8.5 Bungee-Jumping Brides and Quadratic Equations

Bungee jumping isn't usually considered a sport except by those (to wit, the author) who consider it a competition in which each participant tries to prove that he's more insane than (or at least as insane as) the others.

According to a 2013 article in the British newspaper *The Daily Mail*, 1 in 10 people in the United Kingdom would like to do a bungee jump on their honeymoon. Those crazy Brits.

A couple might choose a jump from the Middlesbrough Transporter Bridge, shown in figure 8.12. Courteous as ever, the British man invites his bride to precede him: "Ladies first, m'love!" Barely suppressing a dirty look back at him, she allows herself to be harnessed and stoically dives from the platform 170 ft above the water. When she finally stops bouncing, he peers over and sees that she is 60 ft above the water, that is, 110 ft below him. The guide tells the nervous groom that the bungee cord is 100 ft long when nobody is hanging from it, so it is being stretched 10 ft due to the weight of his new wife.

As she is lowered into a waiting boat, the groom starts asking himself some frantic questions. If his petite 110-lb wife stretches the cord by 10 ft, his 200-lb frame will surely stretch it more. Will the cord be *too* stretchy and let him crash into the water?

Figure 8.12
The Middlesbrough Transporter Bridge in North Yorkshire, England, is a popular site for bungee jumping. A strange-looking bridge, isn't it? Where is the connecting road on either side, high in the air? Actually, the cars, cyclists, and pedestrians cross the River Tees on a gondola hung *below* the structure; it connects the roadway like a ferry in the air. (© *Alan Crawford/Getty Images RF*).

8.5.1 Dangling above the water

Fortunately, he has taken this course, too, so he knows how to quantify the stretchiness of the cord, that is, its Hooke's constant. He imagines a free body diagram of his wife dangling at rest at the end of the cord; this is drawn in the middle panel of figure 8.13. She is not accelerating, so Newton's first law says that the forces on her cancel out. The only forces are her weight pulling down and the cord's tension pulling up:

$$W_y = -T_y$$
$$-110 \text{ lb} = -\left(-k_{\text{cord}}(\Delta y)\right)$$
$$-110 \text{ lb} = -\left(-k_{\text{cord}}(-10 \text{ ft})\right)$$
$$\to k_{\text{cord}} = \frac{110 \text{ lb}}{10 \text{ ft}} = 11 \text{ lb/ft}$$

> **Question**
>
> There were four separate minus signs in the math above! Can you tell why each of them is there?

Our nervous groom's first question is whether *his* head will still be above the water when all the bouncing stops.

To answer this, he just needs to apply the same mathematical treatment to himself to see how much he'll stretch the cord:

$$W_y = -\left(-k_{cord}\left(\Delta y\right)\right)$$
$$\rightarrow \Delta y = \frac{W_y}{+k_{cord}} = \frac{-200 \text{ lb}}{11 \text{ lb/ft}} = -18.2 \text{ ft}$$

The negative sign means simply that he'll be stretching the cord down.

So the stretched cord will be 118.2 ft long, and he'll be dangling 51.8 ft above the water. Whew!

8.5.2 How low will he bounce?

However, even if he'll be above the water when dangling, he will stretch the cord even *more* when he's bouncing! What if he slams into the river on the way down? Our groom desperately wants to know how far down he'll go at the lowest point in the bounce.

This one won't be so easy, since at the bottom of the bounce he will not be in equilibrium, as the upward tension will be larger than his weight, shown in the right panel of figure 8.13. A simple force equation won't be enough. We'll need to look at the energy flow here.

Figure 8.14 shows the same jump at three different times. On the left, the nervous man has just stepped off the platform. His velocity is, for the moment, zero, so KE = 0. The cord is not stretched, so PE_{elast} = 0.

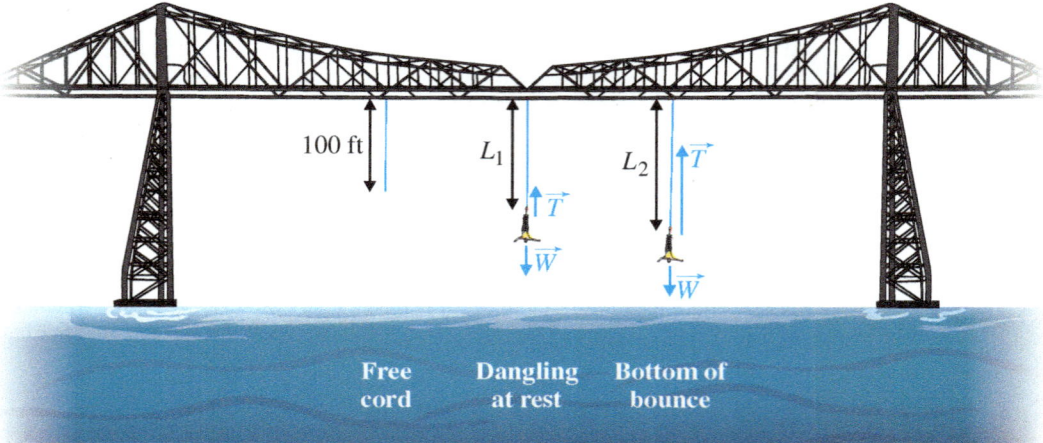

Figure 8.13. Free-body diagrams for a person dangling at rest (middle) and one accelerating upward at the bottom of a bounce (right). In both cases, the length of the cord is larger than its unstretched length of 100 ft (left).

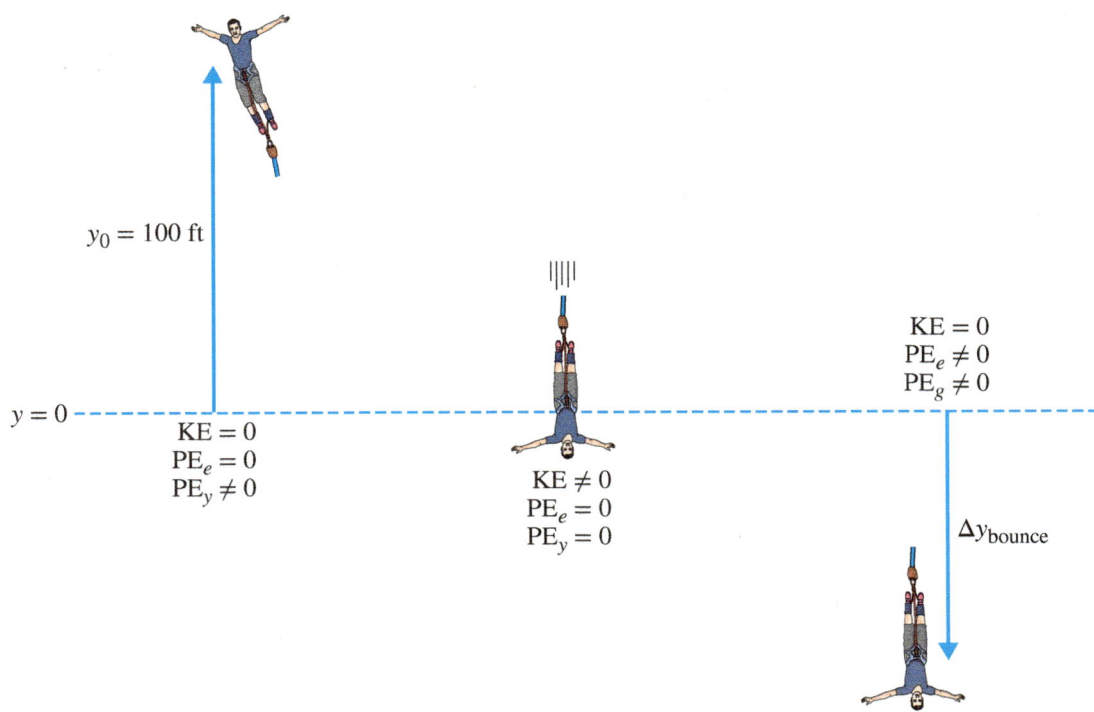

Figure 8.14. Three stages of a bungee jump. **Left**: the jumper has just stepped off the platform. **Middle**: the jumper in free fall. **Right**: the jumper at the bottom of the bounce.

We always have to choose a vertical reference position; here it is 100 ft below the bridge, that is, the unstretched length of the cord. This means that just as he steps off the platform, the groom has positive gravitational potential energy: $PE_{grav} = mgh = (200 \text{ lb})(100 \text{ ft}) = 20{,}000 \text{ lb} \cdot \text{ft}$. We are not going to worry about air drag or heating of the cord, so this must be the jumper's energy at *every* point in the jump.

At the bottom of the bounce, the groom is again at rest momentarily, so $KE = 0$. However, he has positive PE_{elast} because the cord is stretched, and *negative* PE_{grav} because his position is below $y = 0$. Time for equations:

$$KE + PE_{elast} + PE_{grav} = 20{,}000 \text{ lb} \cdot \text{ft}$$
$$0 + \tfrac{1}{2} k_{cord}(\Delta y)^2 + mg\,\Delta y = 20{,}000 \text{ lb} \cdot \text{ft}$$
$$\rightarrow \tfrac{1}{2} k_{cord}(\Delta y)^2 + mg\,\Delta y - 20{,}000 \text{ lb} \cdot \text{ft} = 0$$

The quadratic equation

Wow, we almost made it through a physics course without ever having to discuss the dreaded quadratic equation. Often elementary algebra is enough to solve a problem; that's been the case for most of this book. But the solution to the problem at hand now requires us to introduce and discuss this unique mathematical formula, so let's dive in. It actually comes in handy in many mathematical situations outside of physics, too.

An equation like this

$$ax^2 + bx + c = 0$$

has two possible solutions:

$$x = \begin{cases} \dfrac{-b + \sqrt{b^2 - 4ac}}{2a} \\[1em] \dfrac{-b - \sqrt{b^2 - 4ac}}{2a} \end{cases} \quad (8.8)$$

Note that the difference in the two solutions is the sign before the square root.

Using the quadratic equation: common sense is needed

An aspiring physicist, the groom happened to be carrying a calculator on his honeymoon, (this may not bode well for the marriage, actually), so he can use this formula to figure out how far he will stretch the cord in the first bounce. To proceed, he needs to know what to use for a, b, and c in equation 8.8. In this case, they are

$$a = \tfrac{1}{2} k_{\text{cord}} = \tfrac{1}{2}(11 \text{ lb/ft}) = 5.5 \text{ lb/ft}$$

$$b = mg = 200 \text{ lb}$$

$$c = -20{,}000 \text{ lb} \cdot \text{ft}$$

According to equation 8.8, then, he has two choices:

$$\Delta y = \begin{cases} \dfrac{-b + \sqrt{b^2 - 4ac}}{2a} = \dfrac{-200 \text{ lb} + \sqrt{(200 \text{ lb})^2 - 4(5.5 \text{ lb/ft})(-20{,}000 \text{ lb} \cdot \text{ft})}}{2(5.5 \text{ lb/ft})} \\ \qquad = 44.8 \text{ ft} \\ \\ \dfrac{-b - \sqrt{b^2 - 4ac}}{2a} = \dfrac{-200 \text{ lb} - \sqrt{(200 \text{ lb})^2 - 4(5.5 \text{ lb/ft})(-20{,}000 \text{ lb} \cdot \text{ft})}}{2(5.5 \text{ lb/ft})} \\ \qquad = -81.2 \text{ ft} \end{cases}$$

The quadratic equation gives two possible solutions. Choosing the "right" one requires knowing something else about the situation. And we know this much: at the bottom of the bounce, the groom will be *below* the position $y = 0$ (see figure 8.14), so we need to choose the negative solution, -81.2 ft.

That means that he'll be 181.2 ft below the bridge, that is, under water! In great agitation, he tells the attendant to use either a less flexible cord (higher k_{cord}) or a shorter one. Good thing he brought the calculator! On the other hand, his new bride asks him angrily why he didn't take out the blasted thing before *she* jumped. So much for the honeymoon.

Collected Equations

$$\vec{F}_{\text{elastic}} = -k\,\Delta\vec{x} \tag{8.1}$$

$$k_{\text{bow}} = \frac{\text{draw weight}}{\text{draw length}} \tag{8.2}$$

$$\text{PE}_{\text{elast}} = \tfrac{1}{2} k |\Delta\vec{x}|^2 \tag{8.3}$$

$$T_{\text{one oscillation}} \propto \sqrt{\frac{m}{k}} \tag{8.4}$$

$$T_{\substack{\text{one oscillation} \\ \text{simple mass-spring}}} = 2\pi \sqrt{\frac{m}{k}} \tag{8.5}$$

$$f = \frac{1}{T_{\text{one oscillation}}} \tag{8.6}$$

$$k_{\text{arrow}} = \frac{1.94 \text{ lb}}{\text{spine deflection}} \qquad (8.7)$$

$$x = \begin{cases} \dfrac{-b + \sqrt{b^2 - 4ac}}{2a} \\[2ex] \dfrac{-b - \sqrt{b^2 - 4ac}}{2a} \end{cases} \qquad (8.8)$$

Problems

1. Archers must take care to choose the correct arrow for each shot, depending on the draw weight and size of the bow. On a breezy day, the archer decides to use an arrow twice as massive as his usual arrow, to minimize the effects of wind.

    (a) Will the new arrow leave the bow with more energy than, less energy than, or the same energy as his usual one?

    (b) Will the new arrow leave the bow with more speed than, less speed than, or the same speed as his usual one?

    (c) Should his new arrow have a different spine than his usual one? If so, approximately how much more or less should it be? If not, why not?

2. The pole in the vaulting event is used to transfer the vaulter's initial kinetic energy to gravitational potential energy. In the process, part of the energy is temporarily stored as elastic potential energy in the pole. The flexibility of vaulting poles is quantified much as archery arrows are (figure 8.5): a 50-lb weight is hung from the middle of a pole that is supported at both ends; the number of centimeters it deflects is the pole's "flex number."

    (a) If a pole has a flex number of 19, what is Hooke's constant, in newtons per meter?

    (b) If the pole deflects 2 ft from a straight line (see center panel of figure 8.11, where s is the deflection, for example), how much energy is stored in the pole?

    (c) Assuming the vaulter has a mass of 80 kg and was running at 18 mph down the runway, what fraction of this initial energy has been stored in the pole?

3. In this problem, we assume that diving boards obey Hooke's law, which is only approximately true.

    In diving from a springboard, timing and the direction of the force from the board are crucial. Too much board deflection, and the diver is thrown too much forward. Too fast or slow of a board oscillation, and the diver's timing is thrown off and he wastes energy. Using a roller to set the fulcrum of the board, divers can adjust the board's springiness and deflection. Moving the fulcrum toward the diving end produces a stiffer board. The optimal board setting depends on the diver's weight and technique.

    According to *The Guardian* newspaper, the average mass of a diver at the 2012 Olympics in London was 59.8 kg. The lightest in the Games was China's 16-year-old Yadan Hu, weighing in at a mere 36 kg (79 lb). While her main event is the 10-m platform dive, she also competes in the 3-m springboard.

    When she jumps 0.9 m in the air and lands on the end of the springboard, it deflects downward about 0.7 m and returns to horizontal in 300 ms. These are typical values for a well-tuned board.

    (a) What is Hooke's constant of the board?

    (b) Divers sometimes just stand on the end of the board before a competition, to get a feeling for Hooke's constant. If she simply stood on the end of the board (without having jumped), how far would it deflect?

(c) If she had jumped higher (say, 1.2 m) before landing on the board, would it have taken more, less, or the same amount of time to return to horizontal?

(d) Is she at equilibrium when the board is maximally deflected after she lands on it?

(e) The heaviest diver at that Olympics was Germany's Stephan Feck, at 79 kg. If he also landed on the same board after an initial approach jump 0.9 m high, would the board's force point more forward or less forward than it did with Hu? (There is no need for numbers here.)

(f) How long will it take for the board to return to horizontal?

(g) To properly tune the board's timing for his body mass, should he increase or decrease Hooke's constant?

(h) Does properly tuning Hooke's constant affect the resulting direction of the force of the board on Feck? How?

(i) Besides the mass and approach-jump height, what other factors affect the downward deflection and timing of the board?

4. Bones take a beating in athletics, and their flexibility and strength are studied intensely in biomechanics. The figure in problem 4 shows an experiment in which small pieces from a human femur (donated by a heart attack victim) were compressed by a large cylindrical weight. The experiment reveals that the "springiness" of a bone depends on the rate at which it is compressed.

(a) Is the bone stiffer when compressed quickly or when compressed slowly?

(b) What is Hooke's constant of the sample when it's being compressed at 400 in./s?

(c) How does this compare to Hooke's constant of Zdeno Chára's hockey stick? (Although compressing a bone sample and flexing a stick are rather different things, they both involve a restoring force opposing a distortion of the object. Comparing Hooke's constants gives a feeling for the forces and motions involved.)

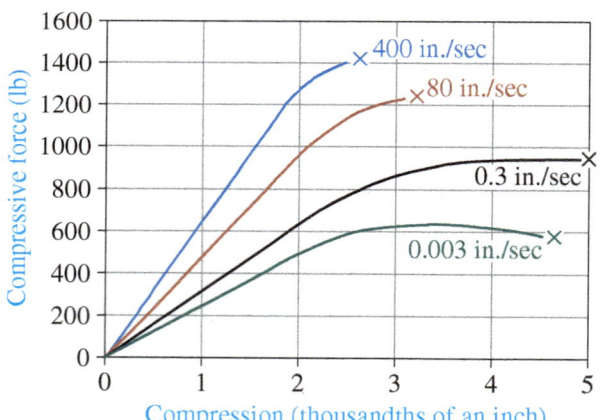

The restoring force of the femur is plotted versus its compression. The curves correspond to crushing the bone slowly (bottom curve) to very rapidly (top curve). At high compression, the bone stops following Hooke's law as it starts to fail; the "×" on each curve indicates when the bone shattered. From J. H. McElhaney, "Dynamic response of bone and muscle tissue", *J. Applied Physiology* **21**: 1233 (1966).

5. Figure 8.14 shows the bungee jump of the 200-lb newlywed groom. We'll assume he and his wife use the $k = 11$ lb/ft cord discussed in section 8.5.

(a) In miles per hour, how fast is the nervous groom moving when he is 100 ft below the bridge (middle panel of the figure)?

(b) How fast is he moving when he is 150 ft below the bridge?

(c) How about his 110-lb wife? How fast is she moving 100 ft below the bridge?

(d) How fast is she moving 150 ft below the bridge?

(e) The bungee cord acts just as a spring does when the person is executing only small oscillations at the bottom (that is, when most of the bouncing is over). What is the frequency of the husband's oscillation at this point?

(f) Does his wife oscillate with greater or lower frequency?

6. (a) When 251-lb Blake Griffin hangs stationary from the front of a rim, it is deflected down 4 in. What is Hooke's constant?

(b) NBA rules prohibit hanging intentionally on the rim, except if it's done to prevent injury to a player. Sometimes, a player must actually lift himself on the rim, to avoid colliding with another player beneath him. How much will the rim be deflected if, instead of hanging stationary, Griffin accelerates his center of mass upward at 10 ft/s^2 by lifting his legs?

7. Resistance bands are long elastic band type of devices that may be used as part of a weight-lifting regimen. One such band has an unstretched length of 4 ft. An athlete who can exert 60 lb of force finds that she can stretch the band to 5.5 ft (that is, 1.5 ft longer than its unstretched length).

(a) What is Hooke's constant of the band?

(b) How much energy has she put into the band? Give the energy in joules and pay attention to units.

(c) How much work has she done by stretching it?

(d) How much *more* work must she do to stretch the band another 6 in.?

(e) How much force will she be exerting when the band is stretched to its new length of 6 ft?

8. Hand grips are springlike contraptions used to build muscle on the forearm. While the springs in these devices are often designed to provide approximately constant resistance, they nevertheless follow Hooke's law to some extent, and we will assume a Hooke spring here.

The handless are 4 in. apart, and the athlete compresses the device in one hand. By squeezing with a force of 10 lb, an athlete finds he compresses the device by 2 in.

(a) How much energy has he put into the spring?

(b) How much force will he have to exert to compress the spring another inch?

A woman works out with a resistance band. (© *McGraw-Hill Education/David Moyer, photographer*)

9. In youth hockey, a typical stick may have a flex rating of 50, meaning Hooke's constant is 50 lb/in.

(a) The athlete holds one end, and the other is on the ice. To flex the stick by 3 in. (see figure 8.11), how much force must he exert with the hand pressing the middle of the stick?

(b) How much force is he exerting with the hand holding the *end* of the stick?

(c) How much work has he done in bending the stick? Give your answer in joules.

10. A bow's draw weight is 40 lb and its draw length is 2 ft.

(a) How much energy is stored in the bow when drawn to its full draw length?

(b) How much force is required to draw it to only one-half of its draw length (1 ft)?

(c) How much energy is stored in the bow when drawn to only one-half of its draw length? Put your answer in joules and pay attention to units.

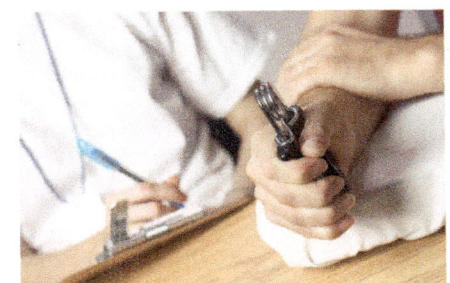

A hand grip is used to evaluate forearm and hand strength. (© *Ingram Publishing/agefotostock RF*)

The Physics of Cycling

Cycling perfectly couples two amazing machines. Long-term power generation of the human engine in its aerobic mode is quantified through new terms such as the VO$_2$max and MET factors. The bike is designed to convert this power efficiently to maximize speed and overcome air drag, hills, and friction. Understanding the details of this seemingly simple machine means learning the physics of rotational motion and torque. There is a *lot* of physics in cycling!

Figure 9.1. Then... Are these guys nuts? Don't they know it's dangerous for your health not to wear a helmet? 1920's-era Tour de France. (© *PresseSports*).

Figure 9.2. ... and now. American cyclist Andrew Talansky in the 2013 Tour de France, almost a century later. (© *Eric Gaillard/Reuters/Corbis*).

People do not "drive" cars, they steer them.
People do not "ride" bicycles, they drive them.

—Unknown

The Physics of Cycling

We've come a long way, baby

Begun in 1903 and held annually except for 11 years during the World Wars, the Tour de France is one of the most iconic sporting events in the world. Over the course of more than 100 Tours, sacred traditions have developed and give the event a familiar feel.

However, so much has changed. The riders at the lead of the peloton in figure 9.1 ride single-gear bikes on wood-rimmed wheels, carrying spare tubes over their shoulders. In figure 9.2, Andrew Talansky rides on carbon fiber and has 22 gears at his disposal; a support car carries spare wheels and bikes, should he have a problem.

The early riders' clothing seems snug, but it is hardly the highly engineered and wind-tunnel-honed racing gear Talansky wore as he placed in the top 10 finishers in his first Tour. Their bikes weighed upward of 25 lb; Talaksy chooses an extrastiff front gear to ensure his bike tips the scale just barely above the 6.8-kg (~15-lb) minimum allowed weight.

More than the bikes have changed, of course. Our knowledge of how the body works and how power is generated and maintained for hours has grown tremendously. Riders regularly check their VO_2max (we'll come to this) and have teams of nutritionists to finely tune their power-to-weight ratio (we'll come to this, too). Smoking is definitely frowned upon.

In short, cycling couples in an ideal way two of the most complex and amazing machines in the world. The science of the body and of the bike is broad and rich. We will explore in detail (1) how the body generates long-term endurance power and how to quantify it; (2) where that power is spent; and (3) the efficiency and details of the machine that convert rotational motion of the pedals into translational ("straight-line") motion of the bike and rider.

Keeping things straight

In cycling as in all sports, energy is constantly being converted from one form to another, and it's easy to get confused about what are the "input" power and "output" power being discussed. Equation 7.16 related the mechanical energy output by the athlete to the chemical energy input to his system in the form of food and oxygen.

We can express this relationship in a pseudo-equation like this:

$$\bar{P}_{\text{chemical input}} \xrightarrow{\eta_{\text{chemical}\to\text{mechanical energy}}} \underbrace{\bar{P}_{\text{output by cyclist}}}_{\text{this is power \textbf{input} to bike that goes to fight air drag, etc.}}.$$

or in proper equation form:

$$\boxed{\bar{P}_{\text{output by cyclist}} = \eta_{\text{chemical}\to\text{mechanical energy}} \times \bar{P}_{\text{chemical input}}} \quad (9.1)$$

The point I'm trying to make clear is that the mechanical energy output by the athlete is the energy that goes "in" to fighting air drag and hills. This might be totally obvious to you, which is great; it's not a super-subtle point. But sometimes it helps to make this point explicit at the outset.

> **Reminder: power, work, and energy**
>
> In long-duration sports, we will often be talking about the power involved, that is, the *rate* of energy converted or the rate of work done. But sometimes we want to know the total *amount* of energy converted or work done, in some period of time. We already discussed this in section 7.5.2, but I'll just repeat the relationship here.
>
> The mechanical work done in some amount of time Δt is
>
> $$\boxed{\text{Work done by cyclist} = W_{\text{cyclist}} = \bar{P}_{\text{output by cyclist}} \cdot \Delta t} \qquad (9.2)$$
>
> and the amount of chemical (food) energy converted in a time Δt is
>
> $$\boxed{\text{Food energy burned} = -\Delta E_{\text{chem}} = \bar{P}_{\text{chemical input}} \cdot \Delta t} \qquad (9.3)$$

Section 9.1 covers the power of sustained effort by a cyclist, and section 9.2 covers how that power is expended fighting wind, hills, and friction. Finally, section 9.3 delves into the physics of the bike itself.

9.1 Input to the Bike: Sustained Human Power

We've already talked about the chemical reactions that go into human power production. For sustained aerobic power, scientists use a variety of formulas and quantities. We'll discuss three of these, starting with MET factors used by nutritionists and medical doctors to describe athletic and nonathletic long-term human activities. Then we'll move on to the treatment of oxygen uptake and VO$_2$max, used to discuss high-power aerobic exertion. Finally, we'll briefly cover the concept of the power-to-weight ratio (PWR) that you'll hear in any conversation about competitive cycling.

9.1.1 Caloric power requirements for long-term effort

Any human activity, athletic or not, requires the consumption of food energy. Naturally, the amount of energy depends on the type and intensity of the activity as well as the size of the person doing it.

We've already talked about very short bursts of energy in chapter 7. Medical researchers and epidemiologists have developed a simple formula for the food energy consumed while performing an extended activity:

$$\boxed{\text{Food energy burned} = -\Delta E_{\text{chem}} = \left(\frac{1 \text{ Cal}}{\text{kg} \cdot \text{hr}}\right) \times \text{MET} \times m_{\text{person}} \times \Delta t.} \qquad (9.4)$$

Here, Δt is the length of time that the person was engaged in the activity. MET is the metabolic equivalent (or metabolic equivalent task (MET)), a number measured for every type of human activity imaginable and published in tables in the scientific literature. Table 9.1 lists the metabolic equivalents for some activities relevant for sports. A more complete list is found in table B.2 and extensive lists can be found in nutritional journals and online. The MET concept is illustrated in examples 9.1 and 9.2.

Table 9.1. Metabolic equivalent (MET) factors and Calories burned when engaged in 1 hr of various physical activities. Values are shown for a 150-lb and 190-lb person. From B. E. Ainsworth et al., "2011 Compendium of physical activities: A second update of codes and MET values," *Medicine and Science in Sports and Exercise* 43:1575 (2011). See table B.2 for a more comprehensive list of sports-related MET values.

		Calories consumed in 1 hr	
Activity	MET	150 lb (62.8 kg)	190 lb (86.4 kg)
Lying quiet, watching TV	1.0	68	86
Walking, 2.5 mph, level firm surface	3.0	205	259
Golf, no cart	4.3	293	371
Softball, baseball	5.0	341	432
Boxing, sparring	7.8	532	674
Basketball game	8.0	545	691
Cycling, 12–14 mph, moderate effort	8.0	545	691
Ice hockey, competitive	10.0	532	864
Cycling, >20 mph racing, not drafting	15.8	1077	1365

If we want to put it in terms of power instead of energy, we can divide both sides of equation 9.4 by Δt and get

$$\bar{P}_{\text{chemical input}} = \frac{1 \text{ Cal}}{\text{hr}} \times \text{MET} \times \frac{m_{\text{person}}}{\text{kg}} = 1.16 \text{ W} \times \text{MET} \times \frac{m_{\text{person}}}{\text{kg}} \quad (9.5)$$

I probably don't need to point out the obvious, but these formulas are simply approximations that mostly work for the "average" person. They are guidelines used by nutritionists. The power consumed by a top athlete will often be a bit less for high MET activities (regular cyclists are more efficient at riding 20 mph, for example) and a bit higher for low MET activities (they may have a "high metabolism," so that they burn more calories than the average person while at rest).

EXAMPLE 9.1 *Resting metabolic rate*

In watts, how much chemical energy does an average adult male require simply to remain alive?

Well, simply surviving means MET = 1, and a 180-lb man has a mass of 82 kg, so equation 9.5 says

$$\bar{P}_{\text{chemical input}} = 1.16 \text{ W} \times 1 \times \frac{82 \text{ kg}}{\text{kg}} = 95 \text{ W}.$$

This is the resting metabolic rate (RMR) that I mentioned in section 7.5.3. Naturally, a larger person burns more and a smaller person less.

So if you are just sitting there watching the game, you are consuming chemical energy at just about the same rate that your television consumes electrical energy, since modern flat-screen TVs operate at about 100 W. All that energy winds up as heat. If the room were sealed, it would get hotter and hotter as the two of you would add 200 J of thermal energy to the room every second. (200 W = 200 J/s.)

EXAMPLE 9.2 *The FDA standard diet*

When you look at the nutrition panels on packaged foods, you see caloric content "based on a 2000-Calorie diet." Will such a diet cause you to gain weight? If so, how much exercise must you do to maintain weight with the FDA standard diet?

Consider the 150-lb (68.2-kg) man who lies down watching TV all day. Over the course of a day, how much food energy does he burn, simply to keep breathing, maintain consciousness, and sustain body temperature?

$$\text{food energy burned} = \left(\frac{1 \text{ Cal}}{\text{kg} \cdot \text{hr}}\right) \times 1 \times (68.2 \text{ kg}) \times \underbrace{(24 \text{ hr})}_{\text{watch units!}} = 1637 \text{ Cal.}$$

If our friend consumes the standard 2000 Cal, he won't be 150 lb for long! He'll immediately begin gaining weight.

Okay then, how much must he exercise to maintain his weight? If just 1.5 hr of his day (not necessarily all at once) is spent walking around at 2.5 mph, and he hops on a stationary bike for a light 0.5 hr (he can keep watching TV while he pedals!), then he becomes essentially an FDA poster child:

$$\text{food energy burned} = \underbrace{\left(\frac{1 \text{ Cal}}{\text{kg} \cdot \text{hr}}\right) \times 1 \times (68.2 \text{ kg}) \times (22 \text{ hr})}_{\text{watching TV}}$$

$$+ \underbrace{\left(\frac{1 \text{ Cal}}{\text{kg} \cdot \text{hr}}\right) \times 3 \times (68.2 \text{ kg}) \times (1.5 \text{ hr})}_{\text{walking 2.5 mph}}$$

$$+ \underbrace{\left(\frac{1 \text{ Cal}}{\text{kg} \cdot \text{hr}}\right) \times 4.8 \times (68.2 \text{ kg}) \times (0.5 \text{ hr})}_{\text{light stationary bike}}$$

$$= 1500 \text{ Cal} + 307 \text{ Cal} + 164 \text{ Cal} = 1971 \text{ Cal}$$

9.1.2 Oxygen uptake, VO₂max, and power

If you read or watch any discussion of sustained athletic performance, you'll surely hear the term $VO_2\text{max}$. There is a lot of confusion as to what this term really means or what it measures. This is partly due to incomplete or incorrect treatment by the experts and partly due to the many different units that people use. Finally, there are so many related terms: VO_2, $VO_2\text{max}$, RMR (from sections 7.5.3 and 9.1.1), MET (section 9.1.1), and PWR (section 9.1.3). Several or all of these terms may be used in a serious sports article.

Let's see if we can make sense of this.

What is VO₂?

Sustained power production is possible only through aerobic reactions, as we discussed in section 7.5.3. Both fat and carbohydrates can be processed aerobically, but for most high-power endurance events such as cycling, we are interested in burning carbohydrates. I gave you the important reaction in equation 7.5, but I'll put it here again:

$$\underbrace{C_6H_{12}O_6}_{\text{eat}} + \underbrace{6O_2}_{\text{inhale}} \rightarrow \underbrace{6H_2O}_{\text{sweat}} + \underbrace{6CO_2}_{\text{exhale}} + 686 \text{ Cal.}$$

Let's assume that you have eaten and stored enough carbohydrate, and it's just sitting there ready to burn. Then the only issue is getting enough oxygen. The more quickly you can absorb oxygen to put into the left side of the reaction, the more quickly you can produce the energy on the right side. That is, the rate at which you uptake oxygen (this is called VO_2) determines the rate at which you burn food Calories (which is $\bar{P}_{\text{chemical input}}$).

We already said that 6 mol of oxygen releases 686 Cal of energy. The "V" in VO_2 means "volume," so usually we talk about liters, rather than moles, of oxygen. It turns out that 1 liter of pure O_2 releases 5.1 Cal.

Figure 9.3 illustrates how VO_2 is measured. The athlete does work on a bicycle while the gases that he inhales and exhales are analyzed to determine the rate at which he is absorbing oxygen. As he is asked to generate more output power (by pedaling harder or faster), he needs to generate more input power by combining the oxygen he absorbs with the food he has stored up. His VO_2 increases as he works harder to produce the needed power.

Figure 9.4 shows results from a test to find the best pilot position for a 4-hr flight of the (pedal-powered) Daedalus aircraft. The test showed no significant difference between pedaling in the upright ("traditional") cycling position and a recumbent (partially lying down) position. The data show nicely the clear relationship between power input, as measured by VO_2 on the y-axis, and power output, measured on the x-axis.

The rate at which oxygen is absorbed by the body (VO_2) determines the rate at which food Calories are burned. Absorbing 1 liter of pure O_2 burns 5.1 Cal.

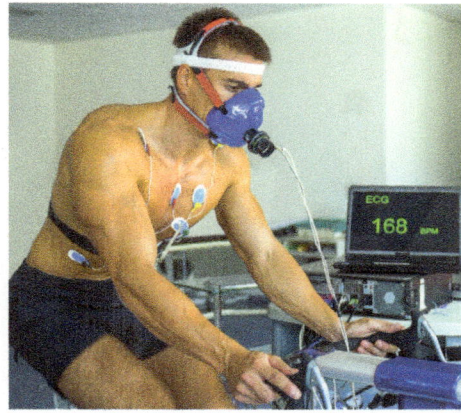

Figure 9.3
A cyclist's oxygen uptake is measured by analyzing the gases inhaled and exhaled, as a function of his power output. (© *technotr/Getty Images RF*).

Formulas and units

Here's one common point of confusion: Sometimes VO_2 is given in units of liters per minute (liter/min), and sometimes it is given in milliliters per kilogram-minute (ml/(kg · min)). If you see it in the second set of units, you have to multiply by the mass of the person to get the rate of oxygen uptake.[1] Both are called VO_2. It can be confusing, but there it is.

In any discussion of VO_2, pay careful attention to the units, as different conventions are in common use.

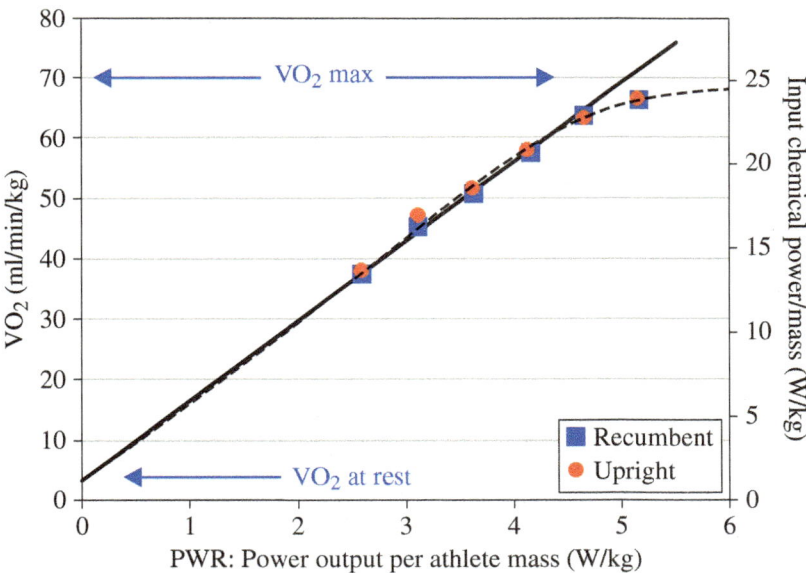

Figure 9.4. Oxygen uptake is measured as the power output by a cyclist is varied, for two different cycling postures. The vertical axis on the right side is obtained by converting the VO_2 value to input chemical power with equation 9.7. On the x-axis is the output power divided by the athlete's mass, that is, the power-to-weight ratio (PWR). Data from S. R. Bussolari and E. R. Nadel, *Human Power* 7, no. 4 (1989).

---

[1] While it's confusing to have two definitions of VO_2, it actually makes sense to look at the rate of oxygen uptake *per kilogram of body mass*, when comparing the fitness of two athletes. After all, a heavier person needs more oxygen simply to support his bulk.

The following formulas convert the oxygen uptake rate to the chemical power released:

$$\bar{P}_{\text{chemical input}} = \left(306 \, \frac{\text{Cal} \cdot \text{min}}{\text{hr} \cdot \text{liter}}\right) \times \text{VO}_2 \qquad \left(\text{if VO}_2 \text{ is in units of } \frac{\text{liter}}{\text{min}}\right)$$
(9.6)

$$\bar{P}_{\text{chemical input}} = \left(0.306 \, \frac{\text{Cal} \cdot \text{min}}{\text{hr} \cdot \text{ml}}\right) \times m_{\text{athlete}} \times \text{VO}_2$$

$$\left(\text{if VO}_2 \text{ is in units of } \frac{\text{ml}}{\text{min} \cdot \text{kg}}\right) \qquad (9.7)$$

These two equations will give the chemical power in the units of Calories per hour (Cal/hr). If we want it in watts or horsepower or anything else, we'll need to convert; conversion factors are given in appendix A. Sorry for the nightmare of the units, but this is what you find if you look at any scientifically oriented articles in *ESPN Magazine* or *Sports Illustrated* or watch a sports science analysis on TV.

The low end: RMR in terms of oxygen uptake

Let's see if we can make a connection to something we said before. In section 9.1.1, we said that a person at rest (MET = 1) burned about 1 Cal/hr for every 1 kg of mass in his body; that was the resting metabolism rate, or RMR. Well, to burn food calories, we have to breathe. As I said above, 1 liter (1000 ml) of oxygen releases 5.1 Cal of energy. So the VO_2 for a person at rest is

$$(\text{VO}_2)_{\text{at rest}} = \left(1 \, \frac{1 \, \text{Cal}}{\text{kg} \cdot \text{hr}}\right) \times \left(\frac{1000 \, \text{ml}}{5.1 \, \text{Cal}}\right) \times \left(\frac{1 \, \text{hr}}{60 \, \text{min}}\right)$$

$$= 3.27 \, \frac{\text{ml}}{\text{min} \cdot \text{kg}}$$

This value is indicated by an arrow on figure 9.4.

The high end: VO$_2$max

The more output mechanical power we generate, the more input chemical power we need, so we increase our oxygen uptake. There is a limit, however. VO$_2$max is the maximum rate at which an athlete can absorb oxygen. In a measurement like the one shown in figure 9.4, the oxygen uptake starts leveling off, even if the mechanical power output increases. This can only mean that anaerobic processes are kicking in, and therefore the effort cannot be sustained for very long.

For the cyclist measured in the experiment of figure 9.4, VO$_2$max is somewhere around 70 ml/(min · kg). (To determine it more accurately, an additional data point at a higher output power would be helpful.) If we look at the data listed in table 9.2, it's clear that the candidate Daedalus pilot is very fit.

An athlete's VO$_2$max is partially determined by the volume of blood his heart can pump and the concentration of his red blood cells. Red-cell concentration can be increased by altitude training, blood doping, or using the hormone erythropoietin (EPO). The latter two methods, of course, are illegal in competitive cycling. EPO use was the scourge of professional cycling in the 1990s. Let's hope that scourge is behind us!

Table 9.2. Approximate reported VO$_2$max values for animals, top athletes, and the rest of us. Check out the horse: 180 ml/(kg · min) means a 600-kg animal can absorb about 100 liters (about 26 gal) of oxygen per minute!

	VO$_2$max (ml/(min · kg))
Iditarod sled dogs	240
Trained healthy racehorse	180
Untrained healthy horse	120
Bjørn Daehlie, cross-country skier (highest human value reported)	~92
Miguel Indurain, Tour de France winner, 1991–1995	88
Steve Prefontaine, mile in 3:54.6	84.4
Bill Rodgers, marathon 2:09:27	78.5
Sebastian Coe, Olympic 1500-m Gold medalist, 1980 and 1984	77
Average male, aged 20–24	44–50
Average female, aged 20–24	37–41
Average male, aged 50–54	31–34
Average female, aged 50–54	26–29

How much *air* must a human breathe?

We've covered a lot in this section, but I want to make one more important point. We've talked about the volume of oxygen *absorbed* by the athlete. But to absorb 1 liter of oxygen, we have to breathe a lot more than 1 liter of air.

The air we breathe is only about 20% oxygen, so to pull 1 liter of oxygen into our lungs, we have to suck in 5 liters of air. And even then, we only *absorb* about 30% of the oxygen that we inhale,[2] so to absorb a 1 liter of oxygen, we have to inhale about 16.7 liters (4.4 gal) of fresh air. Example 9.3 carries this analysis to the next step, relating calories burned to breaths taken.

A human must inhale about 16.7 liters of fresh air in order to absorb 1 liter of oxygen.

EXAMPLE 9.3 *How many breaths are needed to burn 15 Cal?*

To burn 10 Cal of food energy, how many breaths are needed?

We said above that it takes 1 liter of oxygen to burn 5.1 Calories, so the volume of *oxygen* required to burn 15 Cal is

$$\text{volume of oxygen to burn 15 Cal} = \frac{1 \text{ liter of O}_2}{5.1 \text{ Cal}} \times (15 \text{ Cal}) = 2.94 \text{ liters of O}_2$$

We also said that to absorb 1 liter of O$_2$ requires inhaling 16.7 liters of fresh air, so the volume of *air* required is

$$\text{volume of air to burn 15 Cal} = \frac{16.7 \text{ liters air inhaled}}{1 \text{ liter O}_2 \text{ absorbed}} \times \left(2.94 \text{ liters of O}_2\right)$$

$$= 49 \text{ liters of air}$$

Finally, we need to know that with relaxed breathing, we inhale about $\frac{1}{2}$ liter of fresh air each time we inhale. (This is called the "tidal volume," and of course it varies with effort.) So the number of breaths we need is

$$\text{\#breaths} = \frac{1 \text{ breath}}{0.5 \text{ liter of air}} \times 49 \text{ liters of air} = 98$$

---

[2] These numbers vary by person and exertion level, of course.

9.1.3 Power-to-weight ratio

At the end of the day, it's not how much oxygen you breathe and the number of food calories that you burn. Well, it is, if you are looking just to lose weight. But in a competition, it's all about the power you can put out, to move yourself forward.

If you watch a cycling competition on TV, or read about one in articles, you'll eventually hear reference to an athlete's "power-to-weight ratio," or PWR. This is more or less what it sounds like, except that it is usually expressed in watts per kilogram (W/kg), so would more properly be called the power-to-mass ratio.

As we said, VO_2 is often expressed in a "per kilogram" number because of course more body mass will certainly require more oxygen. Similarly, more (muscle) mass will lead to more power output. An interesting question is often whether one athlete generates more power than another *above and beyond* the trivial fact that he has more muscle mass. For this, PWR is a good number for comparing the general fitness of different athletes. It's also relevant because while the bigger athlete might generate more power, he's also got more bulk to move forward on his bike, and to some degree those two effects "cancel." We'll discuss some details on this issue soon.

For a 30-min effort, untrained healthy males may have a PWR of about 2–3 W/kg; females, between 1.5 and 2.5 W/kg. For elite athletes, the numbers are more like 5.6–6.5 W/kg for men and 5.0–5.7 W/kg for women. For shorter efforts, naturally, the PWR will be higher, and fatigued riders show a reduced PWR.

EXAMPLE 9.4 *The efficiency of a Daedalus pilot*

The candidate to fly the pedal-powered aircraft Daedalus puts out more and more power as he increases his rate of oxygen uptake, as shown in figure 9.4. How efficiently is he converting the chemical energy from his food and breath to the mechanical power of the pedals?

If you look really closely at the figure (or if you have the data file as I do), you'll see that the last data point before the curve drops away from the line says that the input power per mass is 22.8 W/kg and the output power per mass (or the PWR) is 4.64 W/kg.

Equation 7.16 gives us a formula for the efficiency:

$$\bar{P}_{\text{work output}} = \eta_{\text{chemical}\rightarrow\text{mechanical energy}} \times \bar{P}_{\text{chemical input}}$$

$$\rightarrow \eta_{\text{chemical}\rightarrow\text{mechanical energy}} = \frac{\bar{P}_{\text{work output}}}{\bar{P}_{\text{chemical input}}}$$

$$= \frac{\bar{P}_{\text{work output}}/m_{\text{athlete}}}{\bar{P}_{\text{chemical input}}/m_{\text{athlete}}}$$

$$= \frac{4.64 \text{ W/kg}}{22.8 \text{ W/kg}}$$

$$= 0.2 = 20\%$$

The pilot's body is 20% efficient at converting food power to pedal power. This is close to the optimum efficiency for any athlete.

9.2 Power Output

Okay, so we have covered some important details regarding where the power is coming from and how chemical (food) power is converted to mechanical power output. Now we focus on that output power.

We've talked about power in terms of work, but there's another very useful formula for the power when a force[3] is exerted on a moving object:

$$P = F_x v_x + F_y v_y = |\vec{F}| \times \left(\text{the velocity component parallel to } \vec{F}\right). \quad (9.8)$$

Actually, in the cycling case, the propulsive force ($\vec{F}$) will always be in the same direction as the cyclist's velocity ($\vec{v}$), unless he's braking (which we'll talk about later). So, for us[4]

$$P_{\text{output cyclist}} = |\vec{F}_{\text{propulsive}}||\vec{v}_{\text{bike}}|. \quad (9.9)$$

The propulsive force is used to climb hills and overcome drag and rolling friction. It can also be used to accelerate the cyclist, but we're going to focus on the situation when the cyclist is moving at a constant velocity for the moment. While the speed certainly changes during a race, there are long stretches where it is reasonably constant, and we can learn a lot by considering this situation.

In that case the net force must be zero, and the propulsive force produced by the cyclist and bike must exactly cancel the opposing forces of hills, rolling resistance, and air drag. In formula form,

$$\text{at constant velocity} \quad \vec{F}_{\text{propulsive}} = -\vec{F}_{\text{hill}} - \vec{f}_R - \vec{F}_D. \quad (9.10)$$

Now we'll go through these three forces, one by one.

9.2.1 Hills

Figure 9.5 shows slopes of various "grades." The grade s is given on the yellow signs on highways; tractor trailers are warned to shift down and take extra care on slopes exceeding 6% ($s = 0.06$). The grade is the increase in vertical height you gain, divided by the distance you travel along the road (that is, along the "hypotenuse" of the hill). In trigonometric terms, s is related to the angle of the slope by

$$s = \sin \theta \quad (9.11)$$

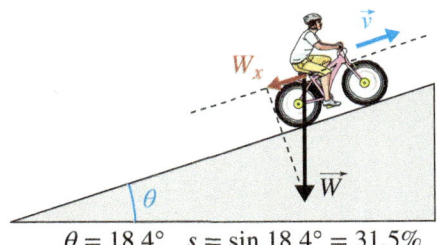

Figure 9.5
Slopes of varying grades don't look nearly as impressive on the page as they are to experience. The fraction of the rider's weight that opposes his forward motion turns out to be the grade s. In the figure, $W_x = |\vec{W}| \times s$.

---

[3] Soon, we'll talk about some of the details of how that forward force is generated by the gears and so on, but for now we use a fact that we all know: when the cyclist churns the pedals, a propulsive force is generated on him and his bike.
[4] There is actually another efficiency that we are ignoring in this book. The "transmission efficiency" of a bike takes into account energy loss due to internal friction, chain stretching, and so on. However, the transmission efficiency of a high-quality bike is greater than 97%, so we'll treat it as 100% and ignore any such losses altogether.

The part of the gravitational force that points along the road is what we're going to call $\vec{F}_{\text{hill}}$. This is the force that opposes you when you go up a hill (or helps when you come down one). It is related to the grade by this formula:

$$\boxed{|\vec{F}_{\text{hill}}| = |\vec{W}| \times s = mgs} \qquad (9.12)$$

So if you are on flat ground, $s = 0\%$ and you are fighting 0% of your weight (that is, none at all). A vertical wall is a 100% grade hill ($s = 1$). If you rode with tires that could stick to the wall, 100% of your weight would oppose your "forward" progress; that is, $|\vec{F}_{\text{hill}}| = |\vec{W}|$.

When you use equation 9.12, keep in mind that the weight (or mass) has to include the bike, which is required to be at least 15 lb (6.8 kg).

The hills represented in the left of figure 9.5 don't look so bad, do they? Well, ask a truck driver or hop on a bike. You will absolutely notice a 4% grade immediately, and a 9% grade can be downright treacherous coming down. I have twice climbed a 14% grade for almost a mile and brag about it to anyone who will listen. (You are listening, right?) There is one block on Filbert Street in San Francisco that has a grade of 31.5%.

The Alpe d'Huez is a legendary slope that often features in the Tour de France. Its average grade is about 8%, though the initial ascent welcoming riders is more than 10%. Example 9.5 explores the forces and power requirements imposed by this slope.

> An downhill slope has a negative value of s.

EXAMPLE 9.5 *Mother Earth tugs Talansky on the Alpe d'Huez*

Top-end cyclists are lean dudes. Often quite short in stature, almost zero fat, powerful legs but typically little upper-body mass, which is essentially "dead weight" to lug up hills and accelerate forward.

At 140 lb (63 kg) Andrew Talansky is a world-class climber and one of the new generation of American cyclists working to bring to an end to the era of EPO doping that stained an otherwise inspiring sport. When he climbs an 8% section of the Alpe d'Huez, gravity is fighting him with a force of only 0.08 × (140 lb + 15 lb) = 12.4 lb. (The extra 15 lb there is the weight of the bike.)

Only 12.4 lb? Try fighting a pull like that for 40 min, and then let me know if "only" is the right word to use. And keep in mind that if you weigh more than 140 lb, the opposing force will be proportionally stronger.

Talansky climbed the 8.6-mi (13.8-km) monster in just about 50 min. In fact, for the first time in Tour history, riders in 2013 climbed the Alpe d'Huez *twice*!

On average, how much power did Talansky have to put out?

First, let's figure out Talansky's average velocity:

$$|\bar{v}_{\text{bike}}| = \frac{\Delta x}{\Delta t} = \frac{13.8 \text{ m} \times 1000 \text{ m/km}}{50 \text{ min} \times 60 \text{ s/min}} = 4.6 \text{ m/s}$$

This is about 10 mph, so wind resistance isn't the major obstacle in such a climb. We'll worry only about gravity in this example.

In order to keep his speed, the bike needs a propulsive force that cancels $\vec{F}_{\text{hill}}$, which we find by equation 9.12.

$$|\vec{F}_{\text{propulsive}}| = |\vec{F}_{\text{hill}}| = mgs = (63 \text{ kg} + 6.8 \text{ kg})(9.8 \text{ m/s}^2)(0.08) = 54.7 \text{ N}.$$

Now we have what we need to calculate the power Talansky is putting out, using equation 9.9:

$$\bar{P}_{\text{output cyclist}} = |\vec{F}_{\text{propulsive}}||\vec{v}_{\text{bike}}| = (54.7 \text{ N}) \times (4.6 \text{ m/s}) = 252 \text{ W}$$

Putting out 252 W of power for 50 min is beyond the limits of a "healthy man," as determined by NASA. (Check out figure 7.10.) And NASA's healthy man doesn't need to keep riding a Tour stage, including a second climb!!

How many food Calories did he burn on each climb?

Climbing a slope like the Alpe d'Huez involves a lot of hard pushing and some out-of-the-saddle cycling. The efficiency of such activity can't be more than 20%. (Remember, 25% is what's possible under optimum circumstances.) So equation 7.16 tells us the rate at which he's consuming chemical energy:

$$\bar{P}_{\text{work output}} = \eta_{\text{chemical} \to \text{mechanical energy}} \times \bar{P}_{\text{chemical input}}$$

$$\to \bar{P}_{\text{chemical input}} = \frac{\bar{P}_{\text{work output}}}{\eta_{\text{chemical} \to \text{mechanical energy}}}$$

$$= \frac{252 \text{ W}}{0.2} = 1260 \text{ W} \times \underbrace{\frac{0.86 \text{ Cal/hr}}{1 \text{ W}}}_{\text{unit conversion}} = 1084 \text{ Cal/hr}$$

To this we should add the resting metabolic rate (RMR) of 63 Cal/hr for a 63-kg man (see section 9.1.1), bringing the input power to 1147 Cal/hr.

Finally, we can see the total energy, in food Calories, that Talansky burned on one climb:

$$\Delta E = \bar{P} \times \Delta t = 1147 \, \frac{\text{Cal}}{\text{hr}} \times \left(50 \text{ min} \times \frac{1 \text{ hr}}{60 \text{ min}}\right) = 956 \text{ Cal}$$

In a single climb, Talansky burned as many Calories as most of us do in half a day.

In this example, we just considered gravity. If we include the effects of rolling friction and air drag that we discuss in the following sections, the power Talansky has to generate is closer to 295 W than 252 W. So gravity is about 85% of the story (252/295 = 85%) on a climb like this.

9.2.2 Rolling resistance

In chapter 3, we talked about kinetic friction, where one surface slides over another, and static friction, where one surface "tries to slide" over another. Both were proportional to the normal force pressing the surfaces together, and both opposed the motion (or "attempted motion").

A very similar force, usually called rolling friction, opposes the motion of a rolling wheel. When we discuss torques shortly, we're going to come back to some details of rolling friction, but for the moment I'll just give the formula for its magnitude:

$$\boxed{|\vec{f}_R| = C_R \times |\vec{F}_N| \approx C_R \times |\vec{W}| = C_R mg} \qquad (9.13)$$

> **Low tire**
>
> On a multiday bike trip with my young son, I found myself exasperated by the boy's sudden slowness one morning. Neither of us could understand what had happened, until after a period of increasing muttering on my part, my son checked and found his rear tire very low on air, even though it didn't look nearly "flat" to the eye. (To be fair, Dad hadn't yet had his morning coffee.)
>
> We didn't measure it, but based on figure 9.6 and considering that his tire was a child's bike fat tire rather than a racer, C_R must have been 0.05.
>
> Now, take a look at equations 9.12 and 9.13. They have exactly the same form—the strength of the force in each case is some number (C_R or s) times the weight of the bike and rider. That means if $C_R = 0.05$ on my son's bike, then asking him to ride was the same as asking him to climb a 5% grade hill!
>
> Dad is occasionally reminded of that morning, even today.

While the magnitude of the normal force is a bit less than the weight on a slope, we'll always use the approximation that $|\vec{F}_N| = |\vec{W}|$ because it's simpler and usually the angle of the slope isn't that big. We call C_R the **coefficient of rolling friction**.[5]

Like kinetic friction, rolling friction converts kinetic energy directly into heat. In this case, internal forces heat the tire as the rubber compresses and decompresses. The less that a tires compresses, the less rolling friction there is.

For that reason, a "hard" wheel that stays round due to high inflation rolls more easily. Figure 9.6 shows rolling friction coefficients for various tires as a function of inflation pressure. High-pressure modern racing tires can have coefficients as low as $C_R \approx 0.004$. With wheels (and rails) that barely deform at all under load, trains can have coefficients as low as $C_R \approx 0.001$.

On the other hand, for automobile tires on concrete roads the coefficient is about 10 times higher: $C_R \approx 0.015$.

9.2.3 Wind drag

Unless he is climbing Alpe d'Huez or riding on a flat tire, the cyclist soon realizes that the biggest obstacle to progress is wind drag. We discussed drag in some detail in chapter 5. Its main features were captured in equation 5.4, copied here:

$$|\vec{F}_D| = \frac{C_D}{2} A \rho_{air} v^2. \tag{5.4}$$

To get maximum speed but low drag force, the cyclist wants to minimize the drag coefficient C_D and the area A facing into the wind. Figure 9.7 shows some machines and postures available to the rider, and table 9.3 lists some parameters for each one. Examples 9.6, 9.7, and 9.8 use these numbers to explore the power and caloric requirements of commuting by bicycle.

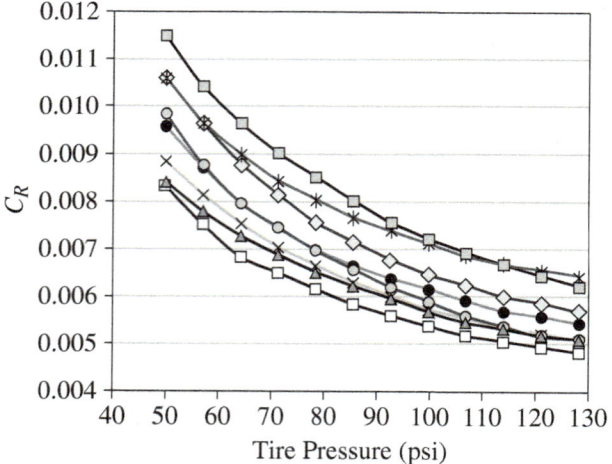

Figure 9.6. The coefficient of rolling resistance as a function of tire inflation for several racing tires. Based on data kindly provided by Jobst Brandt.

---

[5] It's unfortunate that the symbol μ isn't used for rolling friction, as it is for kinetic or static friction.

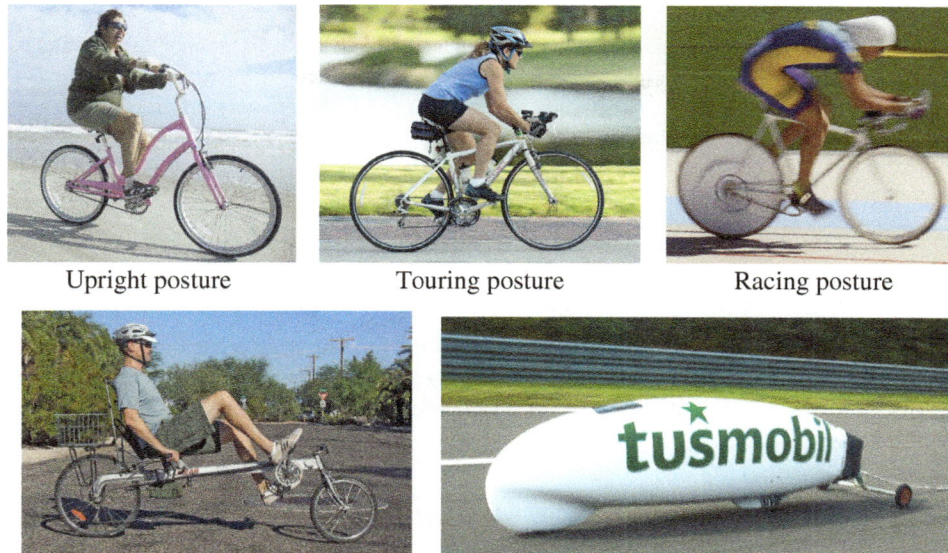

Upright posture   Touring posture   Racing posture

Unfaired recumbent   Faired recumbent

Figure 9.7. Different bicycles and postures lead to very different power requirements from the cyclist, listed in table 9.3. **Top Left:** (© *Granger Wootz/Blend images/Getty Images RF*); **Top Middle:** (© *McGraw-Hill Education/Aaron Roeth, photographer*); **Top Right:** (© *Photodisc/Getty Images RF*); **Bottom Left:** (© *McGraw-Hill Education/David Moyer, photographer*); **Bottom Right:** (© *Thomas Peter/Reuters/Corbis*).

Table 9.3. Different bicycles and postures (illustrated in figure 9.7) can lead to tremendous differences in the power needed to overcome rolling friction and especially air drag. The velocity used in this table is 9 m/s (20 mph). Values for C_D, A, and C_R from David Wilson, *Bicycling Science*, 3rd edition. The power to overcome air drag is given by $\frac{1}{2} C_D \rho_{air} A v^3$, and to overcome rolling friction by $mg C_R v$.

Bike and rider	Power to overcome air drag			Power to overcome rolling friction			Total power
	C_D	A (m²)	P(W)	m (kg)	C_R	P(W)	P(W)
Upright commuting bike	1.15	0.55	277	90	0.0060	48	325
Road bike, touring posture	1.0	0.40	175	86	0.0045	34	209
Racing bike, crouched posture and tight clothes	0.88	0.36	139	81	0.0030	21	160
Unfaired recumbent	0.77	0.35	118	90	0.0045	36	154
Faired recumbent	0.12	0.48	25	95	0.0045	38	63
Automobile	0.3	1.9	250	1,000	0.017	1,500	1,750
Passenger train	1.8	10	7,870	1,500,000	0.0015	198,500	206,370

EXAMPLE 9.6 *Bike commuting—"Should I buy a better bike?"*

Every once in a while a person, interested in beginning to commute via bicycle for environmental or health reasons, approaches me. He often wonders whether it is worth purchasing a road bike, rather than using the trusty around-town upright bike collecting dust in the garage. Let's look at what demands that puts on our potential commuter.

A typical commute to work in Columbus, Ohio, is about 10 mi along flat terrain. For bike commuting to be a viable option, the ride shouldn't take much more than about 0.5 hr.

This means the commuter should go about 20 mph, or 9 m/s, which is a bit on the fast side, but let's use it as a nice round number.

The upright and road bikes will lead to very different drag forces, using the C_D and A values from table 9.3:

$$|\vec{F}_D| = \tfrac{1}{2} C_D A \rho_{\text{air}} v^2$$

upright bike: $\quad |\vec{F}_D| = \tfrac{1}{2}(1.15)(0.55 \text{ m}^2)(1.2 \text{ kg/m}^3)(9 \text{ m/s})^2 = 30.74 \text{ N} = 6.9 \text{ lb}$

road bike: $\quad |\vec{F}_D| = \tfrac{1}{2}(1.0)(0.40 \text{ m}^2)(1.2 \text{ kg/m}^3)(9 \text{ m/s})^2 = 19.44 \text{ N} = 4.4 \text{ lb}$

An around-town bike is a little bit heavier than a road bike, and more importantly, it has softer tires that have a larger rolling friction coefficient. The force of rolling friction on the two bikes, again referring to table 9.3, is

$$|\vec{f}_R| = C_R m g$$

upright bike: $\quad |\vec{f}_R| = 0.0060\,(90 \text{ kg})(9.8 \text{ m/s}^2) = 5.29 \text{ N} = 1.2 \text{ lb}$

road bike: $\quad |\vec{f}_R| = 0.0045\,(86 \text{ kg})(9.8 \text{ m/s}^2) = 3.79 \text{ N} = 0.85 \text{ lb}$

Now we can tell how much power he would need to put out, to overcome these forces while maintaining a speed of 20 mph.

$$P_{\text{output cyclist}} = |\vec{F}_{\text{propulsive}}||\vec{v}_{\text{bike}}| = \left(|\vec{F}_D + \vec{f}_R|\right)|\vec{v}_{\text{bike}}|$$

upright bike: $\quad P = (30.74 \text{ N} + 5.29 \text{ N})(9 \text{ m/s}) = 325 \text{ W}$

road bike: $\quad P = (19.44 \text{ N} + 3.79 \text{ N})(9 \text{ m/s}) = 209 \text{ W}$

Well okay, so what? Does it really matter? Flip back to figure 7.10. Come on, go ahead and do it; it's on page 209.

On that figure we find that NASA's determination for "healthy men" is that they can maintain a power output of 325 W for about...1 min! No wonder you don't see too many around-town bikes traveling at 20 mph! On the other hand, 209 W can be maintained for about 1 hr. The 0.5-hr commute into work will be a vigorous way to start the day, but hardly something that will make a healthy man drop.

So, my answer to my curious friend is that, yes, it is probably worth picking up an inexpensive road bike. Not a carbon-fiber racer, just something with thin tires on which he can lean comfortably forward on the "hoods" of the handle bars, as shown in the second panel of figure 9.7. A person just considering bike commuting shouldn't spend a lot of money on something he or she may give up after a couple of tries. But if an untrained person commutes on an around-town bike, he or she will find that it takes much longer than expected, or the person will be exhausted by the effort, or both. Very likely this person will quickly conclude that bicycle commuting is simply impractical. That is a shame.

EXAMPLE 9.7 *Calories burned during a commute*

In example 9.6, a commuter considered riding his bike in to work. If he rode a road bike, he had to output 209 W of power to maintain a brisk 20-mph pace. How many food calories did he burn on his ride to work?

You know that energy equals power times time, or duration of effort. The issue now is to keep track of which power we're talking about, efficiencies and units. The equations we'll need are 7.15 and 7.16.

Starting with equation 7.16, we can get the input power (the rate at which food energy is converted) from the output power (209 W). We'll also need to assume a reasonable efficiency for the rider. Equation 7.7 tells us that it will be less than about 25%, and in example 9.4 we saw a case where it was 20% (0.2). We'll use that number:

$$\bar{P}_{\text{work output}} = \eta_{\text{chemical} \to \text{mechanical energy}} \times \bar{P}_{\text{chemical input}}$$

$$\to \bar{P}_{\text{chemical input}} = \frac{\bar{P}_{\text{work output}}}{\eta_{\text{chemical} \to \text{mechanical energy}}}$$

$$= \frac{209 \text{ W}}{0.2} = 1045 \text{ W}.$$

Now we turn to equation 7.15 to find the food energy burned in his 30-min ride, being careful about units:

$$\bar{P}_{\text{chemical input}} = \frac{\text{food energy burned}}{\Delta t}$$

$$\to \text{food energy burned} = \bar{P}_{\text{chemical input}} \Delta t$$

$$= \left(1045 \, \frac{\text{J}}{\text{s}}\right)\left(30 \, \text{min} \times \frac{60 \, \text{s}}{\text{min}}\right)$$

$$= 1{,}881{,}000 \, \text{J} \times \frac{1 \text{ Cal}}{4184 \text{ J}}$$

$$= 450 \text{ Cal}.$$

EXAMPLE 9.8 *MET for commuting*

Every activity has its own MET value. What is it for the commuter? Equation 9.5 lets us figure this out, since we know the chemical input power from example 9.7, and we'll just assume his mass is 80 kg:

$$\bar{P}_{\text{chemical input}} = \frac{1 \text{ Cal}}{\text{hr}} \times \text{MET} \times \frac{m_{\text{person}}}{\text{kg}} = 1.16 \text{ W} \times \text{MET} \times \frac{m_{\text{person}}}{\text{kg}}$$

$$\to \text{MET} = \frac{\bar{P}_{\text{chemical input}}}{1.16 \text{ W} \times m_{\text{person}}/\text{kg}} = \frac{1045 \text{ W}}{1.16 \text{ W} \times 80} = 11.3.$$

Headwinds, tailwinds

Now, you intuitively know that it's not really how fast you are going that determines the air's force on you; it's how fast you are going *relative to the wind,* or the "velocity of the wind on your face." Figure 9.8 sketches what you already know. If the wind is blowing to the right ($v_{\text{wind},x} > 0$), then it's much easier to ride to the right. The velocity of the air on your face is

$$\underbrace{\vec{v}_{\text{air on rider}}}_{\substack{\text{velocity of air molecules} \\ \text{as seen by rider}}} = \underbrace{\vec{v}_{\text{wind}}}_{\substack{\text{wind velocity as} \\ \text{seen by a flagpole}}} - \vec{v}_{\text{bike}} \quad (9.14)$$

So if there's wind, we have to take this into account when we think about air drag. First, the *direction* of the drag force will be in the direction of the air relative

Air drag depends on your velocity *relative* to the wind velocity.

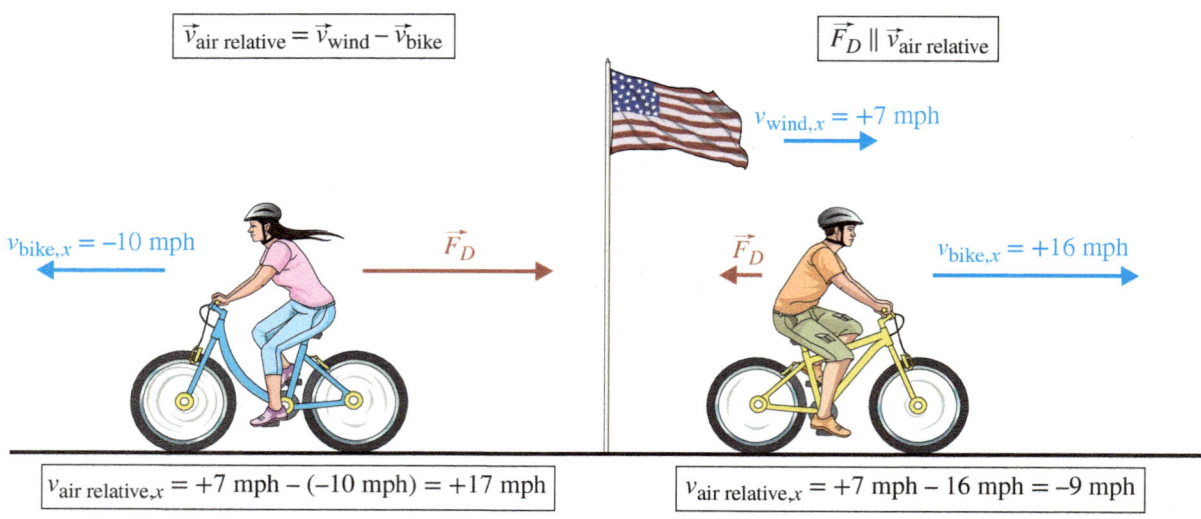

Figure 9.8. Both the *direction* and the *magnitude* of the drag force depend on the velocity of the air relative to the rider. The cyclist riding against the wind experiences far more drag than the cyclist riding with it, because the air molecules slam into her face at a much higher speed.

to the rider ($\vec{v}_{\text{air on rider}}$), not necessarily "opposite to the object's velocity" as before.[6] Second, the formula for the magnitude of the force is

$$|\vec{F}_D| = \frac{C_D}{2} A \rho_{\text{air}} |\vec{v}_{\text{bike}} - \vec{v}_{\text{wind}}|^2 \tag{9.15}$$

9.2.4 The bicycle power equation

Having covered the forces of hills, friction, and drag, we can write a complete equation that relates the power put out by the cyclist to his speed:

$$\bar{P}_{\text{output cyclist}} = \underbrace{C_R m g v_{\text{bike},x}}_{\substack{\text{overcoming} \\ \text{rolling friction}}} + \underbrace{s m g v_{\text{bike},x}}_{\substack{\text{overcoming} \\ \text{hills}}}$$

$$\pm \underbrace{\tfrac{1}{2} C_D A \rho_{\text{air}} \left(v_{\text{bike},x} - v_{\text{wind},x}\right)^2 v_{\text{bike},x}}_{\text{overcoming wind}} + \underbrace{m a_{\text{bike},x} v_{\text{bike},x}}_{\substack{\text{leftover for} \\ \text{acceleration}}}$$

$$= \left((C_R + s) m g \pm \tfrac{1}{2} C_D A \rho_{\text{air}} \left(v_{\text{bike},x} - v_{\text{wind},x}\right)^2 + m a_{\text{bike},x}\right) v_{\text{bike},x}. \tag{9.16}$$

When you are using the bicycle power equation, pay attention to the signs of the hill and wind terms!

Note that when you use equation 9.16, you have to pay attention, okay? If the rider is going *down* the hill, then of course that is going to *reduce* the amount of power he has to put out, so s should be a *negative* number, right? Also, if the wind blows in the same direction as the cyclist, but *faster* than the cyclist, then the wind-force term needs to be negative; that's why there's the "$\pm$" in front of the wind term in the formula. (Usually, the sign will be positive.)

---

[6] Of course, if there is no wind, the velocity of air relative to the rider *is* just opposite to the rider's velocity, as before.

You'll also see that I snuck in an extra term to the bicycle power equation. If the cyclist is putting out *more* power than required to overcome all the opposing forces represented by the first three terms, then he'll accelerate ($a_{bike,x} > 0$). If he puts out less than enough, he'll decelerate ($a_{bike,x} < 0$).

Using the power equation

Riders want to know: for a given course and wind condition, what is the relationship between the power I put out and the time I achieve? The power equation 9.16 is easy to use if you know the speed and you want to find the power. Just "plug and chug." Example 9.6 went through a case like this.

But more often the rider knows the power he can put out and wants to know what his speed or his time will be. That's harder, because equation 9.16 can't be rearranged to give a formula $v_{bike,x}$ = something. (Go ahead and try it. You'll see what I mean.)

This sort of thing happens all the time in science. We know the equation, but simple algebra isn't enough to solve it. So, what to do? Well, of course you can write a computer program to do just about anything, can't you? But we nonprogrammers can use the tools that people invented before computers came along. We can **solve equations iteratively**.

That sounds pretty impressive, doesn't it? It is, and it's another key tool in the scientist's toolbox. Here's how it works. If you know the output power and want to find the velocity:

1. Take your best guess for what you think the velocity is.

2. Plug your number into equation 9.16 to find the output power for that velocity.

3. Is it very close (say, within 1%) to what you know the output power to really be? If so, then you are done!

4. If not, then adjust your guess: Reduce $v_{bike,x}$ a bit if $\bar{P}_{\text{output cyclist}}$ came out too high; increase $v_{bike,x}$ a bit if $\bar{P}_{\text{output cyclist}}$ came out too low.

5. Using your new, adjusted guess, go back to step 2.

The whole thing goes much more quickly than you'd think.

> The iterative procedure shown in example 9.9 looks as if it takes a long time. However in practice, you won't write out each step with explanations, as I do in the example. With modern calculators and the ability to "backspace" to a previous calculation and plug in a new number, the whole iterative procedure should take a minute or so.

EXAMPLE 9.9 *The rarefied air of the London velodrome*

Every Olympics host city wants to be remembered, and London was no different in 2012. Built at a cost of £105 million (about $160 million), London VeloPark is a sleek, beautiful state-of-the-art facility. But the organizers wanted more than a memorable facility—they wanted records to be set at the VeloPark. It worked. In the first day of Olympic competition, world records were set six times in two different events.

As mentioned in chapter 5, the density of air—and therefore air drag—goes down as the temperature goes up. So an underflow heating system lies below the track itself, designed to keep the air temperature at a toasty 82.4°F, though the spectator level was cooled to a more comfortable temperature. (The famous London weather was thought to help, too, since air pressure and density drop even more when a front comes through. While participants in outdoor events might have wanted fair weather, velodrome organizers actually hoped for rain.)

In the individual pursuit, two riders start from opposite sides of the track and race 4 km, each trying to catch the other if he can. Ignoring the effect of the banked turns, **what would be Bradley Wiggins's time in this race, for air at 70°F and at 82.4°F?**

Wiggins has a mass of 69 kg and his 1-hr PWR is reported to be an impressive 6.6 W/kg. The duration of the pursuit is much less than 1 hr, but for the purpose of this example, let's use this PWR, since it is more difficult to find reliable numbers for short races. We know that the race times that we calculate will be somewhat higher than what Wiggins actually achieves.

So Wiggins can put out a power of

$$\bar{P}_{\text{output Wiggins}} = \text{PWR} \times \text{mass} = 6.6 \, \frac{\text{W}}{\text{kg}} \times 69 \, \text{kg} = 455 \, \text{W}$$

At "room temperature"

There are no wind ($v_{\text{wind}} = 0$) or hills ($s = 0$) in our simple race, and at 70°F, the density of air is 1.2 kg/m³. We'll use the racing bike values from table 9.3 for C_D, A, and C_R. The rules demand that track bikes, like Tour de France bikes, have a minimum mass of 6.8 kg (15 lb), so the mass load on the wheels is 69 kg + 6.8 kg = 75.8 kg.

Using all these numbers in equation 9.16, we have

$$\bar{P}_{\text{output cyclist}} = C_R m g v_{\text{bike},x} + \tfrac{1}{2} C_D A \rho_{\text{air}} v_{\text{bike},x}^3$$
$$= (0.003)(75.8 \text{ kg})(9.8 \text{ m/s}^2) \times v_{\text{bike},x}$$
$$+ \tfrac{1}{2}(0.88)(0.36 \text{ m}^2)(1.2 \text{ kg/m}^3) \times v_{\text{bike},x}^3$$
$$= (2.23 \text{ J}) \times v_{\text{bike},x} + (0.190 \text{ kg/m}) \times v_{\text{bike},x}^3$$

Step 1. Guess. It's not crazy to guess that Wiggins could maintain 28 mph (12.5 m/s).

Step 2. Plug in values.

$$\bar{P}_{\text{output cyclist}} = (2.23 \text{ J}) \times (12.5 \text{ m/s}) + (0.190 \text{ kg/m}) \times (12.5 \text{ m/s})^3 = 398 \text{ W}.$$

Steps 3 and 4.

The output power for our guess is too low compared to Wiggins's 455-W output power. So we need to increase our guess. Let's change it to 13.5 m/s and go back to step 2.

Step 2. Plug in (again).

$$\bar{P}_{\text{output cyclist}} = (2.23 \text{ J}) \times (13.5 \text{ m/s}) + (0.190 \text{ kg/m}) \times (13.5 \text{ m/s})^3 = 498 \text{ W}.$$

Steps 3 and 4. Repeat.

This time, the output power for our guess is too high. So, we need to make our guess less than 13.5 m/s, but still larger than 12.5 m/s. Let's use 13.1 m/s and go back again to step 2.

Step 2. Plug in (yet again).

$$\bar{P}_{\text{output cyclist}} = (2.23 \text{ J}) \times (13.1 \text{ m/s}) + (0.190 \text{ kg/m}) \times (13.1 \text{ m/s})^3 = 456 \text{ W}.$$

Steps 3 and 4. Repeat.

Hey, that's almost precisely Wiggins's actual power output—bingo! We're done with this part.

So, we can say that if the temperature is 70°F, Wiggins's time will be

$$\Delta t = \frac{\Delta x}{v_{\text{bike},x}} = \frac{4000 \text{ m}}{13.1 \text{ m/s}} = 305.3 \text{ s} = 5 \text{ min } 5.3 \text{ s}.$$

The velodrome at 82.4°F.

In section 5.7.2, we said that the air density drops 1% for every 5°F increase in temperature. The London velodrome was kept 12.4°F warmer than what we had for the above calculation, so the air density will be

$$\rho_{air} = (1.2 \text{ kg/m}^3) \times \left(100\% - \frac{1\%}{5°F} \times 12.4°\right) = 97.5\% \times (1.2 \text{ kg/m}^3) = 1.17 \text{ kg/m}^3.$$

So the equation that we use now is

$$\bar{P}_{\text{output cyclist}} = (0.003)(75.8 \text{ kg})(9.8 \text{ m/s}^2) \times v_{bike,x}$$
$$+ \tfrac{1}{2}(0.88)(0.36 \text{ m}^2)\underbrace{(1.17 \text{ kg/m}^3)}_{\text{the new value}} \times v_{bike,x}^3$$
$$= (2.23 \text{ J}) \times v_{bike,x} + (0.185 \text{ kg/m}) \times v_{bike,x}^3$$

Hopefully you've gotten the idea of this iterative solution by now, so I won't keep writing down the steps explicitly. If we use this new equation and adjust $v_{bike,x}$, we'll find that a speed of 13.2 m/s gives an output power of exactly 455 W.

So Wiggins's time in the higher-temperature air would be

$$\Delta t = \frac{\Delta x}{v_{bike,x}} = \frac{4000 \text{ m}}{13.2 \text{ m/s}} = 303.0 \text{ s} = 5 \text{ min } 3.0 \text{ s}.$$

The higher temperature leads to more than a 2-s difference. In a sport where times and records are set by *thousandths* of seconds, this is a huge effect. The precisely controlled environment of the velodrome is one key reason that so many track cycling records were set in London.

By the way, these times are significantly worse than Wiggins's typical times, for two reasons. First, as we said at the outset, we were using Wiggins's 1-hr PWR, rather than something more appropriate for a 5-min race. Second, we used numbers for racing road bikes, but track bikes actually have better aerodynamics and lower-resistance wheels than road bikes. On the other hand, we ignored the effect of turns and the fact that the rider must start from rest. So there were plenty of details we could treat better, but the point here was to show that the magnitude of the temperature effect—a few seconds—is quite large.

9.2.5 Cycling versus other modes of transport

After a while, listening to the continuous rapport of commentators during a long stage of the Tour de France can be tiresome. But hey, at least now you know what they mean when they discuss power-to-weight ratios, VO_2max, caloric requirements, and the fact that a 12% grade is fearsome indeed. So that counts for something.

Long stretches of commentary aside, what makes bicycle racing so exciting to the enthusiast are the extreme fitness of the athletes and the amazing speeds that can be achieved, completely powered by the cyclist himself. A human traveling at 20+ mph for hours at a time, or 40+ mph for short stretches—much higher speeds than evolution trained our primitive minds to handle—is amazing, if you think about or experience it. And it's all made possible by the unique efficiency of the bicycle itself.

In this section, I'd like to step away from sports per se, to put the bicycle in the broader context of other modes of human transport.

At city speeds, almost all the cyclist's energy goes into overcoming air drag. Most of the power generated by a car, on the other hand, goes into overcoming rolling friction.

Our two most common ways of getting around town quickly are cars and bikes. In both cases, hydrocarbons (gas, PowerBars) are combined with oxygen to produce energy. In both cases, the efficiency of converting chemical energy into mechanical (propulsive) energy is in the neighborhood of 25%. The propulsive energy goes mostly into overcoming rolling friction and wind drag.[7]

While there are clear similarities between bikes and cars, figure 9.9 shows dramatic differences in how the propulsive power is divided between heating the tires and pushing aside the air. Because bikes are so light and their tires so inflated, the tires deform very little, and the power needed to overcome rolling resistance can just about be ignored, compared to the power used to overcome wind drag. Cars, on the other hand, are much more aerodynamic[8] but are much heavier and have very different wheels. For this reason, *most of the propulsive power in a car goes into overcoming rolling resistance, for speeds less than about 50 mph.* At 20 mph, almost all propulsive power is spent overcoming rolling resistance (that is, heating the tires).

So, let's do a tally for a car moving at 50 mph. Since the engine efficiency is about 25%, out of every 8 gal of fuel in a car's tank 6 gal goes into heating the engine. At a constant speed of 50 mph, 1 gal of the 8 gal goes into overcoming wind drag, and the final 1 gal goes into overcoming rolling resistance. And that's if your tires are at the recommended pressure. Remember my son's experience described in section 9.2.2. When a little boy is fighting a low tire, he complains. When your car is fighting a low tire, it just burns (much) more gas and heats the tire more. Keep your car's tires filled!

For all reasonable speeds, a car takes much more power than a bicycle to propel. If we look at power-to-weight ratios (PWRs), however, the two are much more comparable, and the car wins out at higher speeds. See figure 9.10. Trains, on the other hand, are *much* more efficient at moving mass than road vehicles of any type.

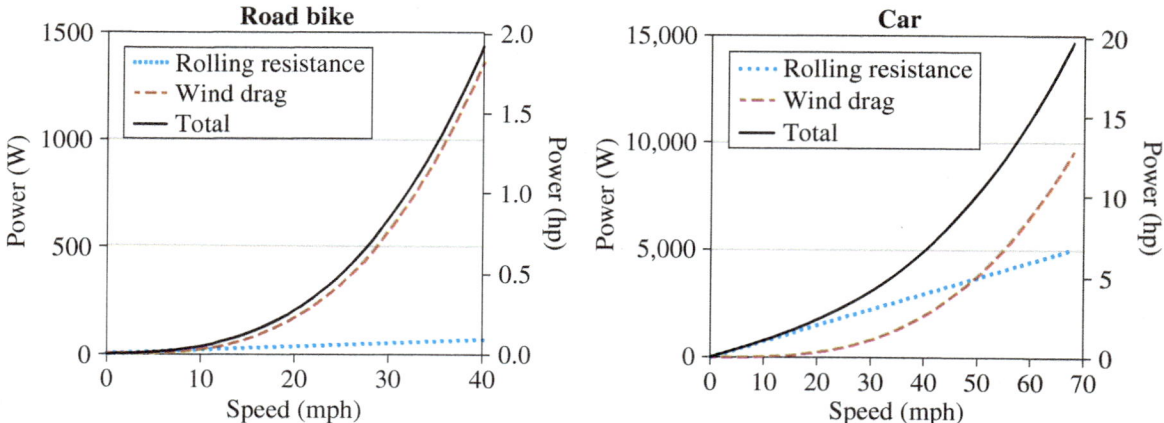

Figure 9.9. The mechanical power to overcome rolling friction and air drag increases with vehicle speed in all cases, but drag becomes increasingly important at high speeds. For bicycles, air drag dominates, while for cars, rolling friction is hugely important at all speeds. On suburban streets, wind resistance hardly matters at all.

---

[7] There are several other energy "losses," including a ~ 5% "transmission inefficiency" (also known as "drivetrain losses") and "parasitic losses" on cars, due to windshield wiper use, and so on. We're just hitting the main points in this book.

[8] That is, cars have a low drag coefficient ($C_D \approx 0.3$) compared to bikes ($C_D \approx 1$). Of course, the frontal area of a car is much larger than that of a bike plus rider, so the drag force is still larger on a car.

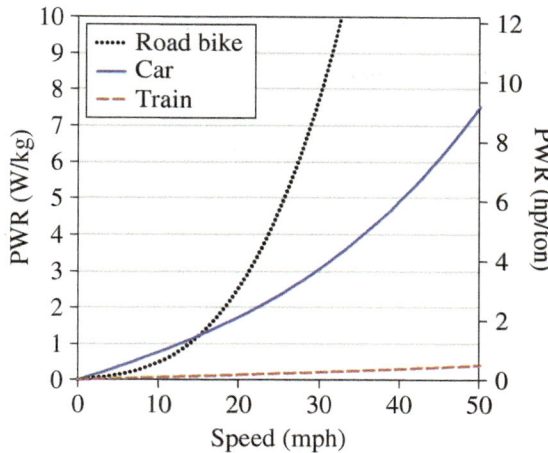

Figure 9.10. The power-to-weight ratio (PWR) for different modes of transport are very different, especially at high speeds.

9.2.6 Drafting

As we've seen, wind drag is the dominant, relentless foe of the high-speed cyclist. Most of the cyclist's energy goes into heating the air as he pushes through it. For this reason, a tremendous research effort has been made to find materials and shapes and postures that limit the drag coefficient of the rider and bike. In the mature, multibillion-dollar sport of bicycle racing, the easy improvements have been made, and the ongoing technical arms race produces an occasional reduction in drag of less than 1%.

Much greater reductions in drag can be achieved by using another cyclist to block the wind, that is, to **draft** behind him. If you've ever ridden in a tight pack of fast cyclists, you intuitively know the difference between leading the pack and riding in the center of it. While you might only be able to maintain speed for a few minutes when leading the pack, when you are inside, it almost seems as if you don't have to pedal at all.

Wind shadow: pace lines

In chapter 5 we talked about objects (mostly spherical balls) moving rapidly through the air. Now we are concerned with one object following another. In a sense, we can think of the lead cyclist as "pushing the air molecules out of the way," so the following cyclist hits fewer of them. But aerodynamics is more complicated than that. There is a low-pressure "wake" that forms behind an object passing through the air, as seen in figures 5.5 and 5.7. If a cyclist follows the lead object closely enough, the "wind on his face" is reduced as he is partially swept along in the wake. The basic idea is shown in figure 9.11.

In the team pursuit event, two teams of four riders start on opposite sides of the velodrome track. Each team tries to overtake the other, or (more commonly) to complete the 4000-m race in the shortest time. On a velodrome track, there are no unexpected obstacles or strange spectators jumping out from behind trees (as on the Tour de France) or hills. Or anything. The only enemy is air, and the entire team strategy is designed to overcome this enemy. Riders on a team form a

Figure 9.11. If he follows a fellow cyclist closely enough (say, one bike-length behind or closer), a following cyclist experiences lower drag. (© *Photodisc/Getty Images RF*).

Figure 9.12. The British men's pursuit team at the 2012 London Olympics. Everything—the helmets, the visors, the tight clothing, the posture, the covered wheel spokes, and especially the group formation—is designed to fight the great enemy, air drag. Even the air is heated to reduce its density. The power measurements were done on another team in 1999 traveling at 37.5 mph. (© *Christophe Ena/AP Photo*).

"pace line," each tucked in the "wind shadow" of the rider in front of him, as seen in figure 9.12.

Detailed studies of riders in these tight pace lines are interesting. Experiments show that the second rider has to put out less power than the lead rider (no surprise there), but the third rider has to put out even *less* power than the second one; the fourth works about as hard as the third. (Often, the change in drag is quantified by a reduced drag coefficient; see example 9.10.) That doesn't necessarily sound "wrong," but aerodynamics are complicated and it can be hard to guess the details just based on simple equations. That's why cars and bikes and even riders spend long hours in wind tunnels, measuring drag.

EXAMPLE 9.10 *What is the drag coefficient of the third member in a team pursuit pace line?*

For track racing, headwinds and hills are not an issue, and rolling friction is tiny compared to wind drag. In this case, the bicycle power equation is simple:

$$P_{\text{output cyclist}} = \tfrac{1}{2} C_D A \rho_{\text{air}} v_{\text{bike},x}^3$$

Clearly, something in the right side of that equation is smaller for the second rider than for the first, but it's not the frontal area A, since all riders have identical positions, as seen in figure 9.12. Also since they ride as a group, $v_{\text{bike},x}$ is the same. The drag coefficients in table 9.3 are for solo riders. When a cyclist drafts another rider or a car, C_D is reduced.

If we assume that the drag coefficient on the lead rider is the same as on a solo rider in race position (which is not entirely correct, as we discuss in a moment), then we can find C_D for rider 3 by taking a ratio:

$$\frac{P_{\text{output cyclist 1}}}{P_{\text{output cyclist 3}}} = \frac{\tfrac{1}{2} C_{D,1} \cancel{A \rho_{\text{air}} v_{\text{bike},x}^3}}{\tfrac{1}{2} C_{D,3} \cancel{A \rho_{\text{air}} v_{\text{bike},x}^3}}$$

$$\rightarrow C_{D,3} = \frac{P_{\text{output cyclist 3}}}{P_{\text{output cyclist 1}}} \times C_{D,1} = \frac{389 \text{ W}}{607 \text{ W}} \times 0.88 = 0.56$$

Simply by drafting close behind his teammates, the third rider reduces his drag coefficient to something between an unfaired and a faired recumbent bike (see table 9.3)!

Echelons

On a still day or inside a velodrome, of course it makes sense to remain immediately behind the cyclist in front of you. But if you watch a professional race outdoors, you'll notice that the best riders are often not in a Indian-file pace line, but are in a diagonal formation known as an **echelon**, as they intuitively seek out the wind shadow formed by other riders.

In figure 9.8, it was already obvious that a breeze can change things. What matters for drag isn't really the rider's motion; it's the velocity at which the air collides with the rider. This is $\vec{v}_{\text{air on rider}}$ as defined in equation 9.14.

Unless the wind comes from directly ahead (headwind) or behind (tailwind), there will be a component of $\vec{v}_{\text{air on rider}}$ that hits the riders from the side. In this case, the wind shadow of the riders does not point directly behind them, but partly to the side, as shown in figure 9.13.

The figure shows an example of a side wind, where the wind makes a right angle with the riders' motion. That figure should remind you of something. Vectors, and trigonometry, from chapter 3. If you clear away the cobwebs with a cup of coffee and look at figure 3.13 and equation 3.14, you'll see that, for pure side-wind situations, the angle of the echelon is

$$\text{side wind: } \theta_{\text{echelon}} = \tan^{-1}\left(\frac{|\vec{v}_{\text{bike}}|}{|\vec{v}_{\text{wind}}|}\right) \qquad (9.17)$$

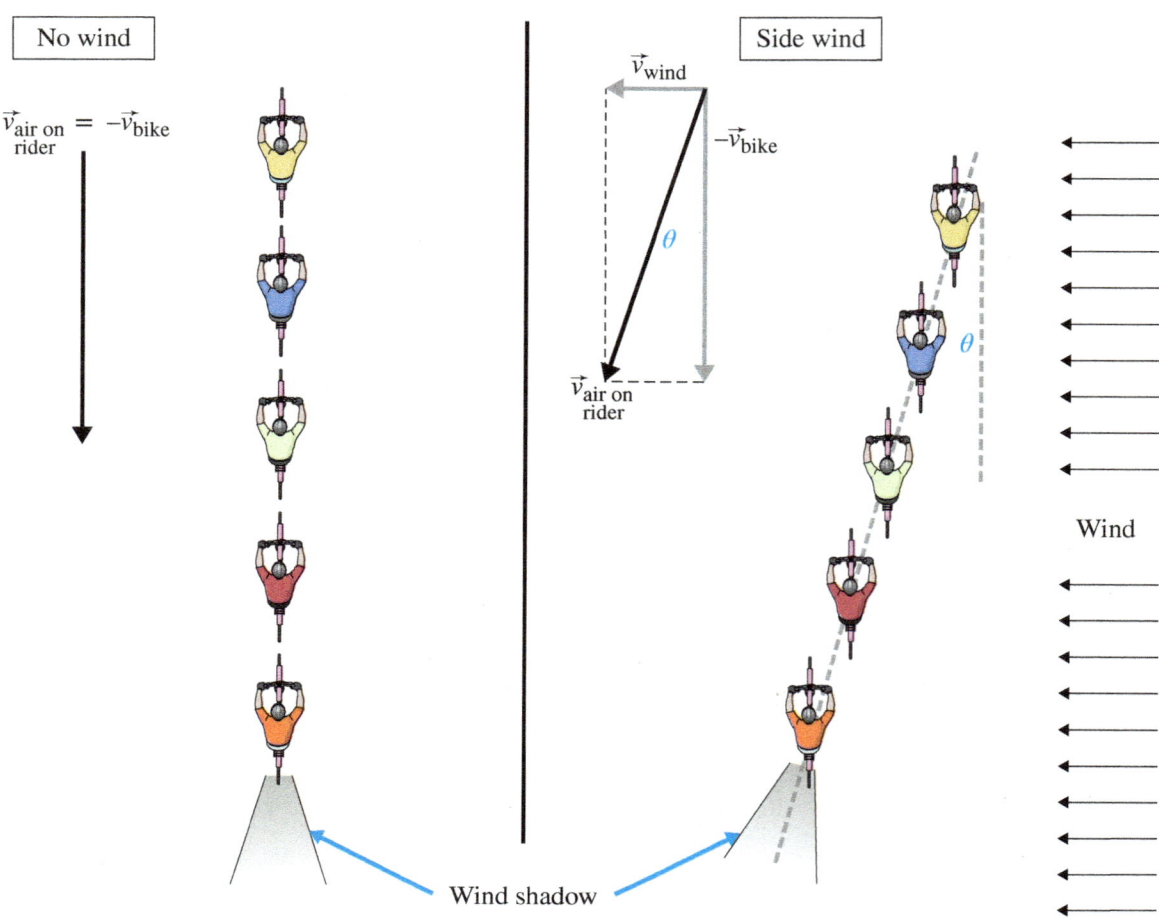

Figure 9.13. The wind shadow of a rider points in the same direction as the velocity at which air hits him, $\vec{v}_{\text{air on rider}}$. If there is no wind, the wind shadow points directly behind him, and the optimum formation for a group of riders is a pace line, as shown on the left. If there is a pace line, the wind shadow points at an angle, and experienced cyclists will shift the formation to make an echelon precisely following the direction of $\vec{v}_{\text{air on rider}}$. To reduce clutter, only the wind shadow of the last rider is drawn.

Question

What if the wind doesn't point at a 90° angle to the rider's motion, but at some other angle? How do you think equation 9.17 would change?

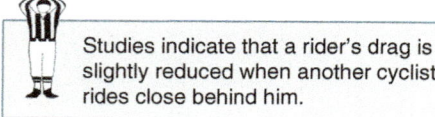

Studies indicate that a rider's drag is slightly reduced when another cyclist rides close behind him.

Effects on the line leader

Here comes the really cool part.

When one cyclist drafts behind another, it's obvious that he gets a significant advantage, compared to riding alone. But what about the guy in front? Does he get a corresponding *dis*advantage, compared to riding alone? Surprisingly, no. It turns out that *both riders benefit when one of them drafts the other*, although the benefit is obviously larger for the rear rider. Different researchers get somewhat different results, and obviously it's going to depend on the postures of the riders and other things, but the reduction in the lead rider's drag coefficient seems to be about 5%.

How can this be? It doesn't take a lot of imagination to understand that when a teammate blocks the wind for him, the slowing force of air on Bradley Wiggins is reduced. But how is the drag on a rider affected by having somebody *behind* him?

Clearly, there must be more to drag than the ballistic picture of pushing air out of the way, but we knew that already by looking at figures 5.3 and 5.10. The wake that follows a ball moving through the air is an important part of drag. It forms a low-pressure zone that in a sense "pulls backward" on the ball.

If we look again at figure 9.11, while the second rider is helped by the low-pressure zone created by the leader, he is at the same time partially filling in that zone with his body. The smaller low-pressure zone leads to slightly less pullback on the lead rider.

The famous teardrop shape of helmets used in velodrome races and time trials is motivated by this phenomenon, as suggested in figure 9.14. These helmets are not used in the regular Tour stages partly because a large unfilled wake region is created if the rider's head is not held in the optimum position. During the regular stages, riders might like to occasionally look down for bumps or look around at other riders!

The filled-in-wake picture is a rough way to think about the physics going on here, though aerodynamics, and especially turbulence, is much more complex than any approximate model that we make up. As we've said before, science often approaches a complex issue by systematically building up a more detailed and nuanced formulation initially based on simpler ideas. We started with the simplest model of drag—knocking molecules out of the way—in section 5.3. That gives us some feeling for what's happening and helped us understand why the force depends on air density and on the square of the speed, rather than the speed itself. Our concept of drag became more detailed as we discussed asymmetric drag as the source of the Magnus force in section 5.4.2; one result was an angled deflection of the turbulent wake of the ball. Now we are focusing even more on the wake and considering what happens as one rider partially occupies the wake of another.

In aerodynamics, going beyond rough descriptions requires experiments, usually in wind tunnels. Figure 9.15 shows the drag coefficient for each of two stock cars. The lead car gets a noticeable benefit if the following car is within about one-half the car's length. The rear car benefits even if the separation is much larger, as confirmed by reports of stock car drivers. From a drag point of view, there is little reason for one car to be almost touching the car in front of it. The main benefit of such a position is that it is easier to remain directly behind the leader and not fall out of his wind shadow. Another benefit is that it looks cool.

There are cool effects if we look at pairs of other drafting objects. Check out figure 9.16, which shows the drag coefficients for two cylinders, one behind the other. Just as with the stock cars, the lead cylinder benefits if the two are close. But check out the rear cylinder. For separations of less than about one diameter, it has a *negative drag coefficient*, meaning that it experiences a *forward* push from the air. It is truly "sucked along" in the wake of the lead cylinder. The bodies of cyclists are very roughly cylinders, so we could ask whether this works for bikes. Alas, no. Even if they were perfect cylinders, the *body* (not the bike) of the following rider would have to be within about 6 in. of the leader's. That just doesn't work with bike geometries.

Okay, just one more. Figure 9.7 shows a sleek fully faired human-powered vehicle (HPV). It is clearly optimized to reduce drag when riding alone. But what if one HPV drafts another? Check out figure 9.17. It shows that when two struts (which have a shape roughly like HPVs[9]) come close, the drag on the rear object hugely *in*creases.

Figure 9.14

The iconic cone-shaped helmet used in velodrome racing and closed-circuit time trials features smaller drag coefficients that shave precious seconds off a racer's time, but only if they are held in the optimum position.

---

[9] Unfortunately, there are no measurements like this on actual HPVs.

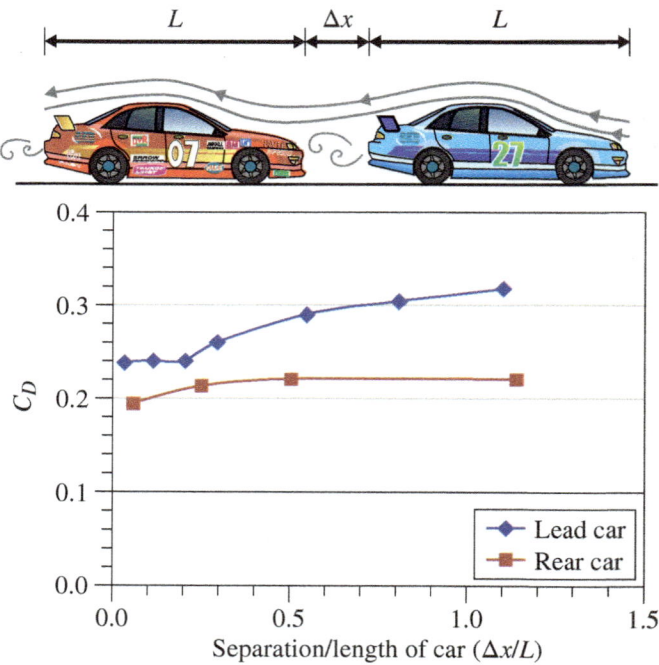

Figure 9.15. The drag coefficient on the front car in a drafting pair is reduced when the cars are within one car length of each other.

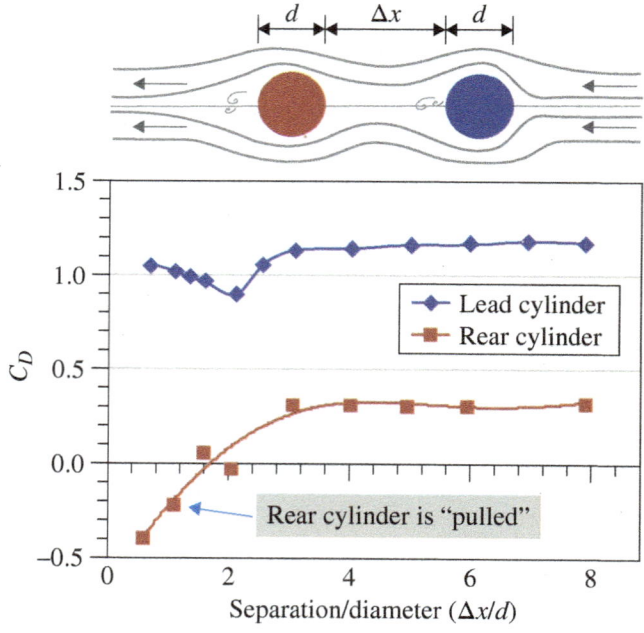

Figure 9.16. With the airflow coming from the right, the left cylinder always has a lower drag coefficient C_D. However, even the right cylinder experiences a reduced drag if the rear cylinder is close.

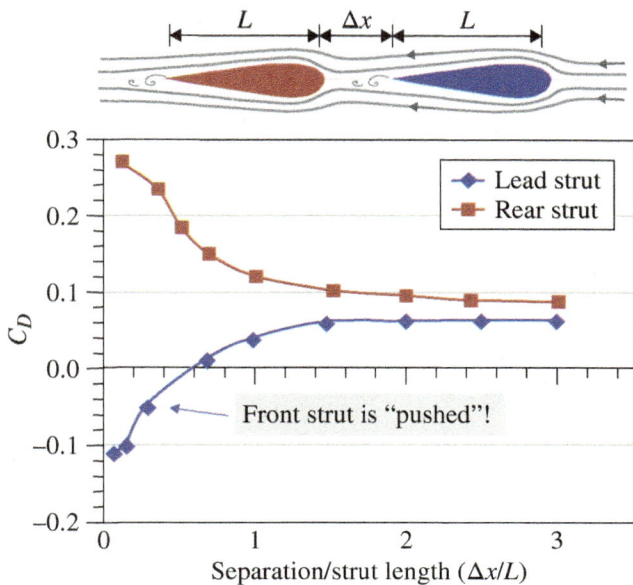

Figure 9.17. Wing struts are very similar in shape to specialized human-powered vehicles. The unique aerodynamic structure of these vehicles makes them particularly ill suited for drafting. When they are very close, the rear strut experiences a much *larger* drag force than it would if it were alone. More surprisingly, the *front* strut experiences such a strong benefit that if the struts are very close, the net aerodynamic force on the front strut is in the *forward* direction (that is, into the wind)!

Even stranger is that the drag on the lead strut *decreases* and C_D even becomes negative if they are really close. Think about it for a second. This means that *the lead object is forced forward and the following object is forced backward!* They repel each other. Okay, that is just weird. Aerodynamics sometimes simply defies simple explanations.

Breaking away

In the classic 1979 film *Breaking Away*, Dave Stoller is a Midwestern boy from Bloomington, Indiana, who dreams of being an Italian cyclist. At one point in the movie, he speeds down a highway at 60 mph by drafting a tractor trailer.

On May 17, 1941, Alfred Letourner took the same approach when he drafted close behind a midget race car on a California freeway near Bakersfield. He broke the motor-paced[10] speed record by reaching 108.9 mph (48.7 m/s) on a fixed-gear wood-rimmed Schwinn bicycle weighing just 20 lb.

Photographs of the feat and the man on his machine are seen in figures 9.18 and 9.24. Letourner's bike differs from most others in several ways. First, it is a fixed-gear bike, meaning that the rider cannot "coast" with his feet stationary on the pedals. This means that the bike need not have brakes—and it does not. Such "fixies" are great for city traffic and quite popular among hipster college students. They are used in velodrome races.

Another difference is that the front wheel is smaller than the rear, and the front fork is shortened, so that the rider leans more forward. This allows Letourner to put his bulk as close to the race car as possible. (No brakes, leaned over, inches away

---

[10] "Motor-paced" simply means he was drafting a motored vehicle.

Figure 9.18. In May, 1941, Alfred Letourner sped at an incredible 108.9 mph, drafting behind a midget race car with a specialized fixture on back. Letourner is seen on his wood-rimmed Schwinn, featuring a huge front chainring. (© *Bettmann/Corbis*).

from the rear of a car at speeds over 100 mph on wood-rimmed wheels. What could possibly go wrong?)

But perhaps the biggest difference is the humungous chain ring in the front. (This is the gear attached to the pedals.) Letourner's chain ring was as large as physically possible. Any larger, and it would touch the ground. Why is this particular feature so important for breaking the speed record?

To really understand the amazing bicycle and elements like this, we have to study the connection between rotational motion of chain rings and wheels and the "translational motion" that we've been studying all along.

9.3 Talansky Drives the Bike

Andrew Talansky, who placed 10th in the 2013 Tour de France, represents a new generation of American cyclists. He is a 139-lb (63-kg) climber whose sustained PWR has been estimated at 5.7 W/kg. The 360 W he generates (5.7 W/kg × 63 kg = 360 W) is input directly to the pedals of his 6.8-kg[11] Cervelo R5 VWD.

The bike then exploits the laws of rotational and translational physics to direct the power output by a cyclist most efficiently into increasing his energy—either kinetic or gravitational—and/or fighting air drag. A bicycle is very nearly the perfect machine.

To fully understand this machine, we're going to work our way "backward" toward the athlete. We'll start where the rubber hits the road, with the wheel. From there, we'll discuss the gears that drive the wheel, then the chain that drives the gears. Finally, we'll cover the pedals pushed by the cyclist to drive the chain.

To discuss the workings of the bike, we'll need to learn some new concepts relating *angular* (or rotational) motion to the *linear* (or translational) motion we've talked about up to this point.

9.3.1 Rolling

To really understand a bike's motion, we need to start with the rolling wheel. The important thing is how fast the wheel is spinning, that is, its **rotational frequency** f_{rot},

---
[11] The minimum mass for a bike permitted on the Tour.

in rotations per second. This is the same quantity we used in section 5.4.2 to quantify how fast a baseball spins when pitched. The rotational frequency is obviously similar to the oscillation frequency of equation 8.6, and we can write

$$f_{\rm rot} = \frac{1}{\text{time required for 1 revolution}}$$

The axle of the wheel (and the bike it's attached to) travels one wheel circumference for 1 revolution (1 rev), so the bike's speed is

$$v_{\rm bike} = \frac{\text{distance traveled in 1 rev}}{\text{time required for 1 rev}} = \frac{2\pi R_{\rm wheel}}{\text{time required for 1 rev}}$$

$$\rightarrow \quad \boxed{v_{\rm bike} = 2\pi R_{\rm wheel} \times f_{\rm wheel}.} \quad (9.18)$$

Traveling fast means a fast-spinning wheel, a large wheel, or both.

One last thing about spinning wheels: Equation 9.18 tells us $v_{\rm bike}$, the speed of the axle, relative to the stationary road. We know that this means the speed of the road, relative to the axle, is also $v_{\rm bike}$. Well, the bottom of the tire is momentarily not moving relative to the road. So the bottom of the wheel is moving with speed $v_{\rm bike}$ relative to the axle. In fact, that's true for every part of the outside of the tire, not just the bottom, so I'll write a boxed equation for it:

$$\boxed{v_{\substack{\text{wheel edge relative} \\ \text{to axle}}} = 2\pi R_{\rm wheel} \times f_{\rm wheel}.} \quad (9.19)$$

Wheel sizes

Bicycle wheels come in a variety of sizes. Unfortunately, there are at least three systems of naming these sizes, and some of the designations really don't make any sense, except in a historical way. An attempt to bring order to rim and tire size designations is to express the diameter of the rim in millimeters (sometimes called the "iso rating") as well as the traditional name. So, sometimes you'll see one, the other, or both, on tires.

Mavic lists the tire for its Cosmic CXR60 C wheel as "23-622 (700x23c)." Looking at table 9.4, we see that "700c" and "622" are just two ways of labeling the same wheel size. The "23" is the additional radius (in millimeters) one gets from the tire itself.

Table 9.4 gives the information we'll need for the most common bike wheels. The important quantities are in the two rightmost columns. The wheel (including tire) radius is the distance between the axle and the ground. This is $R_{\rm wheel}$ in equation 9.18.

Table 9.4. Sizes of the most common road and touring bike wheels. The traditional and iso ratings are usually used to refer to the wheel. The wheel plus tire radius can vary a little, depending on the exact tire used.

Rim designation (traditional)	Rim diameter (mm) (also known as "iso")	Rim radius (m)	Wheel (including tire) radius (m)
700c (or 29er)	622	0.3110	0.3310
650b (or 27.5)	584	0.2920	0.3120
650c	571	0.2855	0.3055
26-in.	559	0.2795	0.2995

Stage 17 of the 2013 Tour

Stage 17 of the 2013 Tour de France was a 31.5-km individual time trial. Talansky finished in 53:14 and averaged about 9.9 m/s (22.1 mph) riding on a 700c Mavic CXR60 rear wheel.

Wheel spin rate

Using equation 9.18 and table 9.4, we can find his wheel's rotational frequency:

$$v_{bike} = 2\pi R_{wheel} \times f_{wheel} \quad \rightarrow \quad f_{wheel} = \frac{v_{wheel}}{2\pi R_{wheel}} = \frac{9.9 \text{ m/s}}{2\pi \, (0.331 \text{ m})} = 4.76/\text{s}.$$

So, for about an hour, Talansky's tire was spinning about one-half as fast as the ball in Roberto Carlos's "impossible" kick (section 5.5.4).

Calories burned

The food Calories burned by Talansky on this stage can be estimated by equation 9.4 and the MET values from table 9.1. Talansky's mass is reported to be 63 kg, so

$$\begin{pmatrix} \text{food} \\ \text{energy} \\ \text{burned} \end{pmatrix} = \left(\frac{1 \text{ Cal}}{\text{kg} \cdot \text{hr}}\right) \times \text{MET} \times m_{person} \times \Delta t$$

$$= \left(\frac{1 \text{ Cal}}{\text{kg} \cdot \text{hr}}\right) \times 15.8 \times (63 \text{ kg}) \times \underbrace{\left(53 \text{ min} \times \frac{1 \text{ hr}}{60 \text{ min}} + 14 \text{ s} \times \frac{1 \text{ hr}}{3600 \text{ s}}\right)}_{\text{must convert 53:14 to hours}}$$

$$= 883 \text{ Cal}.$$

This should be about right for a flat course at a constant speed. But stage 17 was actually hilly for a time trial, and Talansky's speed varied accordingly, as you can see from figure 9.19. So 883 Cal is probably a low-ball estimate of Talansky's Calories burned. In fact, his bicycle computer estimated that he burned about 1000 Cal on the stage.

9.3.2 Drivetrain I: gears

In stage 17, Talansky's wheels spun at 4.76 revolutions per second (rev/s or rps), but his feet certainly didn't turn the pedals that quickly! (Imagine doing that.) Figure 9.20 shows the power output by a typical cyclist as a function of pedaling cadence, that is, the rate of pedal rotation. Most cyclists do best with a cadence of about 90 rpm, or 1.5 rps.

Direct-drive bikes

Bicycle racing as a sport gained popularity toward the end of the 19th century, with the so-called high-wheeler or penny-farthing[12] bicycles. One is shown in figure 9.21. Rather than use a chain as modern bikes do, these bikes had a "direct drive," meaning that the pedals were attached directly to a wheel. One turn of the wheel required one turn of the pedals.[13]

---

[12] The name comes from the British penny and farthing coins, which were of much different size.
[13] And pedaling backward meant traveling backward. This is similar to today's fixed-gear (or "fixie") bikes, though fixies use chains and gear ratios. A better comparison for the penny-farthing is today's unicycle.

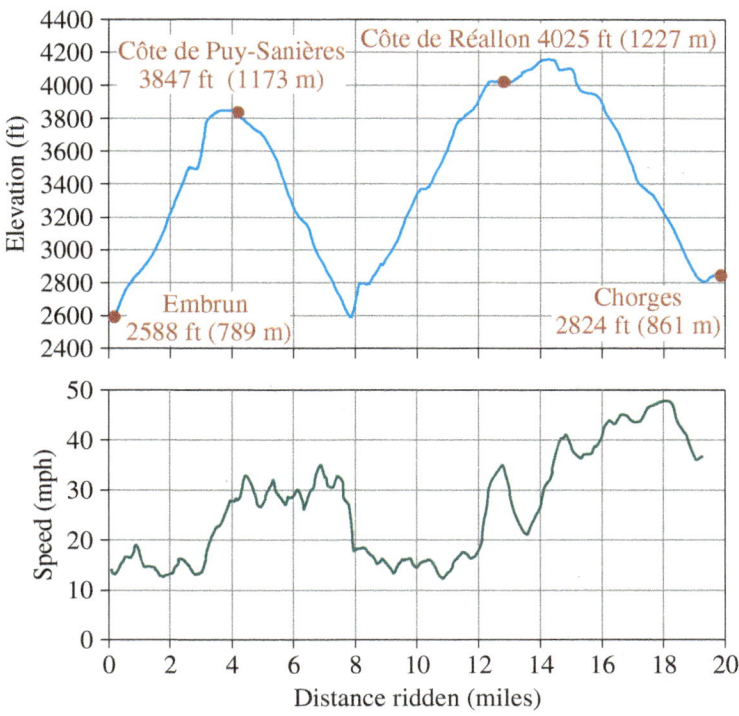

Figure 9.19. Elevation profile for stage 17 of the 2013 Tour de France between the towns of Embun and Chorges. Below, the speed of Andrew Talansky, as measured by his Garmin bicycle computer. Speed data based on http://connect.garmin.com/activity/344028331.

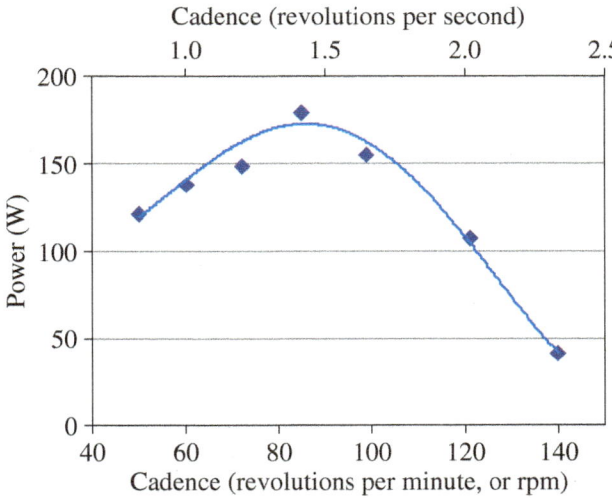

Figure 9.20. The output power of one cyclist as a function of his pedaling cadence. The optimum cadence for a given cyclist varies, but usually falls between 80 and 100 rev/min.

Figure 9.21
William Martin, champion 6-day bicycle rider of the world on a penny-farthing. Circa 1891. (© *Library of Congress Prints & Photographs [LC-USZ62-105442]*).

What if Talansky's bike were direct-drive? Then with a 1.5-rps cadence, his wheel would spin at 1.5 rps, and equation 9.18 says his bike would travel at

$$v_{bike} = 2\pi R_{wheel} \times f_{wheel} = 2\pi (0.331 \text{ m}) \times (1.5/\text{s}) = 3.1 \text{ m/s} = 7 \text{ mph}.$$

That's not very fast at all. Take a look at figure 2.4 (on page 21), giving speeds for foot races: 7 mph doesn't even make it onto the scale! Since stage 17 was about the length of a marathon, a decent long-distance runner would *destroy* Talansky on a modern bike wheel if Talansky used direct drive.

The reason the penny-farthing had such a huge drive wheel (that is, the wheel connected to the pedals) was to achieve greater speeds. Equation 9.18 dictates the speed of the bike. If the human body demands a cadence of 1.5/s, and direct drive means $f_{\text{wheel}} = 1.5/\text{s}$, then the only option is to make R_{wheel} as large as possible. The only limitation really was the distance from the rider's foot to the inseam of his pants (to put it delicately). Front wheels were produced with diameters as large as 60 in., so $R_{\text{wheel}} = 30$ in. $= 0.76$ m. With a 1.5-rps cadence, such a bike would travel at

$$v_{\text{bike}} = 2\pi R_{\text{wheel}} \times f_{\text{wheel}} = 2\pi\,(0.76\text{ m}) \times (1.5/\text{s}) = 7.2 \text{ m/s} = 16 \text{ mph}.$$

That's a much more respectable racing speed. However, it's not much in terms of a "top speed." Furthermore, with this bike, the rider is at sub-optimal power output at lower speeds, due to his lower cadence, as figure 9.20 told us. Climbing steep hills is extremely difficult on a penny-farthing.

Chain drive

The chain drive with variable gears on modern bicycles is *the* key ingredient that allows the cyclist to maintain optimum power output at all speeds. Figure 9.22 shows a modern chain drive, in which a larger front gear is attached to the pedals (so this front gear spins at the rider's pedaling cadence) and the smaller rear gear is attached to the wheel.

The links on the chain are like connected train cars. And just as all train cars move at the same speed,[14] every chain link moves at the same speed v_{link}.

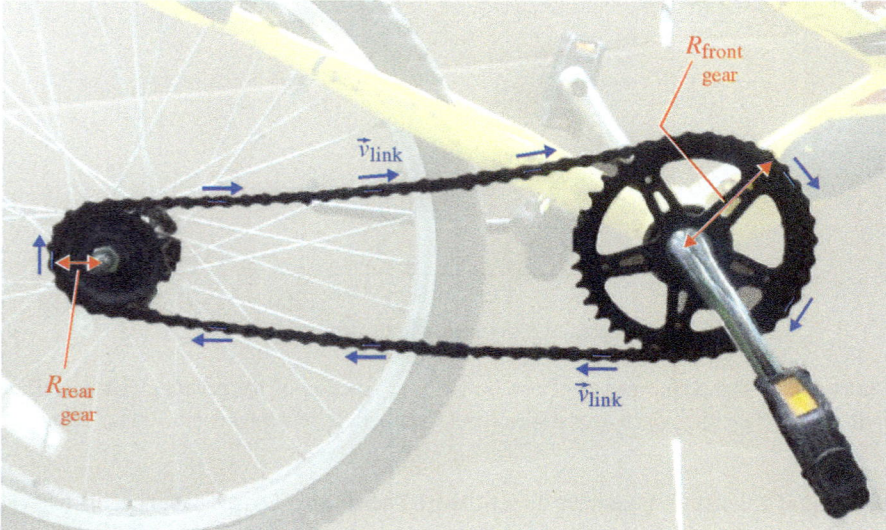

Figure 9.22. Every link in the chain moves at the same speed, v_{link}. This means the teeth on the edges of the gears also move at v_{link}. (*Photo courtesy of Mike Lisa*).

---

[14] Otherwise, a faster car would crash into the car in front of itself.

Now, to get to an important and very cool result, just follow a short train of logic:[15]

1. The gears are "wheels," too, and the edges of those wheels are moving with the *same speed* v_{link} relative to their axles. Looking at figure 9.22 might help make this obvious.

2. If the speeds of the gear edges are the same, then their rotational frequencies are *not* the same. We can relate them by using equation 9.19:

$$v_{\substack{\text{wheel edge relative} \\ \text{to axle}}} = \underbrace{2\pi R_{\substack{\text{front} \\ \text{gear}}} \times f_{\substack{\text{front} \\ \text{gear}}} = 2\pi R_{\substack{\text{rear} \\ \text{gear}}} \times f_{\substack{\text{rear} \\ \text{gear}}}}_{\text{same}}$$

$$\rightarrow f_{\substack{\text{rear} \\ \text{gear}}} = \frac{R_{\substack{\text{front} \\ \text{gear}}}}{R_{\substack{\text{rear} \\ \text{gear}}}} \times f_{\substack{\text{front} \\ \text{gear}}}$$

Since the front gear is bigger than the rear one, this tells us that the rear gear is spinning faster.

The important number is the **gear ratio (GR)**, which is just the ratio of the radii of the front and back gears. It is also the ratio of the number of "teeth" on the front and back gears:

$$\boxed{\text{GR} = \frac{R_{\substack{\text{front} \\ \text{gear}}}}{R_{\substack{\text{rear} \\ \text{gear}}}} = \frac{\#\text{teeth on front gear}}{\#\text{teeth on rear gear}}.} \qquad (9.20)$$

3. The last things to remember are that the wheel is attached to the rear gear (so $f_{\text{wheel}} = f_{\substack{\text{rear} \\ \text{gear}}}$) and the pedals are attached to the front gear (so $f_{\text{pedals}} = f_{\substack{\text{front} \\ \text{gear}}}$). So if you followed step 2, you know how the rotational frequency of the wheel and that of the pedals are related:

$$\boxed{f_{\text{wheel}} = \text{GR} \times f_{\text{pedals}}.} \qquad (9.21)$$

This is important. It lets Talansky's rear wheel rotate at 4.76 rps at the same time his pedal cadence is 1.5 rps. The big limitation of penny-farthings was that they had GR = 1.

4. Finally, by using equation 9.18, we get a formula telling us how the bike's speed is related to the gear ratio, wheel size, and pedaling cadence:

$$v_{\text{bike}} = 2\pi R_{\text{wheel}} \times \underbrace{f_{\text{wheel}}}_{\text{GR} \times f_{\text{pedals}}}$$

$$\rightarrow \boxed{v_{\text{bike}} = 2\pi \times \text{GR} \times R_{\text{wheel}} \times f_{\text{pedals}}.} \qquad (9.22)$$

Adjustable gears make possible maximum power output over a large range of speeds.

Example 9.11 discusses a famous case in which an extreme gear ratio was used to achieve extreme speeds.

---

[15] No complaining allowed. We don't focus on "derivations" in this book, but to understand what makes a bike work, you should understand these four steps.

Front: 54/42 chainring

Rear: 11-28 cassette:
11-12-13-14-15-17-19-21-23-25-28

Figure 9.23
A gear group similar to that used by Andrew Talansky in the 2013 Tour de France. (*Photos courtesy of Mike Lisa*).

Whereas the speed of any given fixed-gear or direct-drive bicycle is determined strictly by the rider's pedal cadence (which is best kept in a limited range) and wheel radius (which cannot be changed during the ride), equation 9.22 for modern bikes includes a gear ratio, which is easily changed over a wide range. It's a huge advance, and it allows optimum power for a range of bike speeds.

Available gear ratios

Talansky used a Shimano Dura-Ace 9070 gear system in the 2013 Tour. Similar (though lower-end) gears are shown in figure 9.23. The two gears on the front chain ring have 54 and 42 teeth, respectively, and the 11 gears on the rear cassette have tooth counts indicated in the figure.

The "highest gear" on a gear system is obtained when the largest gear is used on front (54 teeth, for the setup in figure 9.23) together with the smallest rear gear (11 teeth, in this system). The largest gear ratio here is then

$$GR_{max} = \frac{\text{\#teeth on largest front gear}}{\text{\#teeth on smallest rear gear}} = \frac{54}{11} = 4.91.$$

The lowest ratio with this system is with the smallest front gear and the largest rear:

$$GR_{min} = \frac{\text{\#teeth on smallest front gear}}{\text{\#teeth on largest rear gear}} = \frac{42}{28} = 1.5.$$

What gear combination was Talansky using in stage 17? We'll assume his pedal cadence was $f_{pedals} = 1.5/s$, and we said in section 9.3.1 that his wheel spun at $f_{wheel} = 4.76/s$, so equation 9.21 tells us the gear ratio he used:

$$f_{wheel} = GR \times f_{pedals} \quad \rightarrow \quad GR = \frac{f_{wheel}}{f_{pedals}} = \frac{4.76/s}{1.5/s} = 3.17$$

Talansky could get this ratio by using the large front ring and the 17-tooth rear gear: 54/17 = 3.17. He could get almost the same ratio by using an entirely different combination—the smaller front ring and the 13-tooth rear gear: 42/13 = 3.23. This shows how, even though there are 22 possible gear combinations, many of them give essentially the same gear ratio, so there are really fewer than 22 *different* gears.

EXAMPLE 9.11 *Letourner's big gear*

Figure 9.24
A replica of Letourneur's record-breaking bike, on display at the Bicycle Museum of America, New Bremen, OH. (*Photo courtesy of Mike Lisa*).

Now we see the point behind the huge front gear (57 teeth and a radius larger than the pedal cranks!) and small rear one (only 6 teeth) on Alfred Letourner's record-setting bike (figures 9.18 and 9.24). It allowed his rear wheel to spin at a crazy rate as he flew down the road at 100+ mph, while his feet pedaled at a humanly possible cadence. What was his cadence?

Most likely, Letourner rode a 26-in. wooden-rim wheel. With the tire included, its radius was about 0.2995 m, as listed in table 9.4 Using equation 9.20, we find the gear ratio

$$GR = \frac{\text{\#teeth on front gear}}{\text{\#teeth on rear gear}} = \frac{57}{6} = 9.5.$$

Equation 9.22 can be used to find Letourner's cadence while riding at 108.9 mph (48.7 m/s).

$$v_{bike} = 2\pi GR \times R_{wheel} \times f_{pedals}$$

$$\rightarrow f_{pedals} = \frac{v_{bike}}{2\pi GR \times R_{wheel}} = \frac{48.7 \text{ m/s}}{2\pi \times 9.5 \times 0.2995 \text{ m}} = 2.72 \text{ rps.}$$

A cadence of 2.72 rps (163 rpm) is well beyond the "power peak" shown in figure 9.20. His record-smashing speed was impressive enough, but the fact that Letourner could generate power while churning at 163 rpm for 1 mi is amazing in its own right.

9.3.3 That annoying 2π: angular velocity

By now you might have noticed that as soon as we started talking about rotational motion, formulas started containing factors of 2π. And you know why that is, of course: the circumference of a circle is 2π times its radius.

The factor shows up particularly often when it multiplies f_{rot}, as in equations 9.18, 9.19, and 9.22. At the risk of introducing yet another quantity and boxed formula (this chapter has a lot, but chapter 10 has many fewer), it really turns out to be useful to define the **angular velocity**:

$$\omega = \begin{cases} +2\pi f_{rot} & \text{if the rotation is counterclockwise} \\ -2\pi f_{rot} & \text{if the rotation is clockwise} \end{cases} \quad (9.23)$$

So, in addition to "absorbing" the factor of 2π, as a physicist would say, ω has a sign that tells in which direction something is spinning. This is a lot like the sign on v_x, which tells whether something is moving left or right. Sometimes this sign will matter, and sometimes not. When it does, we'll need to use ω instead of just f_{rot}, which is always positive.

Radian: the "special" unit

The units of angular velocity are sometimes a source of confusion. It turns out that 2π in equation 9.23 is actually 2π **radians**. One radian (abbreviated 1 rad) is an angle of about 57°. So while the rotational frequency f_{rot} is measured in rotations per second, the angular velocity ω is measured in radians per second (rad/s). If the gymnast in figure 9.25 takes 1 s to rotate through the angle shown, his angular velocity will be $\omega = -1$ rad/s.[16]

As it turns out, we're not really going to use the radian directly too much. But it's still a pain in the neck, as you'll see in just a moment.

The speed $|\vec{v}|$ of any part of a rotating object is given by the product of angular velocity and the distance from the axis to the part in question:

$$|\vec{v}| = |\omega| \times \text{(distance to axis)}. \quad (9.24)$$

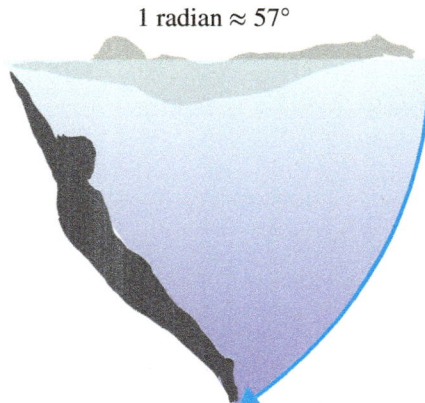

Figure 9.25
A gymnast rotates through an angle of 1 rad.

So if the distance between the gymnast's toes and the high bar in figure 9.25 is 6 ft, then his toes are moving at

$$|\vec{v}_{toes}| = |\omega| \times \text{(distance between toes and bar)}$$
$$= (1 \text{ rad/s}) \times (6 \text{ ft})$$
$$= 6 \text{ ft/s} = 4 \text{ mph}.$$

That seems easy enough, but why isn't the unit rad · ft/s? Basically a radian is a ratio of two lengths (an arc length and a radius) and so isn't really a unit that needs to

 The radian as a "unit" of angle can behave differently from other units when used in formulas.

---

[16] Did you notice the minus sign? That's because he's rotating clockwise. Keep alert!

be canceled algebraically. As such, the radian is the only unit that breaks all the rules I drilled into you in section 1.1, when I said you have to be careful to cancel units, etc. When a radian is multiplied by a length (such as a meter), the result is a length (such as a meter); the radian seems to "disappear" when we are using some formulas.

Don't worry. You'll see that it's not a huge deal when radians are used, and I'll point it out as we go along. Now you are aware that radians appear to act a little differently than other units.

9.3.4 Drivetrain II: chain action

Except for that annoying 2π interlude above, this has all been very nice. Now we understand how all the rotational frequencies and speeds and gearings are related when a bike is cruising down the highway. But how does Talansky get moving in the first place? That is, how does he accelerate?

Ah, well, there it is—that word "accelerate." You suspect that now we're going to start talking about forces, and you are correct. But to understand bicycles, we need to understand not only how to change an object's velocity, which is to induce an acceleration via a force, but also how to change an object's *angular velocity* by inducing an **angular acceleration** via a **torque**.

Torque, which is basically "twisting force," is hugely important in sports, though sometimes the details can get quite involved. We're going to ease into it, still discussing Talansky at constant velocity (that is, *not* accelerating).

We know that there are several forces—predominantly air drag—that push backward on Talansky, even if he's moving at constant speed. So to maintain speed, there must be an equal force pointing forward, to cancel the drag. As you know, this "canceling force" is the frictional force from the road on Talansky's rear wheel, as shown in panel (a) of figure 9.26.

That makes sense, as far as forces go. But let's look *just at the rear wheel about its axle* for a moment. Ask yourself, What will $\vec{F}_{\text{ground on wheel}}$ do to that wheel? Really give it a thought. This is something you want to "get" intuitively.

Well, hopefully you see that $\vec{F}_{\text{ground on wheel}}$ wants to make the wheel rotate counterclockwise about its axle! We say that $\vec{F}_{\text{ground on wheel}}$ *is producing a counterclockwise torque*.

It's time for a formula. The torque depends on the force that is creating it *and* the "lever arm" ℓ.

$$\tau = \begin{cases} +|\vec{F}|\ell & \text{if the torque is counterclockwise} \\ -|\vec{F}|\ell & \text{if the torque is clockwise} \end{cases} \quad (9.25)$$

In the present case, the angle between the force's direction and the line connecting the axle and the application of the force is 90°. In cases like this, ℓ is just the distance between the application of the force and the axle. We'll discuss other cases in a bit.

Torque depends on the force applied and *where* it is applied.

Obviously (since the wheel is rotating at constant angular velocity) something is opposing the torque produced by $\vec{F}_{\text{ground on wheel}}$. And you already know that it is the forward pull from the chain producing a *clockwise* torque on the wheel.[17] Even though $\vec{F}_{\text{ground on wheel}}$ and $\vec{F}_{\text{chain on wheel}}$ point in the same direction, they produce torques in opposite directions, because one is applied above the axle and one below.

---

[17] Since the rear gear and the wheel are connected and move as one unit, it's convenient and fine to treat a force on the gear as a force on the wheel.

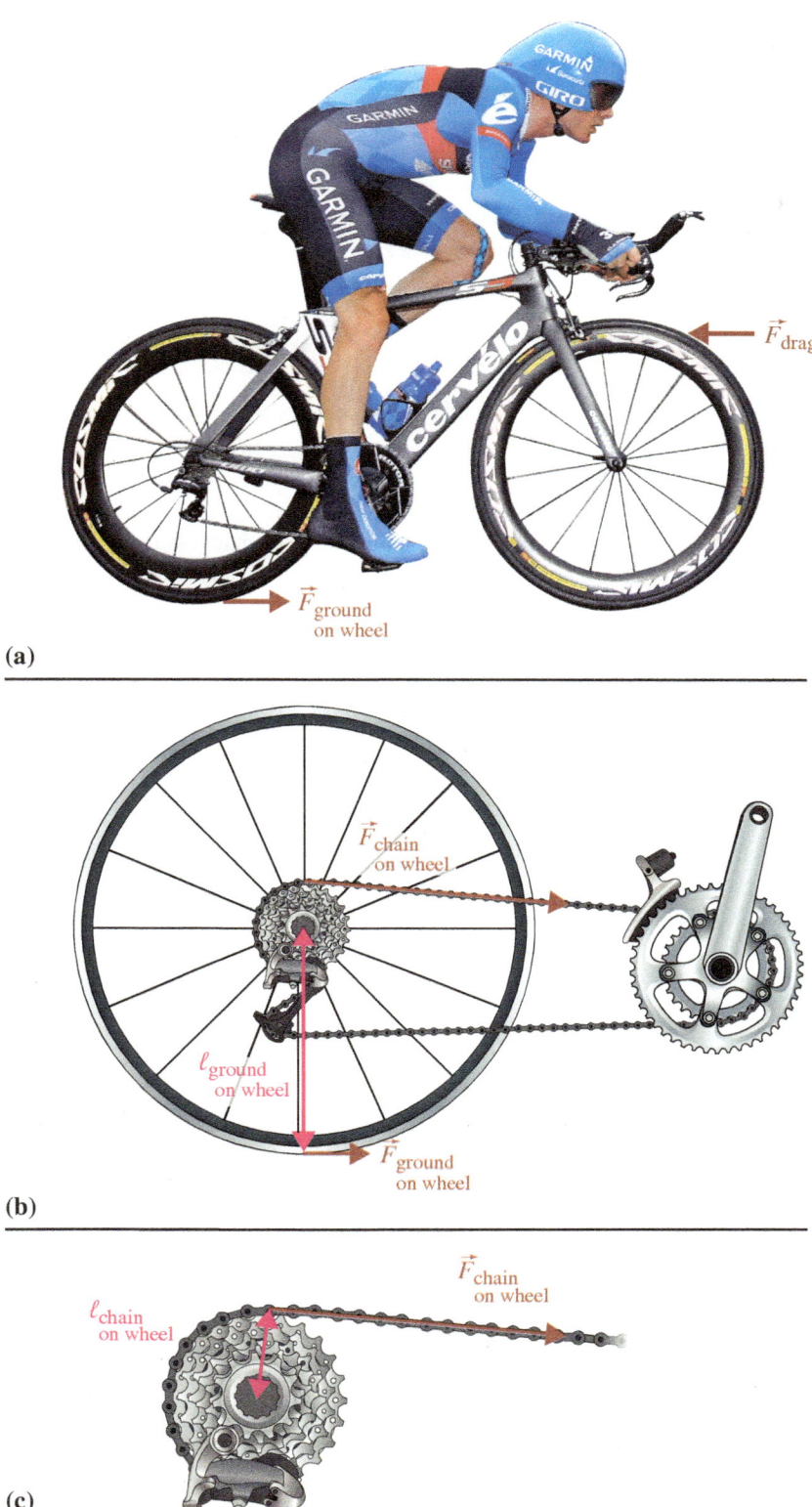

Figure 9.26. The three panels show the same situation, in increasing detail. (a) Talansky in the Tour de France, traveling at a constant velocity. The retarding force of air drag is countered by the forward-directed force of friction from the road. (b) The force from the ground wants to make the wheel rotate counterclockwise. That is, it produces a counterclockwise torque. The force of the chain produces a clockwise torque. (c) The lever arm for the torque from the chain equals the radius of the gear. (© *Eric Gaillard/Reuters/Corbis*).

The connection between all these torques is provided by our hopefully not-forgotten friend Sir Isaac.

Newton's First Law for Angular Motion of a Wheel
A wheel rotates at a constant angular velocity if the *net* torque on the wheel is zero.

$$\sum_i \tau_i = 0 \quad \leftrightarrow \quad \omega = \text{constant} \qquad (9.26)$$

(Actually, this law is for any object that doesn't change shape, not just wheels.)

See that little disclaimer in there about objects that don't change shape? Things get more complicated if something such as a human body changes shape while spinning. We'll come to this in chapter 10.

Talansky's chain

Let's get to some numbers. When a cyclist is putting together a racing bike, the weight of the components is a key issue. However, a light component may not be strong enough to withstand the forces put on it. No rider wants to lose his chain during a race. Let's get a feeling for the tension Talansky puts his chain under while he cruises along at 9.9 m/s in stage 17 of the 2013 Tour.

To figure this out, we have to work through several steps.

The first thing we need to know is the magnitude of the forward-pointing force of the road on Talansky's bike. Since he's moving at constant velocity, this has the same magnitude as the *backward* force of air drag and rolling friction. Using some values from table 9.3 gives us

$$|\vec{F}_{\substack{\text{ground} \\ \text{on wheel}}}| = C_R \left(m_\text{Talansky} + m_\text{bike}\right) g + \tfrac{1}{2} C_D \rho_\text{air} A v^2$$

$$= 0.003 \,(63\text{ kg} + 6.8\text{ kg})\,(9.8\text{ m/s}^2)$$

$$+ \tfrac{1}{2}(0.77)\,(1.2\text{ kg/m}^3)\,(0.36\text{ m}^2)\,(9.9\text{ m/s})^2$$

$$= 2.1\text{ N} + 16.3\text{ N} = 18.4\text{ N}$$

or about 4 lb.

The torque on the wheel from this force is

$$\tau_\text{ground} = |\vec{F}_{\substack{\text{ground} \\ \text{on wheel}}}|\, \ell_{\substack{\text{ground} \\ \text{on wheel}}} = 18.4\text{ N} \times \underbrace{0.331\text{ m}}_{\substack{\text{wheel radius} \\ \text{from table 9.4}}} = +6.1\text{ N}\cdot\text{m}.$$

The plus sign is a reminder that it's a counterclockwise torque.[18] Also notice that the unit of torque is N·m (or ft·lb).

> Even though 1 N·m is mathematically equivalent to a joule (1 J), joules are only used for energy, not torque.

---

[18] Pause and think for a moment: is it clear to you that this is a counterclockwise torque?

The 17-tooth rear cog that Talansky was using has a radius[19] of 1.35 in., or 0.034 m. Equation 9.26 tells us that the torque from the chain has the same magnitude as, but opposite sign to, the torque from the ground:

$$\tau_{chain} = -\tau_{ground} = -6.1 \text{ N} \cdot \text{m}.$$

And equation 9.25 lets us find the force from the chain:

$$\tau_{chain} = -|\vec{F}_{chain}|\ell_{chain} \rightarrow |\vec{F}_{chain}| = -\frac{\tau_{chain}}{\ell_{chain}} = -\frac{-6.1 \text{ N} \cdot \text{m}}{0.034 \text{ m}} = 179 \text{ N}$$

This is about 40 lb, about 10 times the force of the road on the bike.

This tells us that to oppose a 4-lb air drag, Talansky has to pull on the rear cog with the chain with a force of 40 lb! The point is that, **to produce a reasonable torque with a small lever arm** (less than 1.5 in. in the case of the rear cog), **you need a really large force**. If you've ridden a 10-speed, you already know that using the small gears on the rear means you'll have to exert a larger force, compared to using the larger gears. On the upside, the smaller gears allow you to pedal more slowly. It is up to the rider to set the balance between frequency of pedal cycles (cadence) and the force of each pedal cycle.

Most modern chains have no problem handling typical tensions. Specifications of so-called minimum tensile strength are typically 1500 lb or so. Slow wear and stretching over time are a much larger concern.

> **Important point**
>
> In panel (b) of figure 9.26, there are two horizontal forces on the wheel: the force from the chain and the force from the road. If we want to understand the motion of the rider+bike combination, only "external forces" enter into our free-body diagram, as in panel (a) of that figure. The various forces that components of the rider+bike system exert on one another, such as $\vec{F}_{chain \text{ on wheel}}$, are "internal forces" that cannot accelerate the rider+bike system.

> **Question**
>
> In figure 9.26, Talansky is moving at constant velocity. But look at the forces drawn on the wheel—they both point forward. Doesn't that mean that the wheel should accelerate forward? Is there an undrawn force in the horizontal direction? If so, why didn't we include the torque that this force generates in our discussions here?

9.3.5 Drivetrain III: chain reaction

The force of the chain exerts a torque on the rear wheel (through the rear gear). But we know where that force "comes from." The force transmitted by the chain is a *tension*, which means that the left end pulls to the right with the same force that the right end pulls to the left. We saw an example of this in figure 3.20.

For the bike chain, the situation is shown in figure 9.27. Clearly, the chain is exerting a clockwise torque on the rear gear and a counterclockwise torque on the front gear. The two torques have opposite signs. But are they "equal and opposite" torques? *No*.

Equal and opposite forces need not produce equal and opposite torques. This is a key element in the physics of a bicycle.

---

[19] The teeth on a modern bike gear have approximately the same spacing as the links on a chain: 0.5 in. Therefore, the circumference of a gear is 0.5 in. × (#teeth), and the radius is

$$\boxed{R_{gear} = 0.5 \text{ in.} \times (\#teeth)/(2\pi)} \tag{9.27}$$

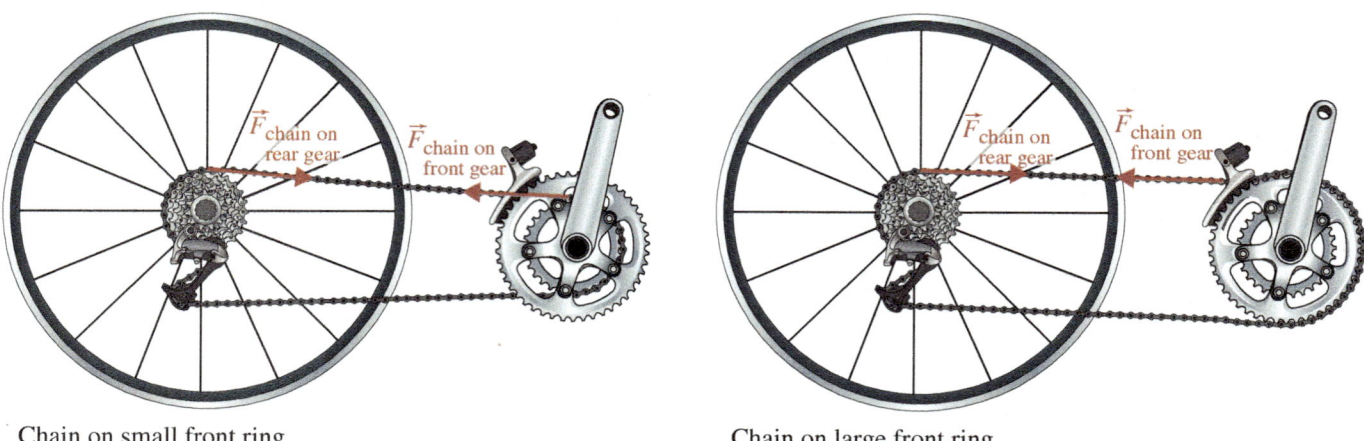

Chain on small front ring   Chain on large front ring

Figure 9.27. All the forces in this figure have the same magnitude (say, 40 lb). In the left panel, a small front gear is selected; in the right a large one is selected. Thus, while the torque exerted on the rear wheel is the same in both panels, the torque on the front gear is larger in the right panel.

The *forces* that the front and rear gears exert on each other are equal and opposite: $\vec{F}_{\text{chain on rear gear}} = -\vec{F}_{\text{chain on front gear}}$. But the counterclockwise torque that the chain exerts on the front gear depends on the lever arm, that is, on the radius of the front gear used.

In the left panel of figure 9.27, a small front gear is used to put a tension in the chain. In the right panel, a larger front gear is being used to exert the *same* tension. That means the force on the rear gear is the same in the two panels. Thus the torque exerted on the rear wheel is the same in the two panels, so the same bike speed could be maintained, and so on.

The *difference* is that the magnitude of the torque exerted on the front gear is less when the small gear is used. This means that the cyclist has to produce less torque, if he uses the smaller front gear rather than the larger one. If you've ridden a multispeed bike, you've surely noticed that it's easier to pedal when the chain is on the small front gear.

So why doesn't a cyclist *always* use the smaller gear then? You know the answer: by choosing the smaller front gear, the gear ratio GR defined in equation 9.20 is reduced, and equation 9.22 tells us that the rider's cadence will need to increase if he wants to maintain speed. That's always the trade-off on a bike: pedal slowly but push hard, or pedal quickly but more easily. If you drive a standard-transmission car or motorcycle, you know that it's the same trade-off in those cases. Both the internal combustion engine and the human body have a narrow range of cadence in which they perform well; this built-in optimal range determines what gear is best in any situation.

9.3.6 Drivetrain IV: pushing the pedals

Finally, starting from the wheel, we have worked our way up the drive chain and now arrive at the pedals, where the rider exerts forces and torques on the machine as he burns chemical fuel in his body. To maintain the bike's speed, the rider needs to exert a *clockwise* torque on the front gearwheel, to exactly cancel the *counterclockwise* torque indicated in figure 9.27. This torque is generated by pushing on pedals located on cranks of length 0.165 m.

A torque is a torque, as they say,[20] so in principle there might be nothing special about a torque generated by a leg. However, understanding how an athlete interfaces with his machine requires a more detailed knowledge of torque. Here's why. Look at panel (a) of figure 9.28. The pedal is at its lowest point, located directly beneath the

---

[20] Actually, *do* they say that? And who are "they"?

axis of the front gear, and the rider is exerting a force pointed directly down. What is the direction of the torque in this case? Clockwise? Counterclockwise? You know the answer: that force will tend to rotate the gear in *neither* direction—there is no torque.

 The lever arm of a torque depends on both *where* the force is applied and the *direction* of the force.

But doesn't that contradict equation 9.25, which says $\tau = \pm|\vec{F}|\ell$? Actually, it does not, because *the lever arm ℓ is not the distance between the application of the force and the axis of rotation, but rather is the shortest distance between the "line of action" of the force and the axis of rotation*. The line of action of a force is an imaginary line that passes through the arrow representing the force, and it is shown in blue in figure 9.28. In panel (a), the line of force passes directly through the axis of rotation, so $\ell = 0$ and that force generates no torque.

In panel (b), the crank is in approximately the 2 o'clock position, and pushing down on the pedal now clearly generates a clockwise torque. The lever arm (shown in white) is no longer zero, but it is not as large as it could be. Panel (c) shows what happens if the rider changes the angle of the applied force: the lever arm increases so that even if the magnitude of the force is kept the same (that is, the rider pushes just as hard as in panel (b)), the torque increases. Road racing shoes clip on to the pedals,[21] allowing the rider to optimize the direction of the applied force.

A little thought and some examination of figure 9.28 should tell you that, for a given force, maximum torque is generated if the force is exerted perpendicular to the crank angle. In that case, $\ell = d$, the crank length. At any other angle, $\ell < d$ and the torque is reduced.

In fact, the torque of a foot pushing on a bicycle pedal is

$$\tau_{\text{pedal}} = |\vec{F}_{\text{foot}}|\ell = |\vec{F}_{\text{foot}}|d\sin\theta, \qquad (9.28)$$

where d is the crank length and θ is the angle between the line of action of the force and the crank, as shown in panel (d) of figure 9.28.

 The torque is largest when the force is perpendicular to a line connecting the point where the force is applied and the axis of rotation. The drive mechanism of a bicycle is designed such that the chain always exerts maximum torque on the gears.

In reality, both the angle of force application and the magnitude of the force change as the cyclist cranks a full cycle. Figure 9.29 shows a measurement of the torque generated as a function of crank angle. Not surprisingly, the torque is much larger on the powerful downstroke. On the upstroke, the torque associated with a

(a)

(b)

(c)

(d)

Figure 9.28. The angle between the rider's leg and the pedal crank changes throughout the pedaling cycle. Even though the distance between the axis of rotation and the point of force application (that is, the pedal) is always equal to the crank length d, the lever arm ℓ changes as the *line of action* of the force changes. The line of action is shown as the yellow dashed line, and the lever arm is the red double-headed arrow. In panel (a), the rider pushes vertically down on the pedal in its lowest position; the line of action passes through the axis of rotation, so no torque is generated. In panel (b), the rider pushes down on a pedal that is in the two o'clock position; a clockwise torque is generated. In panel (c), the pedal position and force magnitude is the same as in panel (b), but the "*direction*" of the force is such that its line of action does not come as close to the axis of rotation; thus, the torque generated in panel (c) is larger than that in panel (b), due to a larger lever arm. Panel (d) is a detailed blow-up of panel (c). (*Photos courtesy of Anna Lisa*).

---

[21] Strangely enough, pedals that clip onto racing shoes are called "clipless" pedals.

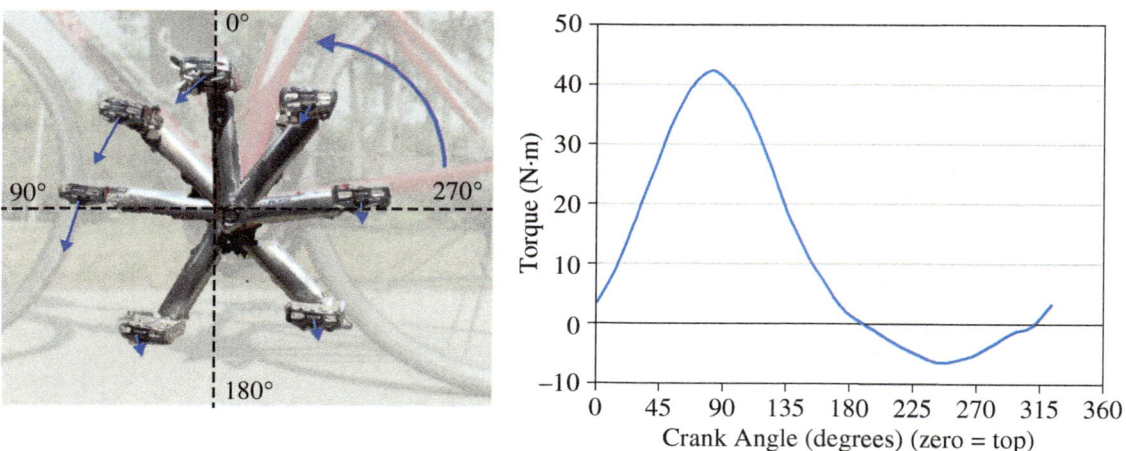

Figure 9.29. The torque generated by a cyclist varies as the crank position changes, for two reasons. First, the amount of force exerted by the rider is greater on the downstroke and reduced on the upstroke. Second, the direction of the applied force is dictated by leg geometry and pedal angle, changing the line of action and lever arm. The figure shows the torque from the force on one of the two pedals. Graph based on data from Sanderson et al., *Journal of Sports Sciences* **18**: 173 (2000). (*Photo courtesy of Mike Lisa*).

given pedal is in fact slightly negative, meaning that the foot is opposing the gear's rotation; it is being "lifted" by the foot pushing on the other pedal.

EXAMPLE 9.12 *The tension in Talansky's chain*

Andrew Talansky is a light guy known for his climbing. On a serious slope, he will sometimes get out of the saddle and stand on his pedals. Let's figure out the tension he puts his chain under, when he rides at constant speed and puts all his weight on the pedal and the crank is parallel to the ground. (That is, we are talking about the 90° position shown in figure 9.29.)

Talansky uses a 170-mm crank on hills, which is just about the limit of a "safe" crank. Longer ones may strike the ground when the rider leans for a curve.

To understand the tension in the chain, we need to think about the forces and torques on the pedals-plus-front-gear object, shown in figure 9.30. Since Talansky's moving at constant speed, this object is spinning at a constant rate, and Newton's first law (equation 9.26) tells us that the clockwise torque generated by Talansky is exactly canceled by the counterclockwise torque generated by the chain.

$$\tau_{\text{Talansky}} = -\tau_{\text{chain}} \quad \rightarrow \quad |\tau_{\text{Talansky}}| = |\tau_{\text{chain}}|$$

$$\rightarrow \quad |\vec{F}_{\text{Talansky}}|\ell_{\text{Talansky}} = |\vec{T}_{\text{chain}}|\ell_{\text{chain}} \quad \rightarrow \quad |\vec{T}_{\text{chain}}| = \frac{\ell_{\text{Talansky}}}{\ell_{\text{chain}}}|\vec{F}_{\text{Talansky}}|.$$

The force Talansky is exerting on the pedal as he stands on it is equal to his weight. (It may be slightly lower if he supports himself somewhat on the opposite pedal or slightly higher if he pulls up on the handlebars.) The lever arm of that force about the rotation point is the crank length d. And the lever arm of the chain's tension is the radius of the gear. To climb, the rider usually uses the small front gear, having 42 teeth in Talansky's case (see figure 9.23). Equation 9.27 gives its radius:

$$R_{\text{gear}} = 0.5 \text{ in.} \times \frac{\#\text{teeth}}{2\pi} = 0.5 \text{ in.} \times \frac{42}{2\pi} = 3.34 \text{ in.} \times \frac{0.0254 \text{ m}}{1 \text{ in.}} = 0.085 \text{ m}.$$

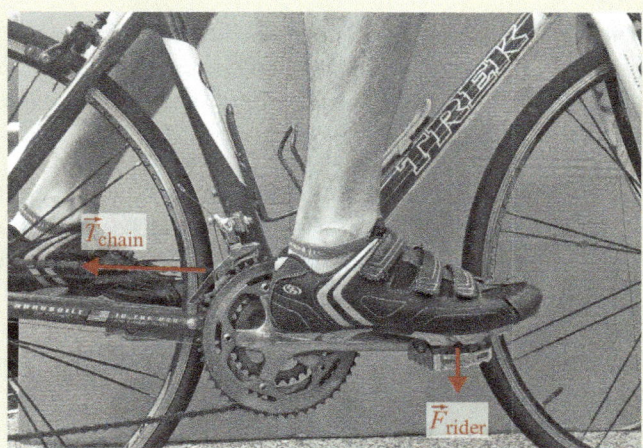

Figure 9.30. Only the forces that generate torques on the pedal-plus-gear ring are shown. (*Photo courtesy of Anna Lisa*).

Finally, we can find the tension:

$$|\vec{T}_{\text{chain}}| = \frac{d}{R_{\text{gear}}}|\vec{W}_{\text{Talansky}}| = \frac{0.170 \text{ m}}{0.085 \text{ m}}(139 \text{ lb}) = 2 \times (139 \text{ lb}) = 278 \text{ lb}.$$

That's the same tension as if two Talansky's were dangling vertically at the end of the chain!

Question

Figure 9.30 shows only forces that generate a torque. Are there other forces? If so, which way do they point? Why do they not generate a torque?

9.3.7 Rolling friction, revisited

I have been a bit coy about the force of rolling friction, first introduced in section 9.2.2. For example, I have not shown any diagrams with an arrow showing $\vec{f}_R$. You've probably been wondering about that. No? Well, you should have been. It's actually an interesting story, and it tells us something about wheels *and* about physics.

If I *had* drawn an arrow for rolling resistance, say, in panel (a) of figure 9.26, where would I draw it, and which way would it point? Well, it's a retarding force, so presumably it would point to the left, opposite to Talansky's motion. And presumably it acts where the wheel meets the road.

But look, the force from the road in the figure points *in* the direction of motion, and it's the *static* friction force. This static friction force is in fact a "reaction" force to the pull of the chain that we discussed earlier (in figures 3.11 and 3.12). It does not slow the wheel unless the bike is being braked. Does the road exert a force both *in* the direction of motion and simultaneously *against* the direction of motion? It turns out that while it's convenient to think of $\vec{f}_R$ as a horizontal force for the purpose of calculations, what's *really* going on has to do more with torque and energy.

A round wheel

Let's take the bike and chain away for a minute and just think of a completely round wheel rolling along the ground. It turns out that, if we neglect air drag, a *completely* round wheel on *completely* flat ground will roll along at constant speed forever, never slowing down![22] These are the only forces on the wheel:

1. The wheel's weight. We've said before that an object's weight can be considered to have effect at its center of mass. That hasn't been important before, but it is now. Since the wheel is symmetric, that means $\vec{W}$ acts right at the center.
2. The normal force from the ground. Since the wheel is perfectly round, only a single tiny point touches the ground. This point is located directly beneath the wheel's center, and $\vec{F}_N$ acts there.

The situation is shown in the left side of figure 9.31. These two forces share a common line of action, and this line passes directly through the axis of rotation. Therefore, the lever arm ℓ is zero, and there is no torque.

"Rolling friction" arises when the line of action of the normal force does not pass through the center of rotation.

A wheel that deforms

All real tires, of course, deform at the interface, meaning that part of the normal force *does* have a nonzero lever arm, as shown in figure 9.31. The important detail is that the bottom of the wheel toward the "front" is being squashed down by the rolling motion, so the normal force is larger there, compared to the part of the wheel touching the ground in the "back." This is indicated by the different-length force arrows in the figure. The upshot is that the larger force on the front generates a torque that opposes the rolling motion—*that's* what's really behind "rolling friction."

Well, there's actually one more thing. The torque slows the wheel, which means the kinetic energy has gone down. Where has the energy gone? You got it—into heat, as the rubber is squashed and released.

Now you know why a deflated tire wastes gas or a little boy's energy: torque and energy issues combine to generate "rolling friction" (and a hot tire). Low-rolling-resistance car tires and racing bicycle tires are hard and deform little.

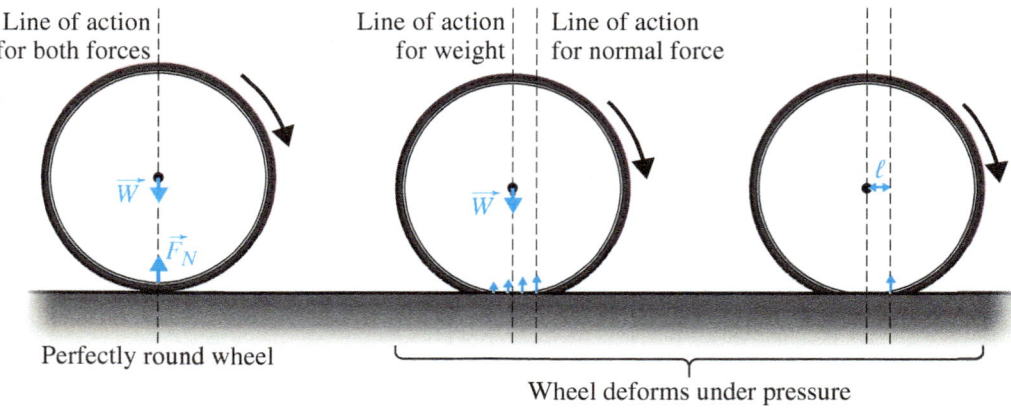

Figure 9.31. If a rolling wheel is deformed where it touches the ground, the normal force on average acts a bit forward of the wheel's axle, generating a torque that tends to slow the wheel.

---

[22] There are interesting caveats to this statement at the atomic level, but we aren't concerned with them.

Section 9.3   *Talansky Drives the Bike*   **289**

> **Question**
>
> What if the wheel didn't heat up at all? What if, when each little segment of tire was squished, it stored all that energy as elastic potential energy, and then released that into kinetic energy when it decompressed, again without heat? Would the wheel slow down?
>
> - If so, then where did the "lost" kinetic energy go?
> - If not, then what's wrong with the picture in figure 9.31, which implies that the wheel *must* slow down because of the torque that opposes motion?

How large is the flattened patch?

It seems pretty obvious that if you want to minimize rolling friction, you want as round a wheel as possible, that is, the smallest possible "flattened patch" where the tire hits the road. For many rolling objects in sports such as bicycle tires and soccer balls (but not bowling balls), internal air pressure is what maintains the shape. The air pushes "out," and the stretched rubber pushes "back in." In this way, air pressure maintains the round shape of the tire or ball.

At the flattened patch where the tire touches the road, however, the rubber is not stretched. Here the air pressure exerts its force directly downward against the road, transmitting the weight of the bike and its rider. Pressure is measured in pounds per square inch (psi). To get the upward (normal) force exerted by the flat patch, multiply the pressure by the area of the patch:

$$|\vec{F}_{\text{normal,tire on road}}| = (\text{tire pressure}) \times (\text{area of flat patch}). \qquad (9.29)$$

The rear wheel of a bike may support 100 lb of weight from the rider-plus-bike. If the tire pressure is 100 psi, the flat patch is 1 in.^{2}. On the other hand, if the tire pressure is low, say, 60 psi, then

$$(\text{area of flat patch}) = \frac{|\vec{F}_{\text{normal,tire on road}}|}{(\text{tire pressure})} = \frac{100 \text{ lb}}{60 \text{ lb/in.}^2} = 1.7 \text{ in.}^2.$$

The increased flat patch size means increased lever arm for the normal force (see figure 9.31) and increased rolling resistance.

Now you understand the physics behind figure 9.6.

Rolling friction for other sports

In cycle racing, the road is much stiffer than the tire, so rolling friction is associated with deformation of the tire.

In some sports, however, the rolling object is hard and the surface is soft, as in billiards. Here it is the table felt that deforms (and slightly heats). In soccer, both the ball and the playing surface distort somewhat.

You intuitively know that the combination of a hard bowling ball on a hard alley surface will lead to very low rolling friction. That's why a little kid can give the tiniest push to the ball and it will slowly roll at almost constant (low) speed down the entire 60-ft lane.

Hybrid wheel and tire: large I → smaller α

Racing wheel and tire: small I → large α

Figure 9.32
The racing wheel on the right has a low spoke count and thin tires. Therefore it has a lower moment of inertia than the heavier, sturdier touring bike wheel on the left. If the same torque is applied to the two wheels, the one with smaller moment of inertia gets a larger angular acceleration. That is, it is easier to spin. (*Photos courtesy of Mike Lisa*).

9.3.8 Back to basics: the wheel again

A tremendous amount of research and development goes into design and construction of bicycle wheels. Strength and durability are obviously crucial, as is aerodynamics.

In cycling, keeping the mass of the wheel to a minimum (while maintaining sufficient strength) is key. However, you might not have guessed that *how the mass is distributed* may also be important. After all, if I'm looking to accelerate a bike, its mass comes into play according to Newton's second law (equation 3.6): $a = \sum F/m$.

Why should it matter where the mass m is? Newton's second law doesn't care if all the mass is in the seat post, or the handlebars, or the top tube. The mass of the bike is all that matters, not how that mass is distributed. While this is true for linear (translational) motion, it turns out to be false when things are spinning. As you may have noticed in this chapter, rotational motion is a little more complicated than linear motion.

Wheels up: angular acceleration and Newton's second law

Naturally, a bike involves both translational and rotational motion simultaneously. To focus on the special features of rotational motion, let's turn the bike upside down on its seat and handlebars, and soon we'll start cranking the pedals[23] as in figure 9.32.

The idea (here as well as in a race when the bike is right side up) is to get the wheel spinning, starting from rest, that is, to give it an **angular acceleration**. This is an important new concept, quantifying how quickly the angular velocity changes from one value (ω_1) to another (ω_2).

$$\bar{\alpha} = \frac{\omega_2 - \omega_1}{t_2 - t_1} = \frac{\Delta \omega}{\Delta t}. \tag{9.30}$$

As usual, the bar over $\bar{\alpha}$ is a reminder that equation 9.30 tells the *average* angular acceleration for the time between t_1 and t_2.

Angular acceleration is a lot like the translational acceleration that we've been using throughout this book.[24] You know intuitively that to give the wheel an angular acceleration (that is, to change the rotational frequency), you're going to have to exert a torque on it, either by cranking the pedals or grabbing the tire with your hand and exerting a force perpendicular to the spokes. More torque, more angular acceleration. So, is there Newton's second law for angular motion, similar to $F = ma$ (equation 3.6)? Of course there is.

Newton's Second Law for Angular Motion

The angular acceleration of an object is the ratio of the net torque on the object to its moment of inertia:

$$\alpha = \frac{\sum_i \tau_i}{I} \tag{9.31}$$

---

[23] An astonishing number of people remember doing precisely this when they were small children, imagining that they were "making ice cream." Nobody seems to know where this concept began or how it spread to all corners of America.

[24] In fact, check out equation 2.4 on page 20. It's essentially the same equation, except now, everything is "angular."

Torque τ is essentially an "angular force." So if you compare the formulas $a = \sum F/m$ and $\alpha = \sum \tau/I$, you can guess that I must be some kind of "rotational mass." And so it is, but it's a little more complicated than the mass we've been talking about until now. The **moment of inertia** (sometimes called rotational inertia) I depends on the shape of an object and the distribution of mass, in addition to the mass itself. It even depends on the axis of rotation; we'll come to this in chapter 10.

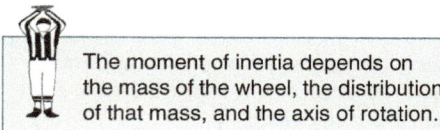

The moment of inertia depends on the mass of the wheel, the distribution of that mass, and the axis of rotation.

Calculating I for realistic shapes can be quite complicated and usually involves some calculus. Table 9.5 lists some moments of inertia for simple shapes that approximate spinning objects in sports. If we look at the formulas in the last column, it's pretty obvious that I grows as more mass is added. That makes sense because it's harder to spin a heavy wheel.

Table 9.5. Moments of inertia relevant to sports. The connections are approximate, since a baseball, for example, is not a completely uniform sphere.

	Shape and objects of approximately that shape	Moment of inertia
	Filled sphere ≈ baseball, billiard ball, golf ball, lacrosse ball, bowling ball Rotation axis passing through the center	$I_{\text{filled sphere}} = \frac{2}{5}mR^2$
	Hollow sphere ≈ soccer ball, table tennis ball, volleyball, basketball, tennis ball, … Rotation axis passing through the center	$I_{\text{hollow sphere}} = \frac{2}{3}mR^2$
	Hoop ≈ bike tire without the wheel Rotation axis through center and perpendicular to plane of hoop	$I_{\text{hoop}} = mR^2$
	Solid disc ≈ frisbee, discus (kind of) Rotation axis through center and perpendicular to plane of disc	$I_{\text{disc}} = \frac{1}{2}mR^2$
	Bike wheel (with tire): between a hoop and a disc Rotation axis through center and perpendicular to plane of wheel	$\frac{1}{2}mR^2 < I_{\text{bike wheel}} < mR^2$
	Thin solid (or hollow) rod ≈ cricket bat, baseball bat Rotation axis passing through one end	$I_{\text{rod}} = \frac{1}{3}mL^2$
	Real baseball bat (or golf club) swung about its end: more mass at end, so larger I than for a simple rod Rotation axis passing through handle	$I_{\text{baseball bat}} > \frac{1}{3}mL^2$
	Heavy mass at the end of a light rope or chain ≈ hammer throw Rotation axis passing through end of rope	$I_{\text{mass on light rope}} = mL^2$

> It's well known that when a baseball player "chokes up" on a bat (that is, grips it a little more toward the middle than the end), he can swing the bat faster. Why? The bat's mass doesn't change, so why should the same effort lead to a faster swing?

> When a tire is connected to a wheel, the moment of inertia of the combined wheel+tire object is just the sum of the moments of inertia of the wheel and the tire separately.

But I also grows as the *square* of the distance of that added mass relative to the axis of rotation. So, for example, you could add 30 lb of material right at the middle of a wheel, and that would not increase the torque required to spin the wheel at all! Add it at the edge of the rim, however, and it gets a lot harder to spin.

Likewise, if you moved all the mass from the inside of a baseball to its surface, the moment of inertia would increase by 66%, from $\frac{2}{5}mR^2$ to $\frac{2}{3}mR^2$ and it would get harder to spin.

> **Question**
>
> A 2011 analysis by cycling guru Lennard Zinn in the magazine *VeloNews* compared two racing wheels. Without an inner tube or tire, the Shimano WH-7850-C24-TU20 rear wheel had a mass of 725 g and moment of inertia of 3.0×10^5 g·cm^2. Meanwhile, the Bontrager Race XXX Lite was 5% lighter (693 g), but at 3.4×10^5 g·cm^2, its moment of inertia was 13% *higher* than Shimano's. Both were 700c wheels, so their radius was the same.
>
> How could this be? What does this tell you about the difference in construction of the wheels?

The kinetic energy of a spinning wheel

Let's keep the bike upside down, so it's not going anywhere. Equation 6.6 for kinetic energy says KE $= \frac{1}{2}mv^2$. If v is the center-of-mass velocity, this formula says that there is no kinetic energy, since $v_{\text{c.m.}} = 0$. But come on, the atoms in the spinning wheel are definitely moving, so the kinetic energy can't really be zero. That's true. *In addition* to the kinetic energy associated with a moving center of mass, there is a rotational kinetic energy. To account for this, we have to add a term to equation 6.6. The energy of a rotating, moving object is

$$\text{KE} = \underbrace{\tfrac{1}{2}mv_{\text{c.m.}}^2}_{\substack{\text{translational} \\ \text{kinetic energy}}} + \underbrace{\tfrac{1}{2}I_{\text{c.m.}}\omega^2}_{\substack{\text{rotational} \\ \text{kinetic energy}}} \qquad (9.32)$$

The "c.m." subscript on I just means we have to use the moment of inertia for the object when the axis passes through the center of mass. In chapter 10, we look at how I depends on the position of the axis. It's not an issue at the moment, as our wheels are rotating about their center of mass.

Understanding the energy of translation and rotation is very useful for shedding some light on an argument that rages in the cycling community, as we discuss next.

EXAMPLE 9.13 *Does I matter in a race?*

One of the most hotly debated topics in online cycling forums is the importance of the wheel moment of inertia for racing performance. Top-end cycling gear is not cheap—racing wheels can easily set you back several thousand dollars—and some manufacturers and enthusiasts claim that a 10% reduction in moment of inertia will improve your race time significantly. There are equally impassioned claims to the contrary; they say that the wheels' moment of inertia has no significant effect on the race.

Who is right? The advantage of a true quantitative understanding of physics is that we don't need to rely on manufacturers' claims or blogs. We can run the numbers ourselves.

First, we'll start by acknowledging that, indeed, any energy stored in the spinning wheels (the second term in equation 9.32) is in a sense "wasted" in terms of getting the rider to the finish line quickly. Making I as low as possible means that all available energy goes into the bike's translational energy—the motion of its center of mass.

In this example, we look at the fraction of a bike's energy that is tied up in rotational kinetic energy of the wheels. We look at a rider+bike combo with a mass of 80 kg, riding at 15 mph (6.7 m/s).

A high-end wheel has a moment of inertia $I = 3.0 \times 10^{-2}$ kg·m². However, this is the moment of inertia for the bare wheel, *without a tire or tube*. Since cyclists generally have a tire on their wheel, we have to account for that. It turns out that, just as the mass of the wheel+tire combined object is simply the mass of the wheel plus the mass of the tire, the moment of inertia of the wheel+tire combined object is simply the sum of the moments of inertia of the tire and the wheel. Moments of inertia add. That makes things easy.

A typical tire has a mass of 250 g or so, and its geometry is pretty well approximated by a hoop. Table 9.5 gives us the formula for the moment of inertia:

$$I_{\text{tire with innertube}} = mR^2 = (0.250 \text{ kg}) \underbrace{(0.33 \text{ m})^2}_{\text{for 700c tire}} = 0.025 \text{ kg} \cdot \text{m}^2.$$

This is almost as big as the 0.030 kg·m² of a typical high-end racing wheel. Accounting for the tire, the rolling wheel has $I_{\text{wheel with tire}} = 5.5 \times 10^{-2}$ kg·m².

Now it's time to compute energies. The translational kinetic energy of the bike+rider combined object is

$$\text{KE}_{\text{trans}} = \tfrac{1}{2}mv_{\text{c.m.}}^2 = \tfrac{1}{2}(80 \text{ kg})(6.7 \text{ m/s})^2 = 1795.6 \text{ J}.$$

To find the rotational kinetic energy of the wheels, we need their angular velocity, which we find by using equation 9.24:

$$\omega = \frac{v_{\text{bike}}}{R_{\text{wheel}}} = \frac{6.7 \text{ m/s}}{0.331 \text{ m}} = 20.4 \text{ rad/s}.$$

So their rotational kinetic energy is

$$\text{KE}_{\text{rot}} = \tfrac{1}{2}I\omega^2 = \tfrac{1}{2}\left(5.5 \times 10^{-2}\right)(20.4/\text{s})^2 = 11.44 \text{ J}.$$

Since there are two wheels, we should double this number and conclude that the energy tied up in the rotation of the wheels is 22.88 J.

Comparing this number to KE_{trans}, we find that the spinning wheels add only about 1.3% to the total energy of the bike! That is nothing! Furthermore, even if the wheel were made out of air, you'd still have the tire, which comprises one-half of the moment of inertia of the spinning object.

On top of all this, remember that almost all the rider's energy goes toward fighting the wind and hills, neither of which has to do with the moment of inertia.

The bottom line is that the moment of inertia for road bike wheels contributes insignificantly to the energy the rider has to put into the bike, and any money spent solely on reducing I_{wheel} is money wasted.

Collected Equations

$$\bar{P}_{\text{output by cyclist}} = \eta_{\text{chemical} \rightarrow \text{mechanical energy}} \times \bar{P}_{\text{chemical input}} \qquad (9.1)$$

$$\text{Work done by cyclist} = W_{\text{cyclist}} = \bar{P}_{\text{output by cyclist}} \cdot \Delta t \qquad (9.2)$$

$$\text{Food energy burned} = -\Delta E_{\text{chem}} = \bar{P}_{\text{chemical input}} \cdot \Delta t \qquad (9.3)$$

$$\text{Food energy burned} = -\Delta E_{\text{chem}} = \left(\frac{1\ \text{Cal}}{\text{kg} \cdot \text{hr}}\right) \times \text{MET} \times m_{\text{person}} \times \Delta t \qquad (9.4)$$

$$\bar{P}_{\text{chemical input}} = \frac{1\ \text{Cal}}{\text{hr}} \times \text{MET} \times \frac{m_{\text{person}}}{\text{kg}} = 1.16\ \text{W} \times \text{MET} \times \frac{m_{\text{person}}}{\text{kg}} \qquad (9.5)$$

$$\bar{P}_{\text{chemical input}} = \left(306\ \frac{\text{Cal} \cdot \text{min}}{\text{hr} \cdot \text{liter}}\right) \times \text{VO}_2 \qquad \left(\text{if VO}_2 \text{ is in units of } \frac{\text{liter}}{\text{min}}\right) \qquad (9.6)$$

$$\bar{P}_{\text{chemical input}} = \left(0.306\ \frac{\text{Cal} \cdot \text{min}}{\text{hr} \cdot \text{ml}}\right) \times m_{\text{athlete}} \times \text{VO}_2 \qquad (9.7)$$
$$\left(\text{if VO}_2 \text{ is in units of } \frac{\text{ml}}{\text{min} \cdot \text{kg}}\right)$$

$$P = F_x v_x + F_y v_y = |\vec{F}| \times \left(\text{the velocity component parallel to } \vec{F}\right) \qquad (9.8)$$

$$P_{\text{output cyclist}} = |\vec{F}_{\text{propulsive}}||\vec{v}_{\text{bike}}| \qquad (9.9)$$

$$\text{at constant velocity} \qquad \vec{F}_{\text{propulsive}} = -\vec{F}_{\text{hill}} - \vec{f}_R - \vec{F}_D \qquad (9.10)$$

$$s = \sin\theta \qquad (9.11)$$

$$|\vec{F}_{\text{hill}}| = |\vec{W}| \times s = mgs \qquad (9.12)$$

$$|\vec{f}_R| = C_R \times |\vec{F}_N| \approx C_R \times |\vec{W}| = C_R mg \qquad (9.13)$$

$$\underbrace{\vec{v}_{\text{air on rider}}}_{\text{velocity of air molecules as seen by rider}} = \underbrace{\vec{v}_{\text{wind}}}_{\text{wind velocity as seen by a flagpole}} - \vec{v}_{\text{bike}} \qquad (9.14)$$

$$|\vec{F}_D| = \frac{C_D}{2} A \rho_{\text{air}} |\vec{v}_{\text{bike}} - \vec{v}_{\text{wind}}|^2 \qquad (9.15)$$

$$\bar{P}_{\text{output cyclist}} = \underbrace{C_R mg v_{\text{bike},x}}_{\text{overcoming rolling friction}} + \underbrace{smg v_{\text{bike},x}}_{\text{overcoming hills}}$$
$$\pm \underbrace{\tfrac{1}{2} C_D A \rho_{\text{air}} \left(v_{\text{bike},x} - v_{\text{wind},x}\right)^2 v_{\text{bike},x}}_{\text{overcoming wind}} + \underbrace{m a_{\text{bike},x} v_{\text{bike},x}}_{\text{leftover for acceleration}} \qquad (9.16)$$

$$= \left((C_R + s)mg \pm \tfrac{1}{2} C_D A \rho_{\text{air}} \left(v_{\text{bike},x} - v_{\text{wind},x}\right)^2 + m a_{\text{bike},x}\right) v_{\text{bike},x}$$

$$\text{side wind: } \theta_{\text{echelon}} = \tan^{-1}\left(\frac{|\vec{v}_{\text{bike}}|}{|\vec{v}_{\text{wind}}|}\right) \tag{9.17}$$

$$v_{\text{bike}} = 2\pi R_{\text{wheel}} \times f_{\text{wheel}} \tag{9.18}$$

$$v_{\substack{\text{wheel edge relative} \\ \text{to axle}}} = 2\pi R_{\text{wheel}} \times f_{\text{wheel}} \tag{9.19}$$

$$\text{GR} = \frac{R_{\text{front gear}}}{R_{\text{rear gear}}} = \frac{\text{\#teeth on front gear}}{\text{\#teeth on rear gear}} \tag{9.20}$$

$$f_{\text{wheel}} = \text{GR} \times f_{\text{pedals}} \tag{9.21}$$

$$v_{\text{bike}} = 2\pi \times \text{GR} \times R_{\text{wheel}} \times f_{\text{pedals}} \tag{9.22}$$

$$\omega = \begin{cases} +2\pi f_{\text{rot}} & \text{if the rotation is counterclockwise} \\ -2\pi f_{\text{rot}} & \text{if the rotation is clockwise} \end{cases} \tag{9.23}$$

$$|\vec{v}| = |\omega| \times (\text{distance to axis}) \tag{9.24}$$

$$\tau = \begin{cases} +|\vec{F}|\ell & \text{if the torque is counterclockwise} \\ -|\vec{F}|\ell & \text{if the torque is clockwise} \end{cases} \tag{9.25}$$

$$\sum_i \tau_i = 0 \quad \leftrightarrow \quad \omega = \text{constant} \tag{9.26}$$

$$R_{\text{gear}} = 0.5 \text{ in.} \times (\text{\#teeth})/(2\pi) \tag{9.27}$$

$$\tau_{\text{pedal}} = |\vec{F}_{\text{foot}}|\ell = |\vec{F}_{\text{foot}}|d\sin\theta \tag{9.28}$$

$$|\vec{F}_{\text{normal,tire on road}}| = (\text{tire pressure}) \times (\text{area of flat patch}) \tag{9.29}$$

$$\bar{\alpha} = \frac{\omega_2 - \omega_1}{t_2 - t_1} = \frac{\Delta\omega}{\Delta t} \tag{9.30}$$

$$\alpha = \frac{\sum_i \tau_i}{I} \tag{9.31}$$

$$\text{KE} = \underbrace{\tfrac{1}{2}mv_{\text{c.m.}}^2}_{\substack{\text{translational} \\ \text{kinetic energy}}} + \underbrace{\tfrac{1}{2}I_{\text{c.m.}}\omega^2}_{\substack{\text{rotational} \\ \text{kinetic energy}}} \tag{9.32}$$

Problems

1. To avoid falling over, a cyclist needs to maintain a minimum speed of about 3 mph. Consider a 180-lb rider on a 20-lb bike attempting the famous 31.5% slope of Filbert Street in San Francisco. Wind drag and rolling friction certainly aren't his main concerns! We'll ignore them.
   (a) In pounds, what is the backward force of the hill $|\vec{F}_{hill}|$?
   (b) How much power must he put out to climb the hill at a constant speed of 3 mph?
   (c) Refer to the NASA estimate in figure 7.10. About how long could a "healthy man" continue such a climb?

2. On a very tough hill, a mountain biker puts out an impressive 376 W of power. Assume his body's efficiency for converting chemical to mechanical energy is 20%.
   (a) At what rate is he burning food energy, in Calories per hour (Cal/hr)?
   (b) What is his oxygen uptake rate VO_2, in liters per minute (liter/min)?

3. It's a windy day with a solid 15-mph breeze. If you don't pedal and just let the wind push you along (from behind) at a constant speed on a level road, how fast will you go?
   Assume you and your bike have a combined mass of 85 kg and you use a commuting bike and an upright posture. You will need to use the values for C_D, A, and C_R from table 9.3.
   (There are several ways to solve this problem. You might end up using the quadratic equation (equation 8.8) or an iterative approach as in example 9.9, though there is a simpler way where neither of these are needed. Just find something that works; the answer should be the same either way.)

4. **What matters more, power or PWR?**
   [This problem will require the iterative technique discussed in section 9.2.4.] In his last two Tour de France wins (1989 and 1990), 67-kg American Greg LeMond averaged 382 W of power. In the following year, 80-kg Spaniard Miguel Indurain averaged an amazing 424 W in his final climb. They both used racing bikes, so use the appropriate values of C_D, A, and C_R from table 9.3. Also, don't forget to include their 6.8-kg bike, where appropriate.
   (a) What are the power-to-weight ratios of these two champions?
   (b) Which one could climb an 8% incline more quickly? That is, what constant speeds could the two riders maintain?
   (c) Which one would be faster on a level road? That is, what constant speeds could the two riders maintain?
   (d) Without doing any calculation, could you say who would be faster on an 8% *down*hill?

5. **Do uphills and downhills cancel out?**
   [This problem will require the iterative technique discussed in section 9.2.4.] For a given level of power output, a cyclist will obviously go more slowly up a slope than down the same slope; and on a level road he'll have a speed somewhere between these two cases. Let's see whether one does any better on a flat road, as opposed to going up and down a hill.
   On a day without wind, Andrew Talansky ($m = 69.8$ kg, including bike) puts out 250 W of power on his racing bike (get values for C_D, A, and C_R from table 9.3).
   (a) How long would it take him to ride 100 km on a level road at constant speed?
   (b) How long would it take him to ride 50 km *up* a hill with a 4% grade at constant speed?

(c) How long would it take him to ride 50 km *down* a hill with a 4% grade at constant speed?

(d) Does the shorter time achieved on the downhill cancel the longer time required for the uphill, or is Talansky's time better on a level road?

6. **Do headwinds and tailwinds cancel?**
You know that when you ride into the wind, you have to work harder to maintain speed; and when the wind is at your back, you work less. If the wind stays constant and you make a round trip, first heading into the wind and then returning with the wind at your back, have you done the same amount of work as if there were never any wind at all?

Assume the road is level, you ride a road bike in touring posture (get values for C_D, A, and C_R from table 9.3), and you keep a steady speed of 15 mph. Assume that you and the bike together have a mass of 90 kg.

(a) At that speed, how long will it take you to each ride a 10-mi leg of the 20-mi round trip?

(b) If there is no wind, how much power must you put out to maintain your speed?

(c) Again if there is no wind, how much total work do you do, for the round trip?

Now let's say there's an 8-mph wind that helps you on the way out and fights you on the way back.

(d) How much power must you put out to maintain speed on the way out?

(e) How much power must you put out to maintain speed on the way back?

(f) How much total work do you do during the round trip?

(g) Now it is time to answer the original question about headwinds and tailwinds canceling. Do you do more, less, or the same amount of work when you do a round trip in the wind?

7. **How long does it take to burn off an apple?**
Here we'll see how long a cyclist (80 kg, including bike) has to ride to burn it off. She rides a commuting bike in an upright position (see table 9.3 for the appropriate values of C_D, A, and C_R) on a level road with no wind.

(a) How much power does she generate, in watts, if she rides at 12 mph?

(b) Assuming her body's efficiency for converting food to mechanical energy is 20%, what is the rate at which she converts energy (that is, $\bar{P}_{\text{chemical input}}$), in Calories per hour?

(c) How long does it take to burn 95 Cal (a typical apple) at this rate?

(d) If she rode instead at 20 mph, how long would it take to burn off the apple?

8. When climbing hills in a major race, riders will often sprint to the top, burning fuel partially anaerobically, like a hockey player on shift. After this, they are winded and look forward to a break on the downslope.

(a) If his speed were initially 10 m/s *and he stopped pedaling*, would a rider speed up, slow down, or coast along at that same speed on a 5% downslope? Use the values for a racing bike and stance from table 9.3, and assume a 85-kg rider+bike combination.

(b) What would be the magnitude of his acceleration (if any)?

(c) Now repeat parts (a) and (b), except use the values from table 9.3 for an upright stance on a commuter bike (with same mass).

9. According to table B.2, the MET value for skipping rope is 12.3. Consider a 70-kg woman skipping rope.

(a) In Calories per hour, at what rate is she burning Calories? That is, what is $\bar{P}_{\text{chemical input}}$?

(b) If that much power were used to turn on several 50-W lightbulbs, how many could bulbs it power?

(c) How many Calories does she burn in 25 min?

(d) How quickly is she absorbing oxygen, in liters per minutes? (That is, what is her VO_2?)

10. How many Calories are burned if a person absorbs 40 gal of oxygen?

11. Alfred Letourneur's bike (figures 9.18 and 9.24) connects a huge front gear to a tiny rear one, so that his legs could spin (relatively) slowly while his wheel spun crazy fast. Of course, the price to pay is that the force that the rider applies is larger for a large drive (that is, front) gear.

    The Space Needle in Seattle, Washington, (figure P9.11) is an extreme example of the opposite situation. The SkyCity restaurant, 500 ft high and 94 ft in diameter, rotates once every hour, giving diners a full view of the city during their meal. The restaurant is on a 125-ton turntable that is driven by a single 1.5-h.p. electric motor, which is essentially a large sewing machine motor! Such motors don't exert a lot of force, but can rotate very rapidly.

    (a) Assume the electric motor turns a gear of radius 0.25 in. If a chain connects this small gear to a 47-ft radius gear connected to the turntable, how many revolutions per minute (rev/min) must the small gear make to rotate the restaurant once per hour?

    (b) Assume that the restaraunt turntable is approximately a 125-ton uniform disk with 47-ft radius. What is its moment of inertia?

    (c) If it rotates once per hour, how much kinetic energy does the turntable have?

    (d) How long does it take a 1.5-h.p. motor to produce that much energy?

    (e) Compare your answer in part (c) with the rotational kinetic energy of a typical bike wheel with $I = 0.1$ kg $\cdot$ m^2 that spins at 5 rps. Which one is larger? By how much?

    (Of course, the drive mechanism for the actual restaurant is considerably more complicated than two gears and a chain.)

Figure P9.11
The Space Needle in downtown Seattle.
(© *Golden Gate Images/Alamy RF*).

12. Remember your free-body diagrams. In figure 9.30, the rider and frame of the bike and center of mass of the front gear are moving at constant velocity, that is, not accelerating. However, the forces drawn on the front gear+pedal combination definitely do not sum to zero. In example 9.12, we found that the downward force of the rider was 139 lb and the chain pulled backward with 278 lb.

    (a) What other components of the bicycle conspire to produce a "compensating force" on the gear+pedal combination? Look at figure 9.12 and sketch these components.

    (b) Where does this compensating force act? (On the edge of the gear+pedal combination? At its center?)

    (c) What is the torque exerted by the compensating force?

    (d) Give the direction and magnitude of the compensating force.

    (e) Draw the compensating force on your sketch from part (a). Does this tell you anything about the shape and design of a bike frame?

13. In example 9.12, we saw that the tension on Talansky's chain was twice his weight (that is, 278 lb) when he stood on the pedals and was using the small front gear.

    (a) Do you think the chain tension would be larger or smaller if he used the large front gear instead?

    (b) Assuming, as in the example, that the bike doesn't accelerate, what would be the tension in the chain if the bike were on the large gear and he stood on the pedal? (As shown in figure 9.23, the large gear has 54 teeth, compared to 42 teeth of the small ring.) Does that agree with your expectation from part (a)?

14. A typical road bike will have 700c tires (see table 9.4).
    (a) If it is ridden at 20 mph, how many times per second are the wheels spinning?
    (b) To keep the pedaling cadence at about 1.5 rev/s (which is about optimum), how many teeth should be used in the rear gear, if the front gear has 54 teeth?

15. What if the manufacturer had gotten his instructions reversed and installed Letourneur's 57-tooth gear on the rear of the bike, and the 6-tooth gear on the front? How fast (revolutions per second) would the rider have to churn his legs to move the bike at 3 mph, about the minimum speed to keep balance? The wheel (+tire) radius is 0.2995 m.

16. In stage 17 of the 2013 Tour de France (figure 9.19), riders hit their second hill at mile 7.9, starting from an altitude of 2600 ft to the town of Côte de Réallon (altitude 4025 ft) at mile 12.6.
    (a) What is the (average) grade of this hill?
    (b) Are such grades very unusual along American highways? (Say, for a "typical" state like Pennsylvania, not extremely hilly ones like Colorado or flat ones like Kansas.)
    (c) What is the backward force of the hill on a 180-lb rider climbing that slope?
    (d) Figure 9.19 shows that Andrew Talansky rode up that hill at about 16 mph. Using the parameters for a racing bike and posture from table 9.3, how much power was Talansky putting out? Assume no wind and constant speed. (His mass including the bike is 69.8 kg.)

17. Consider Dave, the Hoosier cyclist in *Breaking Away* who broke speed limits drafting behind a tractor trailer.
    (a) If Dave were racing at 60 mph, what would be the force of drag on him if he were *not* drafting behind a truck? (Use racing numbers from table 9.3.)
    (b) How much power would he have to produce to keep his speed? You can ignore friction or any hills on the flat Indiana highway. Dave's problem is air!
    (c) For how long do you expect he could sustain that level of effort? (You may want to refer to figure 7.10.)
    (d) Dave was riding a 1970s era 10-speed bike with a 650b wheel. The larger of the two front gears had 50 teeth, and the smallest of the rear gears had 16. What was his pedaling cadence as he flew behind the truck?
    (e) If you look at figure 9.20, how realistic is it that a cyclist would maintain the cadence you obtained in part (d), regardless of any truck to draft behind?

18. See table B.2 for MET values in this problem. After consuming an extra Snickers bar (250 Cal), a 170-lb person decides to work it off by walking the dog at 2.5 mph.
    (a) How far must she walk to burn off the Snickers bar?
    (b) If she were to play singles tennis instead, how long would she have to play to burn off the candy?
    (c) If she plays tennis, what is her VO_2, in liters per minute?

19. Cyclists in the Tour de France require tremendous food (and water) intake.
    (a) How many Calories does a 75-kg cyclist need in 1 day during the Tour? Assume he cycles for 6 hr with an average MET value of 14 and then rests for 18 hr.
    (b) If he starts with the American standard 2000-Cal diet and supplements with 250-Cal energy bars eaten during the ride, how often must he eat an energy bar (for example, once every 2 hr, once every 5 min)?

20. A person consumes the FDA 2000-Cal diet.
    (a) How much pure *oxygen* (in liters) must the person absorb in 1 day to process the energy?

(b) What volume of air must she or he inhale in 1 day to process the energy?

(c) Assuming that the person inhales 0.5 liter of *air* with each breath, how many breaths must he or she take?

(d) Using your answer to part (b), how often must the person inhale during the 24-hr day (for example, once every 2 min, once every 3 s)?

21. Based on the VO_2 max data in table 9.2, what is the maximum number of calories that a 70-kg college female could process in 1 hr?

22. Torque and angular motion are important in several sports. As we said in chapter 5, a basketball is typically shot with a backspin of 2 rps.

    (a) Give a reasonable estimate (not based on formulas, just your best judgment) of the time during which the player's fingers are putting spin on the ball during a foul shot.

    (b) What angular acceleration is needed so that the ball is released with a backspin of 2 rps?

    (c) Estimate the moment of inertia of a men's basketball.

    (d) What torque should the player put on the ball during the shot in order to release it with a backspin of 2 rps?

    (e) If the mass of a basketball were spread uniformly throughout the ball, rather than on the surface, would it be easier or harder to spin? (Assume this modified ball has the same size and total mass as a normal basketball; the only difference lies in how the mass is distributed.)

    (f) If the player exerted the same torque on this modified ball as you found in part (d), for the same time as you estimated in part (a),[25] how much spin would he end up putting on the ball?

    (g) Remember, the mass (or weight) of the modified ball is the same as that of the regulation ball. The bounciness is also the same. *Without putting a spin on it*, do you think an experienced player would be able to distinguish between the two? If so, how?

23. A college student commutes to school at 15 mph on a level road. (See table 9.3 for the appropriate values of C_D, A, and C_R for commuting with an upright posture.) Together, she and her bike have a mass of 85 kg. For this problem, use exact values for the density of air at sea level from table 5.1.

    (a) How much power must she put out in 70°F weather?

    (b) How much power must she put out in 40°F weather?

    (c) Assuming her body's efficiency to do work is 18%, at what rate is she burning Calories on the 40°F day?

24. It's sometimes said that mountain bikes have gear ratios so low that one could ride them straight up (assuming you don't fall off the cliff). With a PWR of 5.7 W/kg, 67-kg Greg Lemond is an excellent climber and candidate for such a feat. When contacted, Lemond said he was too busy to attempt it in person, so we will do it on paper.

    (a) What is the grade *s* for a vertical slope?

    (b) How much power can Lemond put out?

    (c) Wind resistance and rolling friction can be ignored for this problem. The hill is the issue. At full power, at what speed could Lemond ride up the cliff? Give your answer in meters per second and miles per hour, and don't forget to account for his 6.8-kg bike in your calculation.

    (d) If he uses a standard 26-in. tire (see table 9.4), how many times per second is his wheel spinning?

---

[25] That is, he uses the same force and hand motion.

(e) To pedal at the optimum cadence of 1.5 rps, what gear ratio must he use?

(f) If the front gear were as small as possible, say with a radius of 2.5 in., what would be the radius of the rear gear? How does this compare to the radius of the wheel?

25. The rear tire of a touring bicycle, loaded down with gear and rider, may have to support 220 lb.

(a) If the tire is inflated to 85 psi (a typical value for touring tires), what will be the area of tire flatted against the ground?

(b) If the rider loads even more gear onto the bike, the rolling friction will increase. Name two reasons why.

26. As we said at the end of section 9.3.1, Andre Talansky burned 1000 Cal in 53 min on stage 17 of the 2013 Tour. If you use the MET values from table B.2, calculate how long it would take for a 170-lb person to burn that many calories while

(a) Walking the dog.

(b) Playing competitive squash.

27. In example 9.13, we concluded that the rotational energy in the wheel+tire combination comprised only about 1% of the kinetic energy of the bike+rider system, when the bike is moving at 15 mph. How (if at all) does this conclusion change at higher speeds.

(a) What is the translational kinetic energy of an 80-kg bike+rider combination at 30 mph?

(b) What is the rotational kinetic energy of the wheels plus tires on that bike if their radius is 0.331 m and moment of inertia is 0.055 kg · m^2? (Don't forget to account for the fact that there are two wheels.)

(c) What fraction of the energy is tied up in rotational energy of the wheels plus tires?

28. This problem works its way through the complete drivetrain of a bicycle. A bike cruising at constant velocity experiences a propulsive force of 30 N from the ground on its rear wheel. (It also experiences a 30-N retarding force from hills and wind; that's why it moves at constant velocity.) Ignore rolling friction in this problem; that is, $C_R = 0$.

(a) If the bike is moving to the right as you see it (draw a picture for yourself), is the torque that is generated by the propulsive force clockwise or counterclockwise?

(b) What is the magnitude of this torque if the bike uses 700c wheels?

(c) Keeping in mind that the bike moves at constant velocity, what is the magnitude of the torque produced by the chain on the rear gear? Is it clockwise or counterclockwise?

(d) If the rider is using a rear gear with 21 teeth, what is the radius of the rear gear?

(e) What is the tension in the bike's chain?

(f) What is the force that the front gear exerts on the chain?

(g) If the rider is using a front gear with 54 teeth, what is the torque, due to the chain, on the front gear? Is it a clockwise or counterclockwise torque?

(h) What is the magnitude of the torque that the rider is putting on the pedals? Is it clockwise or counterclockwise?

(i) The pedal nearest you is in the 3-o'clock position, and the rider is pushing directly down on the pedal. (On the other pedal, which is "coming up" from behind, he exerts no force.) If the bike uses 165-mm pedal cranks, how much force is the rider exerting on the pedal?

(j) Is the rider exerting more or less force than the original propulsive force that we started with?

10 Twisting Athletes in Flight

We have discussed the physics of spinning balls and wheels. Things get much more interesting when we consider the athlete's body itself, which can change shape and twist about several axes. In this chapter, we'll learn how a gymnast controls her rotation in flight and why she rapidly "windmills" her arms when she starts to topple from a balance beam. Some of the most beautiful and complex motions in sports are based on the possibilities afforded by—and the limitations imposed by—the conservation of angular momentum.

Figure 10.1. The St. Louis Cardinals second baseman Mark Ellis attempts a double play after San Francisco Giants's Hunter Pence is forced out in the ninth inning, July 2, 2014, in San Francisco. As his upper body twists to throw the ball to first base, his lower body reacts by twisting in the opposite direction. Ellis intuitively spreads his legs to maximize their moment of inertia. (© *Eric Risberg/AP Photo*).

At the end of chapter 9, we started talking about rotational motion and how the moment of inertia, which depends on the mass and shape of an object, determines things such as the response to torque and the rotational energy. Going from spinning wheels to swinging bats is already hard enough, but things get *really* interesting when the object is a human body, which can change shape *while* spinning and even start different parts of itself spinning in different directions.

Often without even realizing it, athletes exploit the beautiful laws of rotational physics with their coordinated and intricate body movements. In fact we all do, even if only by swinging our arms as we walk. In this chapter, we will discuss these physical laws and see them in action, probably in some ways that have never occurred to you.

As always, the laws of physics are mathematical. By now you know that numbers run the universe, including sports. However, the calculations to numerically describe the changing moment of inertia of human tissue and muscle, as well as the various translational and rotational motions involved, are extremely complicated. Not hard really, just complicated. And it would be easy to drown in a cascading mass of formulas and details that teach very little but obscure the bigger picture. Therefore, this chapter will be a little less number-based than the preceding ones. However, the concepts here are not trivial and are very cool, so don't take the low formula count as a signal to tune out, okay? Okay.

10.1 Human Rotation

Thinking about the moment of inertia of a bicycle wheel is pretty easy, because most of the mass of the wheel is found at a fixed distance (the wheel radius) away from the axle. Balls and rods are slightly more complicated, but their symmetry allows their moments of inertia to be given by simple formulas; a couple of examples were listed in table 9.5.

The human body is much more complicated, and some of the most reliable measurements of its moments of inertia were made by the Aerospace Medical Research Laboratories of the Air Force Systems Command in the early 1960s, just as manned space missions were beginning. These measurements were performed only on "living[1] male subjects representative of the Air Force population in stature and weight." They were done to allow engineers to design craft that might involve large accelerations, torques, and spins of their contents, including the astronauts. Studies on a more diverse set of subjects periodically show up in sports medicine journals.

Even with all the complications, the moment of inertia I for a human follows rules similar to those for simpler objects:

- I gets larger when more of the body's mass is moved far away from the axis of rotation.

- Even for a fixed body shape, I depends on the axis of rotation. The three "anatomical axes" discussed in medicine, biomechanics, and kinesiology are shown in figure 10.2.

- I depends on where the rotation axis passes through the body. The greater the distance between the axis and the center of mass, the larger I is.

Some postures and rotations relevant to sports are shown in table 10.1.

> **Question**
>
> Think about a "spread-eagle" posture. (This would be the posture if the person in the bottom row of table 10.1 spread apart his legs.) Can you guess, approximately, what I would be for rotations about the longitudinal axis going through his center of mass?
> Now keep our man in this spread-eagle position, but imagine him doing a cartwheel. Now he is rotating about the anteroposterior axis. Would I be larger or smaller than your previous answer? Why?

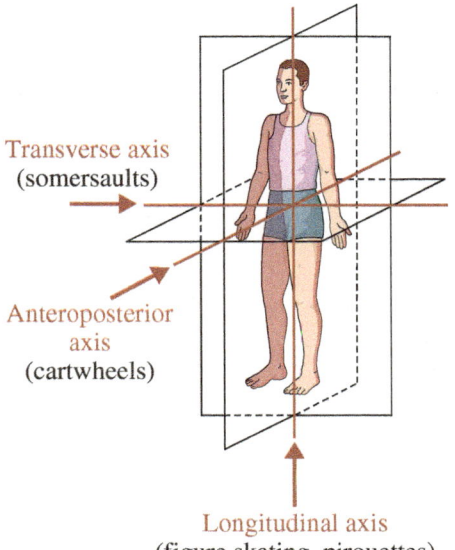

Figure 10.2
The three natural "anatomical axes" are labeled, together with examples of motion that involves rotation about them.

---

[1] The report repeatedly emphasizes that the subjects studied were "living" in the 1963 study. The nature of *previous* measurements is left to the imagination.

Table 10.1. Approximate moments of inertia about various rotational axes for a 70-kg male in different postures. Values from James Hay, *The Biomechanics of Sports Techniques*, and Susan Hall, *Basic Biomechanics*.

Posture and axis	I
Tucked, transverse axis through center of mass	$3.5 \text{ kg} \cdot \text{m}^2$
Pike, transverse axis through center of mass	$6.5 \text{ kg} \cdot \text{m}^2$
Extended, transverse axis through center of mass	$15 \text{ kg} \cdot \text{m}^2$
Extended, axis parallel to transverse axis through hands	$83 \text{ kg} \cdot \text{m}^2$
Standing, arms at side, longitudinal axis through center of mass	$1.1 \text{ kg} \cdot \text{m}^2$
Standing, arms out, longitudinal axis through center of mass	$2 \text{ kg} \cdot \text{m}^2$

10.2 Backward Giant Circle

The horizontal ("high") bar event in the men's artistic gymnastic category is always a highlight of the Olympic games. The somewhat flexible[2] metal bar sits 2.55 m above a 20-cm-thick mat. Gymnasts grip it with their hands (usually inside gloves, called grips, and dusted with chalk) and swing around and over the bar, executing intricate twists and somersaults during the performance.

We're going to focus on one of the basic maneuvers: the backward[3] giant circle. Here the gymnast circles the bar, his body more or less completely extended, as shown in figure 10.3.

The intricate artistry of the twists and flips relies heavily on the gymnast's ability to change his body's shape in precise, complicated ways in flight. For starters, we're going to consider a rather uninspired maneuver in which the athlete starts at the top of the circle from rest and keeps his body rigidly extended for the entire circle. (In reality, he usually bends his body on the upswing; we'll come to this in a moment.)

A crucial component of the routine is the gymnast's speed at each point in the maneuver, especially his speed when he lets go in the final dismount. We'll first discuss his speed while on the bar and then consider what happens when he's in the air.

10.2.1 Torques and spin rate

You already know intuitively that the gymnast will speed up on the downswing and slow again on the upswing. That is, he will experience angular acceleration.

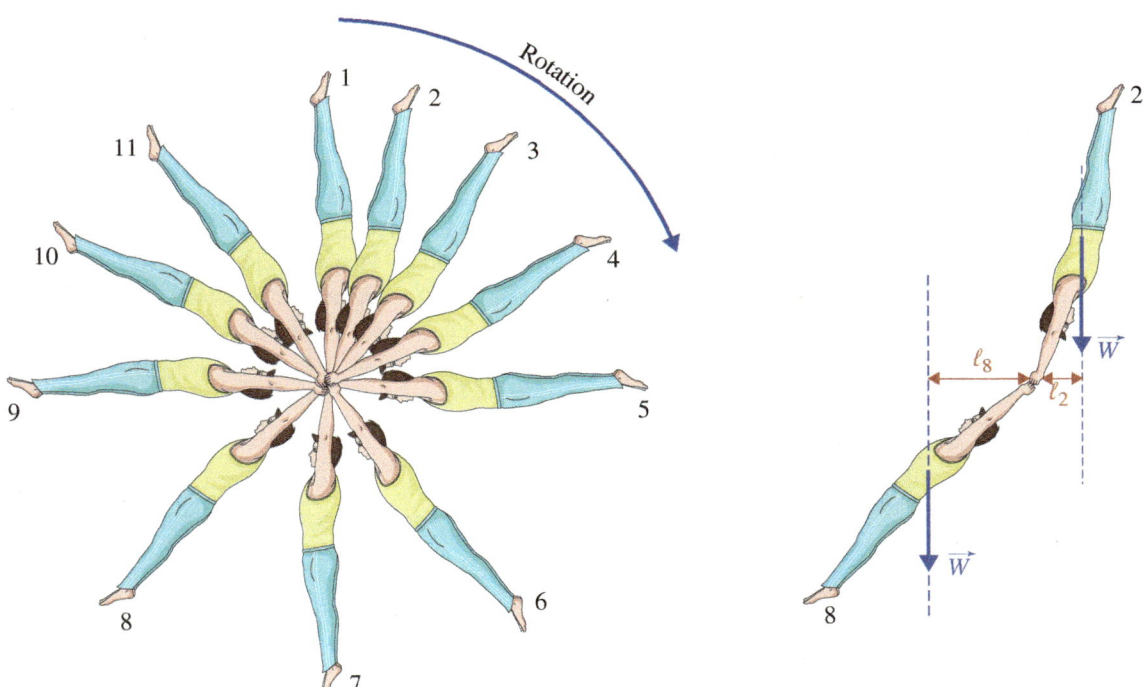

Figure 10.3. The backward giant circle maneuver in the men's artistic gymnastics horizontal bar event. On the right, the torques at two instants of time are analyzed. The weight force is applied at the center of mass. The dashed lines are the "lines of action" of $\vec{W}$.

---

[2] International Federation of Gymnastics regulations stipulate that the bar should deflect by 10 cm when a 2200-N load is placed in its center. Therefore, Hooke's constant is $k = 22{,}000$ N/m ≈ 1500 lb/ft, about the same as a stiff hockey stick and about 75 times that of an archery bow. In this section, we'll treat the bar as perfectly stiff.

[3] The name of this maneuver is strange, since the gymnast is facing forward during this swing.

Newton's second law for rotation (equation 9.31) says that this angular acceleration must be due to a torque. To understand the giant swing, we have to understand the torques.

Torques come from forces, and there are two sources of force on our gymnast. The first is the bar itself, which exerts both pushing/pulling and frictional forces. The magnitude and direction of the forces from the bar change in a complicated way during the giant circle. However, because they are applied *at* the pivot point, the lever arm ℓ is zero, and they exert no torque.[4]

Since **the force of weight can be considered to act *at* the position of the center of mass**, $\vec{W}$ is the only force producing a torque. The gymnast's rotation is powered by gravity.

Torque changes as the gymnast rotates

Unlike the simpler case of a chain pulling on a bicycle gear, even though the torque-producing *force* never changes in direction or magnitude, the *torque it produces* is different at each point in the maneuver. On the right side of figure 10.3 two of the positions are selected. It is clear that the lever arm is much larger in position 8 than it is in position 2. Therefore, the magnitude of the torque is much larger in position 8 than it is in position 2.

We can estimate the lever arms from the figure, assuming the length of gymnast plus his outstretched arms is 2.25 m ($\approx$7 ft 4 in.). Reasonable values are $\ell_8 = 0.7$ m $= 2.3$ ft and $\ell_2 = 0.3$ m $= 1$ ft. The torques on a 160-lb gymnast are then

$$\tau_2 = -(160 \text{ lb})(1 \text{ ft}) = -160 \text{ ft} \cdot \text{lb}$$

$$\tau_8 = +(160 \text{ lb})(2.3 \text{ ft}) = +368 \text{ ft} \cdot \text{lb}$$

In addition to having different magnitudes, these torques have different signs, as we discussed around equation 9.25. The negative torque at position 2 gives a negative angular acceleration at that position ($\alpha_2 = \tau_2/I$). Thus, the negative torque at position 2 is trying to make the already negative angular velocity (because it's a clockwise rotation) even *more* negative. In other words, in simple terms, the gymnast's weight at position 2 is tending to speed him up. You already knew that. But now you know how it is connected mathematically to torque, angular velocity.

The torque at position 8 is a positive number, corresponding to a counterclockwise direction. This means the angular acceleration is positive, so that the change in angular velocity is positive, which means that the already negative angular velocity will tend to get a little less negative. Thus the gymnast's weight at position 8 will tend to slow him down, which you already knew.

10.2.2 Maximal force at the bottom of the swing

An important point in the swing is the very bottom. It is at this point that the forces on the gymnast's arms are maximum, and great strength is required to execute maneuvers here. You can imagine what determines the force on his arm. Obviously, the more he weighs, the greater the force. Also, it makes sense (right?) that the faster he is swinging, the greater the force on his arm at the bottom.

---

[4] Since the horizontal bar has some thickness, the lever arm for the frictional force is not truly zero, and thus there is some slowing torque due to friction. However, the radius of the bar is only about 0.5 in., and friction is small because of chalk on the hands, so we will neglect this small torque.

So, how fast *is* he swinging? Well, he starts at rest at the top, and he speeds up because of the torque from gravity. But what we just found above is that not only is his angular velocity changing at each instant, but even *the rate at which it changes* (that is, the angular acceleration) is changing moment by moment.

If we look at it from this point of view, figuring out the gymnast's rotational rate looks like a hard calculation. We physicists love this sort of thing. Not because we love hard calculations (hey, if we liked hard things, we would have gotten real jobs), but because there is often a different approach that makes things simple. Sometimes looking at nature from different points of view can reveal clear patterns that were not immediately obvious. That's beauty to a physicist.

In the present case, the simpler approach involves looking at the energy, rather than forces and torques. Since we are ignoring the small effects of friction and air drag, **the gymnast's energy is the same at the top of the swing as at the bottom**.

The kinetic energy of a rotating object is given by equation 9.32. For rotation about a fixed axis, an equivalent[5] formula is

$$\text{KE}_{\text{rotation about fixed axis}} = \tfrac{1}{2} I_{\text{axis}} \omega^2 = 2\pi^2 I_{\text{axis}} f_{\text{rot}}^2 \qquad (10.1)$$

It is important to use the appropriate moment of inertia. For the gymnast doing a giant circle, I_{axis} is about 83 kg · m², according to table 10.1.

So the gymnast's total energy is

$$E = \text{KE} + \text{PE}_{\text{grav}} = \tfrac{1}{2} I_{\text{axis}} \omega^2 + mgh_{\text{c.m.}}$$

Now we can figure out the gymnast's angular velocity at the bottom, by equating the energy at the top and bottom of the swing and doing a little algebra:[6]

$$\text{KE}_{\text{bottom}} + \text{PE}_{\text{bottom}} = \text{KE}_{\text{top}} + \text{PE}_{\text{top}}$$

$$\tfrac{1}{2} I_{\text{axis}} \omega^2 + mgh_{\text{c.m., bottom}} = 0 + mgh_{\text{c.m., top}}$$

$$\rightarrow \tfrac{1}{2} I_{\text{axis}} \omega^2 = mg \underbrace{\left(h_{\text{c.m., top}} - h_{\text{c.m., bottom}} \right)}_{2 \times \text{distance to c.m.} = 2L}$$

$$\rightarrow \omega = -2 \times \sqrt{\frac{mgL}{I_{\text{axis}}}} \qquad \text{using the negative root here}$$

$$= -2 \times \sqrt{\frac{(70 \text{ kg}) \left(9.8 \text{ m/s}^2\right) (1 \text{ m})}{83 \text{ kg} \cdot \text{m}^2}}$$

$$= -5.75 \text{ rad/s}$$

There are a couple of things to notice about the math above. First, you'll see that I chose the negative solution when I took the square root of ω because the rotation is clockwise.[7] Second, if you carefully followed all the units in the last step,

---

[5] I won't go into the details, but equation 10.1 can be derived from equation 9.32, using what physicists call the parallel-axis theorem.

[6] Come on, we're at the end of the book. You've seen lots worse by now.

[7] When we take a square root of a number (say, 4), we can choose the positive solution (say, +2) or the negative one (say, −2). As in this case, our choice may be based on our overall understanding of the physical situation being described by the mathematics.

you'd wind up with $\omega = -5.75/\text{s}$ and forget to write down "radians." That is not a disaster; don't worry about it. As I told you in section 9.3.3, when we talk about angular velocity (and angular acceleration actually), we insert the "radian" by hand, as a kind of reminder that it is an angle that is changing. For angles, unless degrees (°) are denoted, 1 means 1 rad, as shown in figure 9.25.

It might help to put the rotation in more familiar terms:

$$f_{\text{rot}} = \left|\frac{\omega}{2\pi}\right| = \frac{5.75 \text{ rad/s}}{2\pi \text{ rad/rot}}$$

$$= 0.92 \text{ rot/s or rps}$$

$$= 55 \text{ rpm}$$

For what follows, the speed of the gymnast's center of mass will be important. This is easy. Equation 9.24 tells us that in this case it is

$$|\vec{v}_{\text{c.m.}}| = |\omega| \times (\text{distance between axis and c.m.}) = (5.75 \text{ rad/s}) \times (1 \text{ m}) = 5.75 \text{ m/s}$$

This is about 13 mph and is the velocity shown in the top panel of figure 10.4.

Okay, we are finally at the point to figure out the force on the gymnast's arms. First, since he's going in a circle, we know that he is accelerating. In fact, when he is at the bottom of the swing, he is accelerating *upward*, toward the middle of the circle. This centripetal acceleration of his center of mass is given by equation 3.16:

$$a_c = \frac{v^2}{r} = \frac{(5.75 \text{ m/s})^2}{1 \text{ m}} = 33.1 \text{ m/s}^2 = 3.37g.$$

He is accelerating upward at more than $3g$!

As shown in the bottom panel of figure 10.4, there are only two forces on him: his weight pulling him down and the bar pulling him straight up. The imbalance between these two forces is the centripetal force causing the centripetal acceleration. Equation 3.17 gives us the imbalance:

$$ma_y = F_{\text{bar},y} + W_y$$

$$m \times (3.37g) = F_{\text{bar},y} - mg$$

$$\rightarrow F_{\text{bar},y} = 3.37mg + mg = 4.37mg = 4.37|\vec{W}|$$

This tells us that the bar is pulling up on the gymnast with a force more than four times his weight![8] By Newton's third law, this is the same force that he must be pulling *down* on the bar. A 170-lb gymnast is hanging on with a force of 740 lb. Furthermore, in our example the gymnast started at the top of the circle at rest, while the gymnast usually already has some speed as he passes the highest point; this increases the velocity at the bottom and pushes the centripetal force even higher, to 850 lb or more.

This amount of upward acceleration and force is on the same level as that experienced by bungee jumpers at the bottom of the bounce. A difference is that in the bungee case, the cord is *attached* to the legs, not held in the hands. It's a small wonder that gymnasts have such strong arms.

---

[8] Even though we got this result for one example of a gymnast with $I_{\text{axis}} = 83 \text{ kg} \cdot \text{m}^2$, the result is approximately true for any gymnast, since athletes with higher weight have proportionally higher moments of inertia.

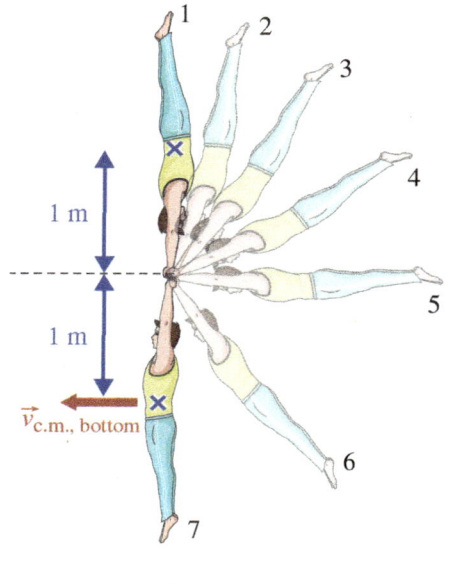

Figure 10.4
Top: The gymnast at two points on the giant circle. At the top of the circle, he is momentarily stationary, so his energy is all potential. At the bottom, he has kinetic and potential energy. **Bottom:** The free-body diagram reveals a large force imbalance, reflecting the fact that the gymnast has a large centripetal acceleration. In fact, the force on his arms is more than four times his weight!

10.2.3 Swinging to speed up

The gymnast started from (almost) a dead stop at the top of the circle. If he holds a constant extended shape, he will speed up on the downswing and slow on the upswing. Since energy is conserved (remember that we're ignoring friction), he will return to the top of the circle (position 1 in figure 10.3) with (almost) zero velocity.

But we know in real life that gymnasts swing their bodies to increase their angular velocity. The gymnast may begin a circle from rest, but by the time he returns to the top of the circle, he's picked up considerable speed. How does that work?

The "secret" (which the athlete knows instinctively) is to change the body's shape at precisely the right time. The gymnast extends his body fully on the downswing, maximizing the torque generated by gravity by maximizing the lever arm. Gravity's torque on the upswing, of course, tends to slow the swing, so the gymnast reduces the lever arm by pulling his body in toward the axis of rotation, that is, the bar.

The left side of figure 10.5 shows an idealized maneuver. At the instant his body reaches the bottom of the circle, the gymnast pulls into a pike position to reduce the lever arm. Upon reaching the top of the circle, he fully extends his body again, maximizing the lever arm for the force of gravity to exert the largest possible torque on the downswing.

A more realistic swing sequence is shown in the right panel of the figure. Rather than quickly drawing up into a tight tuck at the very bottom of the swing, the gymnast slightly draws in his body by bending it; furthermore, he usually performs it a little after reaching the bottom. While the idealized sequence on the left would produce the greatest effect in terms of increasing the rotational frequency, gymnasts simply do not have the strength to quickly accelerate their center of mass *on top of* the 600 lb or so of tension sustained by their arms at the bottom of the swing.

10.2.4 Dismount

After about a minute on the high bar, the gymnast dismounts. This is a high point of the routine, in which he twists and somersaults and then tries to "stick," not shifting

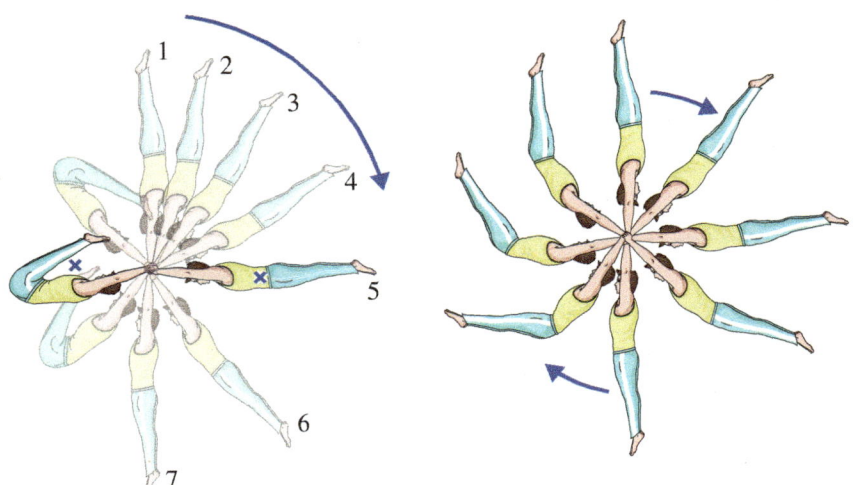

Figure 10.5. A gymnast increases his angular frequency by increasing the lever arm for torque from the gravitational force on the downswing, when torque speeds him, and decreasing it on the upswing, when torque slows him.

his feet from their original landing spot. A dismount is essentially a combination of rotational and projectile motion. To understand the beauty of a dismount, we need to put it all together.

Figure 10.6 shows a basic dismount, where the gymnast executes two somersaults while in flight. There are several lessons to learn here, most of them determined by the fact that *the only force acting on the gymnast in flight is his weight.* (We're ignoring the small effects of air resistance.)

Spinning and flying

The first lesson is something we learned when we discussed the "human projectiles" of Bob Beamon and Blake Griffin in section 4.4. If gravity is the only force in effect, the center of mass will always follow a parabolic trajectory, and all the spinning and twisting shape-changing in the world cannot affect that. The center-of-mass trajectory

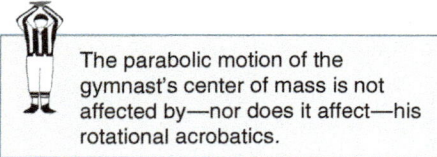
The parabolic motion of the gymnast's center of mass is not affected by—nor does it affect—his rotational acrobatics.

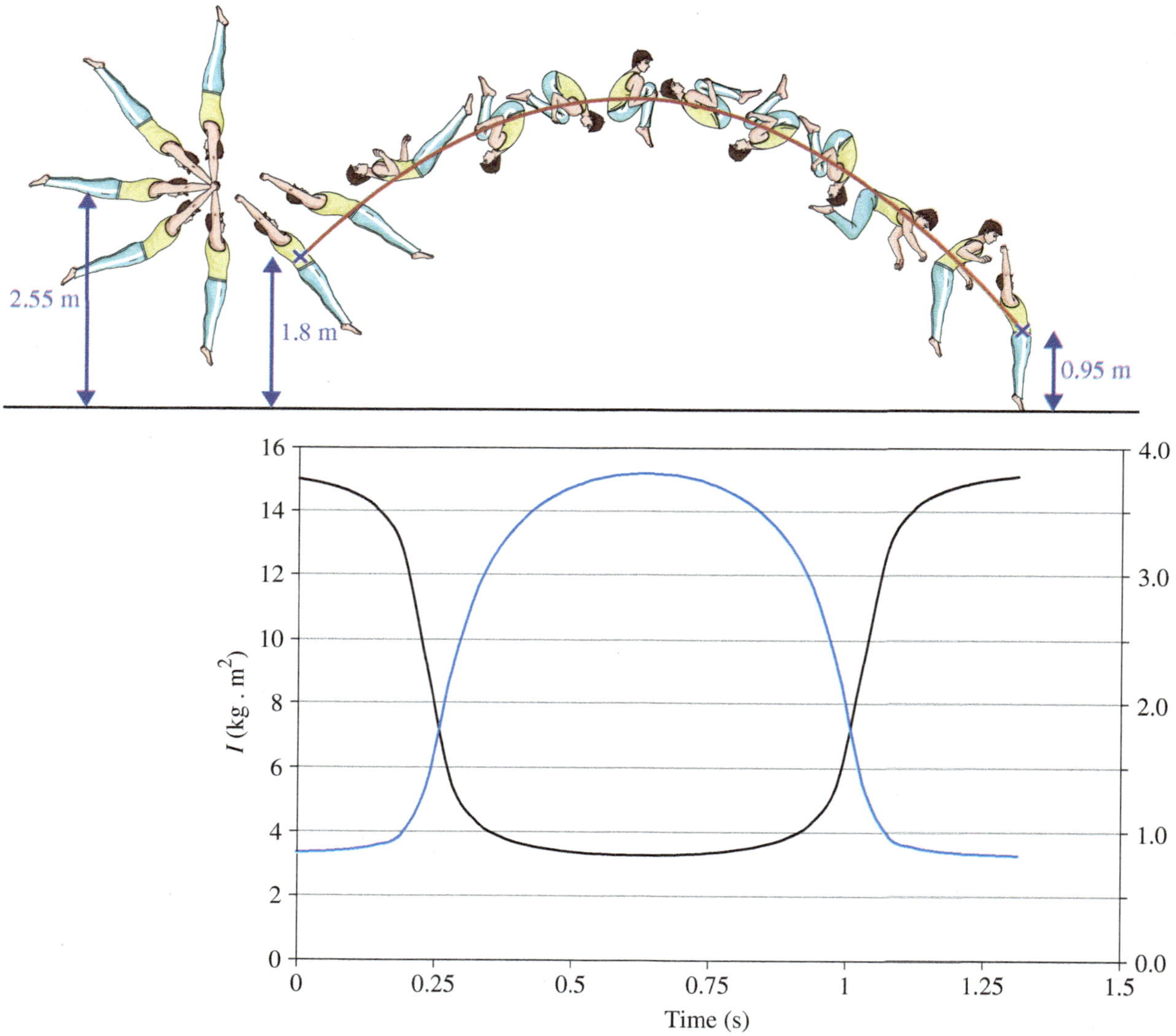

Figure 10.6. When the gymnast releases the bar, the only force on him is gravity. He quickly tucks into a tight position to reduce his moment of inertia (plotted as a function of time with the black line). As a result of angular momentum conservation, his rotational frequency (plotted in blue) increases. All the while, his center of mass follows the familiar parabolic trajectory of all projectiles when air is unimportant. This figure inspired by a similar figure in *The Biomechanics of Sports Techniques* by James Hay.

of the acrobatic gymnast is the same as that of a shot-put launched with the same speed and angle.

And it works the other way, too: just as the motion of the center of mass is unaffected by any rotational motion, the rotational motion is unaffected by how the center of mass moves. The gymnast's change in body shape would result in the same somersaults and twists, no matter if he were dismounting from a high bar, plummeting vertically from a diving platform, or even in the weightlessness of outer space!

Immediately upon release

We've mentioned before an important principle of motion that bears repeating: an object in free flight can only rotate about an axis that passes through its center of mass. Therefore, as soon as he lets go of the bar, the axis of the gymnast's rotation shifts immediately from the bar itself to his center of mass.

In free flight, all rotation takes place about the center of mass.

While the axis of rotation changes upon letting go, the gymnast's angular velocity ω (or rotational frequency f_{rot}) remains the same. For the dismount in figure 10.6, he lets go when $f = 0.8$ rot/s. This corresponds roughly to the giant circle of the previous section.

Tucking in

If you watch the routines of divers, gymnasts, or acrobats, you quickly realize something striking. While they cannot affect the trajectory of their center of mass in any way, *they can significantly change their rotational motion by changing body shape in flight*. After release, the gymnast in the figure goes from an extended posture to a tight tuck. This moves much more of his mass close to the axis of rotation and reduces his moment of inertia from 15 to 3.5 kg · m^2, according to table 10.1.

Remarkably, this simple change causes his spin rate to increase greatly! What drives this effect is one of the most important concepts in all physics. To understand it, the first thing to remember is that the force of gravity (that is, a person's weight) can be considered to act at the center of mass. Since the center of mass is on the axis of rotation, there is no torque exerted on an object in free flight subject to only gravity.

An object in free flight experiences no torque about its center of mass.

No torque? I know what you're thinking. You're thinking, "If there's no torque, then Newton's first law for rotation should apply!" And you are right!

You're also thinking, "And that law says that if there is no torque, the angular velocity shouldn't change!" And you're wrong!

To understand why, we need to go beyond equation 9.26, Newton's first law for angular motion of a wheel, because a person differs from a wheel in the most important way: he can change shape. The complete version of the law applies to wheels, balls, *and* athletes:

Newton's First Law for Angular Motion

The **angular momentum** of an object doesn't change if the *net* torque on it is zero.

$$\sum_i \tau_i = 0 \quad \leftrightarrow \quad L = \text{constant} \qquad (10.2)$$

The important new quantity here is the angular momentum:

$$L = I\omega \qquad (10.3)$$

Since the sign of ω tells the direction of rotation (remember equation 9.23), the angular momentum L has a sign, too.

> **Question**
>
> Is the angular momentum of the gymnast in figure 10.6 positive or negative?

Equation 10.2 is one consequence of the **conservation of angular momentum**. As I've said before, physicists love conservation laws because they are incredibly powerful concepts revealing the deepest symmetries of nature.[9] And there are only a few of them in our universe, so each one is precious. In fact, only three matter in sports: the conservation of momentum (equation 6.4), conservation of energy (equation 7.9), and now the final one: conservation of angular momentum.

Any conservation law says that a quantity (total momentum, etc.) doesn't change even when "something" happens. As we've done earlier, we'll use the subscript b to denote a quantity before that something happens and "a" for after. Here the something is the athlete changing his shape. Here's the conservation law and a useful formula for how the rotational frequency after the change is related to what it was before:

Newton's first law for rotation does **not** say that the rotational frequency stays constant when there is no torque. In fact, it says that if the object changes shape, the rotational frequency **must** change so as to conserve angular momentum.

$$\underbrace{L_a = L_b}_{\text{conservation law}} \quad \rightarrow \quad I_a \omega_a = I_b \omega_b \quad \rightarrow \quad \underbrace{\omega_a = \frac{I_b}{I_a} \omega_b \quad f_{\text{rot},a} = \frac{I_b}{I_a} f_{\text{rot},b}}_{\text{useful relationships}} \quad (10.4)$$

The curves at the bottom of figure 10.6 make this point: when the moment of inertia *decreases* by a factor of 4.2, the spin rate *increases* by a factor of 4.2. This allows the gymnast to complete two complete somersaults in flight.

Prepare for landing

As he approaches the ground, the gymnast extends his body again, achieving two important functions. First, he needs to land feet down and in the "victory" pose. Second, the increased moment of inertia reduces his rotational frequency dramatically, increasing the chance of a controlled finish in which his feet "stick" to the mat upon landing.

Through years of practice, the athlete instinctively knows precisely when to uncurl his body. Uncurl too late, and he'll wind up landing on his back (ouch!). Uncurl too early, and he won't complete the reverse somersault; his shins or toes will strike the mat at a high speed and he'll tumble forward. A successful dismount requires exquisite timing and simultaneous coordination of all three independent motions: horizontal at fixed rate, vertical at fixed acceleration, and rotational at fixed angular momentum (but variable spin rate).

10.3 Figure Skating: Spinning on Ice

Before we get started, allow me to welcome all my fellow physics professors, who have quickly looked at the index and jumped to this section of the book. The figure skater performing an upright scratch spin (in which he stands on one foot on the

---

[9] Yes, in fact, it is as dramatic as it sounds.

ice and twirls about his longitudinal axis) is a huge favorite of professors, for three reasons: (1) the motion is purely rotational without the added complications of projectile motion; (2) the duration of the motion can be several seconds long, unlike for our gymnast who hits the ground after a second or so; (3) many professors can perform the technique in the classroom standing on a commonly used rotating platform. Few of us like to perform somersaults in the air in front of students.

If he supports himself only on the toe of his skate blade, the only forces on the skater in a spin are his weight and the ground reaction force on his skate, as shown in figure 10.7. Both of these forces act on points along the axis of rotation, so neither one can exert a torque about that axis. Kinetic friction with the ice and air drag both exert torques to slow his rotation, but these are small enough to be ignored. His motion is therefore determined by Newton's first law for rotations: his angular momentum does not change.

But his rotational frequency sure can!

The skater begins his spinning motion with arms (and often a leg and trunk) extended away from the rotational axis. As he then pulls his limbs in toward his body, his moment of inertia I goes down, so f_{rot} must increase. The effect can be dramatic. Seemingly effortlessly, he transitions from an almost lazy twirling into a dizzying blur. After a second or two, he again extends his arms to slow his spin and complete or continue his routine.

Question

The camel spin, shown in figure 10.7, maximizes the skater's moment of inertia, so that a slow spin "hides" a large amount of angular momentum that can be dramatically revealed by moving the mass closer. But if the purpose is to maximize the mass that is far away from the axis of rotation, why doesn't Chan extend his hands above his head?

Figure 10.7. Three-time world champion figure skater Patrick Chan of Canada executes a so-called "camel spin." This configuration maximizes the amount of mass located far from the rotation axis. The forces on him act along the rotation axis, so exert no torque. (© *Stephane Reix/For Picture/Corbis*).

Figure 10.8. As a spinning skater pulls in his arms, his moment of inertia I goes down. Since If_{rot} cannot change, that means his rotational frequency goes up.

Kinetic energy of an upright spin

Figure 10.8 shows an upright spin, with typical numbers for its rate and moments of inertia. As we've said, the angular momentum doesn't change, since there is no external torque applied. In each panel, $If_{rot} = 3.8$ kg·m²/s.

But what about the kinetic energy: does that change? Equation 10.1 tells us how to find the energy:

$$KE = 2\pi^2 I f^2_{rot}$$
$$\rightarrow KE_{arms\ extended} = 2\pi^2 \left(1.9\ \text{kg} \cdot \text{m}^2\right) (2/s)^2 = 150\ \text{J}$$
$$KE_{compact} = 2\pi^2 \left(0.8\ \text{kg} \cdot \text{m}^2\right) (4.75/s)^2 = 356\ \text{J}$$

While the angular momentum stayed the same, the kinetic energy *doubled*. If the net force on the skater is zero (which it is), then what did the work to increase the kinetic energy? The skater himself did the work. To move his hands in toward his trunk, he had to "pull against the centrifugal force."[10] The motion of his hands was

---

[10] Actually, of course, it is more correct to say that he had to increase the *centripetal* force to reduce the radius of the circular motion. But this is an example where the concept of centrifugal force is helpful for the intuition.

in the same direction as the net force on them, so according to equation 7.10, positive work was done. This work increased the skater's kinetic energy.

The same principle is at play when the gymnast increases his energy by swinging his body on the high bar, as in figure 10.5. Lifting his legs requires work (that is, energy converted from the chemical energy in his body) to overcome gravity and the "centrifugal force" associated with his swing.

This is another important example of the importance of the malleability of the human body in sports. *No* object can change its *translational* kinetic energy (that is, the speed of its center of mass) without external influences, but a person changing shape can change *rotational* kinetic energy, even without the help of external forces or torques.

> **Question**
>
> When the skater wants to slow his rotation, he usually extends his arms and/or legs. (He may also reach out and slightly scrape a skate along the ice to exert a slowing torque.) Doesn't this *reduce* his rotational energy? If so, how does that jibe with the fact that he has to exert forces on his hands to move them? Is he doing negative work?

10.4 Rotational Action and Reaction

Athletes control their motion by changing body shape. To accelerate along the baseline, Dwight Howard extends his leg at a diagonal angle in figure 3.12 or extends it vertically to jump in figure 3.4. In both cases, the body's shape is being changed to exert a force *on something else*, the floor in this case. This generates a "reaction" force on the athlete, which changes his motion. As we've emphasized, the center of mass of an object only accelerates when *something else* acts on it.

However, even when "there's nothing to push against" in free flight, the way the athlete changes shape is crucial to sports. Although he can't change the motion of his center of mass, we saw in section 4.4 that the athlete can control the flight trajectory of his hand holding a basketball, his feet landing in a sand pit after a long jump, or his back clearing a high-jump bar.

And as we've just seen, it is similar for rotational motion. In free flight, there is no possibility for a person to change his angular momentum, since gravity exerts no torque. Nevertheless, we saw how a gymnast can greatly change his spin rate by changing body shape. In fact, *even if an athlete in flight has no angular momentum at all*, the laws of rotational physics can result in complex maneuvering crucial to the sport, as we discuss next.

10.4.1 Acrobatics of a long-jumper, revisited

We discussed in detail the contortions that the long-jumper makes to move his center of mass from a location high in his body at takeoff to low in his body at landing. But let's look at it in greater detail. Check out figures 4.11, 4.13, and 4.14. In each, we see the legs swinging upward in preparation for landing; that makes sense in terms of (1) raising the lowest point in his body relative to his center of mass and (2) his feet hitting the sand as far forward as possible.

But what about his arms and upper body? Sure, he wants to lower them to move his center of mass to a lower position in his body, but is there a reason to swing the

Figure 10.9

The long-jumper's angular momentum is zero at takeoff and so must remain zero throughout his flight. The clockwise angular momentum of the legs is "canceled" by the angular momentum of the upper body and arms. This can be considered in terms of an "action/reaction" pair of torques. Figure from James Hay, *The Biomechanics of Sports Techniques*.

arms forward and pike about the abdomen as shown in figure 10.9? Indeed there is. Look at it this way: When he leaves the ground, his body has zero angular momentum, and we know that it must remain zero until something (such as the sand) exerts some torque on him.

Swinging up his legs gives them a counterclockwise angular momentum. So, to maintain zero angular momentum of his body as a whole (it's a law, not a choice) some *other* part of his body must get an equal amount of clockwise angular momentum. That other part is his upper body.

But hey, a person has free will, right? What if the jumper decided to break the law and *not* swing his arms and pike about the abdomen? Tough. We can break human laws, but not those of nature. *Some* part of his body would have to rotate forward; angular momentum *will* be zero, like it or not. If he didn't pike and swing his arms, his entire body (including his legs) would rotate counterclockwise, which wouldn't be good for the long jump. The jumper can use his free will to distribute the counterclockwise angular momentum as he wishes, but it's got to go somewhere.

> **Question**
>
> At an intuitive level, the long-jumper in midair has to consider effects of translational motion of the center of mass, rotational motion of various body parts, and, to some extent, aerodynamics. In terms of each of these, would there be a benefit to rotating the arms so that they extend *behind* the athlete? Would there be any disadvantages?

10.4.2 Throwing, kicking, twisting

The long-jumper in the previous section executed rotation and counter-rotation about the transverse axis, the same axis of rotation used in the simple high-bar routine we discussed earlier. More complex counterrotation maneuvers are everywhere in sports, if you know how to look for them.

Double play

Figure 10.1 shows an excellent example of the conservation of angular momentum about the longitudinal axis. To throw the ball hard to first base for the double play, Mark Ellis whips his right arm around. Since he left the ground with little or no angular momentum, *something* must counterrotate about the longitudinal axis. If he lets his body, as a whole, pick up the rotation, his trunk and shoulder will counterrotate; his throw will lose power and may very well be sent off-target.

Ellis intuitively uses his lower body to counter the angular momentum of his arm, twisting at the waist. His legs are more massive than his arm, and he instinctively spreads them to increase their moment of inertia. Therefore, to generate as much angular momentum as his upper body (though in the opposite direction), he need not move them as quickly as he's moving his arm, in order to generate enough angular momentum. This is a good thing, since players can usually whip around their arms more quickly than they can their legs!

Soccer kicks

Another excellent example is found in a hard soccer kick. Figure 10.10 shows the typical body posture of a good soccer player before and after the kick. Here there are rotations and counterrotations about *two* axes simultaneously: the longitudinal and transverse.

Figure 10.10. In executing a hard soccer kick, the rotational motion of a player's leg must be compensated by counterrotations of the trunk and arms. **Left:** (© *McGraw-Hill Education/Jill Braaten, photographer*); **Right:** (© *Koji Aoki/Getty Images RF*).

1. A bird looking down from above sees the right leg rotate quickly counterclockwise about the **longitudinal axis**. To avoid counterrotation of the entire body as a whole, the player swings his arms clockwise. The arms are less massive than the leg, so to maximize their moment of inertia, the player keeps them maximally extended. As seen in the left panel of figure 10.10, the player has intuitively anticipated this "reaction" twist of the arms before executing the kick.

2. From the side, rotation and counterrotation about the **transverse** axis are evident. As the leg swings counterclockwise (up), the player bends at the waist to rotate his trunk and head clockwise (forward and down), similar to the long-jumper's maneuver shown in figure 10.9. The trunk is more massive than the single leg, so need not rotate as far or quickly as the leg does. In their windup for the kick, you'll notice that many players actually arc their back somewhat, in anticipation of the forward swing of the trunk during the kick.

10.4.3 Balance beam

Gymnastic routines on the balance beam can be breathtaking. While performing artistic leaps and flips, the gymnast must be ever aware of the position of her center of mass. As we said in chapter 1, if her center of mass is located to the side of her support position, she will tend to topple. An example is shown in figure 10.11. There are two possible axes that would be natural to think about:

- Rotation about the balance beam: In this case, the ground reaction force $\vec{G}$ generates no torque since it has no lever arm. The only other force is her weight $\vec{W}$, which causes a clockwise torque.

- Rotation about her center of mass: In this case, the torque from her weight is zero, but the ground reaction force causes a clockwise torque.

Either way, unless she is already in motion or does something, the gymnast in the figure is going to tip over to the right.

Figure 10.11

If a gymnast's center of mass is located to the side of her point of support, the resulting torque will tend to make her even more unbalanced and eventually topple her.

318 Twisting Athletes in Flight

Figure 10.12. A gymnast in danger of toppling over intuitively "windmills" her arms counterclockwise. Figure from James Hay, *The Biomechanics of Sports Techniques*.

Figure 10.12 shows the intuitive reaction of a gymnast who finds her center of mass slightly out of line with her support point. She lifts her nonsupporting leg and quickly rotates her arms counterclockwise (in this case); in some instances, she may "windmill" her arms quickly. This action gives her body a counterclockwise angular momentum, and the inevitable "reaction" is a *clockwise* rotation of her body as a whole. If the gymnast is not too far off-center, she will often be able to remain on the beam (though she will certainly lose points from the judges in a competition).

> **Question**
>
> We've said several times that a person cannot accelerate her center of mass—something *else* must exert a force on her. By rapidly rotating her limbs, the gymnast in figure 10.12 does manage to get her center of mass to accelerate to the right. How can you explain this apparent discrepancy?

Collected Equations

$$\text{KE}_{\substack{\text{rotation about} \\ \text{fixed axis}}} = \tfrac{1}{2} I_{\text{axis}} \omega^2 = 2\pi^2 I_{\text{axis}} f_{\text{rot}}^2 \tag{10.1}$$

$$\sum_i \tau_i = 0 \quad \leftrightarrow \quad L = \text{constant} \tag{10.2}$$

$$L = I\omega \tag{10.3}$$

$$\underbrace{L_a = L_b}_{\text{conservation law}} \rightarrow I_a \omega_a = I_b \omega_b \rightarrow \underbrace{\omega_a = \frac{I_b}{I_a} \omega_b \quad f_{\text{rot},a} = \frac{I_b}{I_a} f_{\text{rot},b}}_{\text{useful relationships}} \tag{10.4}$$

Section 10.4   *Rotational Action and Reaction*   **319**

Problems

1. (a) If the gymnast in section 10.2.4 had not tucked in tightly, but had instead remained fully extended and rotating at 0.8 rot/s, how many turns would he have completed before landing? (You'll need to know the flight time; use $\Delta t = 1.4$ s as in figure 10.6.)
   (b) How would he land? (On his feet, on his head ... ?)

2. In a triple Axel (named after Norwegian Axel Paulson, who invented the Axel jump in 1882), a figure skater makes 3.5 rot about the longitudinal axis while in the air with her arms pulled in. We'll estimate her moment of inertia in this stance to be $1.1 \text{ kg} \cdot \text{m}^2$ from table 10.1.
   (a) If the skater rises 80 cm in the jump, what must be her rotational frequency to complete the jump? (You will need to remember stuff from chapter 2.)
   (b) What is her angular momentum?
   (c) What is her rotational kinetic energy?
   (d) What is her *total* kinetic energy just at takeoff?
   (e) The skater wants a jump named after her, too. She leaves the ground exactly as before, but at the peak of her jump she suddenly extends her arms outward (look again at table 10.1). How many rotations will she make before landing?

3. (a) Why do tightrope walkers carry long poles or heavy objects in their outstretched hands? See figure P10.3.
   (b) Around which anatomical axis (see figure 10.2) are rotational "actions and reactions" most relevant in this case?

4. Jumping a hurdle at full speed requires quick acrobatics.
   (a) About which anatomical axis or axes (see figure 10.2) are there rotational actions and reactions?
   (b) Very often the hurdler brings his front arm not only forward, but also across his chest (figure P10.4a). Is there a reason for this that comes from physics?
   (c) How (if at all) do the hurdler's acrobatics change the trajectory of his center of mass?
   (d) Name two reasons why hurdlers generally bend forward at the waist (figure P10.4b) as they pass over the hurdle.

5. The cheerleader in figure P10.5 jumped straight off the ground with her arms above her head and no angular momentum.
   (a) As she changed her posture to the position shown, did the angular momentum of her arms cancel out the angular momentum of her legs?
   (b) If she hadn't brought her arms down but had only lifted her legs, would she begin rotating?
   (c) Besides striking an artistic pose, what might be another reason to bring down her arms?
   (d) If she only raised her left leg and kept her right leg pointed down, which way would her trunk rotate?
   (e) If her goal were to rotate her trunk clockwise by moving only *two* of her limbs to the position shown, leaving the other two in their original position, which two should she move?

Figure P10.3
Tightrope or slackline walkers often hold weighted objects in their outstretched hands. (© *Dynamic Graphics/JupiterImages RF*).

Figure P10.4a
In clearing the hurdle, the athlete often swings his arm not only forward, but across his chest. Why? (© *Purestock/Superstock RF*).

Figure P10.4b
Why does a hurdler bend at the waist when clearing a hurdle? (© *Rubberball/Getty Images RF*).

Figure P10.5 A cheerleader strikes an artistic and symmetric pose in mid-air. (© *Royalty-Free/Corbis*).

6. (a) When running, we bend the knee of the leg we are bringing forward. One reason is simply so that our foot does not drag on the ground, but a more important reason from physics, that bending the knee makes it easier to swing the leg up quickly. Why?

    (b) When we run (or walk), why do we swing the left arm forward and right arm backward as we swing the right leg forward?

7. A diver stands on the edge of a 10-m platform, considering the dive he'd like to perform. But, deep in thought, he loses his balance and topples off. His feet are pointed more or less downward, but he realizes that he's slowly rotating toward a painful belly flop.

    (a) Which way should he windmill his arms, to try to land feet first in the water? (Over his head from behind, or over his head from in front?)

    (b) Would it help if he also "bicycled" his legs in a churning motion?

8. Think of this problem the next time you see beginners at a roller- or ice-skating rink. Stand up and lean backward on your heels until you are just about to lose your balance and fall backward.

    (a) Which way do you "windmill" your arms—over your head from behind, or over your head from in front—to regain your balance? (Don't bother looking in the chapter; just do it. Your body already knows the answer.)

    (b) What happens if you windmill them in the other direction?

9. The skater in figure 10.8 increased his rotational frequency from 2 to 4.75 turns/s.

    (a) What was his angular velocity with arms extended?

    (b) If it took him 0.3 s to fully draw his arms in (that is, to go to the position at the right in the figure), what was his angular acceleration?

    (c) How much torque was required to achieve this angular acceleration?

10. You may want to refer to table 10.1 for moments of inertia. The gymnast in section 10.2.4 was spinning with $f_{rot} = 0.8$ rot/s at the moment he released.

    (a) What was his angular momentum at that point?

    (b) What was his rotational energy at that point?

    (c) After he went into a full tuck, what was his angular momentum?

    (d) How much work did he do while tucking?

Supplementary Chapters

PART II

11 Lines of Action on the Line of Scrimmage: The Torque Wars

Figure 11.1. In this chapter, we revisit the tangle between the Rams defensive end Robert Quinn (on the left) and the Titans tackle Byron Stingily. In the two panels on the right, only the forces that generate torque about the rear foot are drawn. (© *Mark J. Peters/ZUMA Press/Corbis*).

Football is a game played with arms, legs and shoulders, but mostly from the neck up.

—Knute Rockne

Concepts from the Primary Chapters in this chapter:

- Chapter 3: Newton's third law; friction
- Chapter 9: Torque; line of action

In section 3.4.2, we looked at the matchup shown in figure 11.1 where Robert Quinn is coming in low on a block by Byron Stingily of the Tennessee Titans. By Newton's third law, the two players exert equal and opposite forces on each other: $\vec{P}_S = -\vec{P}_Q$. We discussed how, by pushing partially *up* on Stingily while being pushed partially *down*, Quinn gained the advantage by decreasing Stingily's friction and increasing his own.

That's very nice. But, you know, players in a blocking situation are not usually trying to slide their opponent backward. Rather, they are trying to put the other player off-balance so that they must step backward to avoid toppling over altogether. That is, Quinn is trying to rotate Stingily clockwise, tipping him back. Likewise, Stingily would like to rotate Quinn counterclockwise.

Well, we know how you get something to rotate, right? Apply a torque! We know from section 9.3.4 that torque involves a force and a lever arm. Newton's law demands that the two players exert the same force on each other, but they need not exert the same torque on each other. The whole trick in winning the torque battle

Lines of Action on the Line of Scrimmage: The Torque Wars **323**

in blocking lies in positioning your body and that of your opponent to optimize lever arms.

We're going to find that Quinn has the best position in terms of both friction *and* torque. Let's see how.

Pivots and relevant forces

If we're talking about rotations, we need first to identify the pivot, that is, the point about which the players might rotate. Both Quinn and Stingily are supporting themselves almost entirely with their rear foot, and if one of them starts to rotate, it will be about this point.

This knowledge really simplifies our analysis of torques, because we know that any forces that act *at* the pivot exert no torque, because they have no lever arm. Therefore, we don't need to worry about the normal or frictional forces from the ground here.

The only forces we need to worry about are the players' weights and the force they exert on each other. These are highlighted in figure 11.1 and analyzed in figures 11.2 and 11.3.

Lever arms

As we discussed in section 9.3.6, the lever arm ℓ is the closest distance (the perpendicular distance) between the line of action of a force and the pivot point. This can also be related to the distance d between the point of application of the force and the angle formed by the line of action and the line connecting the point of application and the pivot. Oh good grief! Those previous two sentences make it so awkward and complicated-sounding. It is clearer if you just look at panel (d) of figure 9.28 and at figure 11.2.

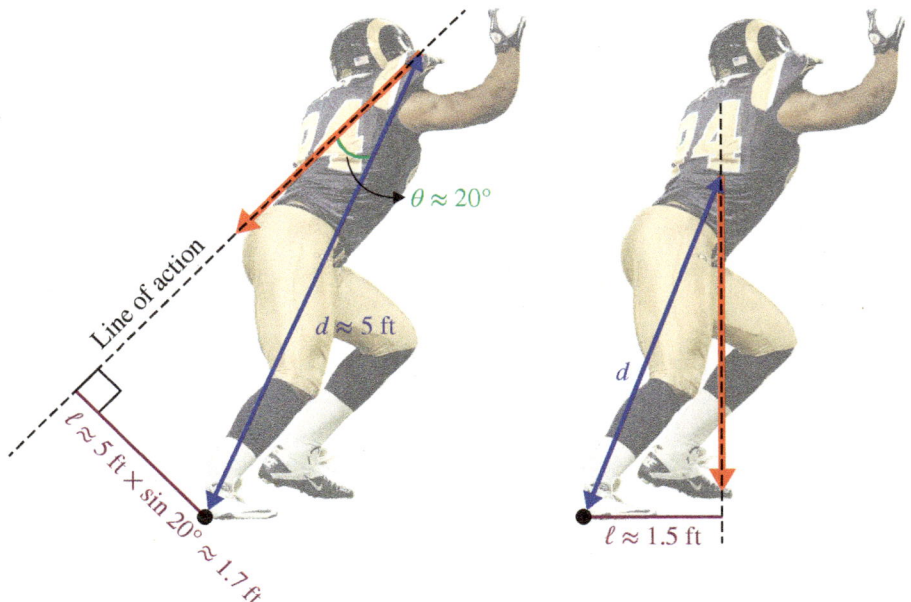

Figure 11.2. Left: The push from Stingily generates a counterclockwise torque on Quinn. **Right:** His weight generates a clockwise torque. (© *Mark J. Peters/ZUMA Press/Corbis*).

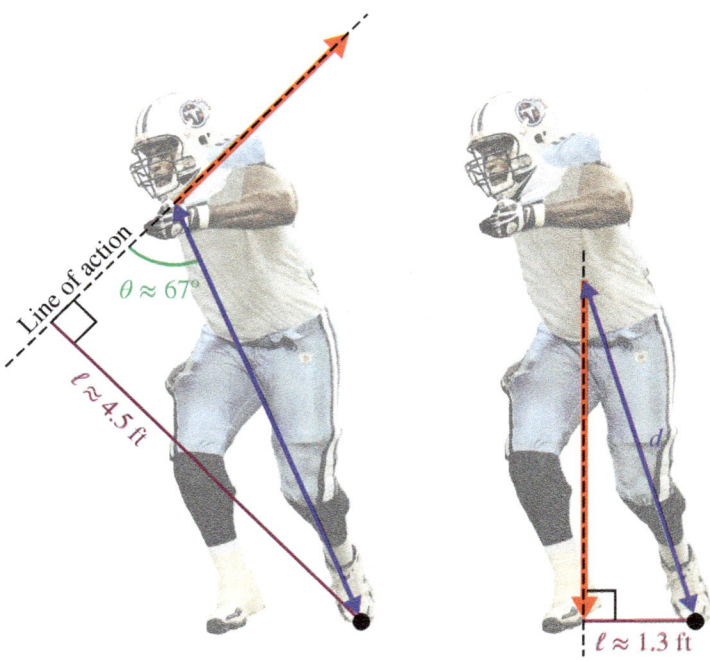

Figure 11.3. Left: The push from Quinn generates a clockwise torque on Stingily. **Right:** His weight generates a counterclockwise torque. (© *Mark J. Peters/ZUMA Press/Corbis*).

The distance between Quinn's pivot foot and the spot where Stingily is applying the pushing force is about 5 ft, and the angle θ is 20°. Therefore the lever arm for that force is $\ell_{P_Q} = d \sin \theta = 5 \text{ ft} \times \sin 20° = 1.7$ ft. Estimated lever arms for the other forces in this situation are shown in figures 11.2 and 11.3.

Quinn's tipping point

Given these lever arms and Quinn's weight, equation 9.25 tells us the total torque on Quinn:

$$\sum \tau_Q = +|\vec{P}_Q|\ell_{P_Q} - |\vec{W}_Q|\ell_{W_Q} = |\vec{P}| \times 1.7 \text{ ft} - 264 \text{ lb} \times 1.5 \text{ ft}$$
$$= +|\vec{P}| \times 1.7 \text{ ft} - 396 \text{ ft} \cdot \text{lb}.$$

In that equation, you'll see that signs matter: the counterclockwise torque from the push got a positive sign, and the clockwise torque from his weight got a negative sign. I also decided to label the magnitude of the pushing force as $|\vec{P}|$ with no subscript, since this is the same for Quinn and Stingily.

If the net torque is negative, then Quinn will tend to rotate clockwise. This won't tip him over; in fact, it will advance him as he shifts weight onto his front foot. If the net torque is positive, however, he'll tend to tip backward, which means he'll have to step back with his nonpivot foot or else topple backward.

Simple math lets us find the push strength $|\vec{P}|$ where this would happen:

$$\sum \tau_Q > 0 \quad \rightarrow \quad +|\vec{P}| \times 1.7 \text{ ft} - 396 \text{ ft} \cdot \text{lb} > 0 \quad \rightarrow \quad |\vec{P}| > \frac{396 \text{ ft} \cdot \text{lb}}{1.7 \text{ ft}} > 233 \text{ lb}$$
$$\rightarrow \quad |\vec{P}| > 233 \text{ lb}$$

Lines of Action on the Line of Scrimmage: The Torque Wars

Quinn is forced backward if he and Stingily are pushing off with a force greater than 233 lb.

Stingily's tipping point

The same procedure can be done for Stingily. The important thing here is that the pushing force—the one that wants to topple him backward—has a large lever arm simply due to geometry, as seen in figure 11.3. This is bad news for Stingily.

$$\sum \tau_S = -|\vec{P}|\ell_{P_S} + |\vec{W}_S|\ell_{W_S} = -|\vec{P}| \times 4.5 \text{ ft} + 318 \text{ lb} \times 1.3 \text{ ft}$$
$$= -|\vec{P}| \times 4.5 \text{ ft} + 413.4 \text{ ft} \cdot \text{lb}.$$

Stingily loses the battle if the net torque on him is negative, that is, clockwise. His tipping point is

$$\sum \tau_S < 0 \quad \rightarrow \quad -|\vec{P}| \times 4.5 \text{ ft} + 413.4 \text{ ft} \cdot \text{lb} < 0 \quad \rightarrow \quad |\vec{P}| > \frac{413.4 \text{ ft} \cdot \text{lb}}{4.5 \text{ ft}}$$
$$\rightarrow \quad |\vec{P}| > 92 \text{ lb}.$$

This is much lower than Quinn's tipping point of 233 lb. Stingily will be pushed backward much more easily than his opponent.

The bottom line: stay low

Byron Stingily is a big man and has 54 lb on Robert Quinn. However, by managing to get his body low, Quinn obtains two key advantages. First, he increases the friction force on him and decreases it on his foe. Second, and more importantly, he wins the torque war by optimizing the lines of action. If the players are pushing off with a force greater than 92 lb, Stingily is forced backward.

12 A Barry Bonds Home Run

Baseball is ninety percent mental. The other half is physical.
—Yogi Berra

Concepts from the Primary Chapters in this chapter:

- Chapter 3: Newton's laws
- Chapter 6: Collisions, impulse
- Chapter 7: Energy
- Chapter 8: Vibrations

 Major League Baseball digitally tracks every pitch and every hit in every game. The data are available to the public.

On August 7, 2007, the great San Francisco Giants slugger Barry Bonds hit his 756th career home run, eclipsing Hank Aaron as the all-time home run leader. We know quite a lot about this home run. For example, we know that the ball traveled a long way, landing in the stands a horizontal distance of 429 ft from home plate and at a height about 12 ft above the field level. And we know that the total "hang time" was 5.12 s.

We also know quite a bit about the initial trajectory of the ball from very precise information gathered using calibrated video cameras. For example, we know that the ball was hit very hard, having left the bat with a speed of 112 mph at the nearly optimal launch angle of 27°. The same cameras[1] were also used to track the pitched baseball, which was moving at about 77 mph when it made contact with the bat. And although we have no direct measurement of the spin rate of the batted ball, we can estimate from the landing point and hang time that it had about 1200 rpm of backspin as it left the bat.[2] In this chapter, using the famous Barry Bonds swat as our example, we are going to examine the physics of hitting a home run.

12.1 Ball–Bat Collision: Speeds, Impulse, Force

As the great baseball philosopher Yogi Berra once said, "You can observe a lot by watching." We can't easily include a video that you can watch in a textbook. However, we can do the next best thing, which is to show you in figure 12.1 a computer rendering of an actual swing, created using high-speed motion analysis equipment to track the flight of the incoming and outgoing baseball as well at the swing of the bat. Each image of the ball or bat is separated in time from the adjacent image by 0.004 s (4 ms). The ball approaches from the right at a slight downward angle, then leaves the collision point at a steeper upward exit angle. The momentum of the ball has clearly changed, initially pointing to the left before the collision, then pointing (primarily) to the right after the collision. The bat has exerted a rightward impulse to the ball $J_{x,\text{ball}} = p_{x,\text{ball},a} - p_{x,\text{ball},b}$, as we discussed in chapter 6.

---

[1] Did you know that these cameras are used to measure the trajectory of every ball pitched in MLB games since the 2007 season? In addition, the data for all these pitches are publicly available online for anyone to download. There are even entire conferences dedicated to analyzing these data.

[2] You can learn how all these things are determined from the camera information at this website: http://baseball.physics.illinois.edu/b756.html

Section 12.1   Ball–Bat Collision: Speeds, Impulse, Force

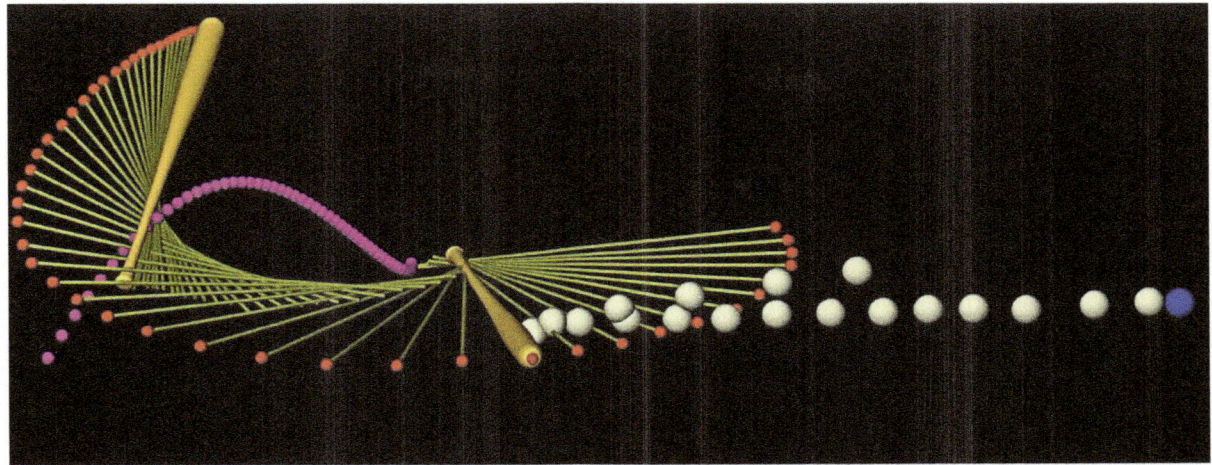

Figure 12.1. Tracking the ball and bat during an actual swing. (*Image courtesy of J.J Trey Crisco Ph.D.*).

The red dots track the position of the tip of the bat for the same 4-ms frames. From the separation between sequential dots, we see the bat speeding up as the ball approaches. Just after the collision, it moves more slowly because of the leftward impulse imposed by the ball ($J_{x,\text{bat}} = -J_{x,\text{ball}}$, remember?). To see this slowing due to momentum conservation in actual high-speed videos of MLB games, check out http://baseball.physics.illinois.edu/HighSpeedClips.html.

Another interesting feature is revealed by comparing the distance between sequential positions of the ball before and after the collision. We can see that not only does the ball change directions, but also it has a higher outgoing than incoming speed. If you are the batter, that is a good thing since higher outgoing speed will greatly improve your chances of getting on base. If you are the pitcher, the news is not so good.

For the Washington Nationals pitcher Mike Bacsik, it certainly was not good news on that August evening! Table 12.1 lists the velocity of Bacsik's pitch and Bonds's response just before and after the hit. The pitch was falling and angling slightly toward the catcher's left, that is, toward the outside of the plate for the left-handed Bonds. After the hit, it was rising and headed toward right field.

Equation 6.2 tells us the impulse that Bonds gave Bacsik's ball:

$$J_{\text{ball,toward pitcher}} = (0.01 \text{ slug}) \times (141.23 \text{ ft/s} - (-112.41 \text{ ft/s})) = 2.53 \text{ slug} \cdot \text{ft/s}$$

$$J_{\text{ball,vertical}} = (0.01 \text{ slug}) \times (74.86 \text{ ft/s} - (-12.96 \text{ ft/s})) = 0.88 \text{ slug} \cdot \text{ft/s}$$

$$J_{\text{ball,to right}} = (0.01 \text{ slug}) \times (40.5 \text{ ft/s} - (-1.33 \text{ ft/s})) = 0.42 \text{ slug} \cdot \text{ft/s}$$

Table 12.1. The velocity of the ball before and after Bonds hit it, courtesy of the MLB camera system. The forces listed are estimates, assuming a 0.7-ms contact time.

	Toward pitcher	Upward	Toward catcher's right
Pitched ball velocity (ft/s)	−112.41	−12.96	−1.33
Batted ball velocity (ft/s)	141.23	74.86	40.50
Average force during hit (lb)	3614	1257	597
Peak force (lb)	7228	2514	1194

Figure 12.2. The ball and bat in contact. (*Reproduced by permission of The News-Gazette, Inc. Permission does not imply endorsement*).

In figure 12.1, the ball is in contact with the bat for exactly one image, telling us that the collision time is no greater than 0.004 s. In fact, when you look with really high-speed cameras, you find that the collision time is really a lot less than that, more like 0.0007 s, or 0.7 ms. Thus the collision is very violent, as you can see from figure 12.2, a rare photograph of opportunity showing the ball and bat in contact. The ball is highly deformed from its usual nearly spherical shape and kind of wraps itself around the cylindrical surface of the bat. You know that if you take a baseball in your hands and try to compress it in the manner shown in the picture, it is not humanly possible to do it. The forces required are much too large.

In fact, we know how to figure out the force that Bonds inflicted on the ball, using equations 6.10 and 6.11. Looking just at the component in the direction of the stunned Bacsik, we find the *average* force on the ball to be

$$F_{\text{ave,toward pitcher}} = \frac{J_{\text{ball,toward pitcher}}}{\Delta t} = \frac{2.53 \text{ slug} \cdot \text{ft/s}}{0.0007 \text{ s}} = 3614 \text{ lb}$$

and the *peak* force is

$$F_{\text{ave,toward pitcher}} = 2 \times \frac{J_{\text{ball,toward pitcher}}}{\Delta t} = 2 \times \frac{2.53 \text{ slug} \cdot \text{ft/s}}{0.0007 \text{ s}} = 7228 \text{ lb}$$

which is more than 3.5 tons! The vertical component and sideways peak force components were also above 1000 lb, as listed in table 12.1. Bonds *crushed* that ball!

Those forces are enormous, much larger than a human being—even Bonds—can exert on a baseball. But a human being doesn't have to exert those forces, since that is the job of the bat. So what does the batter do? Well, the batter is the one who swings the bat in an anaerobic burst of power, as we discussed in chapter 7. The batter puts energy into the bat by applying a force over a relatively long time, about 150 ms. Some of that energy gets transferred to the ball by the bat applying a much larger force but acting over a much shorter time of about 0.7 ms.

To say anything more about the bat, we need to discuss one of the most important equations in baseball.

12.2 Batted Ball Speed (BBS)

For anyone at all familiar with the game of baseball, the batted ball speed (BBS for short) is crucial for a batter to get a hit, whether he is swinging for the fences as Barry Bonds does or just trying to punch the ball through the infield. The faster the ball comes off the bat, the more likely it is that you will get a hit. So let's ask the question: What does BBS depend on?

The most important equation in baseball comes from combining and rearranging two of the equations we've already studied, equation 6.18 for the coefficient of restitution and equation 6.4 for the conservation of momentum:

$$\text{BBS} = q v_{\text{pitch}} + (1+q) v_{\text{bat}}, \qquad (12.1)$$

where the "bounce factor" q is given by

$$q = \frac{e - m_{\text{ball}}/m_{\text{bat,eff}}}{1 + m_{\text{ball}}/m_{\text{bat,eff}}}, \qquad (12.2)$$

where e is the coefficient of restitution that we talked about in chapter 6.

The BBS equation is written such that all speeds are normally positive numbers.

Now, you know that velocity components have signs, so that the horizontal velocity of the pitch and that of the struck ball will have opposite signs because they go in opposite directions. When baseball people do the algebra to write the BBS formula, equation 12.1, they arrange it so that all the v's are positive numbers in the "usual" case. So if the pitch is heading toward home plate, $v_{\text{pitch}} > 0$ and if the ball is hit forward, BBS > 0, and so on. And if the batter swings the bat forward, $v_{\text{bat}} > 0$. Baseball folks are pretty reasonable people.

Baseball hits and other collisions

In chapter 6, we talked about several collisions, but they were all simpler than the baseball hit:

- In the bowling discussion of section 6.3, the ball was moving, but its target (the pin) was initially stationary. For the baseball case, this would correspond to a bunt, and in equation 12.1, we'd use $v_{\text{bat}} = 0$ to find the BBS.

- In the golf discussion of section 6.4.2, the ball was initially stationary until hit by the club. In terms of baseball, this would correspond to hitting a ball off of a practice tee; we'd set $v_{\text{pitch}} = 0$ in equation 12.1.

- Finally, we did discuss the case where *both* colliding partners were initially moving—the football tackle in section 6.2. There, however, the math was simplified by the fact that the players stuck together after the collision; that is, it was completely inelastic. In terms of baseball, this would correspond to maybe a baseball made out of putty that sticks to the bat. We'd use $e = 0$ in equation 12.1.

A collision is a collision, no matter if it's a bat and a ball or two humans crashing together, so the same mathematics applies. But in baseball, we have to think of a case where both objects are moving and do not stick together. It's the most general case.

We'll talk more about the special features of a ball–bat collision shortly. For now, let's try looking at some numbers for a typical ball–bat collision on the sweet spot (which we'll discuss soon) of the bat and see what we can learn.

Here are the numbers:

$$e \approx 0.5 \quad m_{\text{ball}} \approx 5\,\text{oz} \quad m_{\text{bat,eff}} \approx 20\,\text{oz} \quad \rightarrow \quad q \approx \frac{0.5 - \frac{5\,\text{oz}}{20\,\text{oz}}}{1 + \frac{5\,\text{oz}}{20\,\text{oz}}} \approx 0.2,$$

from which we get

$$\boxed{\text{BBS} \approx 0.2\, v_{\text{pitch}} + 1.2\, v_{\text{bat}}.} \tag{12.3}$$

A faster pitch leads to a greater BBS, but bat speed is six times as important as pitch speed.

Note that the number multiplying v_{bat} is six times larger than the number multiplying v_{pitch}. From that we can conclude that bat speed matters *six times* more than pitch speed in determining the speed of the batted ball. That may seem surprising to you. But think about it a bit. You know that you can hit a ball pretty far off a tee ($v_{\text{pitch}} = 0$) but not very far by bunting ($v_{\text{bat}} = 0$). So the fact that bat speed matters much more really shouldn't be all that surprising. What is the reason from physics behind this result? It is due primarily to that extra 1 in the expression multiplying v_{bat}, which arises because the bat is moving in the same direction as the outgoing ball. It is also due to the smallness of the bounce factor q. For a superball ($e \approx 1$) and a very massive bat, q would be close to 1 (see discussion in chapter 6 regarding the ideal golf ball and super-heavy club). Under these conditions, the bat speed would be only twice as important as the pitch speed for BBS. So while the result may have been known to you qualitatively, it is nice to get it quantitatively and to understand the underlying physics behind it.

12.3 Focus on the Bat

A baseball bat is a beautiful thing and more complicated than other objects that we've considered in collisions. Different parts of the bat move at different speeds. It vibrates and it has the famous "sweet spot." We'll discuss these here.

12.3.1 Bonds's swing

First, let's use the formula to estimate Bonds's bat speed for his home run. The pitched ball was moving at 77 mph, and the batted ball was moving at 112 mph. Equation 12.3 tells us

$$\text{BBS} \approx 0.2 v_{\text{pitch}} + 1.2 v_{\text{bat}}$$

$$\rightarrow v_{\text{bat}} = \frac{\text{BBS} - 0.2 v_{\text{pitch}}}{1.2} = \frac{112\,\text{mph} - 0.2 \times 77\,\text{mph}}{1.2} = 80.5\,\text{mph}.$$

This is impressive, especially when compared to other home runs in Major League Baseball, for which the average BBS is about 100 mph and bat speed 70 mph. Bonds's 80.5 mph (118 ft/s) was definitely on the high end of things. For those of you who follow baseball closely, you possibly knew that already, at least intuitively. After all, many people consider him one of the top three batters who ever played the game, so it's not surprising that his bat speed is well above average.

The bat is moving at 80.5 mph? Since the bat is *rotating* about a pivot somewhere near Bonds's shoulders, we know that different parts of the bat will be moving at different speeds, just as the rim of a bike wheel moves faster than the hub. So what *part* of the bat is moving at 80.5 mph? Well, it's the location where the collision takes place, which is about 5.5 ft away from the pivot point.

With that information, it's interesting to see what the bat's rotational frequency is when Bonds hits the ball. Equation 9.24 tells us:

$$v = 2\pi f_{\text{rot}} \times (\text{distance to axis}) \rightarrow f_{\text{rot}} = \frac{v}{2\pi (\text{distance to axis})}$$
$$= \frac{118 \text{ ft/s}}{2\pi \times 5.5 \text{ ft}} = 3.4 \text{ rps},$$

which is about the spin rate of the wheel of a bike going 14 mph.

12.3.2 The bat as an extended object
Effective mass

You may have noticed that in the bowling or golf example, the mass of the pin or club enters in the equation, whereas for the baseball–bat collision, it is the "effective mass" of the bat. The reason is that unlike the golf club, where the mass is entirely concentrated in the head where the impact occurs, the bat mass is distributed along its length. To understand what difference that makes, it is easiest to think about a ball hitting a stationary bat. Suppose the ball hits the bat right at its center of mass, which is typically located near the label of the bat. Then the ball exerts a force on the bat resulting in a "linear recoil." That is, the bat simply recoils backward, a consequence of momentum conservation. Now suppose the ball hits the bat in a different location. Then the force that the rebounding ball exerts on the bat also results in a torque on the bat, causing it to rotate about its center of mass, a consequence of angular momentum conservation. We might call this an "angular recoil." Taking both the linear and angular recoil into account is equivalent to replacing the actual mass with an effective mass, given by this formula:

$$m_{\text{bat,eff}} = \left[\frac{1}{m_{\text{bat}}} + \frac{z^2}{I_{\text{c.m.}}} \right]^{-1}. \tag{12.4}$$

In this equation, z is the distance from the impact location to the center of mass, and $I_{\text{c.m.}}$ is the moment of inertia of the bat about its center of mass. The formula shows that the effective mass depends on z, which is the lever arm for angular recoil. For an impact directly at the center of mass, the effective mass equals the actual mass m_{bat}. For an impact an another location, the effective mass will be smaller than the actual mass.

For a hit like Bonds's, the appropriate numbers are $m_{\text{bat}} = 31$ oz, $I_{\text{c.m.}} = 2600$ oz $\cdot$ in.^{2}, and $z = 6.8$ in. Plugging those numbers into equation 12.4 gives $m_{\text{bat,eff}} = 20$ oz, which we used to get equation 12.3.

BBCOR: the ball's role

But there is more to the story than just the effective mass. We also must consider e, the "ball–bat coefficient of restitution" (BBCOR). The BBCOR is primarily a property of the baseball. As we found from looking at figure 12.2, the ball really gets crushed when it collides with the bat. The forces, which we have found to be in the thousands of pounds in table 12.1, compress the ball just as a spring does. But as the ball compresses and then springs back, the fibers of wool yarn that make up most of the volume of the ball rub together, creating a lot of heat and therefore dissipating a lot of mechanical energy. As a result, the BBCOR for the impact of a baseball with a wood bat is no more than about 0.50 at speeds typical of professional baseball.

As we discussed in section 7.2.1, this means that $1 - e^2 = 75\%$ of the total kinetic energy—the energy that Bonds and Bacsik together worked so hard to generate—is lost to thermal modes. That is a huge effect and the primary reason why the bounce factor q is so low.

BBCOR: the bat's role

The coefficient of restitution e quantifies the fraction of kinetic energy that is converted to thermal modes. In chapter 6, we said that e depended on *both* objects in the collision. While the BBCOR is largely determined by the energy converted to heat in the ball, the bat plays an important role in dissipating energy, in ways you might not suspect.

Rather than ask what the bat does to the ball, we can ask what the ball does to the bat. We have already seen that the forces and torques exerted by the ball on the bat result in a combination of linear and angular recoil. But the ball can do even more than that, once we realize that a bat is not a perfectly rigid object.

What exactly do we mean by "rigid"? In simple language, rigid means that the bat cannot bend. It certainly seems that it is not easily possible to bend a bat. If you hold it in your hand and try to bend it, you will almost surely fail. But it's also not so easy to compress a baseball by using our hands, yet it sure compresses in a hit!

The main point is that the amount of force that a human can exert with the hands is tiny compared to the size of the actual forces that the ball and bat mutually exert on each other, as we see from table 12.1. Those forces can actually bend a bat. Skeptical? Then take a look at figure 12.3, as well as some of the high-speed video at http://baseball.physics.illinois.edu/HighSpeedClips.html, which clearly show the bat

Figure 12.3. A remarkable shot showing the bending of the bat just after collision with a softball. Note that the impact is near the tip of the bat, resulting in the center of the bat bowing toward the pitcher. (*Image courtesy of Illini Media Company*).

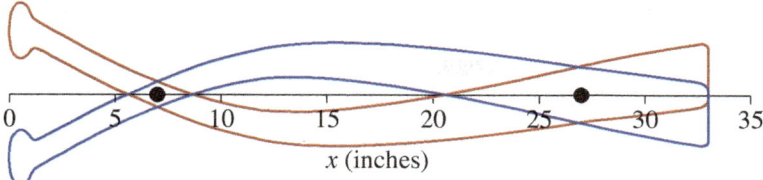

Figure 12.4. The lowest vibrational mode of a typical 33-in. wood bat, which oscillates at a frequency of about 170 Hz. The two black dots are the nodes, one in the handle and one in the barrel. The latter node, located about 7 in. from the barrel tip, is close to the sweet spot of the bat. When the ball strikes near the tip of the bat, beyond the node in the barrel, then the initial bending of the bat will result in a bowing of the center of the bat toward the pitcher, exactly as shown in figure 12.3. If the ball strikes between the two nodes, the center of the bat will bow away from the pitcher.

bending after contact with the ball. The bat is not a rigid object; it can bend and then vibrate. Actually, if you have ever hit a baseball with a bat, you already know this. When the ball hits the bat in the "wrong spot," the bats vibrates and stings your hands. It is not a pleasant feeling. But when the ball hits the bat at just the right spot (the "sweet spot"), the ball flies off the bat and you don't feel a thing in your hands.

The principal mode of vibration of a wood bat (see figure 12.4) looks very similar to the vibrating arrow we discussed in figures 8.4 and 8.8, in which you see the shaft vibrating about fixed points located near the two ends of the arrow. These fixed points are called "nodes." A bat also has nodes that you can easily find. Take your favorite wood bat and suspend it vertically, holding it lightly between your thumb and forefinger, about 6 in. from the knob. Now take a baseball and strike the bat lightly on the barrel end. You should be able to feel and hear the vibrations. Now scan the impact point along the barrel. You should find a point about 7 in. from the barrel tip where the vibrations go away. That is the node of the lowest vibration. If the ball strikes the bat at the location, you will not feel a sting in your hands. That node is the simplest definition of the sweet spot of a bat.

The "sweet spot" of a bat is located near the nodes of the lowest vibrational modes of the bat. Striking this point will produce almost no vibrations in the bat.

So we now know that when the ball hits the bat at (or near) the sweet spot, the impact "feels good," meaning that there is little or no vibration felt in your hands. But that also is the location that makes the BBCOR the largest. Why? When vibrations are excited by the ball striking the bat, it represents energy that might have gone into the rebounding ball but was instead transferred to the bat. As a result, the BBCOR of the bat depends on the impact location. It is largest where the vibrations are minimized, right at the point we call the sweet spot.

The BBCOR depends on where the ball strikes the bat. It is maximum at the sweet spot.

In reality, life is a bit more complicated than shown in figure 12.4, since there are many modes of vibration. Each successive mode has a higher oscillation frequency, and each has a node a little closer to the tip of the barrel. Research has shown that the actual sweet spot—the location that simultaneously feels best and has the largest BBCOR—is at the node of the second vibration, which has a frequency of about 570 Hz. That node is located about 5 in. from the barrel tip.

Aluminum bats

Finally, let's talk about the cursed "A" word—aluminum bats! Why is an aluminum bat generally perceived to be a better hitting instrument? A wood bat has a solid barrel, with the cross section retaining its circular shape during the collision. But a hollow aluminum bat is different because it has a thin, flexible wall that can "give" when the ball hits it. Some of the initial energy that would otherwise have gone into compressing the ball instead goes into compressing the wall of the bat. The more flexible the wall, the more it compresses and the less the ball compresses. With less compression of the ball, there is less energy lost in the collision.

The main point here is that when a baseball compresses, most of the energy goes immediately into thermal modes; it can never be recovered as "useful" energy. When the thin barrel of an aluminum bat compresses, however, that energy goes into elastic potential energy PE_{elast}. And as we discussed in chapters 7 and 8, PE_{elast} is converted back to kinetic energy upon decompression. It is not "lost."

With less energy lost the BBCOR is larger. The increase in BBCOR associated with the flexing of the wall of a hollow bat is popularly called the "trampoline effect." Whereas the BBCOR at the sweet spot of a wood bat is about 0.50, the corresponding BBCOR for an aluminum bat can be as large as 0.60. With the larger BBCOR comes a larger bounce factor, greater batted ball speed, and more hits and home runs. Some organizations, such as the NCAA, regulate the performance of nonwood bats and require "BBCOR-certified" bats, so that the game is more "woodlike" in how it is played. In doing so, they use the same principles we have discussed in this chapter.

Collected Equations

$$\text{BBS} = q v_{\text{pitch}} + (1+q) v_{\text{bat}} \tag{12.1}$$

$$q = \frac{e - m_{\text{ball}}/m_{\text{bat.eff}}}{1 + m_{\text{ball}}/m_{\text{bat.eff}}} \tag{12.2}$$

$$\text{BBS} \approx 0.2\, v_{\text{pitch}} + 1.2\, v_{\text{bat}} \tag{12.3}$$

$$m_{\text{bat.eff}} = \left[\frac{1}{m_{\text{bat}}} + \frac{z^2}{I_{\text{c.m.}}} \right]^{-1} \tag{12.4}$$

The Pole Vault

I love the pole vault because it is a professor's sport. One must not only run and jump, but one must think.

—Sergey Bubka

> Concepts from the Primary Chapters in this chapter:
>
> - Chapter 2: Kinematics
> - Chapter 3: Vector forces; centripetal force
> - Chapter 4: Human center of mass
> - Chapter 6: Impulse
> - Chapter 7: Energy conservation, conversion, and "loss"
> - Chapter 8: Flexible equipment
> - Chapter 9: Torque
> - Chapter 10: Changing moment of inertia through internal work

13.1 Origins

The modern pole vault has its origins in the lowlands of northern Europe, where ordinary people were faced with rivulets and canals in the marshy countryside. People either carried their own poles or used poles kept at common crossing points. As with so many everyday human activities, the practice inspired a competitive sport. Called "Fierljeppen," the event is scored on distance rather than height. In this event, competitors actually climb the pole even as it rotates, to maximize the distance.

In modern pole vaulting, climbing the pole is forbidden; the upper hand must not move higher on the pole. Instead, as we will discuss, athletes rotate into something like a handstand to increase the height of their center of mass, pushing up and off at the last moment.

13.2 The Modern Event

The modern pole-vault event is quite similar to the high jump, with the goal being to pass one's body above a horizontal bar without knocking it down. The bar is supported on pegs such that even a moderate disturbance will cause it to fall, voiding the jump. As in the high jump, the vaulter approaches the bar at a run, attempts to clear the bar, and lands in a foam pad.

13.2.1 Performance progression

Performance improvements in individual sports usually result from (1) technological advances, (2) innovations in technique, or (3) the emergence of a singular champion setting long-standing records, often spurring increased performance in the sport as a whole. A look at the last century of pole-vaulting records (figure 13.1) shows that this event has been driven by all three.

Technological advances

In the first half-century of official vaulting records, the record jump increased by only 75 cm (about 30 in.). Cornelius Warmerdam's 4.77-m vault in 1942 was considered an almost insuperable feat; 15 years passed before fellow American Robert Gutowski

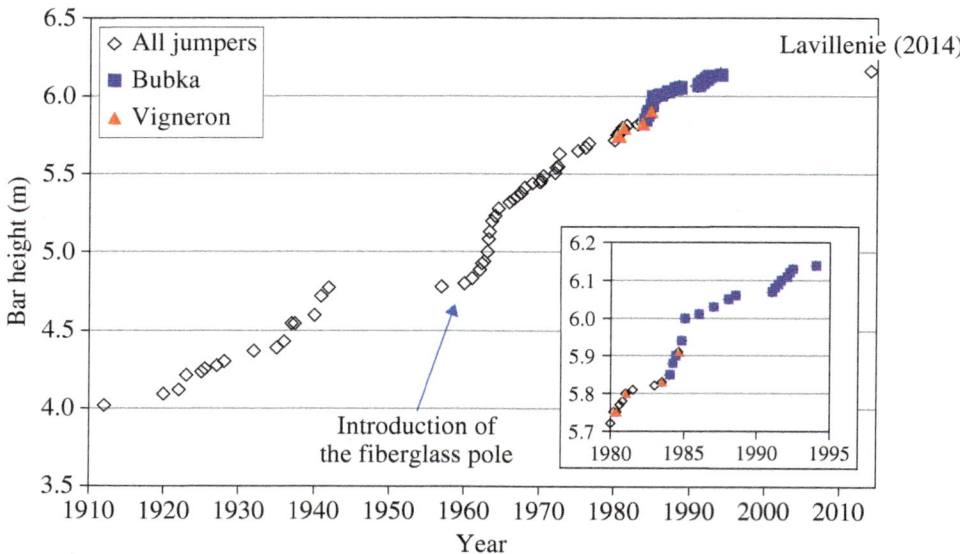

Figure 13.1. The progression of men's pole-vault world records over the past 100 years shows several interesting features. The introduction of fiberglass poles in the early 1960s quickly and dramatically ended a decade-long drought. Ukraine's Sergey Bubka dominated the sport beginning in 1984, repeatedly breaking his own record by tiny increments. His 6.14-m (20-ft 1.7-in.) jump in 1994 stood for 20 years.

beat it by just 1 cm. The introduction of the fiberglass pole in the early 1960s led to rapid advances that blew away these incremental improvements.

We discussed similar technological advances in swimming and javelin events in section 5.6. Each sport handles these advances differently. Swimming banned polyurethane swimsuits but let the records stand (usually with an asterisk). To handle changes in javelin design, the International Association of Athletics Federations (IAAF) created a "new rules" category for throws with the (now-banned) altered javelin design and nullified some results. The rules for pole vaulting, however, are very clear. USA Track and Field (USATF) regulations state that "the pole may be of any material or combination of materials and of any length of diameter." Other regulating bodies for the sport have essentially the same rule.

So while Warmerdam or Gutowski may well have jumped much higher with a modern pole, their records are no more or less valid than those that followed. Pole-vaulting performances have to be considered in the context of the times and equipment available.

We'll say more about the pole in a bit.

Innovations in technique

Fosbury's invention of the Flop in the high jump (section 4.4.2) led to much increased jumping performance; nobody uses the straddle technique anymore in that event. Similarly, underwater swimming (including dolphin kicking, discussed in section 2.1) greatly decreases lap times because of the decreased drag underwater; since the 1950s, regulating bodies have been tweaking rules to limit underwater swimming during meets, to preserve both the integrity of the sport and the health of athletes who can suffer from oxygen starvation.

Together with improvements in pole technology, new vaulting techniques are regularly being developed; coaching and training have become increasingly scientific. You can find a huge number of articles in research journals studying the optimum motion of the body during various stages of the vault. The nuances of these studies are too involved for us to delve into great detail, but we'll point out some general important aspects of modern vaulting technique in this chapter.

Sergey Bubka

First, the statistics. Sergey Bubka was born in 1963 in southeastern Ukraine. He is 6-ft (1.83-m) tall, and during his 20-year career as a pole vaulter, he weighed about 180 lb (80 kg). Since his retirement in 2001, Bubka has devoted most of his time to promoting sport as a positive force in young lives, especially in the former Soviet states.

His pole-vaulting career is legendary. Winning six consecutive gold medals in IAAF World Championships as well as an Olympic gold, he dominated the sport. He set the world record 17 times, almost always superseding his own previous one. He briefly lost his world-record crown only once, to France's Thierry Vigneron in Rome in 1984. Bubka regained it only a few minutes later at the same meet. His 1994 record vault at 6.14 m stood for 20 years, surpassed by 2 cm only in 2014 by Renaud Lavillenie of France.

The following quote, excerpted directly from Wikipedia, is priceless:

> *The fact that most of the time the record he improved was his own demonstrates his absolute dominance in the event. Exactly how high he could have jumped at his best is unknown: because of the large prizes on offer from event promoters for breaking world records, the majority of his world record attempts were made at 1 cm higher than the existing record, and once achieved, he would not attempt another record jump until the next opportunity to collect a prize, even though video footage shows substantial clearance showing he could have achieved a [greater] height.*

Long jumping was transformed forever by Bob Beamon's 8.9-m leap (section 4.4.3) in 1968. It was over 20 years before fellow American Mike Powell broke the record by just 2 in. Powell's record stands today.

In contrast, Bubka continually increased the maximum vault, inspiring increased effort among his competitors as they tried to match his achievements. In this way, Bubka is probably better compared to Usain Bolt or Michael Phelps than to Bob Beamon; their continued dominance tends to "up the game" of their respective events. We'll take a look at the obstacles facing an athlete striving to eke out the extra centimeter of height in a pole vault.

13.2.2 Contributions to height

Figure 13.3 shows various contributions to the center-of-mass height achieved by the vaulter. The ultimate center-of-mass height is given by the sum of (1) its initial height h_0, (2) the distance it rises because of the pole's rotation h_{rise}, and (3) the extra distance achieved h_{push} as the vaulter does work and pushes off from the pole. According to Peter McGinnis of the State University of New York at Cortland, h_{push} is about 35 cm for world-class vaulters.

In a successful vault, the athlete does not touch the horizontal bar (or only does so gently). However, the clearance between the vaulter's center of mass and the bar (labeled "center of mass clearance" in figure 13.3) is often negative, as athletes perform a forward version of the Fosbury Flop (section 4.4.2), as seen in figure 13.2. Top vaulters such as Bubka had negative center-of-mass clearances of several inches.

Figure 13.2
A vaulter performs a belly-down version of the Fosbury Flop. (© *Digital Vision/Getty Images RF*).

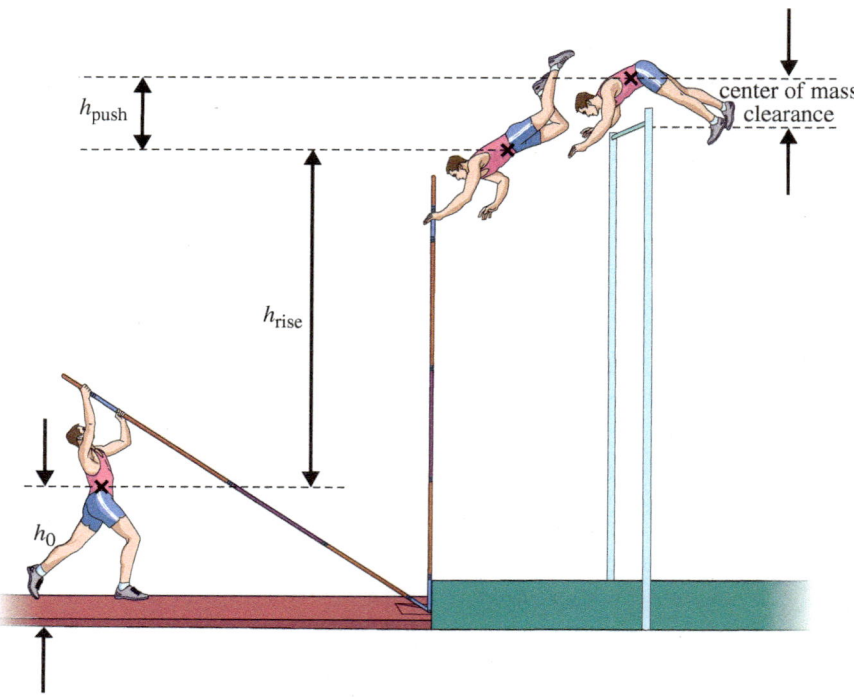

Figure 13.3. Three contributions to the pole vaulter's center-of-mass height.

13.3 Pole Vault 101: Energy Flow

Bubka said that the pole vault is "a professor's sport," and he is right. The event is often used as a classroom example of energy conversion between many of its various forms. As we will discuss, several important details are usually glossed over. Nevertheless, a simplified analysis of the energy flow is instructive as a start. For the moment, we'll ignore the final push at the end of the vault, keeping in the back of our minds that it adds 1 ft or more to the vault height (h_{push} in figure 13.3).

13.3.1 Energy-based estimate of vaulting height

Figure 13.4 sketches various stages of the event.

At position 1, Bubka is at rest ($KE_1 = 0$) but is standing upright, so $h_{c.m.} \approx 1$ m. The energy of the Bubka+pole system is

$$E_1 = KE_1 + PE_{g,1} = 0 + mgh_{c.m.,1} = (80 \text{ kg})(9.8 \text{ m/s}^2)(1 \text{ m}) = 784 \text{ J}$$

Between positions 1 and 2, Bubka takes about 10 strides and does work to reach his top speed of 9.9 m/s.[1] His energy is then

$$E_2 = KE_2 + PE_{g,2} = \tfrac{1}{2}mv_2^2 + mgh_{c.m.,2}$$
$$= \tfrac{1}{2}(80 \text{ kg})(9.9 \text{ m/s})^2 + (80 \text{ kg})(9.8 \text{ m/s}^2)(1 \text{ m})$$
$$= 4704 \text{ J}.$$

Bubka increases his energy sixfold in the approach.

---

[1] Actually, some reports list 9.9 m/s as Bubka's *average* velocity during the approach. This would be extremely high, rivaling Usain Bolt's record-setting 100-m dash. Video analysis suggests that Bubka's peak speed was perhaps a little higher than 9.9 m/s.

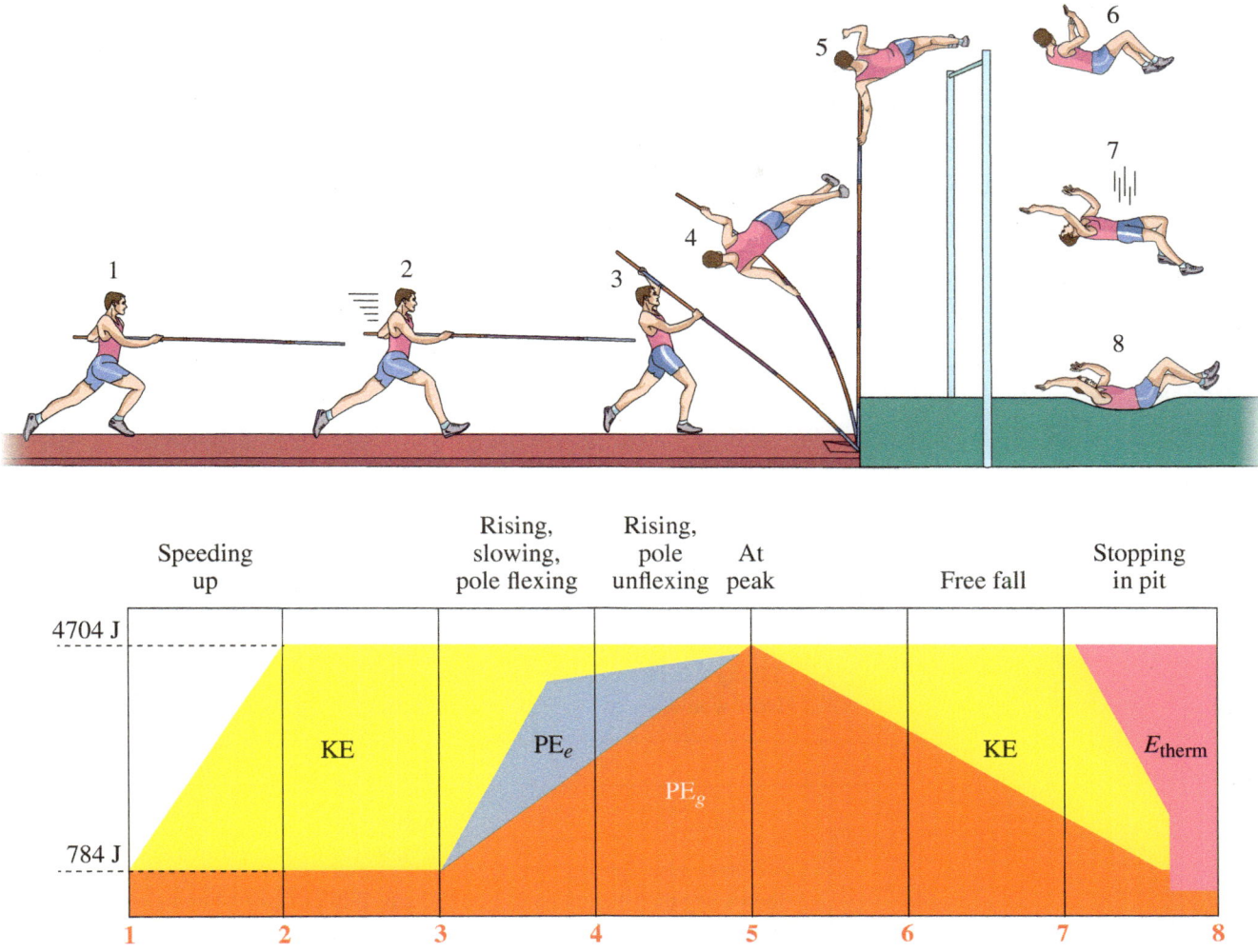

Figure 13.4. A simplified picture of the energy flow in a pole vault. By stage 2, the vaulter is up to top running speed, and the total energy is set. This simple treatment ignores energy losses to air drag, heating of the pole, and inelastic aspects of human motion that are discussed later. The vaulter starts off with some gravitational potential energy because his center of mass is about 1 m off the ground.

The simple picture is that Bubka maintains this top speed for about 8 more strides until he plants the pole at 3. From then on, no energy is "lost" to heat. Kinetic energy is simply converted to gravitational potential energy (that is, he rises) and elastic potential energy (that is, the pole flexes). The *total* energy remains fixed at 4704 J, but the mixture of KE, PE_{grav}, and PE_{elast} varies with time.

Finally, at 4, the pole begins to unbend, so the contribution of PE_{elast} decreases. Bubka also continues to slow, reducing KE. Ideally, the most efficient vault occurs if *all* the available energy is converted into PE_{grav} just at the moment that the pole is upright. (Other situations are illustrated in figure 13.5.) In that case,

$$E_5 = 0 + 0 + mgh_{c.m.,5} \quad \rightarrow \quad h_{c.m.,5} = \frac{E_5}{mg} = \frac{4704 \text{ J}}{(80 \text{ kg})(9.8 \text{ m/s}^2)} = 6 \text{ m}$$

In fact, this is only about 5 in. lower than Bubka's actual record-setting vault.

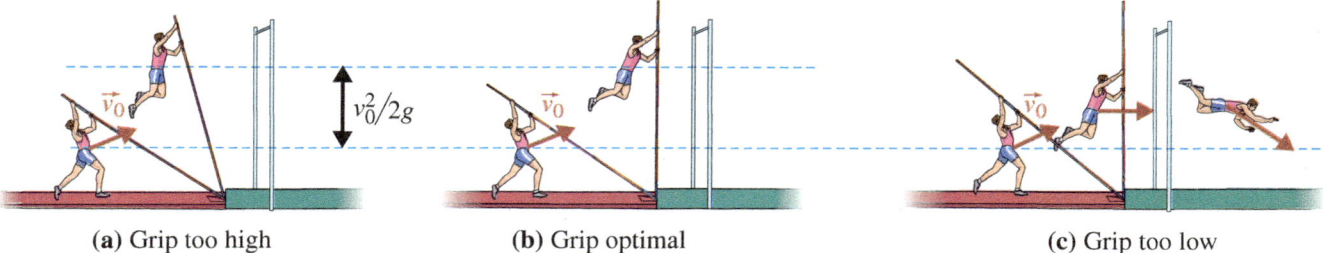

(a) Grip too high   **(b)** Grip optimal   **(c)** Grip too low

Figure 13.5. Selecting the appropriate grip height is important. In all three panels, the vaulter has the same launch velocity and the same initial height. In panels (a) and (b), all of the initial kinetic energy has been converted to gravitational potential, so that the vaulter has risen the same amount. Unfortunately for vaulter (a), however, the pole is not vertical so he is too far back from the horizontal bar, and will simply fall back onto the runway. Vaulters call this "being denied." The vaulter in panel (c) gripped the pole too low. When the pole is vertical, he is lower than (a) and (b), and still has kinetic energy.

> **Question**
>
> One of many things we are ignoring here is the fact that Bubka gives a vertical leap at position 3. It is hard from video footage to estimate the strength of this vertical push, but make a reasonable estimate of his vertical launch speed.
> Taking this into account, how high would he go?
> How would the energy graph at the bottom of figure 13.4 look?

> **Question**
>
> We have also ignored the work that vaulters do at the top of the vault, increasing their height by $h_{\text{push}} \approx 30$ cm. Without going into numbers, how would this effect change the energy graph at the bottom of figure 13.4?

13.3.2 What matters in the simple calculation

According to the simple energy-only treatment that we've given, the only things that really matter are the vaulter's speed and height. Most things don't seem to matter at all.

(A) Mass doesn't matter

Physicists love to do math symbolically, putting in numbers only at the end. While students can find this distracting and abstract, there is sometimes something to learn, so be patient. Going over the last few paragraphs and using only symbols, we find

$$h_{\text{c.m.},6} = \frac{E_5}{mg} + h_{\text{push}} = \frac{mgh_{\text{c.m.},3} + \tfrac{1}{2}mv_3^2}{mg} + h_{\text{push}} = \underbrace{h_{\text{c.m.},3}}_{h_0} + \underbrace{\frac{v_3^2}{2g}}_{h_{\text{rise}}} + \underbrace{h_{\text{push}}}_{\approx 30 \text{ cm}}$$

So, from an energy point of view, the vaulter's mass doesn't matter. Sure, it takes twice as much energy to lift a 100-kg person over the bar as it takes for the 50-kg person. But if he is running at the same speed, the 100-kg person *starts* with twice as much energy.

(B) The pole doesn't matter

When we look at figure 13.4 and the discussion above, it seems the details of the pole don't really matter at all. The pole certainly doesn't *add* any energy to the jump—it just participates in the conversion of energy from the kinetic to the gravitational potential form.[2] The pole's elastic energy rises and falls, but it doesn't even show up in the equations giving the maximum height attained.

In fact, there is no reason that kinetic energy can't be converted to gravitational energy directly, without any flexing of the pole at all. Heck, the poles used by people in Holland were made as rigid as possible. And until the 1960s even professional athletes used bamboo or stiff aluminum poles that flexed relatively little.

Why were flexible fiberglass poles required to break the 5-m (never mind the 6-m!) mark?

(C) The athlete's maneuvers don't matter

Sure, the extra 30 cm that the athlete gains from the final push-off helps. But beyond that, nothing in the energy-only treatment above explains why vaulters and coaches spend so much time on techniques of timing and swinging the body in just the right way. If only energy matters, is that all for show?

Statements A, B, and C above aren't really true

Sergey Bubka is 6 ft tall and can hit 9.9 m/s (22 mph). That's not so far out of reach even for a physics professor. And if all I have to do is hang on to the pole, then why am I *writing* about Bubka instead of breaking his record? Why have fewer than 20 humans on the planet ever cleared 6 m?

To understand more, we have to go beyond Pole Vault 101.

13.4 Pole Vault 102: Beyond Energy Flow

Although the energy-only treatment of Pole Vault 101 is a favorite of professors of introductory physics, coaches and athletes know that there is much more to the story. The pole, maneuvers, and even mass matter. The quantity and level of detail of basic research, kinesiology studies, and advice and commentary on techniques are simply overwhelming. A discussion of the nearly infinite details makes valuable reading for coaches and athletes looking to improve their performance, but can obscure the major physics points.

The motion involved in a vault is composed of several simultaneous and intertwined elements. It is sometimes called a **double pendulum**. The vaulter's center of mass swings about the pivot of his hands at the end of the pole; this is the first pendulum. Meanwhile, the pole itself is pivoting about its fixed end in the vaulting box on the ground; this is the second pendulum. Advanced physics majors and graduate students often are often presented with a double pendulum when the professor wants to convince them that they need to learn an advanced form of mechanics called the Lagrangian approach. At first sight, these students think they can use their old $F = ma$ treatment to calculate the system, but soon find it way too complicated.

And it gets worse! While all this is going on, the pole flexes and straightens, and the athlete changes his shape and exerts torques and forces on the pole itself.

---

[2] As we discussed in chapters 8 and 9, other sports equipment, such as bicycles and hockey sticks, likewise add no energy to an event.

Rather than attempt anything like a detailed treatment here, we'll focus on individual elements of the vault, providing scientific insight into details you'll notice when watching a vaulting competition.

There are four main ingredients for a good vault:

1. Maximizing the initial energy

2. Minimizing mechanical energy "loss"

3. Full conversion of available energy to the gravitational form

4. Work done by the athlete during the vault.

13.4.1 Maximizing initial energy: carry weight

While Pole Vault 101 left out many important details, its main premise remains true:

> **The height achieved by the vaulter depends primarily on his initial energy at the time of the vault.**

As we've seen, the main advice to the vaulter who wants to maximize his initial energy may be summarized thus: "Run fast! Jump hard at takeoff! Be tall!"

An aspect that might not be so obvious has to do with the **carry weight** of the pole and how it is carried down the runway.

Modern poles weigh between 1.6 and 3 kg (3.5 and 6.5 lb). At first, you wouldn't guess that such little extra weight would slow the vaulter much as he accelerates down the runway. However, since the vaulter holds the pole from one end, the carry weight of the pole is much larger. The carry weight is the force the vaulter must exert to keep the pole horizontal. It can be many times the pole's weight.

Figure 13.6 shows a vaulter holding a 5-m-long, 6-lb pole horizontally. For a right-handed vaulter, the left hand is in front and exerts an upward force $\vec{F}_L$ on the pole. The right hand exerts a downward force $\vec{F}_R$ at one end, and the force of weight $\vec{W}$ acts at the pole's center of mass. To keep the pole steady, the net force *and* the net torque must be zero.

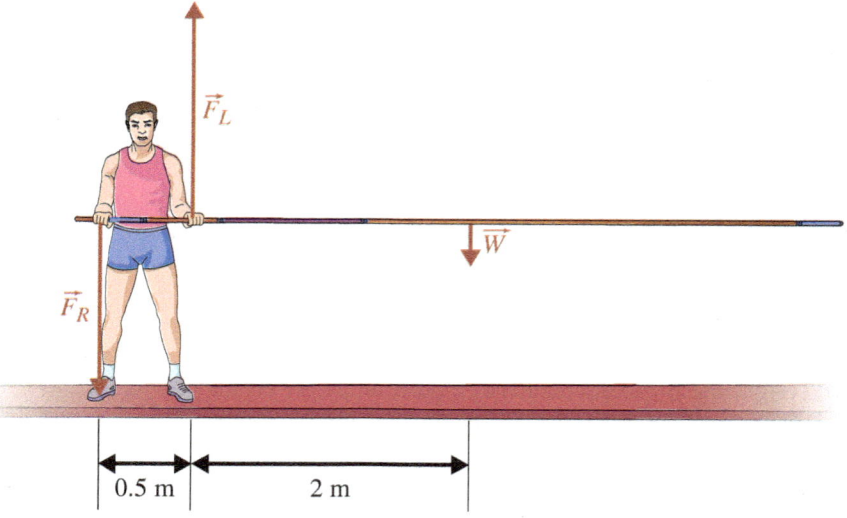

Figure 13.6. A vaulter holds a 5-m-long pole horizontally. The pole's weight $\vec{W}$ can be considered to act at the pole's center of mass, halfway along its length.

The torque around the left hand is

$$\sum \tau_i = \underbrace{-|\vec{W}|(2\text{ m})}_{\substack{\text{torque from}\\\text{gravity}}} + \underbrace{|\vec{F}_R|(0.5\text{ m})}_{\substack{\text{torque from}\\\text{right hand}}}.$$

Setting the net torque to zero lets us find the downward force exerted by the right hand:

$$|\vec{F}_R| = \frac{2\text{ m}}{0.5\text{ m}}|\vec{W}| = 4|\vec{W}| = 24\text{ lb}.$$

Question

Why didn't the force due to the left hand appear in the torque equations above?

Next we look at the net force on the pole:

$$\sum F_{y,i} = -|\vec{F}_R| + |\vec{F}_L| - |\vec{W}|.$$

And setting *that* to zero gives us the force on the left hand:

$$|\vec{F}_L| = |\vec{F}_R| + |\vec{W}| = \underbrace{4|\vec{W}|}_{\text{see above}} + |\vec{W}| = 5|\vec{W}| = 30\text{ lb}.$$

So *each* arm needs to exert the carry weight of about 25 lb of force as the vaulter accelerates down the runway. That's a lot more effort than simply carrying 6 lb extra, and it can significantly inhibit the vaulter's sprinting action. Now you can understand why manufacturers tout any reduction in a vaulting pole's mass. Reducing the weight of a pole by 1 lb translates into about 5 lb *per arm* in carry weight.

The pole angle on approach

If you've watched vaulting competitions, you may have noticed that the vaulters begin their approach with the poles pointed at a steep angle relative to horizontal, as shown in figure 13.7. They then steadily lower the pole as they approach the launching box. This is precisely to reduce the effective carry weight of the pole, as shown in figure 13.8.

If the pole is held at an angle θ relative to horizontal, the torque due to gravity is reduced because its lever arm is reduced. Rather than 2 m, the lever arm is $\ell_W = (2\text{ m})\cos\theta$. For a 60° carry angle, the carry weight is effectively cut in half, since $\cos 60° = \frac{1}{2}$.

Question

Would the carry weight of the pole change if the vaulter brought his or her hands closer together on the pole? If so, how? How might that factor into coaching advice to the vaulter?

Figure 13.7
A vaulter holds her pole at an angle to reduce the carry weight on her arms. (© UpperCut Images/ Getty Images RF).

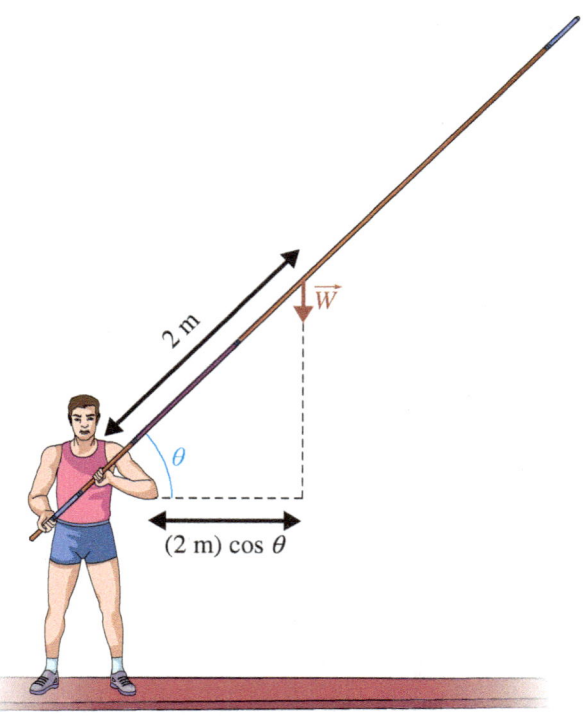

Figure 13.8. If the pole is held at an angle, the lever arm of its weight is reduced, lowering the effective carry weight.

Question

In the discussion above, I simply stated that the forward (left) hand must exert an upward force and the rear hand a downward force. Is that so obvious? Is it possible to change either or both of those directions and still hold the pole steady? If so, how?

13.4.2 Minimizing inelastic energy "loss"

Vaulting expert and biomechanist Peter McGinnis points out that elite vaulters accelerate on their last three strides before launch, while developing vaulters tend to slow down somewhat. This is mostly because they are focusing on planting the pole and positioning their feet for a good jump. But among total neophytes, there can be another reason for reluctance to throw oneself full-force into a just planted pole. Here you are, hurdling down the runway at 20+ mph, when suddenly one end of the pole you are gripping comes to a *dead stop*. Nicholas Linthorne, another vaulting expert, has called this event "an indirect collision" between the athlete and the ground, with the pole as intermediary. If the pole is extremely stiff, the end that you are holding will immediately transmit the force of the collision directly along the length of the pole.

Let's stick with a stiff pole and imagine the worst-case situation, shown in figure 13.9. You plant the pole at a very shallow angle while your velocity is purely horizontal (you forgot to jump!). When the far end of the pole hits the box, a powerful and nearly horizontal shock is sent directly toward your center of mass. Oof! *If* you manage to hang on, this is a completely inelastic collision. Your body will absorb (that is, convert to thermal modes such as inelastic musculoskeletal stretching and perhaps internal damage) essentially all the energy you'd built up during your approach.

Figure 13.9. A vaulter with horizontally directed velocity plants a stiff pole at a very shallow angle. The powerful shock brings him to a dead halt; his body absorbs the energy inelastically.

Any object, including the human body, can withstand only small accelerations before conversion ("loss") of some of the energy to thermal modes. Minimizing these losses is the job of the vaulter, second in importance only to maximizing his speed on the approach. There are three keys to this task, each based on physics concepts we've discussed in this book.

Key 1: large plant angle

Figure 13.5 showed the importance of gripping the pole at the optimal distance from the far end. Having chosen where to grip the pole, good vaulters hold the pole high (head level or a bit higher) to maximize the plant angle θ_P, as shown in figure 13.10. The vector nature of forces (section 3.4.1) is crucial here. Since to a good approximation **the force of a stiff pole on the vaulter always points along the pole itself**, this maximizes the upward component of the force. Up sounds good.

In another difference from the worst-case scenario of figure 13.9, the vaulter leaps in the final stride, maximizing θ_V and the upward component of velocity. Again, up sounds good.

Vaulting experts advise as high a hold as possible and a strong leap at takeoff.

Question

1. Use geometry to estimate the plant angle for Sergey Bubka.

2. What would you expect for θ_V for a world-class athlete such as Bubka?

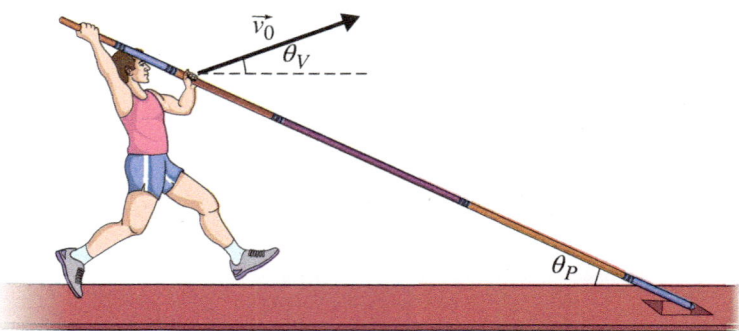

Figure 13.10. To maximize the vertical components of velocity and acceleration, the vaulter tries to maximize the plant angle θ_P and launch angle θ_V.

Key 2: swinging about the grip

We'll stick with a rigid pole for a bit more.

Now the vaulter has some upward-directed velocity (from his launching leap) and one source of upward-directed force from the pole planted with large angle θ_P. Still, if we account for gravity, the net force on the vaulter points roughly horizontally to the left. This means one thing: **he *must* have a large left-pointing acceleration.** Hmm... that doesn't sound too promising, does it? It looks as if it's going to be a slightly less painful version of the "oof" of figure 13.9.

All that energy he built up running horizontally, and now a sudden leftward acceleration is going to slow him down!

But wait, scratch that. A left-facing net force actually means that **his *center of mass* must accelerate to the left**. The fact that the human body is an "extended object" (physicists love that term) makes all the difference.[3] The launch position in figure 13.10 leads naturally to a swinging motion pivoted about the hand gripping the pole. Assuming that the pole stays fixed in place (of course it doesn't really) and doesn't bend (it does really), the vaulter begins circular motion. Great! We learned in section 3.5.3 that an athlete can accelerate to the left without changing speed (that is, losing kinetic energy), as long as it is *centripetal* acceleration at work.

In figure 13.11 the pole is kept rigid and fixed in place, to focus on this point. If the vaulter keeps his upper arm rigid to avoid inelastic stretching and negative work (remember figure 7.8 and section 7.6), the loss of mechanical energy to heat is minimized.[4] The force of gravity (his weight) does slow the vaulter in the figure, but this converts kinetic to gravitational potential energy; this mechanical energy is not lost.

Key 3: a flexible pole

Qualitatively, that all sounds fine. But we find that there is still a problem when we do a numerical evaluation of the forces on the vaulter's arm as he undergoes the swinging motion.

Using equation 3.17 with numbers appropriate for Bubka at launch, we find

$$F_{\text{arms}} = m\frac{v^2}{r_{\text{c.m.}}} = (80 \text{ km})\frac{(10 \text{ m/s})^2}{1 \text{ m}} = 8000 \text{ N} \times \frac{1 \text{ lb}}{4.45 \text{ N}} = 1800 \text{ lb}.$$

---

[3] This is one reason why this aspect of the event is never raised in introductory courses. Those courses deal as much as possible with point-like objects.

[4] Experts such as McGinnis stress the importance of keeping limbs rigid in a pole vault, precisely to avoid energy loss.

Figure 13.11. The "man pendulum." By executing circular motion about his hands fixed on the pole, the horizontal component of the vaulter's center of mass points opposite his initial velocity, but without significant loss of energy. In this figure, the vaulting pole is fixed in space, just to make the point clear.

Gripping a pole via friction, against the sudden application of almost a ton of force, would not be possible. Even if his hands were glued to the pole, his arms would certainly experience inelastic (and painful) stretching.

A flexible pole helps solve this problem, basically spreading the impulse over a longer time, reducing the peak force (cf. equation 6.11). The resisting force is initially small,[5] gradually increasing as the pole flexes. Not only does this lead to a gentler "collision," but there's an extra bonus: The larger restoring force from the pole occurs after the pole has rotated more toward a more vertical orientation; so the biggest push from the pole is directed at a larger angle than the plant angle θ_P (figure 13.10).

In terms of minimizing the energy loss due to shock, the vaulter should choose as flexible a pole as possible. However, Bubka and other world-class vaulters use relatively stiff poles. The reason is related to the third goal mentioned on page 342, which we discuss next.

13.4.3 Fully exploiting the energy: flexibility and timing

Assuming that the vaulter has maximized the initial amount of energy (section 13.4.1) and minimized inelastic energy losses (section 13.4.2), his goal is to ensure that no energy is "left over" in kinetic or elastic modes when he crosses the plane of the crossbar.

One aspect of this goal is important even for a nonflexible pole, as already sketched in figure 13.5: Grip the pole too low, and you'll never convert all your kinetic energy into gravitational energy.

When a flexible pole is considered, another factor comes into play: timing.

Flex numbers and pole weight

Modern vaulting poles are hollow columns of carbon fiber or fiberglass. For small bends, a vaulting pole follows Hooke's law, introduced in section 8.1. The flexibility

---

[5] In fact, modern poles are built slightly "prebent," to ensure that the pole immediately starts to bend (in the right direction!) when planted, avoiding "straight-pole shock."

of a vaulting pole is measured in a similar fashion as it is for archery arrows. The pole is supported near each end, and a weight (usually 50 lb) is hung from its center. The number of centimeters the pole deflects (see figure 8.5 for the same measurement with arrows) is the **flex number** of a pole.[6] A large flex number means a flexible pole.[7] The Hooke spring constant is

$$k_{\text{pole}} = \frac{50 \text{ lb/cm}}{\text{flex number}}$$

At maximal flexing during the vault (about position 4 in figure 13.4), modern vaulting poles store an amount of elastic potential energy equivalent to about one-half of the vaulter's initial kinetic energy.

> **Question**
>
> A 14-ft pole might have a flex number of 20. How does the Hooke constant for such a pole compare to each of the following?
>
> - A recurve bow (section 8.1.2)
> - An archery arrow (section 8.3.1)
> - A hockey stick (section 8.4)

In section 8.2 we said that any oscillatory system, that is, a system with a springlike restoring force coupled with a mass, oscillates with a characteristic period. In our case, the pole oscillates between fully extended (at launch), and compressed (midvault) to extended again (ideally, when the pole is vertical). According to equation 8.4, the time to compress and uncompress the pole, if the system is left to itself,[8] is given by the proportionality

$$T_{\text{flex then unflex}} \propto \sqrt{\frac{m_{\text{vaulter}}}{k_{\text{pole}}}}.$$

As we said in section 8.2, it is complicated to calculate the constant of proportionality for any realistic system, but it's usually between 3 and 10.

For a typical 160-lb vaulter ($m_{\text{vaulter}} = 5$ slug) and pole (say flex number of 20),

$$k_{\text{pole}} = \frac{50 \text{ lb/cm}}{20} \times \frac{30.48 \text{ cm}}{1 \text{ ft}} = 76.2 \text{ lb/ft}$$

$$\rightarrow T_{\text{flex then unflex}} \approx \left(\begin{array}{c}\text{number between}\\ 3 \text{ and } 10\end{array}\right) \times \sqrt{\frac{5 \text{ slug}}{76.2 \text{ lb/ft}}}$$

$$= \text{between } 0.8 \text{ and } 2.56 \text{ s}$$

---

[6] Different manufacturers use slightly different ways to measure flex numbers. The method I'm describing comes from Jeffrey Watry, an engineer at Gill Athletics.

[7] Related to flex number and to the underlying tensile strength of the material the pole is made from is a **weight rating**. This specifies the maximum weight of a vaulter who can use the pole safely, with little chance of it breaking during the vault. Amazingly, modern poles can be bent by 170° without breaking. Poles with higher weight ratings are generally stiffer, but two poles with identical weight ratings don't necessarily have the same stiffness or flex number, because of different material composition, length, and so on. The weight rating is not important for the physics we'll discuss here.

[8] Of course, the pole+vaulter system is *not* left to oscillate by itself, since it is simultaneously rotating about the vaulting box, subject to the Earth's pull, and experiencing forces from the vaulter's acrobatics. Indeed, for the large flex angles of a realistic pole vault, the system doesn't exactly follow Hooke's law. Therefore, like much else in this section, our discussion here simply focuses on one aspect of the complex vaulting motion to get a rough idea of the time scales involved and what drives them.

Ideally, the vaulter wants the pole to completely extend (return all its stored energy) as the pole rotates to vertical. In high-level competitions, this rotation indeed takes about 1 s, which is in the range indicated above. So it is up to the vaulter to fine-tune his pole selection according to run-up speed, launch angle, pole plant angle, and hand placement.

If k_{pole} is too small and/or $m_{vaulter}$ is too large, the pole won't unflex until the vaulter has passed under (or through!) the bar. Conversely, if the pole is too stiff or the vaulter too light, the pole will unflex too early and throw the vaulter backward onto the runway; in this case, the vaulter is said to have been "denied." Elite vaulters bring several poles to a meet, selecting a pole for each jump according to the attempted height, technique, and environmental factors (for example, wind).

13.4.4 Work done by the athlete

Obviously, all energy in the vault originates in work done by the athlete; the pole adds no energy on its own. The majority of this work is done prior to launch. Nevertheless, there are more opportunities for the athlete to add energy to the system and gain precious centimeters.

Loading the pole

Figure 13.12 shows several stages of the vault. The forces involved in stage *A* are shown in Figure 13.13. For a short time (until about 0.2 s after launch, according to McGinnis), the lower hand pushes the pole, increasing its elastic potential energy; this is an example of actively "loading" the pole. The "reaction" forces of the pole on the vaulter are shown in the right panel of the figure. While the forces on the left and right hands point in almost opposite directions, they each produce a torque about the center of mass in the *same* direction, tending to rotate the vaulter clockwise into position *B*. This starts to position the vaulter for the final handstand maneuver at the end of the vault.

Figure 13.12. Detailed sketch of a vaulter's position and pole orientation throughout the vault.

350 The Pole Vault

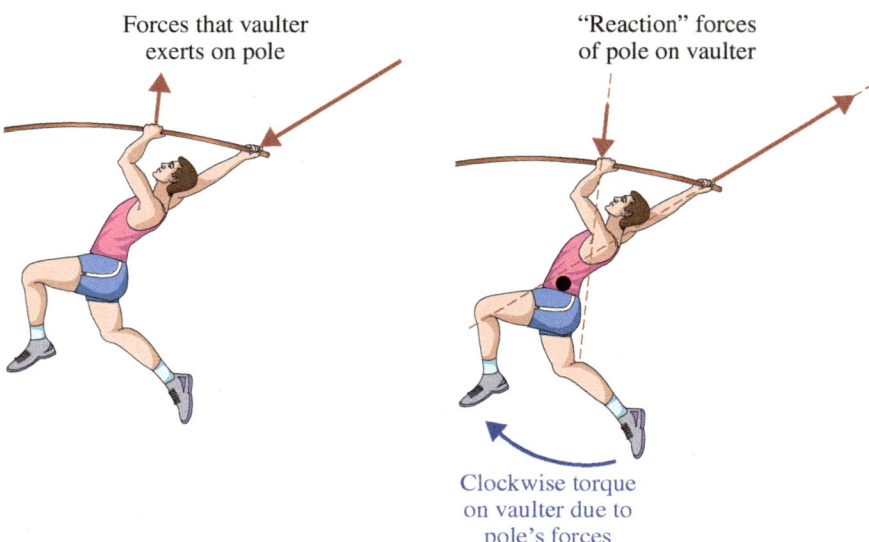

Figure 13.13. Early in the vault, the vaulter pushes with his lower arm, "loading" the pole. The forces of the pole on him produce a clockwise torque. Dashed red lines show the "lines of action." A similar figure appears in *Biometrics of Sport and Exercise* by Peter McGinnis.

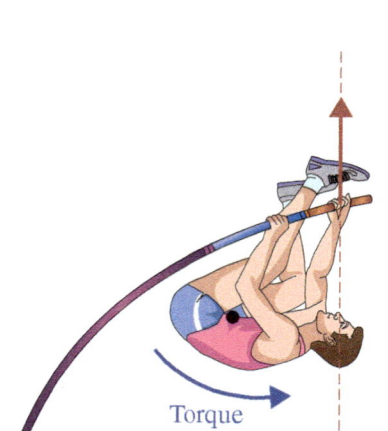

Figure 13.14
As he rotates to position *B*, the vaulter pulls in his legs, increasing the kinetic energy of rotation. The line of action of the pole's forces on him changes as well, producing a counterclockwise torque that slows his rotation.

Pulling in the legs

The clockwise rotation continues as the vaulter progresses from stage *A* to *B*. Importantly, however, the lines of action of the forces on his hands change, as shown in figure 13.14. Because the torque from the force on his upper hand has switched from clockwise (position *A*) to counterclockwise (position *B*) his somersaulting rotation begins to slow. This is a good thing, or else he'd "overrotate" and find himself on the wrong side of the pole when he reached the top.

Another thing that has changed while progressing from *A* to *B* is that the vaulter has drawn in his legs, thus decreasing his moment of inertia about his center of mass (see, for example, table 10.1). As we discussed with the figure skater in section 10.3, "pulling in" while spinning requires positive work, so the vaulter has put additional energy into the system by this action.

Extending and pushing off

The vaulter extends his body in to a handstand in position *C*. The vaulter is again doing work by lifting his center of mass, infusing even more energy into the system. This extension also kills off any remaining rotation of the vaulter about his center of mass by increasing his moment of inertia. Likewise, the extension increases the moment of inertia of the pole–vaulter system about the planted end of the pole, slowing any remaining rotation.

Finally, as indicated in figure 13.15, the vaulter pushes directly down on the pole to lift himself up and (hopefully) over the bar, injecting the final shot of energy before gravity takes over.

Figure 13.15
Positions *C* and *D*: Extending the body increases the moment of inertia and further slows rotation. The positive work from the lift and push in the vertical direction injects a final burst of energy into the system.

Is It Better to Run through First Base or to Dive?

14

If a batter hits a pitched ball in American baseball, he must run and touch first base before someone (usually the first baseman) on the fielding team touches the base while holding the ball. If the batter makes it, he's "safe." Otherwise, he's "out."[1]

Once he touches first base, the runner is declared safe; he is allowed to run through and past the base without risk of being tagged out. This rule allows the runner to maintain top speed without having to decelerate before reaching the base. This is different from the situation on second or third base, where the runner can be tagged out if he doesn't remain on the base. In those cases, the runner often slides or dives the last few feet before reaching the bag, relying on friction to rapidly slow him so at least one part of his body remains in contact with the bag.

Periodically the question arises: Would it perhaps be better to dive into first base, instead of running through it?

It's not a crazy question. In the high jump or long jump (sections 4.4.2 and 4.4.3), the trajectory of the athlete's center of mass is beyond his control, but he is not scored on the height or distance achieved by his center of mass. Rather, the athlete manipulates the shape of his body and the laws of physics to optimize his performance. Beamon's center of mass did not need to travel 8.9 m, nor did Fosbury's need to achieve a height of 2.24 m in Mexico City in 1968.

Similarly in baseball, what matters is that *some part* of the batter's body must touch the bag as soon as possible. By manipulating his body orientation, a batter who dives into first base touches the bag before his center of mass arrives. By contrast, the batter who runs through first base touches the bag essentially when his center of mass arrives.

Most players run through first base, but not all. Before his retirement in the 1980s, the legendary Pete Rose (also known as "Charlie Hustle") eked out many a play by diving into bases including occasionally first. In recent years, Nick Punto is perhaps the most notorious example of a head-first-into-first player.

Punto and the run-versus-dive question itself are frequent topics in magazines, newspaper articles, and endless online baseball discussion forums and comment sections. Especially in the latter format, one study is by far the most frequently cited as conclusive "scientific proof" that running through first is faster than diving into it. Let's discuss this before continuing.

Concepts in this chapter:

- Chapter 2: Kinematics
- Chapter 3: Newton's laws friction
- Chapter 4: Changing human center of mass
- Chapter 5: Air drag
- Chapter 6: Impulse
- Chapter 9: Torque

14.1 The Story according to *Sport Science*

Originally aired by Fox Sports Network in 2008 and later picked up by ESPN, *Sport Science* is a series of television vignettes exploring scientific aspects of sports, frequently featuring big-name athletes. While it doesn't exactly "teach" the science, it

---

[1] A less common way to be out is for a fielding player to tag the runner with the ball in hand, before the runner makes it to first.

is always entertaining and of high production value. Sometimes, however, it doesn't get things exactly right. The diving-versus-running episode contains errors that are so glaring and so frequently propagated in arguments that they deserve discussion. The physics that we have already learned will help us spot the error and identify the correct approach to the question.

In the episode, college baseball player Jake Wood runs to first base. He is running "at top speed of almost 18 mph" (to quote the show) as he approaches the bag, and the analysis compares what happens as he covers the final 10 ft. If he runs through the bag, it takes 260 ms to cover this distance whereas diving takes 270 ms.

After going through the material in this book, you are now qualified as an expert in mechanics,[2] the branch of physics that quantifies and explains motion. Therefore, the explanation of Jake Wood's motion is of interest. To quote:

> *What's surprising is that when we overlay the high-speed footage, we see that as Jake dives, he actually pulls ahead. [see figure 14.1]. But the moment his feet leave the ground, he decelerates by 50%! By remaining upright [i.e. not diving], Jake takes one more propulsive step forward and maintains his velocity through the bag.*

The show concludes by pointing out that the 10-ms difference between the two methods corresponds to about the diameter of a baseball.

Question

Can you confirm that last statement?

The program is intriguing and makes some interesting points. As you probably noticed right away, one of the *physics* points is stunningly incorrect; we'll discuss this in section 14.3. First, however, we briefly discuss the roughness of the statements made.

Figure 14.1. Overlaid high-speed video footage where the diving Jake Wood is described as "pulling ahead" of the running Jake Wood in the program.

---

[2] To receive your official certificate, send $20 to me via my editor.

14.2 Too Close to Call

The width of a baseball is a small difference compared with the 90-ft base path. But are the measurements made in the program precise enough to determine such a difference?

A numerical check

Let's check out some of the numbers used in the analysis. If Jake maintains his "top speed" of 18 mph, then how long would it take him to travel the final 10 ft?

$$\Delta t = \frac{\Delta x}{v_x} = \frac{10 \text{ ft}}{18 \text{ mph} \times \frac{146.7 \text{ ft/s}}{100 \text{ mph}}} = 0.379 \text{ s} = 379 \text{ ms}.$$

That's a *lot* more than the 260 ms quoted in the program. A 120-ms discrepancy in something simple like this makes one wonder about how well these measurements can determine a 10-ms difference between diving and running through.

Obviously something is strange here. What happened? There's no way to know for sure without more information about the measurement process, but it may be related to the way that the "start" location is defined.

Where to start?

The time to stop the stopwatch seems very clear: just as soon as some part of Jake touches the bag. But when should one *start* the watch? (See figure 14.2.) The answer seems to be when Jake is 10 ft from the bag. But which part of Jake? His center of mass? his hand? his front foot? When running through the base, his front foot touches the bag, so perhaps we should start the clock when his front foot is 10 ft from the base, as it looks in the figure. But this is not so precisely determined and will depend sensitively on where he is in his "stride pattern" for any given attempt.

The "starting positions" at $t = 0$, shown in figure 14.2, are only approximately the same. Certainly, the distance traveled by Jake's outstretched hand in the diving case (indicated by the blue line on the ground) is much shorter than the distance his foot must travel in the running case.[3] It apparently isn't so clear—at least to within the diameter of a baseball—precisely when to start the clock. In a "fair" race (in which the clock was started when Jake begins running from home plate, rather than when he is within a vaguely defined 10 ft from first base) the time difference might be *more* than 10 ms, corresponding to a distance much greater than a baseball diameter.

Implications (and hypocrisy on my part?)

The numerical check and considerations above lead one to question how precisely these measurements distinguish time differences between running and diving. Physicists would likely consider a 10-ms difference as "within the noise" with these methods.

I can already hear you: "Hey! You've been teaching us physics in this book, based on estimates and methods just like the ones you are now criticizing! The pot is calling the kettle black." Not really. Our use of estimates and approximations has been

---

[3] On the other hand, their center-of-mass positions are similar. It cannot be determined whether they are within 3 in. of each other.

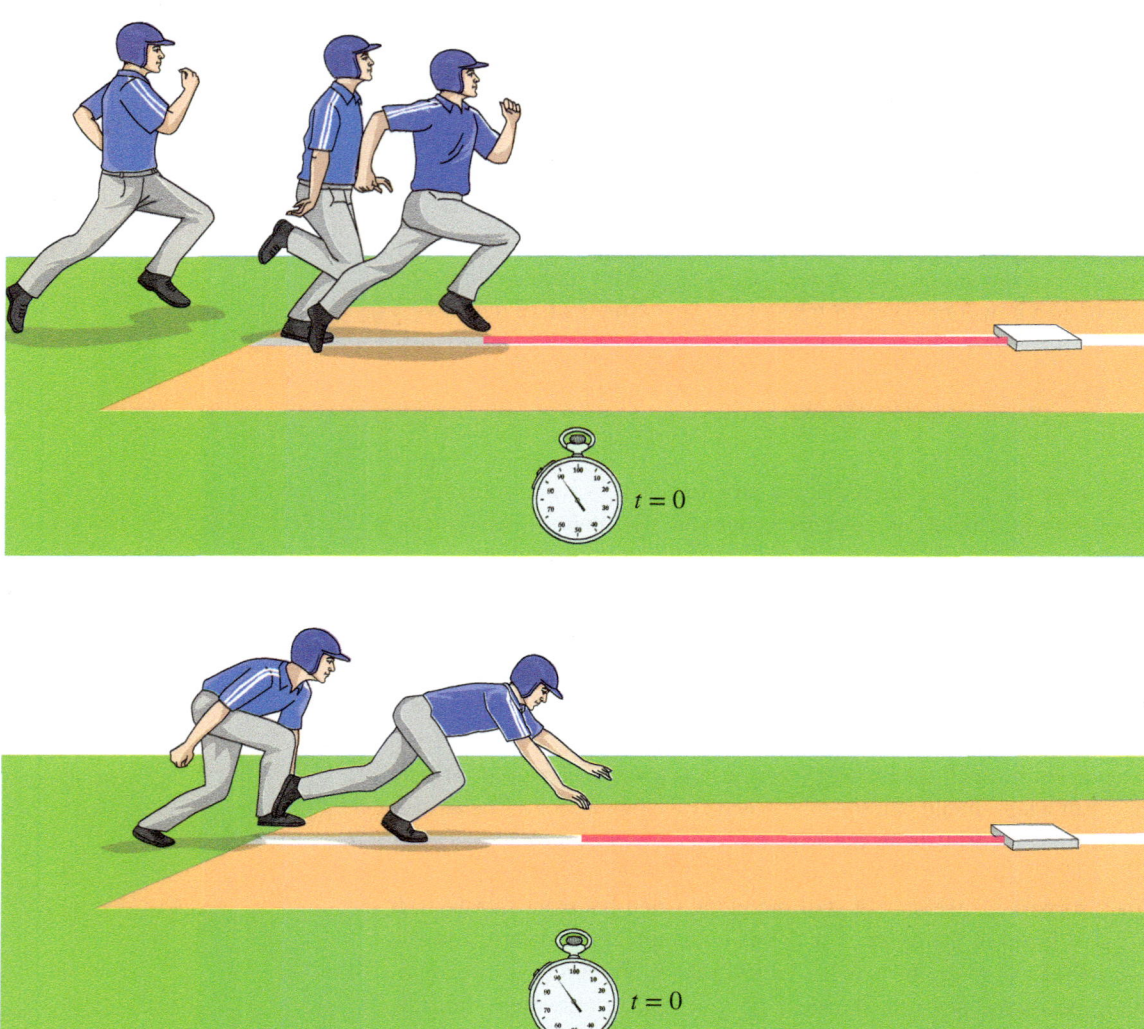

Figure 14.2. The clock starts when Jake Wood is 10 ft from first base. Is his distance from the bag sufficiently well defined to within an inch or two? Is the distance traveled by the front part of his body (pink line on the ground) exactly the same in the two cases?

valid and extremely useful to gain insight into the forces, torques, and speeds and to understand the *principles* driving sports.

However, if the focus is not on principles but on a tiny effect, the estimates and approximations have to be good enough to measure that effect before claiming a result that will echo endlessly on the Internet.

14.3 Diving Speed

Speaking of principles, that's where the *real* problem lies in this discussion. If we put aside any problems associated with timing measurements that we've just discussed, the video footage shows very clearly that Jake *is* moving more slowly during the dive than he does if he runs through the bag. Understanding the physics behind that difference is our goal now.

14.3.1 "50% Deceleration"

We've seen repeatedly that understanding sports usually starts by quantifying the motion. The key description in this regard would be the statement "the moment his feet leave the ground, he decelerates by 50%!"

While "decelerates by 50%" sounds pretty scientific, as we learned in chapter 2, acceleration is measured in ft/s^2, not percentage. We'll make a reasonable guess that "decelerates by 50%" means that Jake's speed drops to one-half its original value, from 18 to 9 mph ($\Delta v_x = v_x - v_{x,0} = -9$ mph). To figure out the acceleration, then, we have to know how long it takes for this change to happen.

Possibility 1: Jake is torn apart

If the change in velocity occurs "the moment his feet leave the ground," then the change takes essentially no time at all.[4] In this case, the acceleration would be infinite ($a_x = \Delta v_x/\Delta t = \infty$ if $\Delta t = 0$), implying that the force on Jake would be infinite. Poor Jake could never survive a change in speed that takes place "the moment his feet leave the ground!"

"Come on," I hear you say, "you know that's not what they meant." Really? Well, if you say so. But regardless of what they meant, most people would have no problem at all with a change in speed that takes zero time. It's up to those of us (and hey that includes you now) who know a little physics to keep discussions of scientific issues on track. Baseball is one thing, but these little misunderstandings and "intuitive misconceptions" build up and affect much more substantive issues in our larger world.

Possibility 2: Jake is yanked back

Perhaps they mean that Jake's velocity *starts to change* when his feet leave the ground, but that the change occurs over the course of his flight. That would be more reasonable. In that case

$$a_x = \frac{\Delta v_x}{\Delta t} = \frac{9 \text{ mph} \times \frac{146.7 \text{ ft/s}}{100 \text{ mph}}}{0.27 \text{ s}} = 49 \text{ ft/s}^2.$$

That's 1.5 g's! Jake is being yanked backward with a force that is 50% larger than his weight, or about 260 lb! There is little chance that Jake would agree to a second dive attempt if that meant launching head first into a 260-lb opposing force.

There is no hope. Whether we interpret the statement to mean that Jake's speed changes "in the moment" that his feet leave the ground, or that it changes in flight, we come to conclusions that make little scientific sense. And yet, it doesn't sound so bad on TV.

14.3.2 Newton's first law and air drag

Even worse than the huge accelerations implied by the show's description is the fact that it perpetuates a misconception that seems to reside deep in our bones: that a "propulsive step" is required to "maintain velocity." We experts know that the horizontal component of velocity (which is all that concerns us here) won't change unless there is a horizontal force.

---

[4] After all, a "moment" is just an instant; it has no duration. Sometimes it is easier to think of a moment as a very, very, very short amount of time, to avoid things such as dividing by zero.

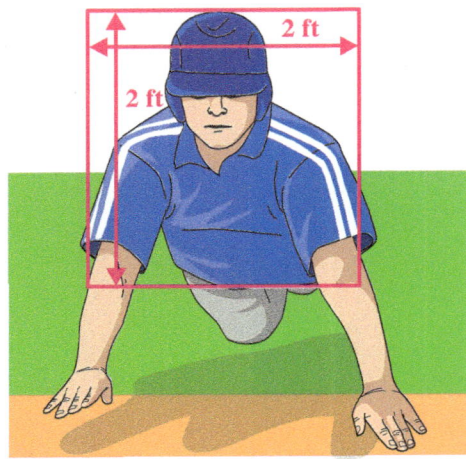

Figure 14.3
A very rough estimate of the cross-sectional area facing into the wind as Jake dives.

We also know that only a few forces matter for sports. Examining the list in table 3.1, we see that the only horizontal force for Jake while in flight is air drag. Equation 5.4 lets us calculate its magnitude:

$$|\vec{F}_D| = \tfrac{1}{2} C_D A \rho_{\text{air}} v^2 = \tfrac{1}{2} \times \underbrace{1.2}_{\substack{\text{estimate from} \\ \text{skydiving}}} \times \underbrace{4 \text{ ft}^2}_{\text{figure 14.3}} \times (0.0024 \text{ slug/ft}^3) \times (26.4 \text{ ft/s})^2$$

$$= 4 \text{ lb}$$

This tiny force will produce a proportionally tiny deceleration of Jake when he dives. It is negligible for this analysis.

To experience the 260-lb drag we mentioned in section 14.3.1, Jake would need to dive into a 145-mph headwind!

14.4 What's Really Happening: Torque and Impulse

So if the whole 50% deceleration thing is nonsense, then why *does* Jake move more slowly (as he clearly does in the video) when he dives? The answer lies in something else we've studied: torque.

To dive, Jake must rotate his body such that his feet move behind his center of mass, and his hands move in front of it. Initiating this counterclockwise rotation requires a counterclockwise torque. This can be produced either by a leftward-pointing force to the top of his body or by a rightward-pointing force to the bottom of his body. The latter choice is all we have, since the ground is the only thing exerting forces on Jake as he prepares to dive.

And this, finally, is it: To rotate his body to dive, Jake purposely executes a "braking step" to produce friction between his shoe and the ground. Netwon's third law tells us that the ground will exert a right-directed force on Jake that does two things:

1. It generates a counterclockwise torque, the desired effect.

2. It slows his overall horizontal velocity, an undesirable but inevitable effect of a force.

So there *is* deceleration after all! But it is not happening "the moment his feet leave the ground" or "after his feet leave the ground," but *before* his feet leave the ground. It's an important point if we really want to understand the motion and physics and not perpetuate common misconceptions.

Analogy with long jump

A similar situation is found in the long jump, where the athlete must plant his foot on the board and rotate his body such that his center of mass is positioned directly above his coiled leg, to achieve maximum vertical force. In a sense, he is trading some horizontal velocity for a vertical velocity component.

Figure 14.4 shows Ohio State long-jumper Jordan Foster as he executes his launch. The "collision" with the board lasts only about 150 ms, and the horizontal force components are large and complex. After an initial force spike corresponding

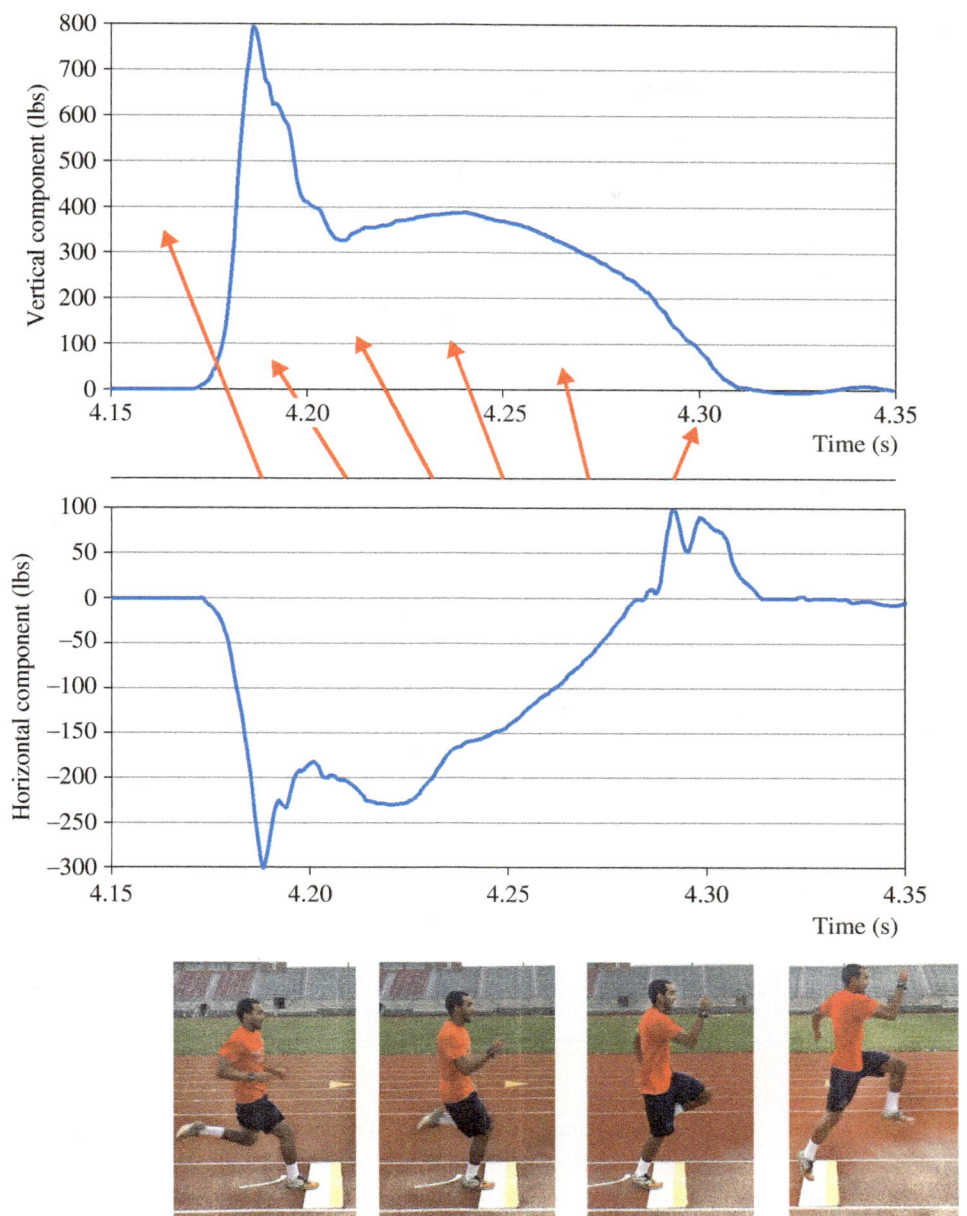

Figure 14.4. Ohio State long-jumper Jordan Foster plants his foot and leaps from the board. The graphs show the rapid evolution of the horizontal and vertical components of the ground's force on him. The arrows indicate the magnitude and direction of this force every 20 ms. (*Photos courtesy of Corey Phifer*).

to the shock of impact with the board, a backward-pointing force reaching 300 lb is exerted on his feet, providing the torque to rotate his body for launch. A forward-directed force is exerted at the end of the launch. However, by this time Foster's leg has largely extended, and he is leaving the board, so this force is relatively weaker.

Long-jumpers focus on launching technique to minimize braking to some extent (for example, by landing flat on the launch foot instead of on the heel), but some amount of backward force is needed to rotate the body for the launch.

EXAMPLE 14.1 *Impulse on Foster*

The jump in figure 14.4 is not Foster's best, as he needed some concentration to launch directly from an unfamiliar force plate used to make the measurement. Nevertheless, we can use the data to estimate the slowing effect of the horizontal force, using our knowledge of collisions from chapter 6.

If we focus just on horizontal components, the change in Foster's momentum due to his launch step is given by the *impulse* (equation 6.2). In this case, we can consider the launch as two "triangle impacts" (see figure 6.4), one pointing backward (negative impulse) and one forward.

The first impulse (between 4.18 and 4.28 s) has a peak force of about −250 lb and duration of about 0.10 s. According to equation 6.11, the impulse is then

$$J_{\text{backward},x} = \tfrac{1}{2} F_{\text{peak},x}\, \Delta t = \tfrac{1}{2}(-250\text{ lb})(0.1\text{ s}) = -12.5 \text{ lb}\cdot\text{s}.$$

This is followed by a smaller forward impulse:

$$J_{\text{forward},x} = \tfrac{1}{2}(+100\text{ lb})(0.02\text{ s}) = +1\text{ lb}\cdot\text{s}.$$

The total impulse is the sum of these two.

Jordan's weight is 160 lb ($m = 5$ slug) and his initial speed is $v_{b,x} = 17.2$ mph = 25.2 ft/s. The connection between momentum and impulse (equation 6.2) lets us figure out how much he slowed:

$$p_{a,x} - p_{b,x} = J_x \quad\rightarrow\quad mv_{a,x} = J_x + mv_{b,x}$$

$$\rightarrow\quad v_{a,x} = \frac{J_x}{m} + v_{b,x} = \frac{-12.5\text{ lb}\cdot\text{s} + 1\text{ lb}\cdot\text{s}}{5\text{ slug}} + 25.2\text{ ft/s} = 22.9\text{ ft/s},$$

which is 15.6 mph. So, this particular launch slowed Foster from 17.2 to 15.6 mph.

Question

Continuing the analysis of example 14.1, what is the vertical component of his velocity when he jumps? What is his launch angle?

14.5 Other Issues

Those who ridicule the very concept of diving into first base point out that Olympic sprinters do not dive across the finish line. To this the "diver camp" responds: "Well, maybe they would try it if they didn't have a rough and hard track to land on." It's good point. And it's related to another argument against diving: the possibility of injury to the player's valuable hands. Indeed, even slides into second or third base are safer when done feet first, in this regard.

There is also the unpredictability factor. In a *perfect* dive, the player's outstretched hand touches the bag while he is still in the air, so friction won't slow him. Even if a player's perfect dive is better than running through the bag, that doesn't mean his *average* dive will be better.

Furthermore, a player who plans to dive will often make one or two "stutter steps," in anticipation of his leap. Pole vaulters and long-jumpers very carefully measure their approach distance and calibrate their strides to avoid such stutter steps. A baseball player who has just made a hit, however, does not have this luxury. After hitting the ball, he begins from an awkward position and scrambles in the dirt to begin his sprint to first base. Unless he is very lucky, there is little chance that his feet will be in the perfect position to make an optimal dive when he reaches first base.

14.6 Concluding Remarks

Diving may be better for some athletes

Now we have a more complete and correct idea of what happens when an athlete needs to rotate his body forward while running. It should be clear that this is a complex maneuver, and you can hopefully see that **no law of physics precludes the possibility that diving might be the better option for a player**. After all, an outstretched hand *does* provide a benefit, reducing the distance the batter must cover. This is counterbalanced by the fact that he must slow to rotate his body.

Which of these two effects "wins" depends on the runner and the strength of his technique. A "good diver," that is, one who consistently pushes off especially hard after rotating (think of the second part of Jordan Foster's jump in figure 14.4), may well be better off diving.

Respect

In science, as in sports, we can often learn as much from our errors as from our successes. ESPN's *Sport Science* series is generally excellent and informative. Laying an egg every once in a while is understandable, and my intention here was not to harp on an honest mistake, but to use it as a valuable point of discussion.

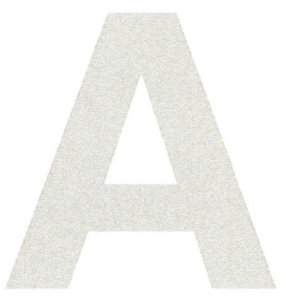

Unit Conversions

units of length

1 m = 100 cm = 39.4 in. = 3.28 ft = 1.09 yd
1 mi = 1.61 km = 5280 ft = 1760 yd   1 in. = 2.54 cm

units of volume

1 liter = 0.264 gal   1 gal = 3.785 liters   1 m^3 = 1000 liters

units of speed

100 mph = 44.7 m/s = 146.7 ft/s

units of force

1 N = 0.225 lb = 3.6 oz   (1 lb = 4.45 N)   (1 lb = 16 oz)
1 ton = 2000 lb

units of energy

1 J = 1 N · m = 0.738 lb · ft
1 kWh (kilowatt-hour) = 3.6×10^6 J
1 Cal ("food calorie") = 4184 J

units of power

1 W = 1 J/s = 0.86 Cal/hr   1 h.p. (horsepower) = 746 W

units of torque

1 m · N = 0.738 ft · lb

units of angle

1 rot = 2π rad = 360°   1 rad = 57.3°

mass and weight

1 slug = 14.5 kg   (1 kg = 0.0685 slug) (For reference, Shaq's mass is just around 10 slug.)
    1 gr = 64.799 mg (This is used in archery.)

We cannot say that 1 kg *equals* 2.2 lb, since mass and force are different quantities. It would be like saying that 1 in. equals 25 s. So there is no "conversion factor" between kilograms and pounds. Nevertheless, one often wants to know the weight (force due to gravity) of an object of given mass, on the surface of the Earth, where sports are currently played.

1 kg weighs 2.2 lb   1 kg weighs 9.8 N
1 slug weighs 32 lb   1 slug weighs 142 N

It can be handy to remember that 1 N is the weight of a typical apple.

Units Abbreviations

Cal = (food) calorie
cm = centimeter
ft = foot
g = gram
gal = gallon
gr = grain
h.p. = horsepower
hr = hour
in = inch
J = joule
kg = kilogram
km = kilometer
kW = kilowatt
lb = pound
m = meter
mg = milligram
mi = mile

min = minute
ml = milliliter
mm = millimeter
mph = miles per hour (mi/hr)
N = newton
oz = ounce
rad = radian
rev = revolution
rot = rotation
rpm = rotations (or revolutions) per minute
rps = rotations (or revolutions) per second
s = second
W = watt
yd = yard

B Tables of Relevant Physical Properties

Table B.1. Typical times for which a ball is in contact with a striking surface for common sports situations. For the football tackle, this is the approximate time required for the two humans to achieve a common velocity.

Event	Contact time (ms)
Billiards	0.2–0.5
Football kickoff	8
Baseball hit	0.7–1.3
Softball hit	3.5
Golf drive	0.45–1.0
Handball	12.5
Soccer head	23
Squash	3.3
Tennis	5.0–8.0
Volleyball	10.0
Basketball dribble	10.0
Football tackle	100–250

Table B.2. Metabolic equivalent factors for various activities. See section 9.1.1 for details on their use. From B.E. Ainsworth et al., "2011 Compendium of physical activities: A second update of codes and MET values," Medicine & Science in Sports & Exercise, 43: 1575 (2011).

Activity	MET	Activity	MET
Lying quietly and watching TV	**1.0**	Football, competitive	8.0
Standing quietly, standing in a line	1.3	Football or baseball, playing catch	2.5
Standing, light work	2.0	Frisbee, ultimate	8.0
Walking the dog	3.0	Golf, using power cart	3.5
Bicycling, mountain, uphill, vigorous	14.0	Golf, walking, carrying clubs	4.3
Bicycling, mountain, competitive	16.0	Hockey, field	7.8
Bicycling, mountain, general	8.5	Hockey, ice, competitive	10.0
Bicycling, <10 mph, leisure	4.0	Lacrosse	8.0
Bicycling, 16–19 mph, racing, not drafting	12.0	Motocross, off-road motor sports, ATV, general	4.0
Bicycling, >20 mph, racing, not drafting	15.8	Racquetball, competitive	10.0
Bicycling, stationary, 51–89 W, light-to-moderate effort	4.8	Racquetball, general	7.9
Bicycling, stationary, 201–270 W, very vigorous effort	14.0	Rugby, union, team, competitive	8.3
Resistance training (weight lifting, free weight, nautilus, universal)	6.0	Soccer, competitive	10.0
Rope skipping, general	12.3	Soccer, casual, general	7.0
Rowing, stationary ergometer, general, vigorous effort	6.0	Softball, pitching	6.0
Rowing, stationary, 150 W, vigorous effort	8.5	Squash (competitive)	12.0
Jogging, general	7.0	Squash, general	7.3
Running, 5 mph (12 min/mi)	8.3	Sports spectator, very excited, emotional, physically moving	3.3
Running, 7 mph (8.5 min/mi)	11.0	Table tennis	4.0
Running, 9 mph (6.5 min/mi)	12.8	Tennis, hitting balls, nongame play, moderate effort	5.0
Running, 14 mph (4.3 min/mi)	23.0	Tennis, singles	8.0
Running, cross-country	9.0	Volleyball, competitive, in gymnasium	6.0
Running, marathon	13.3	Volleyball, beach, in sand	8.0
Bowling, indoor, bowling alley	3.8	Wrestling (one match = 5 min)	6.0
Boxing, in ring, general	12.8	Track and field: shot, discus, hammer throw	4.0
Boxing, punching bag	5.5	Track and field: high jump, long jump, triple jump, javelin, pole vault	6.0
Cricket, batting, bowling, fielding	4.8	Track and field: steeplechase, hurdles	10.0

Table B.3. The size and mass of common balls are listed, together with typical values of speed and spin relevant to play.

Radius		Mass		Ball	v		Spin
cm	ft	kg	slug		m/s	mph	rps
2	6.56×10^{-2}	2.7×10^{-3}	1.86×10^{-4}	Table tennis	9–25	20–55	10–25
2.13	6.99×10^{-2}	0.045	3.10×10^{-3}	Golf	Up to 65	Up to 145	Up to 180!
3.18	1.04×10^{-1}	0.057	3.93×10^{-3}	Tennis ball	30	65	30
3.2	1.05×10^{-1}	0.145	1.00×10^{-2}	Lacrosse ball	30–50	65–110	?
3.74	1.23×10^{-1}	0.15	1.03×10^{-2}	Baseball	25–45	55–100	up to 20
2.86	9.38×10^{-2}	0.156	1.08×10^{-2}	Numbered billiard ball	—		—
2.86	9.38×10^{-2}	0.17	1.17×10^{-2}	Billiard cue ball (heavy)	0.5–8	1–15	—
3.58	1.17×10^{-1}	0.16	1.10×10^{-2}	Cricket	—	—	10
3.81	1.25×10^{-1}	0.17	1.17×10^{-2}	Hockey puck ($h = 2.54$ cm)	20–43	45–95	—
4.85	1.59×10^{-1}	0.185	1.28×10^{-2}	Softball	18–32	40–70	?
10.5	3.44×10^{-1}	0.27	1.86×10^{-2}	Volleyball	15–25	35–55	3–6 (est.)
8.9	2.92×10^{-1}	0.42	2.90×10^{-2}	Football ($l = 28$ cm)	18–26	40–60	10 (spiral)
11.1	3.64×10^{-1}	0.425	2.93×10^{-2}	Soccer ball	15–25	35–55	10
10.5	3.44×10^{-1}	0.425	2.93×10^{-2}	Women's water polo	—	—	—
11.1	3.64×10^{-1}	0.425	2.93×10^{-2}	Men's water polo	—	—	—
11.6	3.80×10^{-1}	0.54	3.72×10^{-2}	Women's basketball (size 6)	5–10	10–25	2
12	3.94×10^{-1}	0.62	4.28×10^{-2}	Men's basketball (size 7)	5–10	10–25	2
18	5.90×10^{-1}	1	6.90×10^{-2}	Women's discus	15–25	34–56	—
22	7.22×10^{-1}	2	1.38×10^{-1}	Men's discus	15–25	34–56	—
5	1.64×10^{-1}	4	2.76×10^{-1}	Women's 4-kg shot	10–14	22–32	—
6	1.97×10^{-1}	7.26	5.01×10^{-1}	Men's 16-lb shot	10–14	22–32	—
10.9	3.58×10^{-1}	7.26	5.01×10^{-1}	Bowling ball, 16 lb	6–10	15–25	—
—	—	7.3	5.03×10^{-1}	Carbon bike	8–20	18–45	—
14	4.59×10^{-1}	18.2	1.26	Curling stone	3	7	—

Table B.4. Coefficients of static and kinetic friction. We use these values in this book. They are "typical" values, approximate only. For example, μ_S may be 0.55 for one football sled on dry grass and 0.43 for another.

Surface 1	Surface 2	μ_S	μ_K
Football sled	Dry grass	0.5	0.4
Basketball shoes	Wooden floor	1.1	0.8
Basketball shoes	Dusty wooden floor	0.3–0.6	
Cleated shoes	Astroturf	1.5	
Cleated shoes	Grass	0.9	
Skates	Ice		0.003–0.007
Skis	Snow		0.05–0.2
Ice	Ice	0.1	0.03
Frozen puck	Ice	0.1	0.03
Room-temperature puck	Ice	>0.1	>0.03
Rubber tire	Dry concrete	1.0	0.8
Rubber tire	Wet concrete	0.7	0.5
Granite curling stone	Swept ice		0.05
Synthetic basketball	Glass (backboard)	0.47	
Synthetic basketball	Acrylic	0.55	
Synthetic basketball	Hardwood	0.57	
Leather basketball	Glass (backboard)	0.47	
Leather basketball	Acrylic	0.40	
Leather basketball	Hardwood	0.42	
NCAA basketball	Glass (backboard)	0.48	
NCAA basketball	Acrylic	0.43	
NCAA basketball	Hardwood	0.70	
Playground basketball	Glass (backboard)	0.27	
Playground basketball	Acrylic	0.95	
Playground basketball	Hardwood	0.72	
Tennis ball	Wooden basketball court		0.25
Tennis ball	Acrylic-coated court		0.5
Tennis ball	Supreme Court (synthetic)		0.61

Table B.5. Fluid densities relevant to sports. The density of air depends on altitude, temperature, and weather conditions; see chapter 5.

Fluid	Density
Water	1000 kg/m^3 = 1 kg/liter ≈ 1.94 slug/ft^3
	(62 lb/ft^3 ≈ 1 lb/pint)
Air (sea level, 70°F, clear skies)	1.2 kg/m^3 ≈ 0.0024 slug/ft^3
	(0.075 lb/ft^3)
Air (Denver, 70°F, clear skies)	0.97 kg/m^3 ≈ 0.0019 slug/ft^3
	(0.061 lb/ft^3)

Table B.6. Some typical drag coefficients. These will vary somewhat with surface roughness and speed. Online tables abound; most of these are from http://www.engineeringtoolbox.com/drag-coefficient-d_627.html

Object	C_D
Smooth sphere (low speeds)	0.47
Solid hemisphere	0.42
Solid hemisphere flow normal to flat side	1.17
Bike racer	0.88
Upright person, wind in face	1.0–1.3
Ski jumper	1.2–1.3
Head-down skydiver	0.4–0.6
Dolphin	0.0036
Tractor trailer truck	0.96
Model-T Ford	0.8
Toyota Prius	0.26

Further Reading

I have benefited greatly from the huge literature on the science of sports. Many of these books were written by physics professors who are enthusiasts of a given sport and want to share with the reader the excitement of a deeper scientific understanding of the game. Here are some of the most helpful that I've found.

Textbooks

Textbooks are primarily intended for use in a course. In most physics courses, knowledge of one topic builds upon another. A hallmark is that they will have problem sets for the student to work.

- *The Biomechanics of Sports Techniques*, 4th ed., James G. Hay, Prentice Hall, Englewood Cliffs, NJ, 1993.
  This classic text was first published in 1973. It is perhaps the earliest textbook with a similar spirit to the book you are reading now. It is well written and has great illustrations.

- *The Dynamics of Sports*, 4th ed., David F. Griffing, Dalog Company, Oxford, OH, 1995.
  First published in 1982, this was one of the earliest proper textbooks on the physics of sports and served as the basis for Prof. Griffing's course at Miami University. It is a concise treatment with crisp examples.

- *Biomechanics of Sport and Exercise*, 2nd ed., Peter M. McGinnis, Human Kinetics, Champaign, IL, 2005.
  This is a comprehensive and clearly written textbook with a focus on the human body in sport, but it also includes the mechanics of objects.

- *Physics of the Human Body*, Irving P. Herman, Springer-Verlag, Berlin, 2007.
  The aim of this incredibly comprehensive book is not to teach physics, but to use it to understand the human body. With 571 figures, 135 tables, hundreds of equations, and an assumed background in physics, the target audience for Prof. Herman's book is quite different from the audience for mine. Instructors, however, will find this book a treasure trove of information and ideas.

Popularizations

Popularizations are written to bring a scientific discussion of sports to the general public. Although not necessarily "light reading," they are quite readable and often humorous. The author's job is made much easier by the fact that he need not *teach* the topic, but simply *refer* to torques, forces, and momenta. The reader is educated, but not as completely as she or he would be in a course.

- *Bicycling Science*, 3rd ed., David Gordon Wilson, MIT Press, Cambridge, MA, 2004.
  This book is a masterpiece. The first edition was published in 1974. An MIT professor combines brilliant writing and technical expertise to explore the history, mechanics, and science of every aspect of the bicycle and the body of its rider. It is more rigorous than most popularizations, but not overbearing. It is highly recommended to anybody interested in two-wheeled transportation. There is more to know than you think.

- *Sport Science*, Peter J. Brancazio, Simon and Schuster, New York, 1984.
  This classic monograph by Professor Brancazio contains clear textual descriptions with less of a focus on math. It is a favorite of physicists. Topics are arranged along the lines of a physics textbook, but the work is intended for the general public, outside of the classroom.

- *Newton at the Bat*, Eric Schrier and William Allman, Charles Schribner's Sons, New York, 1984.
  This wonderful collection of 36 three- to four-page essays first appeared in *Science* magazine, covering topics as diverse as weight loss, shoe soles, artificial turf, and darts. It is hard to find but recommended.

- *The Physics of Baseball*, 3rd ed., Robert K. Adair, Harper Collins, New York, 2002.
  First published in 1990, Professor Adair's superb book inspired a series of similarly minded books by physics professors on almost every imaginable sport. Many are mentioned here.

- *Keep Your Eye on the Ball*, 2nd ed., Robert Watts and A. Terry Bahill, W. H. Freeman and Company, New York, 2000.
  This book gives a beautiful treatment of balls and bats, and especially the aerodynamics of a baseball, including measurements by the authors (both engineering professors). It has an interesting historical perspective as well.

- *Newton on the Tee*, John Zumerchik, Simon and Schuster, New York, 2002.
  This nonmathematical book provides valuable insights into the science of golf in an easy-to-digest manner.

- *The Physics of Hockey*, Alain Haché, Johns Hopkins Press, Baltimore, MD, 2002.
  This is a readable introduction to the forces of sticks, pucks, and colliding players. There is a nice chapter on the science of ice and an excellent discussion of the slap shot.

- *The Physics of Football*, Timothy Gay, Harper Collins, New York, 2005.
  This book showed me how fun a popularization could be while still educating the unsuspecting reader. I read it through and then read it again. Aerodynamics, nutrition, and of course collisional physics are interspersed with anecdotes and historical context about the game.

- *The Physics of Basketball*, John Fontanella, Johns Hopkins Press, Baltimore, MD, 2006.
  A former college player, Professor Fontalla uses the interaction of the ball with the air, floor, and hoop to illustrate classical mechanics at play.

- *The Knucklebook*, Dave Clark, Ivan R. Dee, Chicago, 2006.
  The subtitle says it all: "Everything you need to know about baseball's strangest pitch." It is fun, readable, and full of interesting facts about the science, history, and people associated with the knuckleball.

- *Why a Curveball Curves*, Frank Vizard, ed., Hearst Books, New York, 2008.
  This lively, fun collection of essays is put out by *Popular Mechanics* magazine. Contributing authors include professors, professional athletes, and veteran coaches.

- *Gold Medal Physics*, John Eric Goff, Johns Hopkins Press, Baltimore, MD, 2010.
  Professor Goff uses famous athletes and events, from the Olympics to Sumo wrestlers, as a vehicle to explain how a scientist understands the world around him. It uses introductory mathematics and is quite readable.

- *Gliding for Gold*, Mark Denny, Johns Hopkins University Press, Baltimore, MD, 2011.
  This book is a very readable explanation of physical principles behind winter Olympic events.

- *100 Essential Things You Didn't Know You Didn't Know about Sport*, John D. Barrow, The Bodley Head, London, 2012.
  Released just in time for the 2012 Summer Olympics in London, mathematics professor Barrow's book comprises 100 one- to two-page well-written essays of interesting tidbits. Important physical effects play a role in about 20 of these. All are fun.

Manuals

Manuals are books aimed largely at coaches or athletes looking to improve their performance. Many authors find that providing a scientific understanding of the sport is an important part of the process.

- *Search for the Perfect Swing*, Alastair Cochran and John Stobbs, Triumph Books, Chicago, 1968.
  Reprinted essentially unchanged in 2005, this classic contains superb strobe photographs of major golfers' swings and clear explanations with no math. It is the best discussion of the double-pendulum and two-spring model I've seen.

- *Gymnastics: A Mechanical Understanding*, Tony Smith, Holmes & Meier Publishers, Inc., New York, 1982.
  Clear and concise, this handbook is intended largely to give gymnastic coaches a solid foundation in the science of body mechanics. It was an unexpected gem of a find. Simple illustrations emphasize forces and torques.

- *The Illustrated Principles of Pool and Billiards*, Dr. Dave Alciatore, Sterling Publishing Company, New York, 2004.
  This surprisingly comprehensive book is aimed largely at the player but written by an engineering professor who cannot help but show the technical beauty and physics of the game. A supplementary DVD contains many fascinating high-frame-rate videos of collisions. I was surprised at the level of complexity involved.

- *Technical Tennis*, Rod Cross and Crawford Lindsey, Raquet Tech Publishing, Vista, CA, 2005.
  This small, valuable book is packed with an incredible amount of practical information on strings, raquets, balls, and their interactions.

- *Physics and the Art of Dance*, 2nd ed., Kenneth Laws, Oxford University Press, New York, 2008.
  An enjoyable discussion of how physical principles affect the decisions and esthetics of ballet can be found in this book.

- *Advanced Sports Nutrition*, 2nd ed., Dan Benardot, Human Kinetics, Champaign, IL, 2012.
  This book contains detailed, sport-specific nutritional strategies for athletes, with clear justifications bolstered by the latest research.

Focused studies

Focused studies are technically in-depth explorations of a topic, "heavier" than a casual popularization.

- *Determination of the Moments of Inertia of the Human Body and Its Limbs*, Wilhelm Braune and Otto Fischer, Springer-Verlag, Berlin, 1988.
  This 1988 issue is a reissue of the original work published in 1892. It is a highly focused and technical study (the title says it all) of a surprisingly difficult topic. I wound up incorporating little from this book into mine, but instructors with special interest in the mass distribution of the human body are recommended to read this book.

- *Fluid-dynamic Drag*, S. F. Hoerner, self-published, 1965.
  This is an incredible labor of love by an aerodynamical engineer trained in war-time Germany. While publishers at the time did not see the value in such an exhaustive treatment of this specialized topic, the data and systematics compiled in this self-published book are continually referenced by industrial and scientific practitioners today.

- *Physics of Baseball and Softball*, Rod Cross, Springer, New York, 2011.
  This is a well-organized, pedagogical and exhaustive exploration of every aspect of the game, often in great technical detail. It is probably too "heavy" to be considered a popularization, but this book is a trove for instructors and could be used as a textbook, if augmented with exercises.

- *Projectile Dynamics in Sport*, Colin White, Routledge, New York, 2011.
  A highly systematic and mathematical treatment of every type of projectile that flies through the air is found in this book. Launch, aerodynamics, and bounce physics are discussed, followed by a sport-by-sport set of chapters.

Collections

These are **collections** of articles by researchers or leaders in the field.

- *The Physics of Sports*, edited by Angeno Armenti, Jr., AIP Press, Springer-Verlag, New York, 1992.
  This is an excellent collection of basic research journal articles by physicists, commissioned by the American Institute for Physics. It is a favorite reference.

- *Biomechanics and Biology of Movement*, Benno Nigg, Brian MacIntosh, and Joachim Mester, eds., Human Kinetics, Champaign, IL, 2000.
  This is a collection of excellent monographs on the human body in motion, by prominent researchers in the field. It could be used as an advanced-level undergraduate or graduate course textbook, if augmented with exercises and examples.

Other

- *The Physics of Baseball*, Alan Nathan, http://baseball.physics.illinois.edu, 2011.
  The Internet is full of excellent websites on the science and sports. (It is at least as full of incredibly bad ones unfortunately.) Because of the temporary nature of websites, I've decided not to list them. Nevertheless, Prof. Alan Nathan's site is an exception. If it vanishes from the location I've listed, simply search for his name and baseball and you will be treated to video, figures, popularizations, and research articles. His work on baseball physics (and nuclear physics!) has been widely published in research journals but not yet in a book.

- *Resource Letter PS-2: Physics of Sports*, Cliff Frohlich, *American Journal of Physics*, **79**: 565 (2011).
  While I have benefited from many hundred research articles and will not list them all here, I must mention one metaresource. A prolific researcher and writer of research articles, Professor Frohlich has compiled an incredibly useful list of books, periodicals, and research papers related to the physics of sports. I have worked my way up and down this list.

Answers to Odd-Numbered Problems

Numerical answers may vary by up to 20%, depending on units and conversions used during the calculation.

Chapter 1

1. (a) 9.17×10^1 m
   (b) 4.22×10^4 m
   (c) 1.02 m
3. (a) 3.52 min
   (b) 211.4 s
5. 1037 N
7. (a) 0.087 h.p.
   (b) 5.7 bulbs (6 bulbs)
9. Wrong answers: a, b, c, f, g
11. About 2.5 ft
13. 17.6 fields
15. between 100 and 750
17. (a) Yes, it is important
    (b) Heavier
    (c) My opinion: on!
    (d) More difficult with hat on

Chapter 2

1. 50 ms
3. (a) +10.4 m/s
   (b) −9.2 m/s
5. (a) 0.866 s
   (b) 0.433 s
   (c) 50%
7. 5.36 m/s
9. (a) 1.4 g's
   (b) 332 g's
11. (a) 282.8 ft/s
    (b) 146.7 ft/s at best (100 mph)
    (c) 336 ft
13. 0.147 s (assuming a 3-ft drop) this is roughly human reaction time
15. (a) My impression is always that the springboard diver spends more time in the air.
    (b) 10-m platform: 1.43 s
    (c) 3-m springboard: 1.35 s
17. (a) 434 ms
    (b) 458 ms
    (c) Difference in pitches is 24 ms which is larger than bat crossing time of 9.7 ms; so yes, it matters very much to the timing of the swing
19. (a) 4.9 yd/s^2 (14.8 ft/s^2)
    (b) 1.7 s
21. (a) 5.35 m/s^2 (0.55 g's)
    (b) Ford: 0 to 60 in 5 s gives almost identical acceleration!

Chapter 3

1. (a) 78.88 mph
   (b) 1.36 mi
3. 156.5 lb in forward direction
5. (a) 48 ft/s^2
   (b) Same acceleration
   (c) Burst upward as well as forward
7. (a) 2.8 lb (small, no?)
   (b) No, doubling the pushing force will not double a, because it does not double the *net* force on the stone
9. (a) 230 lb
   (b) 230 lb
   (c) 36.8 ft/s^2
11. 446 lb
13. 137 ft/s^2 41.4° relative to horizontal
15. (a) 14.2 ft/s^2
    (b) 71 lb
    (c) Up
    (d) Weight, force from rope
    (e) 231 lb
17. (b) 0.8 lb, backward
    (c) 10.9 ft/s (7.4 mph)
    (d) 1.6 ft (19.2 in.) away; that is, the stone will travel 88.4 ft
19. (a) To the left—*away* from the sled
    (b) 396 lb (!) (1764 N)
21. 2350 lb (more than a ton!) (10,440 N)

Answers to Odd-Numbered Problems

Chapter 4

1. (a) 144 ft
   (b) 144 ft
3. (a) 16 yd/s = 33 mph
   (b) No
5. (a) 4.56 s
   (b) 1.22 s
7. (a) $v_{\text{sideline}} = 10.6$ yd/s
       $v_{\text{downfield}} = 10.6$ yd/s
   (b) 12 yd/s at 62°
9. (a) 144 ft
   (b) 116.9 ft/s
   (c) 55°
11. (a) 34.3 ft/s
    (b) 2.01 s
    (c) 27.6 ft/s
13. (a) Not as high
    (b) Same height
    (c) Higher
    (d) Not as high
15. (a) 46.6 ft
    (b) Two possible correct answers: 28° and 59.8°

Chapter 5

1. (a) 3.11 ft^3
   (b) 193 lb
   (c) Yes
   (d) 452 ft
   (e) 2.84 ft^3
   (f) 176 lb
   (g) No
   (h) N/A (due to buoyancy alone, he will *never* come to rest!)
3. (a) 3 in.^{2} = 2.08×10^{-2} ft^2
   (b) 0.158 lb
   (c) 89.3 ft/s = 61 mph
5. (a) 2.02 s
   (b) v_x does not change; v_y changes
   (c) Parabola
   (d) 16.09 m/s
   (e) 1.408 m/s
   (f) -1.4 m/s^2
   (g) 0.58 N
   (h) 2.25 rps
7. No
9. (a) 25 ft/s
   (b) Up and backward; up; up and forward
   (c) 0.072 lb = 1.2 oz
11. (a) For no air:
       x: change fixed amount
       y: change diff amount
       v_x: no change
       v_y: change fixed amount
       a_x: no change
       a_y: no change
    (b) With air:
       x: change diff amount
       y: change diff amount
       v_x: change diff amount
       v_y: change diff amount
       a_x: change diff amount
       a_y: change diff amount
13. (a) clockwise
    (b) 2.56 ft
    (c) 17.3 rps
    (d) No
    (e) 2.3 in. (0.19 ft)
15. 65.5 rps
17. (a) 0.4330 s
    (b) 0.4336 s
    (c) Larger
19. (a) 25.6 ft/s^2 down
    (b) 37.5 ft/s^2 down
21. (a) 0.214 lb = 3.42 oz
    (b) About 10 in.
    (c) Less
23. (a) About 10.4 lb (46.1 N) at sea level; 7.8 lb (34.9 N) in Mexico City
    (b) About 7.6 cm or 3 in.
    (c) About 12.6 N, or 2.8 lb less drag due to tailwind
    (d) About 8.4 cm or 3.3 in.

Chapter 6

1. 2900 mph
3. (a) 11.9 kg m/s
   (b) 232 lb
   (c) 232 lb
5. (a) +5.2 kg m/s
   (b) 28.3 m/s
   (c) 0.366
   (d) 1749 N = 393 lb
7. (a) 9 in.
   (b) 4.4 in.
   (c) Slower
9. 7.96 m/s
11. 53 ft
13. 2060 lb (a ton of force, from a little cue ball!)
15. (a) 101 kg (222 lb)
    (b) 1790 lb
    (c) 0
17. (a) 9.57 yd/s
    (b) 0
    (c) roughly 10 to 70 J
    (d) No. Only a tiny amount (less than 1%) of the energy was converted into heat or damage.

Answers to Odd-Numbered Problems

Chapter 7

1. (a) 9.8×10^6 J
   (b) 2.58 MW (!) or 3450 h.p.
   (c) 8.55×10^6 J
   (d) 19.1 m/s (43 mph)
   (e) 4.5 m/s (10 mph)

3. (a) 2242 J
   (b) It takes about 2 s to use all the ATP and PCr stored in the cells; any shorter time, and you don't use all the energy readily available; any longer time, and you are waiting for ATP to be replenished from carbohydrate stores.
   (c) 1.2 h.p. (896.8 W)
   (d) 7.6 Cal

5. (a) 225 W
   (b) 271 Cal

7. The diver has 5.3 times the energy of Lewis (diver: 7840 J; Lewis: 1475 J).

9. 2.4 m (7.9 ft)

11. −2560 J

13. (a) −1200 J
    (b) −2400 J
    (c) 3600 J
    (d) No.

15. (a) 791 J
    (b) 7.9 kW (!) (10.6 h.p.)
    (c) They spin and build up energy more slowly.

17. (a) 200 J
    (b) $e = 1 \to \eta = 0.6$; $e = 0.83 \to \eta = 0.5$
    (c) $KE_{ball} = 119.9$ J; $KE_{head} = 80.1$ J; $E_{therm} = 0$
    (d) $KE_{ball} = 100.4$ J; $KE_{head} = 88.2$ J; $E_{therm} = 11.4$ J
    (e) 9% faster

19. (a) 3600 J
    (b) 291 mph (130.2 m/s)
    (c) 0.77 s
    (d) 1.4 s

Chapter 8

1. (a) Slightly more
   (b) Less
   (c) About the same spine, because the period of the drawstring+arrow will increase by $\sqrt{2}$ *and* the period of the arrow's oscillation will increase by the same amount, if the spine is unchanged.

3. (a) 2304 N/m
   (b) 15.3 cm
   (c) Same

5. (a) 80 ft/s (54.5 mph)
   (b) 72.1 ft/s (49.2 mph)
   (c) 80 ft/s (54.5 mph)
   (d) 40.1 ft/s (27.3 mph)
   (e) 0.2 bounce/s
   (d) No (momentarily at rest, but accelerating up)
   (e) More forward
   (f) About 445 ms
   (g) Increase
   (h) Yes, it makes it less forward

   (f) She'll bounce with greater frequency.

7. (a) 40 lb/ft
   (b) 61 J
   (c) 61 J
   (d) 47.5 J more work (that is, 108.5 J total)
   (e) 80 lb

9. (a) 150 lb
   (b) 75 lb
   (c) 25.4 J

Chapter 9

1. (a) 63 lb (280 N)
   (b) 376 W
   (c) 1 to 2 min

3. 7 mph (3.1 m/s)

5. (a) 9434 s (2:37:14)
   (b) 7558 s (2:05:58)
   (c) 3355 s (0:55:55)
   (d) He is faster with no hills. The hills don't "cancel."

7. (a) 83.7 W
   (b) 368 Cal/hr

   (c) 0.258 hr (about 16 min)
   (d) 0.069 hr (4 min)

9. (a) 861 Cal/hr
   (b) 20 bulbs
   (c) 359 Cal
   (d) 2.8 liters/min

11. (a) 37.6 rpm
    (b) 1.17×10^7 kg · m² (8.62×10^6 slug · ft²)
    (c) *Only* 17.8 J
    (d) 0.016 s (16 ms)

   (e) The spinning bike wheel has 49.4 J of energy, 2.8 times that of the rotating restaurant!

13. (a) Torque must be the same, but now there is a larger lever arm, so expect less force (tension).
    (b) 216 lb

15. 6.8 rps!

17. (a) 136.7 N (30.7 lb)
    (b) 3666 W
    (c) Not even 1 s!

(d) 4.38 rps!

(e) No

19. (a) 7650 Cal!

(b) 3.8 bars/hr (1 bar every 15 min!)

21. 878 Cal, going *all out* for an hour

23. (a) 148 W

(b) 154.5 W

(c) 739 Cal/hr

25. (a) 2.59 in.^{2}

(b) Retarding torque increases because (i) area of flat patch increases, increasing lever arm for the torque; (ii) upward force of road increases, increasing the force for the torque.

27. (a) 7193 J

(b) 90.2 J

(c) Still only 1.2%

Chapter 10

1. (a) 1.12 rot

(b) He would land more or less on his back.

3. (a) (i) If the pole sags, it lowers the center of mass, reducing the lever arm for gravitational torque, adding stability; (ii) increases moment of inertia, so any rotation is slowed; (iii) the person can "counterrotate" the pole in order to regain balance.

(b) Anteroposterior axis

5. (a) No

(b) No

(c) Increase the height reached by other parts of her body (head, legs)

(d) Rotate to her left (clockwise in the photo)

(e) Left leg and right arm

7. (a) Over his head from behind

(b) Yes

9. (a) 12.56 rad/s

(b) 57.6 rad/s^2

(c) 0

INDEX

A

acceleration, 20, 25
    centripetal, 68, 71
    gravitational, 29
Adams, Phillip, 155
aerobic energy conversion, 208
air density, 135
air resistance, *see* force, drag
anaerobic energy conversion, 207
anatomical axes, 303
angular acceleration, 280, 290
angular momentum, 311
    conservation of, 312, 314
angular velocity, 279
approximation, 106, 122, 125
Archemides, 107
archery, 224
    archer's paradox, 231
    arrow spine, 230, 233
atmospheric pressure, 115
ATP, 197
auto racing
    drafting, 270
    IndyCar airfoil, 64

B

Bacsik, Mike, 327
ballet, 91
Barnes, Lynda, 156
basal metabolic rate, *see* RMR
baseball
    bat, 332
        aluminum, 333
        sweet spot, 333
    BBCOR, 331, 334
    BBS, 329
    curveball, 120, 122
    diving into first, 351
    double play, 302
    hit, 326
    knuckleball, 129
    slide, 26
basketball
    bounce, 170, 178, 186, 191
    bounce pass, 178
    dribble, 170
    dunk, 40, 92
    shot, 126, 170
    synthetic ball, 169
    vertical leap, 52
Beamon, Bob, 94

Beasley, Cole, 155
billiards, 159, 165
    90-degree rule, 165
Bolt, Usain, 21
Bonds, Barry, 326
bowling, 156, 165
    birthday party, 161
    lily, 162
Brees, Drew, 2
Brown Trafton, Stephanie, 73
Bubka, Sergey, 337
bungee jumping, 237

C

Calorie, 194
Carlos, Roberto, 124
center of gravity, *see* center of mass
center of mass, 8, 52, 91
Chara, Zdeno, 234
cheerleading, 13, 319
coefficient
    drag, 111, 366
        dependence on speed, 111
        dependence on surface, 112
    friction, 57, 171, 365
    Magnus, 118
    restitution, 168, 170, 174, 331
        dependence on speed, 175
    rolling friction, 256
collision, 145
    2D, 155, 162, 165
    contact time, 151, 362
    elastic, 157
    inelastic, 147, 234
    isolated system, 161, 166
    partially inelastic, 168
    types of, 168
component, *see* vector component
Conkle, Jeff, 216
constant acceleration kinematics, 23
cos(), *see* trigonometry
cricket, 116
Cuban, Mark, 169
curling, 68
cycling, 245
    bicycle power equation, 260
    drafting, 265
        effect on leader, 268
    echelon, 267
    gearing, 274, 277, 278, 283, 284
    paceline, 265

    penny-farthing, 274
    pursuit, 265
    velodrome, 261
    wheels, 273, 290

D

deceleration, 25
density, 108, 365
discus, 69, 130
displacement, 16
diving, 31
Dong-Hyun, Im, 224

E

Earnhardt, Jr., Dale, 67, 74
efficiency, 191, 197, 198, 202, 226, 236, 264
    golf drive, 192
Ellis, Mark, 302, 316
endothermic & exothermic reactions, 196, 197
energy
    chemical, 194, 198
        storage, 198, 209
    conservation of, 199, 201, 307
    conversion to other forms, 150, 187, 188, 190, 196, 202
    elastic potential, 187, 225, 226, 235
    forms relevant to sports, 194
    gravitational potential, 188
    in a collision, 150, 161, 234
    kinetic, 149, 161, 292, 307
    mechanical, 199
    thermal, 151, 187
        heat death, 193
    total, 188, 199
    water analogy, 200
EPO, 250
equilibrium, 40, 42
estimation
    with video, 6

F

fat, 198
Federer, Roger, 6
figure skating
    spin, 312
football
    air drag, 113
    blocking, 63
    fair catch, 87

helmet, 153
linemen, 64
pass, 82
punt, 83, 86
sled, 56, 62
sound on the line, 19
tackle, 147, 155
force, 40
"centrifugal", 70, 72, 74
action-reaction, 48, 72
buoyancy, 107, 128
centripetal, 71
drag, 110, 134
head/tailwind, 260
ground reaction (GRF), 41, 48, 53, 60
imaginary, 67
in a collision, 151, 153
lift, 130, 133
Magnus, 117, 120, 179
net, 42, 47
normal force, 40
peak, 48, 153
tension, 72, 173
types of, 40
weight, 40, 43
acts at center of mass, 306
Fosbury, Richard, 92
Foster, Jordan, 356
free body diagram, 41, 59
freefall, see gravity, freefall
frequency
oscillation, 228
rotational, 273
friction, 55, 59, 64
anti-lock brakes, 57
rolling, 255, 287

G

glucose, 197
glycogen, 198, 207
glycolysis
anaerobic, 207
golf
angle of attack, 181
backspin, 180
ball, 112
coefficient of restitution, 174
drive, 172, 176, 192
efficiency, 192
loft, 173
graph, 15, 22, 24
gravity
constant g, 29, 43
freefall, 29, 52
GRF, see force, ground reaction

Griffin, Blake, 91, 243
gymnastics
balance beam, 317
Giant Circle, 305
high bar, 305

H

hangtime, 32, 83, 84
Harrington, Pádraig, 172
heat, see energy, thermal
high jump, 92
hockey
flex rating, 235
slapshot, 234
wrist shot, 33, 55
Hooke's Law, 225
horseshoes, 103
Howard, Dwight, 39, 59
hurdle, 319

I

impulse, 146, 147, 151
iterative solution, 261

J

javelin, 133
Johnson, Chris, 147
jumping, 46, 50, 92, 94
landing, 53

K

kilocalorie, see Calorie
kilogram (kg), 43
knuckling, 129

L

lactic acid, 207
Letourner, Alfred, 271, 278
Lewis, Ray, 24, 153
lift, see force, lift
long jump, 94, 103, 315, 356

M

Magnus, see force, Magnus
mass, 43
MET (metabolic equivalent task), 246, 247, 363
Micham, Matthew, 31
mole, 197
moment of inertia, 291
addition of, 293
human, 303, 304, 310, 314
momentum, 146
conservation of, 148, 163

transfer, 148, 174
vector, 155, 162

N

NASCAR, 67
Nash, Steve, 178
Newton (unit N), 40
Newton's First Law
rotational motion, 282, 311
translational motion, 42
Newton's Second Law
rotational motion, 290
translational motion, 47
Newton's Third Law
translational motion, 48, 146

O

Olympics
Beijing 2008, 15, 73
Helsinki 1952, 43
London 2012, 213, 261
Mexico 1968, 92, 94
Mexico City 1968, 43
Nagano 1998, 68
Sochi 2014, 68
origin, 16
oscillations, 230

P

Paul, Chris, 170
Paxson, John, 126
PCr, 197
period (oscillation), 227
Perry, Don, 27
Phelps, Michael, 15
pole vault, 335
pit, 152
polyurethane swimsuits, 134
pool, see billiards
position, 15
pound (lb), 40
power, 205, 207, 227, 246, 253, 260
power-to-weight ratio (PWR), 252
projectile motion, 81
Punto, Nick, 351

Q

quadratic equation, 239
Quinn, Robert, 63, 323

R

radian, 279
range, 83, 175
golf drive, 176
reaction time, 19, 21

resting metabolic rate (RMR), 210, 247, 250
rifle, 81
Robinson, Denard, 21
rolling resistance, *see* friction, rolling
rope climbing, 27
rotational frequency, *see* frequency, rotational
rugby line-out, 13

S

Salimikordasiabi, Behdad, 211
scalar, 17
scientific notation, 3
shoe
    cushioning, 54
shot put, 88
sin(), *see* trigonometry
ski jump, 133
slug, 43
Smith, Garvin, 27
soccer
    bent kick, 118, 124
    twisting during kick, 316
speed
    average, 15
    of sound, 19
    running records, 20
spin, 117, 118, 177, 179, 180
spiral, 121, 125
Sport Science television show, 351
Stern, David, 169
Stingily, Byron, 63, 323

subatomic physics, 120
swimming, 15, 134

T

table tennis, 179
Talansky, Andrew, 245, 254, 272, 275
tennis
    serve, 6
    spin, 143
terminal velocity, 114
throwing, 65
torque, 177, 280, 356
    lever arm, 280, 283, 285, 306, 323
    line of action, 177, 285, 323
    zero in free flight, 311
trampoline effect, 334
triangle impact, 153
trigonometry, 61
tug-of-war, 69
typical scales, 7

U

Uecker, Bob, 105
Uncle Rico, 203
units, 3
    angles, 279
    conversion, 4, 360
    energy, 149, 194
    frequency, 228
    mass, 43
    metric, 4
    of force, 40

    of momentum or impulse, 147
    power, 205
    SI versus Imperial, 4
    VO_2, 249

V

Van Treese, Ben, 133
vector, 17, 61
    component, 16, 17, 61
velocity
    average, 17
    instantaneous, 18
    relative, 157
Verlander, Justin, 115, 137
vertical motion, 27
Vlasic, Blanka, 92
VO_2, 249
VO_2max, 250
volleyball, 103

W

water polo, 108
weightlifting, 211, 216
Wells, Beanie, 24
Williams, Chandler, 147
wind shadow, 265
wind tunnel measurements, 111, 116, 118, 130, 270, 271
work, 203, 211
    done by spinning skater, 315
    negative, 213
work-energy theorem, 203